Classics in Mathematics

Daniel W. Stroock
S. R. Srinivasa Varadhan

Multidimensional Diffusion Processes

Daniel W. Stroock
S. R. Srinivasa Varadhan

Multidimensional Diffusion Processes

Reprint of the 1997 Edition

 Springer

Daniel W. Stroock
Massachusetts Institute of Technology
Department of Mathematics
77 Massachusetts Ave
Cambridge, MA 02139-4307
USA

S. R. Srinivasa Varadhan
New York University
Courant Institute of Mathematical Sciences
251 Mercer Street
New York, NY 10012
USA

Originally published as Vol. 233 in the series
Grundlehren der mathematischen Wissenschaften

Mathematics Subject Classification (2000): 60J60, 28A65

Library of Congress Control Number: 2005934787

ISSN 1431-0821
ISBN 978-3-662-22201-0   ISBN 978-3-540-28999-9 (eBook)
DOI 10.1007/978-3-540-28999-9

Springer is a part of Springer Science+Business Media

springeronline.com

© Springer-Verlag Berlin Heidelberg 1997
Originally published by Springer-Verlag Berlin Heidelberg New York in 1997

Production: LE-TeX Jelonek, Schmidt & Vöckler GbR, Leipzig
Cover design: *design & production* GmbH, Heidelberg

Printed on acid-free paper   41/3142/YL - 5 4 3 2 1 0

Grundlehren der mathematischen Wissenschaften 233

A Series of Comprehensive Studies in Mathematics

Springer-Verlag Berlin Heidelberg GmbH

Daniel W. Stroock
S.R. Srinivasa Varadhan

Multidimensional Diffusion Processes

 Springer

Daniel W. Stroock
Massachusetts Institute of Technology
Department of Mathematics
77 Massachusetts Ave
Cambridge, MA 02139-4307
USA

e-mail: dws@math.mit.edu

S.R. Srinivasa Varadhan
New York University
Courant Institute of Mathematical Sciences
251 Mercer Street
New York, NY 10012
USA

e-mail: varadhan@cims.nyu.edu

Cataloging-in-Publication Data applied for

A catalog record for this book is available from the Library of Congress.

Bibliographic information published by Die Deutsche Bibliothek
Die Deutsche Bibliothek lists this publication in the Deutsche Nationalbibliografie;
detailed bibliographic data is available in the Internet at http://dnb.ddb.de

Mathematics Subject Classification (2000): 60J60, 28A65

ISSN 0072-7830
ISBN 978-3-662-22201-0   ISBN 978-3-540-28999-9 (eBook)
DOI 10.1007/978-3-540-28999-9

Springer-Verlag Berlin Heidelberg New York
a member of BertelsmannSpringer Science+Business Media GmbH

http://www.springer.de

© Springer-Verlag Berlin Heidelberg 1979, 1997
Originally published by Springer-Verlag Berlin Heidelberg New York in 1997

Cover design: *design & production* GmbH, Heidelberg

Printed on acid-free paper   41/3142/db - 5 4 3 2 1 0

To our parents:
    Katherine W. Stroock
    Alan M. Stroock
    S.R. Janaki
    S.V. Ranga Ayyangar

Contents

Frequently Used Notation

I. Topological Notation. Let (X, ρ) be a separable metric space.

1) B° is the interior of $B \subseteq X$.
2) $\bar{B}$ is the closure of $B \subseteq X$.
3) ∂B is the boundary of $B \subseteq X$.
4) $\mathscr{B}_X$ is the Borel field of subsets of X.
5) $C_b(X)$ is the set of bounded continuous functions $f: X \to R$.
6) $B(X)$ is the set of bounded $\mathscr{B}_X$-measurable $f: X \to R$.
7) $U_\rho(X)$ is the set of bounded ρ-uniformly continuous $f: X \to R$.
8) $M(X)$ is the set of probability measures on $(X, \mathscr{B}_X)$.
9) $\|f\| = \sup\limits_{x \in X} |f(x)|$ for $f \in B(X)$.

II. Special Notation for Euclidean Spaces

1) R^d is d-dimensional Euclidean space.
2) $|x| = (\sum\limits_1^d x_j^2)^{1/2}$ for $x \in R^d$.
3) $B(x, r) = \{y \in R^d: |x - y| < r\}$.
4) $\langle x, y \rangle = \sum\limits_1^d x_j y_j$ for $x, y \in R^d$.
5) $S^{d-1} = \{x \in R^d: |x| = 1\}$.
6) $\hat{C}(R^d) = \{f \in C_b(R^d): \lim\limits_{|x| \to \infty} f(x) = 0\}$.
7) $C_0(\mathscr{G})$ is the set of $f \in C_b(\mathscr{G})$ having compact support.
8) $C_b^m(\mathscr{G})$ is the set of $f: \mathscr{G} \to R$ possessing bounded continuous derivatives of order up to and including m.
9) $C_b^\infty(\mathscr{G}) = \bigcap\limits_{m=0}^\infty C_b^m(\mathscr{G})$.
10) $C^\infty(\mathscr{G})$ is the set of $f: \mathscr{G} \to R$ possessing continuous derivatives of all orders.
11) $C_0^\infty(\mathscr{G}) = C^\infty(\mathscr{G}) \cap C_0(\mathscr{G})$.
12) $C_b^{m,n}(\mathscr{G})$ for $\mathscr{G} \subseteq [0, \infty) \times R^d$ is the set of $f: \mathscr{G} \to R$ such that f has m bounded continuous time derivatives and bounded continuous spacial derivatives of order less than or equal to n.
13) $L^p(\mathscr{G})$, $1 \leq p \leq \infty$, is the usual L^p-space defined in terms of Lebesgue measure on $\mathscr{G}$.

14) $L^p_{loc}(\mathscr{G})$ is the set of $f\colon \mathscr{G} \to R$ (or $\mathbb{C}$) such that $f \in L^p(K)$ for all compact K in $\mathscr{G}$.

III. Path Spaces Notation

1) $C(I, R^d)$ for $I \subseteq [0, \infty)$ is the set of R^d-valued functions on I into R^d.
2) $\Omega_d(\Omega)$ (see p. 30).
3) $\mathscr{M}_t(\mathscr{M})$ (see p. 30).
4) $x(t, \omega)$ (see p. 30).

IV. Miscellaneous Notation

1) $a \wedge b$ is the smaller of the numbers $a, b \in R$.
2) $a \vee b$ is the larger of the numbers $a, b \in R$.
3) S_d is the set of symmetric non-negative definite $d \times d$ real matrices.
4) S_d^+ is the set of nondegenerate elements of S_d.
5) $\|A\|$, where A is a square matrix, and is the operator norm of A.
6) $\sigma(\mathscr{C})$, where $\mathscr{C}$ is a collection of subsets of X, and is the smallest σ-algebra over X containing $\mathscr{C}$.
7) $\sigma(\mathscr{F})$, where $\mathscr{F}$ is a set of functions on X into a measurable space, and is the smallest σ-algebra over X with respect to which every element of $\mathscr{F}$ is measurable.
8) $[\lambda]$, $\lambda \in R$, is the integral part of λ.
9) $\xi \sim I^s_d(a, b)$ (see p. 92).

Chapter 0

Introduction

The main purpose of this book is to elucidate the martingale approach to the theory of Markov processes. Needless to say, we believe that the approach has many advantages over previous ones, and it is our hope that the present book will convince some other people that this is indeed so. When we began this project we were uncertain whether to proselytize by intimidating the reader with myriad examples demonstrating the full scope of the techniques or by persuading him with a careful treatment of just one problem to which they apply. We have decided on the latter plan in the belief that it is preferable to bore than to batter. The result is that we have devoted what may seem like an inordinate number of pages to a rather special topic. On the other hand, we have endeavoured to present our proofs in such a way that the techniques involved should lend themselves to easy adaptation in other contexts. Only time will tell if we have succeeded.

The topic which we have chosen is that of diffusion theory in R^d. In order to understand how this subject fits into the general theory of Markov processes, it is best to return to Lévy's ideas about "stochastic differentials." Let $x(\cdot)$ be a Markov process with values in R^d and suppose that for $t \geq 0$ and test functions $\varphi \in C_0^\infty(R^d)$

$$(0.1) \quad E[\varphi(x(t + h)) - \varphi(x(t)) \,|\, x(s), s \leq t] = h L_t \varphi(x(t)) + o(h), \qquad h > 0,$$

where, for each $t \geq 0$, L_t is a linear operator on $C_0^\infty(R^d)$ into $C_b(R^d)$. It is obvious that L_t must satisfy the weak maximum principle, since if φ achieves its maximum at x^0 then $E[\varphi(x(t + h)) - \varphi(x(t)) \,|\, x(t) = x^0] \leq 0$. Moreover, if $x^0 \in R^d$ and $\eta \in C_0^\infty(R^d)$ is such that $\eta(x^0) = 1$, $0 \leq \eta \leq 1$, and η is identically zero outside an ε-neighborhood of x^0, then $P(|x(t + h) - x^0| \geq \varepsilon \,|\, x(t) = x^0) \leq E[(1 - \eta(x(t + h))) \,|\, x(t) = x^0] \leq h |L_t \eta(x^0)| + o(h)$. Thus if $\varphi \in C_0^\infty(R^d)$ vanishes in a neighborhood of x^0, then $|L_t \varphi(x^0)| \leq C_\varepsilon \|\varphi\|$. That is to say, L_t is "quasilocal." Therefore, if we now define the operator L_{t, x^0} by the relation

$$L_{t, x^0}\varphi(x) = [L_t(\tau_{x - x^0}\varphi)](x^0), \qquad x \in R^d,$$

where $\tau_y \varphi(\cdot) = \varphi(\cdot + y)$, then L_{t, x^0} is a quasi-local, translation invariant operator satisfying the maximum principle. This class of operators is well-known and can be shown to coincide with the class of generators of time homogeneous independent increment processes (cf. Hille and Philips [1957]). In particular, we can conclude that

$$L_t\varphi(x^0) = \frac{1}{2} \sum_{i,j=1}^d a^{ij}(t, x^0) \frac{\partial^2 \varphi}{\partial x_i \, \partial x_j}(x^0) + \sum_{i=1}^d b^i(t, x^0) \frac{\partial \varphi}{\partial x_i}(x^0)$$

$$+ \int \left(\varphi(x^0 + y) - \varphi(x^0) + \frac{\langle y, \nabla\varphi(x^0) \rangle}{1 + |y|^2} \right) M(t, x^0; dy)$$

where $((a^{ij}(t, x^0)))$ is non-negative definite and $M(t, x^0; \cdot)$ is a Lévy measure. More important, we develop from these considerations the intuitive picture of the process $x(\cdot)$ leaving $x(t)$ like the independent increment process with characteristics $a(t, x(t))$, $b(t, x(t))$, and $M(t, x(t); \cdot)$. Throughout this book we will be restricting ourselves to continuous Markov processes. For a continuous process, the Lévy measure M must be absent. That is, if $x(\cdot)$ is a continuous Markov process and (0.1) obtains, then

$$(0.2) \qquad L_t = \frac{1}{2} \sum_{i,j=1}^d a^{ij}(t, x) \frac{\partial^2}{\partial x_i \, \partial x_j} + \sum_1^d b^i(t, x) \frac{\partial}{\partial x_i}$$

and for small $h > 0$, $x(t + h) - x(t)$ is like the Gaussian independent increment process having mean $b(t, x(t))$ and covariance $a(t, x(t))$. (A slightly different presentation of these ideas is given in the introduction to Itô [1951]. We recommend Itô's discussion to the interested reader.)

The structure of this book can now be explained in terms of the ideas introduced in the preceding paragraph. Starting from (0.1), various tacks toward an understanding of the process $x(\cdot)$ suggest themselves. The most analytic of these is the following. Let $P(s, x; t, \cdot)$ denote the transition probability function determined by $x(\cdot)$ (i.e., $P(s, x; t, \Gamma) = P(x(t) \in \Gamma \,|\, x(s) = x)$). From (0.1), we see that

$$\frac{\partial}{\partial t} \int P(s, x; t, dy)\varphi(y) = \lim_{h\downarrow 0} \int P(s, x; t, dz)$$

$$\times \int P(t, z; t + h, dy)(\varphi(y) - \varphi(z))$$

$$= \int P(s, x; t, dy)L_t\varphi(y).$$

Of course, we have used the Chapman-Kolmogorov equation. From this we derive the formal relation:

$$(0.3) \qquad \frac{\partial}{\partial t} P(s, x; t, \cdot) = L_t^* P(s, x; t, \cdot), \qquad t > s,$$

where L_t^* is the formal adjoint of L_t. Equation (0.3) is called the *forward equation* (in physics and engineering literature it is often referred to as the *Fokker-Planck equation*). Since it is clear that

$$(0.3') \qquad \lim_{t\downarrow s} P(s, x; t, \cdot) = \delta_x(\cdot),$$

it is reasonable to suppose that one can recapture $P(s, x; t, \cdot)$ from (0.3) and (0.3'). Indeed, this was done with great success by Kolmogorov [1931] and Feller [1936] in their pioneering work on this subject. However, there are severe technical problems with (0.3). In particular, one must tacitly assume that $P(s, x; t, \cdot)$ admits a density $p(s, x; t, y)$ and think of (0.3) as being an equation for $p(s, x; t, y)$ as a function of t and y; and even when such an assumption is justified, there remain inherent difficulties in the interpretation of L_t^* unless the coefficients are smooth. For this reason, people turned their attention to the *backward equation*. Namely, starting once again from (0.1) we have

$$-\frac{\partial u}{\partial s}(s, x) = -\frac{\partial}{\partial s}\int P(s, x; t, dy)\varphi(y)$$

$$= \lim_{h\downarrow 0}\frac{1}{h}\int P(s - h, x; s, dy)(u(s, y) - u(s, x))$$

$$= L_s u(s, x),$$

where $u(s, x) = \int P(s, x; t, dy)\varphi(y)$, $0 \leq s < t$. (Notice that the preceding computation is not fully justified since we do not know that $u(s, \cdot)$ is in our test-function space. Nonetheless, the argument is correct in spirit.) Hence we arrive at

(0.4) $$\qquad \frac{\partial}{\partial s}P(s, x; t, \cdot) + L_s P(s, x; t, \cdot) = 0, \qquad 0 \leq s < t,$$

(0.4') $$\qquad \lim_{s\uparrow t} P(s, x; t, \cdot) = \delta_x(\cdot).$$

(It should be clear why (0.3) is the forward equation and (0.4) is the backward equation: (0.3) involves the forward (i.e., future) variables whereas (0.4) involves the backward (i.e., past) variables.) Again one might suspect that (0.4) and (0.4') determine $P(s, x; t, \cdot)$ and now there are no problems about interpretation. The study of diffusion theory via the backward equation has been one of the more powerful and successful approaches to the subject and we have included a sketch of this procedure in Chapters 2 and 3.

The major objection to the study of diffusion theory by the method just described is that the hard machinery used comes from the theory of partial differential equations and the probabilistic input is relatively small. A more probabilistically satisfactory approach was suggested by Lévy and carried out by Itô [1951]. The idea here is to return to the intuitive picture of $x(t + h) - x(t)$, for small $h > 0$, looking like the Gaussian independent increment process with drift $b(t, x(t))$ and covariance $a(t, x(t))$. In differential form, this intuitive picture means that

(0.5) $$\qquad dx(t) = \sigma(t, x(t))\, d\beta(t) + b(t, x(t))\, dt$$

where $\beta(\cdot)$ is a d-dimensional Brownian motion and σ is a square root of a. Indeed, $\sigma(t, x(t))(\beta(t + h) - \beta(t)) + b(t, x(t))$ will be just such a Gaussian process;

and if $\{x(s), \ 0 \le s < t\}$ is $\{\beta(s): 0 \le s \le t\}$-measurable, then $\sigma(t, x(t)) \times$ $(\beta(t + h) - \beta(t)) + b(t, x(t))$ will be conditionally independent of $\{x(s): 0 \le s < t\}$ given $x(t)$. There are two problems of considerable technical magnitude raised by (0.5). First and foremost is the question of interpretation. Since a Brownian path is nowhere differentiable it is by no means obvious that sense can be made out of a differential equation like (0.5). Secondly, even if one knows what (0.5) means, one still has to learn how to solve such an equation before it can be considered to be useful. Both these problems were masterfully handled by Itô, a measure of the success of his solution is the extent to which it is still used. We develop Itô's theory of stochastic integration in Chapter 4 and apply it to equations like (0.5) in Chapter 5.

With Chapter 6 we begin the study of diffusion theory along the lines initiated by us in Stroock and Varadhan [1969]. In order to understand this approach, we return once again to (0.1). From (0.1), it is easy to deduce that:

$$\frac{d}{dt_2} E[\varphi(x(t_2)) \,|\, x(s), \ s \le t_1]$$

$$= \lim_{h \downarrow 0} \frac{1}{h} E[E[\varphi(x(t_2 + h)) - \varphi(x(t_2))] \,|\, x(s), \ s \le t_2] \,|\, x(s), \ s \le t_1]$$

$$= E[L_{t_2}\varphi(x(t_2)) \,|\, x(s), \ s \le t_1].$$

Thus

$$E[\varphi(x(t_2)) - \varphi(x(t_1)) - \int_{t_1}^{t_2} L_t\varphi(x(t)) \, dt \,|\, x(s), \ s \le t_1] = 0;$$

or in other words

(0.6) $$X_\varphi(t) \equiv \varphi(x(t)) - \int_0^t L_s\varphi(x(s)) \, ds$$

is a martingale for all test functions φ. (Notice that the line of reasoning leading from (0.1) to (0.6) is essentially the same as that from (0.1) to the forward equation.) One can now ask if the property that $X_\varphi(\cdot)$ is a martingale for all test functions φ uniquely characterizes the process $x(\cdot)$ apart from specifying $x(0)$. To be more precise, given L_t, consider the following problems:

    (i) Is there for each $x \in R^d$ a probability measure P on $C([0, \infty), R^d)$ such that $P(x(0) = x) = 1$ and $X_\varphi(\cdot)$ is a martingale for all test functions φ?
and
    (ii) Is there at most one such P for each x?

    Problems (i) and (ii) constitute what we call the *martingale problem for* L_t. Of course problems (i) and (ii) are interesting only if one can also answer

    (iii) If (i) and (ii) have affirmative answers, what conclusions can be drawn?

To convince oneself that these are reasonable questions, one should recall that in the case when $d = 1$ and $L_t = \frac{1}{2}d^2/dx^2$, Lévy (cf. Doob [1953] or Exercise 4.6.6) characterized Wiener measure as the unique probability measure P on

$C([0, \infty), R^1)$ such that $P(x(0) = 0) = 1$ and $x(t)$ and $x^2(t) - t$ are martingales. That is, he showed that in this case one only needs the functions $\varphi(x) = x$ and $\psi(x) = x^2$. (Actually [cf. Exercise 4.6.6], this is a general phenomenon, since under general conditions one can show that $X_\varphi(\cdot)$ is a martingale for all test functions φ if $X_{\varphi_j}(\cdot)$ and $X_{\psi_{ij}}(\cdot)$ are martingales for $\varphi_j(x) = x_j$ and $\psi_{ij}(x) = x_i x_j$, $1 \le i, j \le d$. When $d = 1$, this general phenomenon was already pointed out in Chapter 9 of Doob [1953].) Furthermore, one should remember that much of Doob's beautiful work in potential theory relies heavily on the observation that $X_\varphi(\cdot)$ is a martingale when P is the process associated with L_t. Thus what if anything is truly new about our approach is that we have made this observation the cornerstone of our theory and asked if in fact it does not underlie the whole structure of diffusion theory.

In Chapter 6 we lay the foundation for everything that follows. In particular, we prove there a basic existence theorem for solutions to the martingale problem. Once this is done, we start laying the groundwork for out attack on the question of uniqueness by deriving general conclusions that can be drawn about a P on $C([0, \infty), R^d)$ under the assumption that $X_\varphi(\cdot)$ is a martingale for all test functions φ. These include the relationship between the martingale problem and the (strong) Markov property, as well as the formula of Cameron, Martin and Girsanov.

Chapter 7 contains a proof of our best general theorem about uniqueness for the martingale problem. What we show is that if the coefficients a and b in L_t are bounded and measurable and for each $T > 0$ and $R > 0$, a satisfies

$$(0.7) \qquad \inf_{\substack{0 \le s \le T \\ |x| \le R}} \inf_{\theta \in R^d \setminus \{0\}} \frac{\langle \theta, a(s, x)\theta \rangle}{|\theta|^2} > 0$$

and

$$(0.8) \qquad \lim_{\delta \downarrow 0} \sup_{\substack{0 \le s \le T \\ |x^1|, |x^2| \le R \\ |x^1 - x^2| < \delta}} \|a(s, x^1) - a(s, x^2)\| = 0,$$

then the martingale problem for L_t is well-posed (i.e., existence and uniqueness hold). As a dividend of our proof, we show that L_t determines a strong Markov, strongly Feller continuous process.

The contents of Chapter 8 are somewhat tangential to the main thrust of our development. What we do there is expand on the theme initiated in Watanabe and Yamada [1971] in their investigation of the relationship between Itô's approach and the martingale problem.

In Chapter 9 we return to L_t's having coefficients of the sort studied in Chapter 7. Here we take advantage of certain analytic relations and estimates upon which our proof of uniqueness in Chapter 7 turns. In brief, the results of these considerations are various L^p-estimates for the transition probability function of the process determined by L_t.

Chapter 10 extends the martingale problem approach to unbounded coefficients. The point made here is that this extension is elementary, provided one can show that the diffusion process does not "explode." We give some standard conditions that can be used to test for explosion.

Again in Chapter 11 we deal with L_t's of the sort studied in Chapters 7 and 10. This time we are interested in stability results for the associated processes. These results can be naturally divided into two categories: convergence of Markov chains to diffusions (i.e., invariance principles of the sort initiated by Erdös and Kac and perfected by Donsker) and convergence of diffusions to other diffusions. Both categories are surprisingly easy to handle given the results of Chapters 7 and 9.

The final chapter, Chapter 12, takes up the question of what can be done in those circumstances when existence of solutions to a martingale problem can be proved but uniqueness cannot. The idea here, is to make a careful "selection" of solutions so that they fit together into a Markov family. The procedure that we use goes back to Krylov [1973]. We also show in Chapter 12 that every solution to a given martingale problem can, in some sense, be built out of those solutions which are part of a Markov family.

The only parts of the book which we have not yet discussed are the beginning and the end. Chapter 1 provides an introduction to those parts of measure and probability theories which we consider most important for an understanding of this book. Although the material here is not new, much of it has been reworked. In particular, our criteria for compactness in Section 1.4 strikes us as a useful variation on the ideas of Prohorov.

Finally, in spite of our attempt to make it look as if it were, the appendix is not probability theory. Instead, it is that part of the theory of singular integrals on which we rely in Chapters 7 and 9. At the present time, one has to depend on these results from outside probability theory and we have provided a proof in the Appendix in order to make the book self-contained.

It is now time for us to thank the many people and organizations to whom we are deeply indebted. The original work out of which this book grew was performed while both of us were at the Courant Institute of Mathematical Sciences. During that period we were encouraged and stimulated by many people, particularly: M. Kac, H. P. McKean Jr., S. Sawyer, M. D. Donsker and L. Nirenberg; and we were supported by grants from the Air Force, the Sloan and Ford foundations as well as general C.I.M.S. funds. Whether this book would ever see the light of day was cast into considerable doubt by the departure from C.I.M.S. of one of us to the Rocky Mountains in 1972. At that time not a sentence of it had been written. However, in 1976 we had the good fortune to visit Paris together under the auspices of Professors Neveu and Revuz; and it was at that time (much to the dismay of an accompanying wife) that we actually began to write this book. Progress from that point on has been slow but steady. During the interim we have incurred a considerable debt of gratitude to several people: wives Lucy and Vasu; secretaries Janice Morgenstern, Gloria Lee, Susan Parris and Helen Samoraj; students Marty Day and Pedro Echeverria; colleagues Richard Holley, G. Papanicolaou, M. D. Donsker, and E. Fabes; gadfly J. Doob, and publisher Springer Verlag. To all these we extend our heart felt thanks along with the promise that they do not necessarily have to read what we have written.

Chapter 1

Preliminary Material: Extension Theorems, Martingales, and Compactness

1.0. Introduction

As mentioned in the Introduction, the point of view that we take will involve us in a detailed study of measures on function spaces. There are a few basic tools which are necessary for the construction of such measures. The purpose of this chapter is to develop these tools. In the process, we will introduce some notions (e.g., conditioning and martingales) which will play an important role in what follows. Section 1.1 contains the basic theorem of Prohorov and Varadarajan characterizing weakly compact families of measures on a Polish space. Using their results, we prove the existence of conditional probability distributions. The final topics in Section 1.1 are the extension theorems of Tulcea and Kolmogorov. Section 1.2 introduces the notions of progressively measurable functions and martingales. In connection with martingales we prove Doob's inequality, his stopping time theorem and a useful integration by parts formula. Finally we prove a result connecting martingale theory and conditioning.

In Section 1.3 we specialize the results of Section 1.1 to the case when our Polish space is $C([0, \infty); R^d)$ (i.e., the space of R^d valued continuous functions on $[0, \infty)$ with the natural topology induced by uniform convergence on bounded intervals). Section 1.4 contains a useful sufficient condition for compactness of a family of measures on $C([0, \infty); R^d)$ in terms of certain martingales associated with them.

1.1. Weak Convergence, Conditional Probability Distributions, and Extension Theorems

Throughout this section (X, D) will stand for a Polish space (i.e., a complete separable metric space) and $\mathcal{B} = \mathcal{B}_X$ its Borel σ-field. We denote by $M(X)$ the set of all probability measures on $(X, \mathcal{B})$ and by $C_b(X)$ the set of all bounded continuous functions on X. We will view $M(X)$ as a subset of the dual space of $C_b(X)$ and give it the inherited weak* topology. It will turn out that this topology makes $M(X)$ into a metric space.

1.1.1 Theorem. *Let $\mu_n \in M(X)$ for each $n \geq 1$. Given $\mu \in M(X)$, the following are equivalent:*

(i) $\lim_{n \to \infty} \int f \, d\mu_n = \int f \, d\mu$ for every $f \in C_b(X)$

(ii) $\lim_{n \to \infty} \int f \, d\mu_n = \int f \, d\mu$ for every $f \in U_\rho(X)$ where $U_\rho(X)$ is the set of bounded uniformly continuous functions on (X, ρ) and ρ is any equivalent metric on X.

(iii) $\lim \sup_{n \to \infty} \mu_n(C) \le \mu(C)$ for any closed set C in X

(iv) $\lim \inf_{n \to \infty} \mu_n(G) \ge \mu(G)$ for any open set G in X

(v) $\lim_{n \to \infty} \mu_n(B) = \mu(B)$ for any $B \in \mathscr{B}$ such that $\mu(\partial B) = 0$.

Proof. That (i) implies (ii) is obvious, and (iii) is equivalent to (iv) by complementation. Also the following simple argument shows that (iii) and (iv) together imply (v):

$$\lim_{n \to \infty} \sup \mu_n(B) \le \lim_{n \to \infty} \sup \mu_n(\bar{B}) \le \mu(\bar{B}) = \mu(B^0) \le \lim_{n \to \infty} \inf \mu_n(B^0) \le \lim_{n \to \infty} \inf \mu_n(B).$$

It remains to show that (ii) implies (iii) and (v) implies (i). To prove that (ii) implies (iii), let C be a closed set and choose

$$f_k(x) = \left[\frac{1}{1 + \rho(x, C)} \right]^k, \quad k \ge 1$$

where $\rho(x, C) = \inf\{\rho(x, y): y \in C\}$. Then $f_k \in U_\rho(X)$ for each $k \ge 1$ and $f_k(x) \downarrow \chi_C(x)$ for each $x \in X$. Therefore

$$\mu(C) = \lim_{k \to \infty} \int f_k \, d\mu = \lim_{k \to \infty} \lim_{n \to \infty} \int f_k \, d\mu_n \ge \lim_{n \to \infty} \sup \mu_n(C).$$

Finally, to see that (v) implies (i), take $f \in C_b(X)$ and, given $\varepsilon > 0$, choose N and $\{a_i\}_1^{N-1}$ so that

$$- \|f\| - 1 = a_0 < a_1 \cdots < a_{N-1} < a_N = \|f\| + 1,$$

$\mu(\{x: f(x) = a_i\}) = 0$ for $1 \le i \le N - 1$, and $a_i - a_{i-1} < \varepsilon$ for $1 \le i \le N$. Set $B_i = \{x: a_{i-1} \le f(x) < a_i\}$ for $i = 1, 2, \ldots, N$. Then the B_i's are disjoint, their union is X and $\mu(\partial B_i) = 0$ for each i. Also

$$\left\| f - \sum_{i=1}^{N} a_i \chi_{B_i}(x) \right\| < \varepsilon.$$

Hence,

$$\lim_{n \to \infty} \sup \left| \int f \, d\mu_n - \int f \, d\mu \right|$$

$$\le 2\varepsilon + \lim_{n \to \infty} \sup \left| \int \left(\sum_{1}^{N} a_i \chi_{B_i} \right) d\mu_n - \int \left(\sum_{1}^{N} a_i \chi_{B_i} \right) d\mu \right|$$

$$\le 2\varepsilon + \sum_{1}^{N} |a_i| \lim_{n \to \infty} \sup |\mu_n(B_i) - \mu(B_i)|$$

$$\le 2\varepsilon.$$

Since $\varepsilon > 0$ was arbitrary, the proof is complete. $\quad\square$

Remark. The equivalence of (*i*) and (*ii*) implies that for any λ, $\mu \in M(X)$: if $\int f \, d\mu = \int f \, d\lambda$ for $f \in U_\rho(X)$, then $\int f \, d\mu = \int f \, d\lambda$ for all functions f in $C_b(X)$ and in fact $\lambda = \mu$ on $\mathscr{B}$.

Since X is a separable metric space, by Tychonoff's embedding theorem, X is homeomorphic to a subset of a compact metric space. Thus X admits an equivalent metric ρ with respect to which it is totally bounded. Choose such a ρ and let $\bar{X}$ denote the completion of (X, ρ). Then $\bar{X}$ is compact and $U_\rho(X)$ is isomorphic to $C(\bar{X})$, which is separable. With these remarks we will prove that the weak* topology on $M(X)$ is metrizable. In fact, define for μ, $\lambda \in M(X)$

$$(1.1) \qquad \Delta(\mu, \lambda) = \sum_{k=0}^{\infty} \frac{1}{2^k} \frac{\left| \int \varphi_k \, d\mu - \int \varphi_k \, d\lambda \right|}{\|\varphi_k\|}$$

where $\{\varphi_k : k \geq 0\}$ is dense in $U_\rho(X)$. By the remark following Theorem 1.1.1, Δ is clearly a metric on $M(X)$.

1.1.2 Theorem. *If Δ is defined on $M(X)$ by (1.1), then the topology induced on $M(X)$ by Δ is the weak* topology.*

Proof. Obviously the topology induced by Δ is weaker than the weak* topology. We now show that for any $\mu \in M(X)$ and any weak* neighborhood N of μ, there is a $\delta > 0$ such that

$$\{\lambda : \Delta(\lambda, \mu) < \delta\} \subseteq N.$$

If not, there is a sequence $\{\lambda_n\}_1^\infty \subseteq M(X)$ such that $\Delta(\lambda_n, \mu) \to 0$ as $n \to \infty$ and yet $\lambda_n \notin N$ for any $n \geq 1$. On the other hand, from the definition of Δ in (1.1), the denseness of $\{\varphi_k : k \geq 0\}$, and the equivalence of (*i*) and (*ii*) of Theorem 1.1.1, it would follow that

$$\lim_{n \to \infty} \int f \, d\lambda_n = \int f \, d\mu$$

for all $f \in C_b(X)$. Since N must contain a set of the form

$$\left\{ \lambda : \max_{1 \leq j \leq l} \left| \int f_j \, d\lambda - \int f_j \, d\mu \right| < \varepsilon \right\}$$

for some choice of ε, l and $\{f_j\}_1^l \subseteq C_b(X)$, we have a contradiction. $\square$

Remark. With a little more work it is possible to show that there is a metric $\tilde{\Delta}$ on $M(X)$, equivalent to Δ, such that $(M(X), \tilde{\Delta})$ is a Polish space. We will not be needing this fact in what follows.

1.1.3 Theorem. *Let Γ be a compact subset of $M(X)$. Then for each $\varepsilon > 0$ there is a compact $K \subseteq X$ such that $\mu(X \backslash K) \leq \varepsilon$ for all $\mu \in \Gamma$. In particular for any $\mu \in M(X)$ and $\varepsilon > 0$, there is a compact $K \subseteq X$ such that $\mu(X \backslash K) \leq \varepsilon$.*

Proof. Let $\{x_j : j \geq 1\}$ be dense in X, and for $k \geq 1$ and $n \geq 1$ put

$$G_k^n = \bigcup_{j=1}^{n} B\left(x_j, \frac{1}{k}\right).$$

By Theorem 1.1.1, the map $\mu \to \mu(G_k^n)$ is lower semicontinuous, and clearly

$$\mu(G_k^n) \uparrow 1$$

as $n \to \infty$ for each k. Since Γ is compact, it follows from Dini's theorem that for each $\varepsilon > 0$ and $k \geq 1$, there is an n_k such that

$$\inf_{\mu \in \Gamma} \mu(G_k^{n_k}) \geq 1 - \frac{\varepsilon}{2^k}.$$

Set $K = \bigcap_{k=1}^{\infty} \overline{G_k^{n_k}}$. Then

$$\inf_{\mu \in \Gamma} \mu(K) \geq 1 - \varepsilon.$$

Moreover, K is closed in X and is therefore complete. Finally, for any $k \geq 1$

$$K \subset \bigcup_{1}^{n_k} B\left(x_i, \frac{2}{k}\right)$$

and so K is totally bounded. Thus K is compact and the proof is finished. $\quad\square$

1.1.4 Theorem. *Let $\Gamma \subseteq M(X)$ be given and assume that for each $\varepsilon > 0$ there is a compact set $K \subseteq X$ such that*

$$\inf_{\mu \in \Gamma} \mu(K) \geq 1 - \varepsilon.$$

Then Γ is precompact in $M(X)$ (i.e., $\bar{\Gamma}$ is compact).

Proof. We first recall that if X itself is compact, then by Riesz's theorem and standard results in elementary functional analysis, $M(X)$ is compact. In general we proceed as follows. Choose a metric ρ on X equivalent to the original one such that (X, ρ) is totally bounded and denote by $\bar{X}$ its completion. Then $\bar{X}$ is compact and we can think of $M(X)$ as being a subset of $M(\bar{X})$. Thus it remains to show that if $\{\mu_n\}_1^{\infty} \subseteq \Gamma$ and $\mu_n \to \bar{\mu}$ in $M(\bar{X})$, then $\bar{\mu}$ can be restricted to X as a probability measure μ, and $\mu_n \to \mu$ in $M(X)$. But, by our assumption on Γ, there is a sequence of compact sets $\{K_l\}$, $l \geq 1$ in X such that $\mu_n(K_l) \geq 1 - 1/l$ for $n \geq 1$. Since K_l is compact and therefore closed in $\bar{X}$, it follows from Theorem 1.1.1 that $\bar{\mu}(K_l) \geq \underline{\lim}_{n \to \infty} \mu_n(K_l) \geq 1 - 1/l$. Thus

$$\bar{\mu}\left(\bigcup_{1}^{\infty} K_l\right) = 1$$

and so we can restrict $\bar{\mu}$ to $(X, \mathscr{B}_X)$ as a probability measure μ. Finally since $\mu_n \to \bar{\mu}$ in $M(\bar{X})$

$$\lim_{n \to \infty} \int \varphi \, d\mu_n = \int \varphi \, d\mu$$

for all $\varphi \in U_\rho(X)$, and by Theorem 1.1.1 this implies that $\mu_n \to \mu$ in $M(X)$.

Remark. Note that whereas Theorem 1.1.3 relies very heavily on the completeness of X, Theorem 1.1.4 does not use it at all.

1.1.5 Corollary. *Let $F \subseteq C_b(X)$ be a uniformly bounded set of functions which are equicontinuous at each point of X. Given a sequence $\{\mu_n\}_1^\infty \subseteq M(X)$ such that $\mu_n \to \mu$ in $M(X)$, one has*

$$\lim_{n \to \infty} \sup_{\varphi \in F} \left| \int \varphi \, d\mu_n - \int \varphi \, d\mu \right| = 0.$$

Proof. Suppose there exists $\varepsilon > 0$ such that

$$\limsup_{n \to \infty} \sup_{\varphi \in F} \left| \int \varphi \, d\mu_n - \int \varphi \, d\mu \right| > \varepsilon.$$

By choosing a subsequence if necessary, we can assume that for each n there is a φ_n in F such that

$$\left| \int \varphi_n \, d\mu_n - \int \varphi_n \, d\mu \right| \geq \varepsilon > 0.$$

Let $M = \sup_{\varphi \in F} \|\varphi\|$ and choose a compact set K in X such that $\sup_{n \geq 1} \mu_n(X \backslash K) \leq \varepsilon/8M$. It follows from this that $\mu(X \backslash K) \leq \varepsilon/8M$. From the Ascoli-Arzela theorem and the Tietze extension theorem, we can find a ψ in $C_b(X)$ such that $|\psi| \leq M$, and a subsequence $\varphi_{n_j} = \psi_j$ of $\{\varphi_n\}$ converging to ψ uniformly on K. But then

$$\left| \int \psi_j \, d\mu_{n_j} - \int \psi_j \, d\mu \right| \leq \left| \int (\psi_j - \psi) \, d\mu_{n_j} \right| + \left| \int (\psi_j - \psi) \, d\mu \right|$$

$$+ \left| \int \psi \, d\mu_{n_j} - \int \psi \, d\mu \right|$$

$$= \left| \int_K (\psi_j - \psi) \, d\mu_{n_j} \right| + \left| \int_{X \backslash K} (\psi_j - \psi) \, d\mu_{n_j} \right|$$

$$+ \left| \int_K (\psi_j - \psi) \, d\mu \right| + \left| \int_{X \backslash K} (\psi_j - \psi) \, d\mu \right|$$

$$+ \left| \int \psi \, d\mu_{n_j} - \int \psi \, d\mu \right|.$$

Therefore

$$0 < \varepsilon \le \overline{\lim_{j \to \infty}} \left| \int \psi_j \, d\mu_{n_j} - \int \psi_j \, d\mu \right|$$

$$\le 0 + 2M \cdot \frac{\varepsilon}{8M} + 0 + 2M \cdot \frac{\varepsilon}{8M} + 0$$

$$= \frac{\varepsilon}{2}$$

which is a contradiction. $\square$

We now turn to the study of conditional probability distributions. Let $(E, \mathscr{F}, P)$ be a probability space and $\Sigma \subseteq \mathscr{F}$ a sub σ-field. Then the conditional expectation of an integrable function $f(\cdot)$ is another integrable function $g(\cdot)$ which is Σ-measurable and satisfies:

$$(1.2) \qquad \int_A g(q)P(dq) = \int_A f(q)P(dq) \qquad \text{for all } A \in \Sigma.$$

The function $g(\cdot)$ exists and is unique in the sense that any two choices agree almost surely with respect to P. (See Halmos [1950], for instance, to find a proof of the existence and elementary properties of conditional expectations.) The function g is denoted by $E[f|\Sigma]$. If we want to call attention to the measure P that is used, we use $E^P[f|\Sigma]$ in place of $E[f|\Sigma]$. In the special case when the function $f(\cdot)$ is the indicator function $\chi_B(q)$ of a set B in $\mathscr{F}$ we refer to the conditional expectation as the conditional probability and denote it by $P(B|\Sigma)$. It has some elementary properties inherited from the properties of conditional expectations. For instance if B_1 and B_2 are disjoint sets in $\mathscr{F}$

$$P(B_1 \cup B_2|\Sigma) = P(B_1|\Sigma) + P(B_2|\Sigma) \qquad \text{a.s.}$$

Since $P(B|\Sigma)$ can be altered on a set of measure zero for each $B \in \mathscr{F}$ it is important to know if one can choose a "nice" version of $P(B|\Sigma)$ such that $P(B|\Sigma)$ is a countably additive probability measure on $\mathscr{F}$ for each $q \in E$. Such a choice, if it can be made, will be called a conditional probability distribution and will be denoted by $\{Q_q(B)\}$.

Definition. A *conditional probability distribution* of P given Σ is a family Q_q of probability measures on $(E, \mathscr{F})$ indexed by $q \in E$ such that
(i) For each $B \in \mathscr{F}$, $Q_q(B)$ is Σ-measurable as a function of q
(ii) For every $A \in \Sigma$ and $B \in \mathscr{F}$

$$P(A \cap B) = \int_A Q_q(B)P(dq).$$

In general a conditional probability distribution need not exist. However if we replace $(E, \mathscr{F})$ by a Polish space X and its Borel σ-field $\mathscr{B}$, then for any P on $(X, \mathscr{B})$ and any sub σ-field $\Sigma \subseteq \mathscr{B}$ a conditional probability distribution of P given Σ exists. We state and prove this as a theorem.

1.1.6 Theorem. *Let X be a Polish space and $\mathscr{B}$ its Borel σ-field. Let P be a probability distribution on $(X, \mathscr{B})$ and $\Sigma \subseteq \mathscr{B}$ a sub σ-field of $\mathscr{B}$. Then a conditional probability distribution $\{Q_x\}$ of P given Σ always exists.*

Proof. Since X is a Polish space, there exists on X an equivalent metric ρ such that (X, ρ) is totally bounded. Therefore the space $U_\rho(X)$ of uniformly continuous bounded functions on (X, ρ) is separable. Let $\{f_j\}_1^\infty$ be a countable subset of $U_\rho(X)$ such that $f_1(\cdot) \equiv 1$, $\{f_j\}_1^\infty$ are linearly independent and the linear span W of $\{f_j\}_1^\infty$ is dense in $U_\rho(X)$. We denote by g_i some version of $E[f_i | \Sigma]$. We can assume, without loss of generality, that $g_1(\cdot) \equiv 1$. For $n \geq 1$ let Λ_n be the set of n-tuples $(r_1, r_2, \ldots, r_n)$ with rational entries such that

$$r_1 f_1(x) + \cdots + r_n f_n(x) \geq 0 \qquad \text{for all} \quad x \in X.$$

The set Λ_n is clearly countable. From the properties of conditional expectations,

$$r_1 g_1(x) + r_2 g_2(x) + \cdots + r_n g_n(x) \geq 0$$

for almost all x. Thus if we let, for each $(r_1, \ldots, r_n) \in \Lambda_n$,

$$F_{(r_1, \ldots, r_n)} = \{x : r_1 g_1(x) + \cdots + r_n g_n(x) < 0\}$$

then $F_{(r_1, \ldots, r_n)} \in \Sigma$ and $P[F_{(r_1, \ldots, r_n)}] = 0$. Therefore if

$$F = \bigcup_{n=1}^\infty \bigcup_{(r_1, \ldots, r_n) \in \Lambda_n} F_{(r_1, \ldots, r_n)}.$$

Then $F \in \Sigma$ and $P(F) = 0$.

Choose $x \in X \backslash F$ and fix it. We now define the linear functional L_x on W by

$$L_x(f) = t_1 g_1(x) + \cdots + t_n g_n(x)$$

where $f \in W$ is written uniquely as

$$f = t_1 f_1 + \cdots + t_n f_n$$

for some n and real numbers $t_1, t_2, \ldots, t_n$. We want to show now that $L_x(f)$ is a nonnegative linear functional on W. Suppose $f \in W$ is non-negative. Then $t_1 f_1 + \cdots + t_n f_n = f \geq 0$. Given any rational number $\varepsilon > 0$ we can find rationals $(r_1, \ldots, r_n)$ such that

$$r_1 f_1 + \cdots + r_n f_n \geq -\varepsilon = -\varepsilon f_1$$

and $|r_j - t_j| < \varepsilon$ for $1 \le j \le n$. Therefore

$$(r_1 + \varepsilon)f_1 + \cdots + r_n f_n \ge 0.$$

From the definition of the set F, it follows that

$$(r_1 + \varepsilon)g_1(x) + \cdots + r_n g_n(x) \ge 0.$$

By letting $\varepsilon \to 0$ over the rationals, we conclude that

$$t_1 g_1(x) + \cdots + t_n g_n(x) \ge 0,$$

or equivalently $L_x(f) \ge 0$. Since $L_x(f_1) = 1$ and W is dense in $U_\rho(X)$, $L_x(f)$ defined on W extends uniquely as a non-negative linear functional on $U_\rho(X)$. We continue to call this extension $L_x(f)$. We can view the space $U_\rho(X)$ as the space $C(\bar{X})$, where $\bar{X}$ is the completion of the totally bounded space (X, ρ). Note that $\bar{X}$ is compact. By the Riesz Representation Theorem, there is a probability measure $\bar{Q}_x$ on $(\bar{X}, \mathscr{B}_{\bar{X}})$ such that

$$L_x(\bar{f}) = \int_{\bar{X}} \bar{f}(y)\bar{Q}_x(dy)$$

for all $\bar{f} \in C(\bar{X})$.

Thus we have shown that for all $x \in X \backslash F$, there is a probability measure $\bar{Q}_x$ on $(\bar{X}, \mathscr{B}_{\bar{X}})$ such that

(1.3) $$g_i(x) = \int \bar{f}_i(y)\bar{Q}_x(dy)$$

for all i. (Here we use the notation $\bar{f}$ to denote the extension of an $f \in U_\rho(X)$ to $\bar{X}$.) This shows that the mapping

$$x \to \int \bar{f}(y)\bar{Q}_x(dy)$$

on $X \backslash F$ is $\Sigma[X \backslash F]$-measurable for all $f \in W$, and therefore for all $f \in U_\rho(X)$. Moreover, it is easy to see that

(1.4) $$E^P\left[\int \bar{f}(y)\bar{Q}_.(dy), A \cap (X \backslash F)\right] = E^P[f(\cdot), A]$$

for all $f \in U_\rho(X)$ and $A \in \Sigma$. Given a compact set K in X, choose $\{\varphi_n\} \subseteq U_\rho(X)$ so that $\bar{\varphi}_n \searrow \chi_K$. Then (1.4) yields:

(1.5) $$E^P[\bar{Q}_.(K), X \backslash F] = P(K).$$

Next choose compacts K_n in X so that $K_n \subseteq K_{n+1}$ and $P(K_n) \geq 1 - (1/n)$. Then $D = \bigcup_1^\infty K_n$ is in $\mathscr{B}_{\bar{x}}$ and, from (1.5),

$$E^P[\bar{Q}.(D), X \backslash F] = 1.$$

Thus there is a $\Sigma[X \backslash F]$-measurable P-null set F' such that

(1.6) $\bar{Q}_x(D) = 1$ for $x \in X \backslash (F \cup F')$.

In other words, for $x \in X \backslash (F \cup F')$, $\bar{Q}_x$ can be considered as a probability measure Q_x on $(X, \mathscr{B}_X)$. For $x \in F \cup F'$, define $Q_x \equiv P$. Then

$$x \rightarrow \int f(y) Q_x(dy)$$

is a Σ-measurable mapping on X for all $f \in U_\rho(X)$. Moreover, by (1.4) and the fact that $P(F \cup F') = 0$, we have

$$E^P\left[\int f(y) Q.(dy), A \right] = E^P[f(\cdot), A], \qquad A \in \Sigma,$$

for all $f \in U_\rho(X)$. Thus

(1.7) $\int f(y) Q.(dy) = E[f \,|\, \Sigma]$ a.s.

for all $f \in U_\rho(X)$. Since the class of bounded $\mathscr{B}_X$-measurable f for which

$$x \rightarrow \int f(y) Q_x(dy)$$

is Σ-measurable and (1.7) holds is linear and is closed under bounded point-wise convergence, it follows from Exercise 1.5.2 that this class coincides with the bounded $\mathscr{B}_X$-measurable functions. In particular, $Q.(B)$ is Σ-measurable and is equal to $P(B \,|\, \Sigma)$ a.s. for all $B \in \mathscr{B}_X$. This completes the proof that $\{Q_x\}$ is a conditional probability distribution. $\square$

One can now use the conditional probability distribution to construct a version of the conditional expectation for any integrable function $f(\cdot)$ on $(X, \mathscr{B}, P)$.

1.1.7 Corollary. *Let $f(\cdot)$ be any integrable function on $(X, \mathscr{B}, P)$. Then for almost all x,*

$$\int_X |f(y)| Q_x(dy) < \infty \quad \text{and} \quad \int_X f(y) Q_x(dy) = E[f \,|\, \Sigma] \qquad \text{a.s.}$$

Proof. Clearly the class of non-negative functions for which (1.7) holds is a monotone class containing bounded measurable functions on $(X, \mathscr{B})$. A routine application of the monotone convergence theorem yields Corollary 1.1.7 for non-negative integrable functions. By separating any integrable function into its positive and negative parts, we can complete the proof of the corollary.   □

If the sub σ-field Σ is countably generated, then we can say something more about the conditional probability distribution. If $x \in X$ is any point we define the atom $A(x)$ containing x by

$$A(x) = \cap \{A: x \in A, A \in \Sigma\}.$$

It follows from Exercise 1.5.3 that $A(x)$ is in fact an element of the σ-field Σ.

1.1.8 Theorem. *If Σ is countably generated then there is a P-null set N (i.e., $P(N) = 0$) in Σ such that for $x \in X \setminus N$*

$$Q_x(A(x)) = 1.$$

Proof. If $A \in \Sigma$, then

$$Q_x(A) = P(A \mid \Sigma) = \chi_A(x) \qquad \text{a.s.}$$

If Σ_0 is a countable field generating Σ, then outside a single null set $N \in \Sigma$

$$(1.8) \qquad\qquad Q_x(A) = \chi_A(x) \quad \text{for all} \quad A \in \Sigma_0.$$

Since both sides of (1.8) are probability measures, it follows from (1.8) that

$$Q_x(A) = \chi_A(x) \quad \text{for} \quad x \in X \setminus N \quad \text{and} \quad A \in \Sigma.$$

In particular, if $A = A(x)$, then for $x \in X \setminus N$

$$Q_x(A(x)) = \chi_{A(x)}(x) = 1.$$

This completes the proof.   □

A conditional probability distribution satisfying $Q_x(A(x)) = 1$ a.s. is called a *regular conditional probability distribution.*

Let $(E, \mathscr{F})$ be a measurable space and $\mathscr{F}_n$, $n \geq 0$, an increasing family of sub σ-fields of $\mathscr{F}$ such that $\mathscr{F}$ is generated by $\bigcup_n \mathscr{F}_n$. If we have a consistent family $\{P_n\}$ of probability measures on $(E, \mathscr{F}_n)$ (i.e., $P_{n-1} = P_n \upharpoonright \mathscr{F}_{n-1}$), under suitable conditions one can obtain a probability measure on $(E, \mathscr{F})$. The problem of course is to show that P, which can be defined naturally on $\bigcup_n \mathscr{F}_n$, is countably additive on it, and therefore can be extended uniquely to $\mathscr{F}$. We need a basic assumption on the nature of the σ-fields $\mathscr{F}_n \subset \mathscr{F}$. For each $q \in E$ and n we define

$$A_n(q) = \cap \{B: B \in \mathscr{F}_n \quad \text{and} \quad q \in B\}.$$

We make the following hypothesis:

(1.9)   For every sequence $\{q_n\}_1^\infty \subseteq E$ such that $\bigcap_{n=0}^{N} A_n(q_n) \neq \emptyset$

for every N, we have $\bigcap_{n=0}^{\infty} A_n(q_n) \neq \emptyset$.

If E is a product space of the form $E = \prod_{i=1}^{\infty} E_i$ and $\mathscr{F}_n$ is the σ-field generated by the first n coordinates, the condition is always satisfied. On the other hand, if $E = C[0, 1]$, the space of continuous functions on $[0, 1]$, and $\mathscr{F}_n$ is the σ-field generated by the path on $[0, 1 - (1/n)]$, then the condition is not fulfilled. This is because a consistent sequence of continuous functions on $[0, 1 - (1/n)]$ determine a continuous function on $[0, 1)$, but not necessarily on $[0, 1]$. The first theorem we prove is a variation of Tulcea's extension theorem. It is particularly well suited for Markov processes.

1.1.9 Theorem *Let $(E, \mathscr{F})$ be a measurable space and $\{\mathscr{F}_n : n \geq 0\}$ a non-decreasing sequence of sub σ-algebras whose union generates $\mathscr{F}$. Let P_0 be a probability measure on $(E, \mathscr{F}_0)$, and, for each $n \geq 1$, let $\pi^n(q, dq')$ be a transition function from $(E, \mathscr{F}_{n-1})$ to $(E, \mathscr{F}_n)$. Define P_n on $(E, \mathscr{F}_n)$ for $n \geq 1$ so that*

$$P_n(B) = \int \pi^n(q, B) P_{n-1}(dq) \quad \text{for} \quad B \in \mathscr{F}_n;$$

and assume that, for each $n \geq 0$, there is a P_n-null set $N_n \in \mathscr{F}_n$ such that

$$q \notin N_n, \; B \in \mathscr{F}_{n+1}, \; \text{and} \; B \cap A_n(q) = \emptyset \Rightarrow \pi^{n+1}(q, B) = 0.$$

Then there is a unique probability measure P on $(E, \mathscr{F})$ with the properties that $P \upharpoonright \mathscr{F}_0 = P_0$ and, for each $n \geq 1$,

$$P(B) = \int \pi^n(q, B) P(dq') \quad \text{for} \quad B \in \mathscr{F}_n.$$

For each n, P_n is probability measure on $\mathscr{F}_n$ and P_{n+1} agrees with P_n on $\mathscr{F}_n$. Hence there is a unique finitely additive P on $\bigcup_n \mathscr{F}_n$ such that P agrees with P_n on $\mathscr{F}_n$. We will show that P is countably additive on $\bigcup_n \mathscr{F}_n$ and therefore extends uniquely as a probability measure to $\mathscr{F}$. We have to show that if $B_n \in \bigcup_n \mathscr{F}_n$ and $B_n \downarrow \emptyset$ then $P(B_n) \downarrow 0$. Assume that $P(B_n) \geq \varepsilon > 0$ for all n. We will produce a point $q \in \bigcap_n B_n$. We can assume, without loss of generality, that $B_n \in \mathscr{F}_n$. For $0 \leq n \leq m$ and $B \in \mathscr{F}_n$ we define $\pi^{m,n}(q, B)$ to be $\chi_B(q)$. For $n > m$ and $B \in \mathscr{F}_n$ we define $\pi^{m,n}(q, B)$ inductively by

$$\pi^{m,n}(q, B) = \int \pi^n(q', B) \pi^{m,n-1}(q, dq').$$

Clearly

$$P(B) = \int \pi^{0,\,n}(q,\,B) P_0(dq) \quad \text{for} \quad B \in \mathscr{F}_n.$$

We also have for $n > m$

$$\pi^{m,\,n}(q,\,B) = \int \pi^{m+1,\,n}(q',\,B) \pi^{m+1}(q,\,dq')$$

for $B \in \mathscr{F}_n$. Define

$$F_n^0 = \left\{ q \colon \pi^{0,\,n}(q,\,B_n) \geq \frac{\varepsilon}{2} \right\} \qquad n \geq 0.$$

Then $F_{n+1}^0 \subseteq F_n^0$ and

$$\varepsilon \leq P(B_n) \leq \frac{\varepsilon}{2} + P_0(F_n^0).$$

Therefore $P_0(\bigcap_n F_n^0) \geq \varepsilon/2$ and we can find $q_0 \notin N_0$ such that $\pi^{0,n}(q_0, B_n) \geq \varepsilon/2$ for all $n \geq 0$. Suppose we have found $q_0, q_1, \ldots, q_m \in E$ such that $q_k \in A_{k-1}(q_{k-1}) \backslash N_{k-1}$ for $1 \leq k \leq m$ and $\pi^{k,n}(q_k, B_n) \geq \varepsilon/2^{k+1}$ for $1 \leq k \leq m$ and $n \geq 0$. Let

$$F_n^{m+1} = \{ q \colon \pi^{m+1,\,n}(q,\,B_n) \geq \varepsilon/2^{m+2} \}.$$

Then $F_{n+1}^{m+1} \subseteq F_n^{m+1}$ and for $n > m$

$$\frac{\varepsilon}{2^{m+1}} \leq \pi^{m,\,n}(q_m,\,B_n) \leq \frac{\varepsilon}{2^{m+2}} + \pi^{m+1}(q_m,\,F_n^{m+1}).$$

Hence

$$\pi^{m+1}\left(q_m,\, \bigcap_{n=0}^{\infty} F_n^{m+1} \right) \geq \varepsilon/2^{m+2}.$$

We can therefore conclude that $A_m(q_m) \cap \bigcap_0^\infty F_n^{m+1} \neq \varnothing$; and so there is a $q_{n+1} \in A_n(q_n) \backslash N_n$ with the property that $\pi^{m+1,n}(q_{m+1}, B_n) \geq \varepsilon/2^{m+2}$ for all $n \geq 0$. By induction on m, we have now shown that a sequence $\{q_m\}_0^\infty$ exists with the properties that $q_{m+1} \in A_m(q_m) \backslash N_m$ and $\inf_n \pi^{m,n}(q_m, B_n) > 0$ for all $n \geq 0$. In particular $\chi_{B_m}(q_m) = \pi^{m,m}(q_m, B_m) > 0$ and therefore $A_m(q_m) \subset B_m$, since $B_m \in \mathscr{F}_m$. Thus $\bigcap_m A_m(q_m) \subset \bigcap_m B_m$. Finally, $q_N \in \bigcap_0^N A_m(q_m)$ for all $N \geq 0$, and so (1.9) implies that $\bigcap_0^\infty A_m(q_m) \neq \varnothing$. This completes the proof. $\square$

We will now establish Kolmogorov's extension theorem for product spaces. Let I be an infinite index set and for each $\alpha \in I$ let X_α be a Polish space with its Borel

σ-field $\mathscr{B}_\alpha$. For sets $F \subset I$ we denote by X_F the product space $\prod_{\alpha \in F} X_\alpha$ and by $\mathscr{B}_F$ the product σ-field $\prod_{\alpha \in F} \mathscr{B}_\alpha$ on X_F. For sets $G \supset F \neq \varnothing$ let σ_F^G denote the canonical projection from X_G onto X_F. We denote σ_F^I by σ_F.

1.1.10 Theorem. *Suppose that for each finite set F we are given a probability measure P_F on $(X_F, \mathscr{B}_F)$ such that, for any two finite sets $\varnothing \neq F \subset G$, $P_F = P_G(\sigma_F^G)^{-1}$. Then there is a unique probability measure P on $(X_I, \mathscr{B}_I)$ such that $P_F = P\sigma_F^{-1}$ for all finite $F \neq \varnothing$.*

Proof. Uniqueness is obvious from the fact that $\mathscr{B}_I$ is the smallest σ-field generated by $\mathscr{A} = \bigcup_F \sigma_F^{-1}(\mathscr{B}_F)$. To prove existence, we observe that there exists a finitely additive P on $\mathscr{A}$ such that $P_F = P\sigma_F^{-1}$. Now suppose that $\{A_n\}_0^\infty \in \mathscr{A}$ is a non-decreasing sequence such that $A_n \downarrow \varnothing$. Without loss of generality, we will assume that $A_n \in \mathscr{B}_{F_n}$, $n \geq 0$ where $\varnothing \neq F_0 \subset F_1 \cdots \subset F_n \cdots$ and $\{F_n\}$ are strictly increasing finite sets. Define $\mathscr{F}_n = \sigma_{F_n}^{-1}(\mathscr{B}_{F_n})$ for $n \geq 0$. It is easy to check that $\{\mathscr{F}_n, n \geq 0\}$ satisfy (1.9). Next, let $\{Q^n\}$ be a regular conditional probability distribution of P_{F_n} given $(\sigma_{F_{n-1}}^{F_n})^{-1}(\mathscr{B}_{F_{n-1}})$ and define

$$\pi^n(q, B) = Q^n_{\sigma_{F_{n(q)}}}(\sigma_{F_n}^{-1}(B))$$

for all $q \in X_I$ and $B \in \mathscr{F}_n$. It is easily checked that π^n is a transition function from $(X_I, \mathscr{F}_{n-1})$ to $(X_I, \mathscr{F}_n)$ and that it satisfies the condition of Theorem 1.1.9. Thus by that theorem there is a unique probability measure $\bar{P}$ on $(X_I, \sigma(\bigcup_0^n \mathscr{F}_n))$ such that $\bar{P}$ equals $P_{F_0}\sigma_{F_0}^{-1}$ on $\mathscr{F}_0$ and

$$\bar{P}(B) = \int \pi^n(q, B)\bar{P}(dq)$$

for all $B \in \mathscr{F}_n$. By induction, we see that $\bar{P}$ equals P on $\bigcup_0^\infty \mathscr{F}_n$. In particular $P(A_n) = \bar{P}(A_n)$, the countable additivity of $\bar{P}$ implies that $P(A_n)\downarrow 0$, and the theorem is proved. $\square$

1.2. Martingales

Throughout this section, E will denote a non-empty set of points q. $\mathscr{F}$ is a σ-algebra of subsets of E, and $\{\mathscr{F}_t : t \geq 0\}$ is a non-decreasing family of sub σ-algebras of $\mathscr{F}$.

Given $s \geq 0$ and a map θ on $[s, \infty) \times E$ into some separable metric space (X, D), we will say that θ is *(right-) continuous* if $\theta(\cdot, q)$ is (right-) continuous for all $q \in E$. If P is a probability measure on $(E, \mathscr{F})$ and $\theta: [s, \infty) \times E \to X$, we will say that θ is *P-almost surely (right-) continuous* if there is a P-null set $N \in \mathscr{F}$ such that $\theta(\cdot, q)$ is (right-) continuous for $q \notin N$.

Given $s \geq 0$ and θ on $[s, \infty) \times E$ into a measureable space, θ is said to be *progressively measurable with respect to $\{\mathscr{F}_t : t \geq 0\}$ after time s* if for each $t \geq s$ the restriction of θ to $[s, t] \times E$ is $\mathscr{B}_{[s, t]} \times \mathscr{F}_t$-measurable. Usually there is no need to mention s or $\{\mathscr{F}_t: t \geq 0\}$, and one simply says that θ is *progressively measurable*. Note that θ is progressively measurable with respect to $\{\mathscr{F}_t: t \geq 0\}$ after time s if and only if θ_s, defined by $\theta_s(t, q) = \theta(s + t, q)$, is progressively measurable with respect to $\{\mathscr{F}_{t+s}: t \geq 0\}$ after time 0. Thus any statement about a progressively measurable function after time s can be reduced to one about a progressively measurable function after time 0. This remark makes it possible to provide proofs of statements under the assumption that $s = 0$, even though s need not be zero in the statement itself. Exercises 1.5.11–1.5.13 deal with progressive measurability and the reader should work them out. The following lemma is often useful.

1.2.1 Lemma. *If θ is a right-continuous function from $[s, \infty) \times E$ into (X, D) and if $\theta(t, \cdot)$ is $\mathscr{F}_t$-measurable for all $t \geq s$, then θ is progressively measurable.*

Proof Assume that $s = 0$. Let $t \geq 0$ be given and for $n \geq 1$ define

$$\theta_n(u, q) = \theta\left(\frac{[nu] + 1}{n} \wedge t, q\right).$$

Clearly θ_n is $\mathscr{B}_{[0, t]} \times \mathscr{F}_t$-measurable for all n. Moreover, as $n \to \infty$, θ_n tends to θ on $[0, t] \times E$. Hence θ restricted to $[0, t] \times E$ is $\mathscr{B}_{[0, t]} \times \mathscr{F}_t$-measurable. $\square$

Given a probability measure P on $(E, \mathscr{F})$, $s \geq 0$, and a function θ on $[s, \infty) \times E$ into $\mathbb{C}$, we will say that $(\theta(t), \mathscr{F}_t, P)$ is a *martingale after time s* if θ is a progressively measurable, P-almost surely right-continuous function such that $\theta(t) = \theta(t, \cdot)$ is P-integrable for all $t \geq s$ and

$$(2.1) \qquad E^P[\theta(t_2) | \mathscr{F}_{t_1}] = \theta(t_1) \qquad (\text{a.s.}, P), \qquad s \leq t_1 < t_2.$$

The triple $(\theta(t), \mathscr{F}_t, P)$ is called a *submartingale after time s* if θ is a real-valued, progressively measurable, P-almost surely right continuous function such that $\theta(t)$ is P-integrable for all $t \geq s$ and

$$(2.2) \qquad E^P[\theta(t_2) | \mathscr{F}_{t_1}] \geq \theta(t_1) \qquad (\text{a.s.}, P), \qquad s \leq t_1 < t_2.$$

In keeping with the remarks following the definition of progressive measurability, we point out that s plays a rather trivial role here and that any statement proved for the case $s = 0$ can be proved in general simply by replacing θ by θ_s and $\{\mathscr{F}_t: t \geq 0\}$ by $\{\mathscr{F}_{t+s}: t \geq 0\}$. Thus, although theorems will be stated for general s, they will be proved under the assumption that $s = 0$. Usually, it will not even be necessary to mention when the (sub-) martingale begins and we will simply say the $(\theta(t), \mathscr{F}_t, P)$ is a (sub-) martingale.

We begin our study of martingales and submartingales with the following lemma. Like nearly everything in this theory, the original version is due to J. L. Doob. Thus we will be somewhat lax in the assignment of credit before each theorem.

1.2.2 Lemma. *Let* $(\theta(t), \mathscr{F}_t, P)$ *be a submartingale after time* s *with values in a closed interval* $I \subseteq R$. *Assume that* g *is a continuous, non-decreasing, convex function on* I *into* $[0, \infty)$ *such that* $g \circ \theta(t)$ *is* P-*integrable for all* $t \geq s$. *Then* $(g \circ \theta(t), \mathscr{F}_t, P)$ *is a submartingale after time* s. *In particular, if* $(\theta(t), \mathscr{F}_t, P)$ *is a martingale or a non-negative submartingale after time* s *and* r *is a number greater than or equal one such that* $|\theta(t)|^r$ *is* P-*integrable for all* $t \geq s$, *then* $(|\theta(t)|^r, \mathscr{F}_t, P)$ *is a submartingale after time* s.

Proof. Assume that $s = 0$. By the version of Jenssen's inequality for conditional expectation values:

$$E[g(\theta(t_2))\,|\,\mathscr{F}_{t_1}] \geq g(E[\theta(t_2)\,|\,\mathscr{F}_{t_1}]) \qquad \text{a.s.}$$

Thus, if $0 \leq t_1 < t_2$, then

$$E[g(\theta(t_2))\,|\,\mathscr{F}_{t_1}] \geq g(E[\theta(t_2)\,|\,\mathscr{F}_{t_1}]) \geq g(\theta(t_1)) \qquad \text{a.s.}$$

This completes the proof of the first assertion. The second assertion is immediate in the case when $(\theta(t), \mathscr{F}_t, P)$ is a non-negative submartingale; simply take $g(x) = x^r$ on $[0, \infty)$. Thus the proof will be complete once we show that $(|\theta(t)|, \mathscr{F}_t, P)$ is a submartingale if $(\theta(t), \mathscr{F}_t, P)$ is a martingale. But if $(\theta(t), \mathscr{F}_t, P)$ is a martingale, then

$$E[\,|\theta(t_2)|\,|\,\mathscr{F}_{t_1}] \geq |E[\theta(t_2)\,|\,\mathscr{F}_{t_1}]| = |\theta(t_1)| \qquad \text{(a.s., } P\text{)},$$

and so $(|\theta(t)|, \mathscr{F}_t, P)$ is a submartingale. $\square$

1.2.3 Theorem. *If* $(\theta(t), \mathscr{F}_t, P)$ *is a submartingale after time* s, *then for any* $\lambda > 0$ *and all* $T > s$:

$$(2.3) \qquad P\left(\sup_{s \leq t \leq T} \theta(t) \geq \lambda \right) \leq \frac{1}{\lambda} E\left[\theta(T), \sup_{s \leq t \leq T} \theta(t) \geq \lambda \right].$$

In particular, if θ *is non-negative, then*

$$(2.4) \qquad P\left(\sup_{s \leq t \leq T} \theta(t) \geq \lambda \right) \leq \frac{1}{\lambda} E[\theta(T)],$$

and for all $r > 1$

$$(2.5) \qquad E\left[\left(\sup_{s \leq t \leq T} \theta(t) \right)^r \right]^{1/r} \leq \frac{r}{r-1} E[\theta(T)^r]^{1/r}$$

(in the sense that the right-hand side is infinite if the left-hand side is).

Proof. Assume that $s = 0$. Relation (2.4) is immediate from (2.3) and (2.5) follows from (2.3) by the nice real-variable theory lemma mentioned in Exercise 1.5.4. Since θ is P-almost surely right-continuous, (2.3) will be proved once we show that for any $n \geq 1$ and $0 = t_0 < \cdots < t_n = T$:

$$P\left(\max_{0 \leq k \leq n} \theta(t_k) \geq \lambda \right) \leq \frac{1}{\lambda} E\left[\theta(T), \max_{0 \leq k \leq n} \theta(t_k) \geq \lambda \right].$$

To this end, define $A_0 = \{\theta(t_0) \geq \lambda\}$, and for $1 \leq k \leq n$ set

$$A_k = \left\{ \theta(t_k) \geq \lambda \quad \text{and} \quad \max_{0 \leq i \leq k} \theta(t_i) < \lambda \right\}.$$

Clearly $A_i \cap A_j = \varnothing$ if $i \neq j$, $\{\max_{0 \leq k \leq n} \theta(t_k) \geq \lambda\} = \bigcup_0^n A_k$, and $A_k \in \mathscr{F}_{t_k}$ for $0 \leq k \leq n$. Hence

$$P\left(\max_{0 \leq k \leq n} \theta(t_k) \geq \lambda \right) = \sum_0^n P(A_k) \leq \frac{1}{\lambda} \sum_0^n E[\theta(t_k), A_k]$$

$$\leq \frac{1}{\lambda} \sum_0^n E[\theta(T), A_k] = \frac{1}{\lambda} E\left[\theta(T), \max_{0 \leq k \leq n} \theta(t_k) \geq \lambda \right],$$

and the proof is complete. $\square$

We now want to introduce the important concept of a stopping time. A function $\tau\colon E \to [0, \infty) \cup \{\infty\}$ is called a *stopping time (relative to $\{\mathscr{F}_t\colon t \geq 0\}$)* if, for all $t \geq 0$, $\{\tau \leq t\} \in \mathscr{F}_t$. Examples of stopping times are plentiful. Certainly if $\tau \equiv t$ for some $t \geq 0$, then τ is a stopping time. A not quite so trivial example is obtained by considering the first time that a right-continuous, progressively measurable θ comes close to a closed set C:

$$\tau = \inf\{t \geq 0\colon \overline{[\theta(u)\colon 0 \leq u \leq t]} \cap C \neq \varnothing\}.$$

In this connection, observe that the last time that θ leaves C is *not* a stopping time since one has to know the entire history of the path $\theta(\cdot, q)$ in order to determine if it never visits C after time t. The question of when the entrance time

$$\tau_A = \{\inf t \geq 0\colon \theta(t) \in A\}$$

is a stopping time is a difficult but interesting problem. If the trajectories are continuous and A is a closed set, then τ_A is easily seen to be a stopping time. For the general case, see Dynkin [1965], Chapter IV, Section 1 for a discussion of the measurability of various entrance times relative to the completed σ-algebras. Since we will be working exclusively with processes that are right continuous and almost surely continuous for every closed set C, the contact time

$$\tau = \{\inf t \geq 0\colon \overline{[\theta(u)\colon 0 \leq u \leq t]} \cap C \neq \varnothing\}$$

is a stopping time and agrees almost surely with the entrance time τ_C.

Given a stopping time τ, we define

$$\mathscr{F}_\tau = \{A \in \mathscr{F}: A \cap \{\tau \le t\} \in \mathscr{F}_t \quad \text{for all} \quad t \ge 0\}.$$

It is easy to check that $\mathscr{F}_\tau$ is a sub σ-algebra of $\mathscr{F}$ and that $\mathscr{F}_\tau = \mathscr{F}_t$ if $\tau \equiv t$. Intuitively, $\mathscr{F}_\tau$ should be thought of as the set of events "before time τ". (Lemma 1.3.2 in the next section makes this intuitive picture precise in the case of a path space.) The following lemma collects together some elementary facts about stopping times.

1.2.4 Lemma. *If τ is a stopping time, then τ is $\mathscr{F}_\tau$-measurable. If τ is a stopping time and θ is a progressively measurable function, then $\theta(\tau) \equiv \theta(\tau(\cdot), \cdot)$ is $\mathscr{F}_\tau$-measurable on $\{\tau < \infty\}$. Finally, given stopping times σ and τ:*

(i) *$\sigma + \tau$, $\sigma \vee \tau$, and $\sigma \wedge \tau$ are stopping times,*
(ii) *if $A \in \mathscr{F}_\sigma$, then $A \cap \{\sigma < \tau\}$ and $A \cap \{\sigma \le \tau\}$ are in $\mathscr{F}_{\sigma \wedge \tau}$,*
(iii) *if $\sigma \le \tau$, then $\mathscr{F}_\sigma \subseteq \mathscr{F}_\tau$.*

Proof. The proof that τ is $\mathscr{F}_\tau$-measurable is left to the reader. Suppose that θ is progressively measurable and let $t \ge 0$ be given. Define f_t on $(\{\tau \le t\}, \mathscr{F}_t[\{\tau \le t\}])$ by $f_t(q) = (\tau(q) \wedge t, q)$. Then f_t is measurable into $([0, t] \times E, \mathscr{B}_{[0, t]} \times \mathscr{F}_t)$. Since θ restricted to $[0, t] \times E$ is $\mathscr{B}_{[0, t]} \times \mathscr{F}_t$-measurable, it follows that $\theta \circ f_t$ is $\mathscr{F}_t[\{\tau \le t\}]$-measurable on $\{\tau \le t\}$. But $\theta \circ f_t$ is just the restriction of θ to $\{\tau \le t\}$, and so the assertion that $\theta(\tau)$ is $\mathscr{F}_\tau$-measurable on $\{\tau < \infty\}$ has been proved.

The proof of (i) is easy and is left to the reader. To prove (ii), we first show that $\{\sigma < \tau\}$ and $\{\tau \le \sigma\}$ are in $\mathscr{F}_\sigma \cap \mathscr{F}_\tau$. By an obvious complementation argument, it suffices to prove that $\{\sigma < \tau\} \in \mathscr{F}_\sigma \cap \mathscr{F}_\tau$. Given $t \ge 0$, let Q_t denote the rational numbers in $[0, t]$. Then

$$\{\sigma < \tau\} \cap \{\tau \le t\} = \bigcup_{s \in Q_t} \{\sigma \le s\} \cap \{\tau > s\} \cap \{\tau \le t\}$$

and the right-hand side is certainly a member of $\mathscr{F}_t$. This proves that $\{\sigma < \tau\} \in \mathscr{F}_\tau$. To show that $\{\sigma < \tau\} \in \mathscr{F}_\sigma$, note that

$$\{\sigma < \tau\} \cap \{\sigma \le t\} = (\{\sigma \le t\} \cap \{\tau > t\}) \cup (\{\sigma < \tau\} \cap \{\tau \le t\});$$

and $(\{\sigma \le t\} \cap \{\tau > t\}) \in \mathscr{F}_t$ by definition, whereas $(\{\sigma < \tau\} \cap \{\tau \le t\}) \in \mathscr{F}_t$, since we have just seen that $\{\sigma < \tau\} \in \mathscr{F}_\tau$. We now know that both $\{\sigma < \tau\}$ and $\{\sigma \le \tau\}$ are in $\mathscr{F}_\sigma \cap \mathscr{F}_\tau$. To complete the proof of (ii), let $A \in \mathscr{F}_\sigma$ and $t \ge 0$ be given. Then

$$(A \cap \{\sigma < \tau\}) \cap \{\sigma \wedge \tau \le t\} = (A \cap \{\sigma \le t\}) \cap (\{\sigma < \tau\} \cap \{\sigma \le t\}) \in \mathscr{F}_t$$

because $\{\sigma < \tau\} \in \mathscr{F}_\tau$, and

$$(A \cap \{\sigma \le \tau\}) \cap \{\sigma \wedge \tau \le t\} = (A \cap \{\sigma \le t\}) \cap (\{\sigma \le \tau\} \cap \{\tau \le t\}) \in \mathscr{F}_t$$

since $\{\sigma \leq \tau\} \in \mathscr{F}_\tau$. Now (ii) is proved. Finally, (iii) is obviously just a special case of (ii). $\square$

We now turn to Doob's optional stopping time theorem. Let σ and τ be stopping times with values in the finite set $\{t_0, \ldots, t_N\}$, where $0 = t_0 < \cdots < t_N = T$, and assume that $\sigma \leq \tau$. Given a martingale $(\theta(t), \mathscr{F}_t, P)$ and $A \in \mathscr{F}_\sigma$, we have:

$$E[\theta(\tau), A] = \sum_{k=0}^{N} E[\theta(t_k), A \cap \{\tau = t_k\}]$$

$$= \sum_{k=0}^{N} E[\theta(T), A \cap \{\tau = t_k\}]$$

$$= \sum_{k=0}^{N} E[\theta(t_k), A \cap \{\sigma = t_k\}] = E[\theta(\sigma), A],$$

since $A \in \mathscr{F}_\sigma \subseteq \mathscr{F}_\tau$ and therefore $A \cap \{\tau = t_k\}$ and $A \cap \{\sigma = t_k\}$ are in $\mathscr{F}_{t_k}$. This proves that

(2.6) $$E^P[\theta(\tau) | \mathscr{F}_\sigma] = \theta(\sigma) \qquad \text{a.s.}$$

Next suppose that $(\theta(t), \mathscr{F}_t, P)$ is a submartingale and define $A(0) = 0$ and $A(t_k) - A(t_{k-1}) = (E[\theta(t_k) | \mathscr{F}_{t_{k-1}}] - \theta(t_{k-1}))$ for $1 \leq k \leq N$. Define $\eta(T) = \theta(T)$ and

$$\eta(t) = \theta(t_k), \qquad A(t) = A(t_k), \qquad \text{and} \quad \mathscr{G}_t = \mathscr{F}_{t_k} \quad \text{if } t \in [t_k, t_{k+1}),$$

for $0 \leq k < N$. Finally, set $M(t) = \eta(t \wedge T) - A(t \wedge T)$. Then $(M(t), \mathscr{F}_t, P)$ is a martingale, and so, by (2.6),

$$E[M(\tau) | \mathscr{F}_\sigma] = M(\sigma) \qquad \text{a.s.}$$

Since $A(\tau) \geq A(\sigma)$, it follows that

(2.7) $$E[\theta(\tau) | \mathscr{F}_\sigma] \geq \theta(\sigma) \qquad \text{a.s.}$$

1.2.5 Theorem. *Let σ and τ be bounded stopping times such that $s \leq \sigma \leq \tau$. If $(\theta(t), \mathscr{F}_t, P)$ is a non-negative submartingale after time s, then*

(2.8) $$E[\theta(\tau) | \mathscr{F}_\sigma] \geq \theta(\sigma) \qquad \text{a.s.}$$

and if $(\theta(t), \mathscr{F}_t, P)$ is a martingale, then

(2.9) $$E[\theta(\tau) | \mathscr{F}_\sigma] = \theta(\sigma) \qquad \text{a.s.}$$

Finally, if $(\theta(t), \mathscr{F}_t, P)$ *is a non-negative submartingale after time s and* $T > s$, *then* $\{\theta(\tau \vee s): \tau$ *is a stopping time bounded by* $T\}$ *is a uniformly P-integrable family.*

Proof. Assume that $s = 0$. Let $(\theta(t), \mathscr{F}_t, P)$ be a non-negative submartingale. Choose $T > 0$ so that $\sigma \leq \tau \leq T - 1$. Given $n \geq 1$, define

$$\sigma_n = \frac{[n\sigma] + 1}{n} \quad \text{and} \quad \tau_n = \frac{[n\tau] + 1}{n}.$$

Then σ_n and τ_n are stopping times and $\sigma_n \leq \tau_n \leq T$. Since σ_n and τ_n take only a finite number of values, we have from (2.7) that:

(2.10) $$E[\theta(\tau_n) | \mathscr{F}_{\sigma_n}] \geq \theta(\sigma_n) \quad \text{a.s.}$$

(2.11) $$E[\theta(T) | \mathscr{F}_{\tau_n}] \geq \theta(\tau_n) \quad \text{a.s.}$$

(2.12) $$E[\theta(T) | \mathscr{F}_{\sigma_n}] \geq \theta(\sigma_n) \quad \text{a.s.}$$

From (2.11) and (2.12), respectively, we know that

$$E[\theta(\tau_n), \theta(\tau_n) \geq \lambda] \leq E[\theta(T), \theta(\tau_n) \geq \lambda]$$

and

$$E[\theta(\sigma_n), \theta(\sigma_n) \geq \lambda] \leq E[\theta(T), \theta(\sigma_n) \geq \lambda].$$

Since $\theta(T) \geq 0$, these imply:

$$E[\theta(\tau_n), \theta(\tau_n) \geq \lambda] \leq E[\theta(T), \sup_{0 \leq t \leq T} \theta(t) \geq \lambda]$$

and

$$E[\theta(\sigma_n), \theta(\sigma_n) \geq \lambda] \leq E\left[\theta(T), \sup_{0 \leq t \leq T} \theta(t) \geq \lambda\right].$$

But, by (2.4),

$$P\left(\sup_{0 \leq t \leq T} \theta(t) \geq \lambda\right) \leq \frac{1}{\lambda} E[\theta(T)],$$

and so we have proved that $\{\theta(\sigma_n): n \geq 1\}$ and $\{\theta(\tau_n): n \geq 1\}$ are uniformly P-integrable. Since θ is P-almost surely right continuous, we now know that $\theta(\sigma_n) \to \theta(\sigma)$ and $\theta(\tau_n) \to \theta(\tau)$ in $L^1(P)$. Finally, if $A \in \mathscr{F}_\sigma$, then $A \in \mathscr{F}_{\sigma_n}$, since $\sigma \leq \sigma_n$; and so by (2.10):

$$E[\theta(\tau_n), A] \geq E[\theta(\sigma_n), A].$$

Letting $n \to \infty$, we now get (2.8).

The proof that $\{\theta(\tau): \tau$ a stopping time bounded by $T\}$ is uniformly integrable when $(\theta(t), \mathcal{F}_t, P)$ is a non-negative submartingale is accomplished in exactly the same way as we just proved that $\{\theta(\tau_n): n \geq 1\}$ is uniformly integrable. The details are left to the reader.

Finally, suppose that $(\theta(t), \mathcal{F}_t, P)$ is a martingale. Then $(|\theta(t)|, \mathcal{F}_t, P)$ is a non-negative submartingale. Thus if σ_n and τ_n are defined as in the preceding, $\{|\theta(\tau_n)|: n \geq 1\}$ and $\{|\theta(\sigma_n)|: n \geq 1\}$ are uniformly P-integrable families and so $\theta(\tau_n) \to \theta(\tau)$ and $\theta(\sigma_n) \to \theta(\sigma)$ in $L^1(P)$. Since, by (2.6),

$$E[\theta(\tau_n)|\mathcal{F}_{\sigma_n}] = \theta(\sigma_n) \qquad \text{a.s.}$$

for all $n \geq 1$, the rest of the argument is exactly like the one given at the end of the submartingale case. $\square$

1.2.6 Corollary. *Let* $\tau: [0, \infty) \times E \to [s, \infty)$ *be a right-continuous function such that* $\tau(t, \cdot)$ *is a bounded stopping time for all* $t \geq 0$ *and* $\tau(\cdot, q)$ *is a non-decreasing function for each* $q \in E$. *If* $(\theta(t), \mathcal{F}_t, P)$ *is a (non-negative sub-) martingale after time* s, *then* $(\theta(\tau(t)), \mathcal{F}_{\tau(t)}, P)$ *is a (non-negative sub-) martingale after time* 0.

1.2.7 Corollary. *If* $\tau \geq s$ *is a stopping time and* $(\theta(t), \mathcal{F}_t, P)$ *is a (non-negative sub-) martingale after time* s, *then* $(\theta(t \wedge \tau), \mathcal{F}_t, P)$ *is a (non-negative sub-) martingale after time* s.

Proof. By Corollary 1.2.6 $(\theta(t \wedge \tau), \mathcal{F}_{t \wedge \tau}, P)$ is a (sub-)martingale. Thus if $A \in \mathcal{F}_{t_1}$, then for $t_2 > t_1$:

$$E[\theta(t_2 \wedge \tau), A \cap \{\tau > t_1\}] \overset{(\geq)}{=} E[\theta(t_1 \wedge \tau), A \cap \{\tau > t_1\}],$$

since $A \cap \{\tau > t_1\} \in \mathcal{F}_{t_1 \wedge \tau}$. On the other hand, $\theta(t_2 \wedge \tau) = \theta(\tau) = \theta(t_1 \wedge \tau)$ on $\{\tau \leq t_1\}$, and so:

$$E[\theta(t_2 \wedge \tau), A \cap \{\tau \leq t_1\}] \overset{(\geq)}{=} E[\theta(t_1 \wedge \tau), A \cap \{\tau \leq t_1\}].$$

Combining these, we get our result. $\square$

The next theorem is extremely elementary but amazingly useful. It should be viewed as the "integration by parts" formula for martingale theory.

1.2.8 Theorem. *Let* $(\theta(t), \mathcal{F}_t, P)$ *be a martingale after time* s *and* $\eta: [s, \infty) \times E \to \mathbb{C}$ *a continuous, progressively measurable function with the property that the variation* $|\eta|(t, q)$ *of* $\eta(\cdot, q)$ *on* $[s, t]$ *is finite for all* $t \geq s$ *and* $q \in E$. *If for all* $t \geq s$

(2.13) $$E\left[\sup_{s \leq u \leq t} |\theta(u)|(|\eta|(t) + |\eta(s)|)\right] < \infty,$$

then $(\theta(t)\eta(t) - \int_s^t \theta(s)\eta(ds), \mathcal{F}_t, P)$ *is a martingale after time* s.

Proof. Assume $s = 0$. Using Exercise 1.5.5, one can easily see that $\int_0^t \theta(u)\eta(du)$ can be defined as a progressively measurable function. Moreover, (2.13) certainly implies that $\theta(t)\eta(t) - \int_0^t \theta(u)\eta(du)$ is P-integrable. Now suppose that $0 \le t_1 < t_2$ and that $A \in \mathscr{F}_{t_1}$. Then

$$E\left[\theta(t_2)\eta(t_2) - \theta(t_1)\eta(t_1) - \int_{t_1}^{t_2} \theta(u)\eta(du),\ A\right] = E\left[\int_{t_1}^{t_2} (\theta(t_2) - \theta(u))\eta(du),\ A\right]$$

since

$$E[(\theta(t_2) - \theta(t_1))\eta(t_1),\ A] = 0.$$

But if $\Delta = t_2 - t_1$, then

$$\int_{t_1}^{t_2} (\theta(t_2) - \theta(u))\eta(du) = \lim_{n \to \infty} \sum_{k=1}^{n} \left(\theta(t_2) - \theta\left(t_1 + \frac{k}{n}\Delta\right)\right)$$
$$\times \left(\eta\left(t_1 + \frac{k}{n}\Delta\right) - \eta\left(t_1 + \frac{k-1}{n}\Delta\right)\right) \qquad \text{a.s.,}$$

and, by (2.13) and the Lebesgue dominated convergence theorem, the convergence is in $L^1(P)$. Finally,

$$E\left[\left(\theta(t_2) - \theta\left(t_1 + \frac{k}{n}\Delta\right)\right)\left(\eta\left(t_1 + \frac{k}{n}\Delta\right) - \eta\left(t_1 + \frac{k-1}{n}\Delta\right)\right),\ A\right] = 0$$

for all $n \ge 0$ and $1 \le k \le n$. Thus

$$E\left[\int_{t_1}^{t_2} (\theta(t_2) - \theta(u))\eta(du),\ A\right] = 0$$

and the proof is complete. $\square$

The final topic of the present section is a theorem which will serve us well in what follows. Basically, this result shows that the martingale property is invariant under certain ways of conditioning a measure. Before we state the theorem, we need the following lemma, which is often useful.

1.2.9 Lemma. *Let θ: $[s, \infty) \times E \to \mathbb{C}$ be a progressively measurable, P-almost surely right continuous function such that $\theta(t)$ is P-integrable for all $t \ge s$. Let $D \subseteq [s, \infty)$ be a countable dense set. If θ is non-negative and*

$$(2.14) \qquad\qquad E[\theta(t_2)\,|\,\mathscr{F}_{t_1}] \ge \theta(t_1) \qquad \text{a.s.}$$

for all $t_1, t_2 \in D$ such that $t_1 < t_2$, then $(\theta(t), \mathscr{F}_t, P)$ is a non-negative submartingale after time s. If

$$(2.15) \qquad\qquad E[\theta(t_2)\,|\,\mathscr{F}_{t_1}] = \theta(t_1) \qquad \text{a.s.}$$

for all $t_1, t_2 \in D$ such that $t_1 < t_2$, then $(\theta(t), \mathscr{F}_t, P)$ is a martingale after time s.

Proof. Assume that $s = 0$. Clearly the proof boils down to showing that in either case the family $\{|\theta(t)|: t \in [0, T] \cap D\}$ is uniformly P-integrable for all $T \in D$. Since (2.15) implies (2.14) with $|\theta(\cdot)|$ replacing $\theta(\cdot)$, we need only show that non-negativity plus (2.14) implies that $\{\theta(t): t \in [0, T] \cap D\}$ is uniformly P-integrable. To this end, we mimic the proof of (2.4), and thereby conclude that

$$P\left(\sup_{t \in [0, T] \cap D} \theta(t) \geq \lambda\right) \leq \frac{1}{\lambda} E[\theta(T)], \qquad \lambda > 0.$$

Combining this with

$$E[\theta(t), \theta(t) \geq \lambda] \leq E[\theta(T), \theta(t) \geq \lambda]$$

$$\leq E\left[\theta(T), \sup_{t \in [0, T] \cap D} \theta(t) \geq \lambda\right], t \in [0, T] \cap D,$$

we conclude that $\{\theta(t): t \in [0, T] \cap D\}$ is uniformly P-integrable. $\square$

1.2.10 Theorem. *Assume that for all $t \geq 0$ the σ-algebra $\mathscr{F}_t$ is countably generated. Let $\tau \geq s$ be a stopping time and assume that there exists a conditional probability distribution $\{Q_{q'}\}$ of P given $\mathscr{F}_\tau$. Let $\theta: [s, \infty) \times E \to R^1$ be a progressively measurable, P-almost surely right-continuous function such that $\theta(t)$ is P-integrable for all $t \geq s$. Then $(\theta(t), \mathscr{F}_t, P)$ is a non-negative submartingale after time s if and only if $(\theta(t \wedge \tau), \mathscr{F}_t, P)$ is a non-negative submartingale after time s and there exists a P-null set $N \in \mathscr{F}_\tau$ such that for all $q' \notin N$, $(\theta(t)\chi_{[s, t]}(\tau), \mathscr{F}_t, Q_{q'})$ is a non-negative submartingale after time s. Next suppose that $\theta: [s, \infty) \times E \to \mathbb{C}$ is a progressively measurable, P-almost surely right-continuous function such that $\theta(t)$ is P-integrable for all $t \geq s$. Then $(\theta(t), \mathscr{F}_t, P)$ is a martingale after time s if and only if $(\theta(t \wedge \tau), \mathscr{F}_t, P)$ is and there is a P-null set N such that $(\theta(t) - \theta(t \wedge \tau), \mathscr{F}_t, Q_{q'})$ is a martingale after time s for all $q' \notin N$.*

Proof. Assume that $s = 0$. We suppose that $(\theta(t), \mathscr{F}_t, P)$ is a martingale. Then by Corollary 1.2.7 so is $(\theta(t \wedge \tau), \mathscr{F}_t, P)$. Let $0 \leq t_1 < t_2$, $B \in \mathscr{F}_\tau$ and $A \in \mathscr{F}_{t_1}$ be given. Then

$$E^P[E^Q[\theta(t_2), A], B \cap \{\tau \leq t_1\}] = E^P[\theta(t_2), A \cap B \cap \{\tau \leq t_1\}]$$

$$= E^P[\theta(t_1), A \cap B \cap \{\tau \leq t_1\}]$$

$$= E^P[E^Q[\theta(t_1), A], B \cap \{\tau \leq t_1\}].$$

Here we have used the fact that $A \cap B \cap \{\tau \leq t_1\}$ is in $\mathscr{F}_{t_1}$. Since $B \in \mathscr{F}_\tau$ is arbitrary this implies that for P—almost all q' satisfying $\tau(q') \leq t_1$

$$E^{Q_{q'}}[\theta(t_2), A] = E^{Q_{q'}}[\theta(t_1), A].$$

Taking a single null set for a countable subalgebra of sets A generating $\mathscr{F}_{t_1}$ we obtain a null set N_{t_1, t_2} such that, for $q' \notin N_{t_1, t_2}$ and $\tau(q') \leq t_1$

$$E^{Q_{q'}}[\theta(t_2) | \mathscr{F}_{t_1}] = \theta(t_1) \quad \text{a.s.} \quad Q_{q'}.$$

We now take a countable dense set D in $[0, \infty)$. We can then find a single null set N such that for $q' \notin N$

$$E^{Q_{q'}}[\theta(t_2) | \mathscr{F}_{t_1}] = \theta(t_1) \quad \text{a.s.} \quad Q_{q'}$$

provided $t_1, t_2 \in D$ and $t_2 \geq t_1 \geq \tau(q')$.

From Lemma 1.2.9 we can now conclude that for $q' \notin N$, $(\theta(t), \mathscr{F}_t, Q_{q'})$ is a martingale for $t \geq \tau(q')$. This can of course be restated as $(\theta(t) - \theta(t \wedge \tau(q')), \mathscr{F}_t, Q_{q'})$ is a martingale for $t \geq 0$. Since

(2.16) $P[q': Q_q[q: \tau(q) = \tau(q')] = 1] = 1$

we are done. Now suppose $(\theta(t), \mathscr{F}_t, P)$ is a non-negative sub-martingale. Then by Corollary 1.2.7 so is $(\theta(t \wedge \tau), \mathscr{F}_t, P)$. By replacing equalities by the obvious inequalities we can conclude that there is a null set N in $\mathscr{F}_\tau$ such that $(\theta(t), \mathscr{F}_t, Q_{q'})$ is a non-negative submartingale for $t \geq \tau(q')$ provided $q' \notin N$. We note that this is equivalent to $\chi_{[0, t]}(\tau(q'))\theta(t)$ being a non-negative submartingale for $t \geq 0$. Again by (2.16) we are done. We now turn to the converse proposition. If $0 \leq t_1 < t_2$ and $A \in \mathscr{F}_{t_1}$ are given, in the martingale case

$$\begin{aligned}
E^P[\theta(t_2), A] &= E^P[E^Q[\theta(t_2), A]] \\
&= E^P[E^{Q_{q'}}[\theta((t_2 \wedge \tau(q')) \vee t_1), A]] \\
&= E^P[E^Q[\theta((t_2 \wedge \tau) \vee t_1, A]] \\
&= E^P[\theta(t_1), A \cap \{\tau \leq t_1\}] + E^P[\theta(\tau \wedge t_2), A \cap \{\tau > t_1\}] \\
&= E^P[\theta(t_1), A \cap \{\tau \leq t_1\}] + E^P[\theta(\tau \wedge t_1), A \cap \{\tau > t_1\}] \\
&= E^P[\theta(t_1); A].
\end{aligned}$$

The submartingale case is proved in the same manner by replacing the equalities by inequalities at the relevant steps. $\square$

1.2.11 Remark. It is hardly necessary to mention, but for the sake of completeness we point out that everything we have said about almost surely right-continuous martingales and submartingales is trivially true for discrete parameter martingales and submartingales. That is, if $(E, \mathscr{F}, P)$ is a probability space, $\{\mathscr{F}_n: n \geq 0\}$ a non-decreasing family of sub σ-algebras, $\{\theta_n: n \geq 0\}$ a sequence of P-integrable complex valued random variables, such that θ_n is $\mathscr{F}_n$-measurable, then $(\theta_n, \mathscr{F}_n, P)$ is a *martingale (submartingale)* if (θ_n is real-valued) and

$$E[\theta(n + 1) | \mathscr{F}_n] \overset{(\geq)}{=} \theta(n) \quad \text{a.s.}$$

for all $n \geq 0$. The obvious analogues of (1.2.4) through (1.2.10) now hold in this context. Indeed, if one wishes, it is obviously possible to think of this set-up as a special case of the continuous parameter situation in which everything has been made constant on intervals of the form $[n, n + 1)$.

1.3. The Space $C([0, \infty); R^d)$

In this section we want to see what the theorems in Section 1.1 say when the Polish space is $C([0, \infty); R^d)$. The notation used in this section will be used throughout the rest of this book.

Let $\Omega = \Omega_d = C([0, \infty); R^d)$ be the space of continuous trajectories from $[0, \infty)$ into R^d. Given $t \geq 0$ and $\omega \in \Omega$ let $x(t, \omega)$ denote the position of ω in R^d at time t. Define

$$D(\omega, \omega') = \sum_{n=1}^{\infty} \frac{1}{2^n} \frac{\sup_{0 \leq t \leq n} |x(t, \omega) - x(t, \omega')|}{1 + \sup_{0 \leq t \leq n} |x(t, \omega) - x(t, \omega')|}$$

on $\Omega \times \Omega$. Then it is easy to check that D is a metric on Ω and (Ω, D) is a Polish space. The convergence induced by D is uniform convergence on bounded t-intervals. We will use $\mathcal{M}$ to denote the Borel σ-field of subsets of (Ω, D). Clearly the map $x(t)$ given by $\omega \rightarrow x(t, \omega)$ is D continuous and therefore measurable, for each $t \geq 0$. Thus

$$\sigma[x(t) : t \geq 0] \subseteq \mathcal{M}.$$

On the other hand, if $\omega^0 \in \Omega$, $t \geq 0$ and $\varepsilon > 0$ are given then

$$\{\omega : \sup_{0 \leq s \leq t} |x(s, \omega) - x(s, \omega^0)| < \varepsilon\}$$

$$= \bigcup_{n=1}^{\infty} \{\omega : |x(s, \omega) - x(s, \omega^0)| \leq \varepsilon \left(1 - \frac{1}{n}\right)$$

$$\text{for all rational } s \text{ in } [0, t]\}.$$

The set in question is therefore clearly in $\sigma[x(t) : t \geq 0]$. Since sets of the form

$$\{\omega : \sup_{0 \leq s \leq t} |x(s, \omega) - x(s, \omega^0)| < \varepsilon\}$$

generate the topology of Ω, we conclude that

(3.1) $\mathcal{M} = \sigma[x(t) : t \geq 0].$

Next we define $\mathcal{M}_t$ for $t \geq 0$ by

(3.2) $\mathcal{M}_t = \sigma[x(s) : 0 \leq s \leq t].$

Clearly $\mathcal{M}_s \subseteq \mathcal{M}_t$ for $s \leq t$ and

$$(3.3) \qquad \mathcal{M}_t = \sigma\left(\bigcup_{s < t} \mathcal{M}_s\right) \quad \text{for } t > 0.$$

By (3.1) we also have that

$$(3.4) \qquad \mathcal{M} = \sigma\left(\bigcup_{t \geq 0} \mathcal{M}_t\right).$$

The following theorem is a handy form in which to have Theorems 1.1.3 and 1.1.4.

1.3.1 Theorem. *A family $\mathcal{P}$ of probability measures P on $(\Omega, \mathcal{M})$ is precompact if and only if*

$$(3.5) \qquad \lim_{A \uparrow \infty} \inf_{P \in \mathcal{P}} P[\,|x(0)| \leq A] = 1$$

and for each $\rho > 0$ and $T < \infty$

$$(3.6) \qquad \lim_{\delta \downarrow 0} \inf_{P \in \mathcal{P}} P\left[\sup_{\substack{0 \leq s \leq t \leq T \\ t - s \leq \delta}} |x(t) - x(s)| \leq \rho\right] = 1.$$

Proof. By the Ascoli–Arzela theorem a closed set $K \subset \Omega$ is compact if and only if

$$(3.7) \qquad \sup_{\omega \in K} |x(0, \omega)| < \infty$$

and for each $T < \infty$

$$(3.8) \qquad \lim_{\delta \downarrow 0} \sup_{\omega \in K} \sup_{\substack{0 \leq s \leq t \leq T \\ t - s \leq \delta}} |x(s, \omega) - x(t, \omega)| = 0.$$

We can therefore obtain immediately the necessity of (3.5) and (3.6) from Theorem 1.1.3. We now prove the sufficiency. Let us pick $\rho = 1/n$ and $T = n$ and find $\delta_n = \delta_n(\varepsilon)$ such that for given $\varepsilon > 0$

$$\inf_{P \in \mathcal{P}} P\left[\sup_{\substack{0 \leq s \leq t \leq n \\ t - s \leq \delta_n}} |x(t) - x(s)| \leq 1/n\right] \geq 1 - \frac{\varepsilon}{2^{n+1}}.$$

We pick A from (3.5) such that

$$\inf_{P \in \mathcal{P}} P[\,|x(0)| \leq A] \geq 1 - \frac{\varepsilon}{2}.$$

If we define

$$K_\varepsilon = \bigcap_n \left[\omega: \sup_{\substack{0 \leq s \leq t \leq n \\ t - s \leq \delta_n}} |x(t, \omega) - x(s, \omega)| \leq \frac{1}{n}\right] \cap [\omega: |x(0, \omega)| \leq A]$$

then clearly $P(K_\varepsilon) \geq 1 - \varepsilon$ for all $P \in \mathscr{P}$. Moreover by the Ascoli–Arzela theorem K_ε is compact in Ω. So the sufficiency part of our theorem now follows from Theorem 1.1.4. $\square$

For a sequence $\{P_n\}$ of probability measures on $(\Omega, \mathscr{M})$ it is sometimes more convenient to have the following alternate form of the condition for precompactness.

1.3.2 Theorem. *Let P_n on $(\Omega, \mathscr{M})$ satisfy the following*

$$\liminf_{A \uparrow \infty} P_n[\,|x(0)| \leq A] = 1$$

and for any $\rho > 0$ and $T < \infty$

$$\lim_{\delta \downarrow 0} \limsup_{n \to \infty} P_n \left[\sup_{\substack{0 \leq s \leq t \leq T \\ t - s < \delta}} |x(t) - x(s)| \geq \rho \right] = 0.$$

Then $\{P_n\}$ is precompact.

Proof. Define for each fixed $T < \infty$ and $\rho > 0$

$$\psi_n(\delta) = P_n \left[\sup_{\substack{0 \leq s \leq t \leq T \\ t - s \leq \delta}} |x(t) - x(s)| \geq \rho \right]$$

We then have for each n, $\psi_n(\delta) \downarrow 0$ as $\delta \downarrow 0$ and

$$\lim_{\delta \downarrow 0} \limsup_{n \to \infty} \psi_n(\delta) = 0.$$

Let $\varepsilon > 0$ be given. There exists $\delta(\varepsilon) > 0$ and $n_0(\varepsilon) < \infty$ such that for $n \geq n_0(\varepsilon)$

$$\psi_n(\delta(\varepsilon)) \leq \varepsilon.$$

Since each $\psi_n(\delta) \to 0$ as $\delta \to 0$, we can find $\delta_1(\varepsilon)$ such that for $\delta \leq \delta_1(\varepsilon)$ and $n \leq n_0(\varepsilon)$

$$\psi_n(\delta) \leq \varepsilon.$$

If we now take $\delta_2(\varepsilon) = \delta(\varepsilon) \wedge \delta_1(\varepsilon)$ we conclude that $\psi_n(\delta) \leq \varepsilon$ for all n and $\delta \leq \delta_2(\varepsilon)$. Therefore

$$\limsup_{\delta \downarrow 0} \psi_n(\delta) = 0$$

and Theorem 1.3.1 is now applicable. $\square$

We next want to see what Theorem 1.1.8 looks like when the conditioning σ-algebra is $\mathcal{M}_\tau$, τ being any stopping time. In order to do this we need the next lemma.

1.3.3 Lemma. *Let τ be any stopping time. Then*

(3.10) $$\mathcal{M}_\tau = \sigma[x(s \wedge \tau): s \geq 0].$$

In particular, $\mathcal{M}_\tau$ is countably generated.

Proof. We first note that $x(s \wedge \tau)$ is $\mathcal{M}_\tau$ measurable for each $s \geq 0$. In fact, by Lemma 1.2.4, $x(s \wedge \tau)$ is $\mathcal{M}_{s \wedge \tau}$-measurable, and $\mathcal{M}_{s \wedge \tau} \subseteq \mathcal{M}_\tau$. To prove $\mathcal{M}_\tau \subseteq \sigma\{x(s \wedge \tau): s \geq 0\}$, we introduce the following notation. Given $\omega \in \Omega$ and $t \geq 0$ let ω_t be the unique element of Ω such that $x(s, \omega_t) = x(s \wedge t, \omega)$ for all $s \geq 0$. By Exercise 1.5.6 we know that $f: \Omega \to R$ is $\mathcal{M}$ measurable if and only if there is a $(\mathscr{B}(R^d))^{Z^+}$ measurable function $F: (R^d)^{Z^+} \to R$ and a sequence $\{t_n\}_1^\infty$ in $[0, \infty)$ such that

(3.11) $$f(\omega) = F(x(t_1, \omega), x(t_2, \omega), \ldots, x(t_n, \omega), \ldots).$$

Moreover, f is $\mathcal{M}_t$ measurable if and only if the $\{t_n\}$ can be chosen from $[0, t]$. Thus f is $\mathcal{M}_t$ measurable if and only if f is $\mathcal{M}$ measurable and $f(\omega) = f(\omega_t)$ for all ω. Now suppose that f is $\mathcal{M}_\tau$ measurable. For $t \geq 0$, define $f_t(\omega) = \chi_{\{t\}}(\tau(\omega)) \cdot f(\omega)$. Then f_t is $\mathcal{M}_t$ measurable and therefore

$$f_t(\omega) = f_t(\omega_t).$$

In particular

(3.12) $$f(\omega) = f_{\tau(\omega)}(\omega) = f_{\tau(\omega)}(\omega_{\tau(\omega)}) \quad \text{for all} \quad \omega \in \Omega.$$

Applying this to $f(\cdot) = \chi_{\{t\}}(\tau(\cdot))$, we get

$$\chi_{\{t\}}(\tau(\omega)) = \chi_{\tau(\omega)}(\tau(\omega_{\tau(\omega)}))\chi_{\{t\}}(\tau(\omega_{\tau(\omega)})).$$

If we now take $t = \tau(\omega)$ in this, we see that $\tau(\omega) = \tau(\omega_{\tau(\omega)})$. Using this fact in (3.12) we conclude that for any $\mathcal{M}_\tau$ measurable f,

(3.13) $$f(\omega) = f(\omega_{\tau(\omega)}) \quad \text{for all} \quad \omega \in \Omega.$$

Finally given a $\mathcal{M}_\tau$ measurable function f, choose F so that (3.11) holds. Then by (3.12)

$$f(\omega) = F(x(t_1 \wedge \tau(\omega), \omega), \ldots, x(t_n \wedge \tau(\omega), \omega), \ldots) \quad \text{for} \quad \omega \in \Omega,$$

which explicitly displays f as a $\sigma[x(s \wedge \tau): s \geq 0]$ measurable function. $\qquad \square$

Given a stopping time τ and a measure P on $(\Omega, \mathcal{M})$ we can now use Theorem 1.1.8 to find a conditional probability distribution $\{Q_\omega\}$ of P given $\mathcal{M}_\tau$ such that off of some P-null set $N \in \mathcal{M}_\tau$, we have

$$(3.14) \qquad\qquad Q_\omega(A) = \chi_A(\omega') \quad \text{for all} \quad A \in \mathcal{M}_\tau.$$

Suppose we define $\tilde{Q}_{\omega'}$ so that outside N, $\tilde{Q}_{\omega'} = Q_{\omega'}$ and for $\omega' \in N$ and $A \in \mathcal{M}$

$$\tilde{Q}_\omega(A) = \chi_A(\omega'_{\tau(\omega')})$$

where $\omega'_{\tau(\omega')}$ is defined as in Lemma 1.3.2. Then it is clear that $\tilde{Q}_{\omega'}$ is again a conditional probability distribution of P given $\mathcal{M}_\tau$. In addition, for the new version (3.14) holds for all ω'.

1.3.4 Theorem. *If P is a probability measure on $(\Omega, \mathcal{M})$ and τ is a stopping time, then there exists a conditional probability distribution $\{Q_\omega\}$ of P given $\mathcal{M}_\tau$ such that (3.14) holds for all ω'.*

If $\{Q_\omega\}$ is as in the preceding theorem we will call it a *regular conditional probability distribution of P given $\mathcal{M}_\tau$* and we will abbreviate this phrase by *r.c.p.d. of $P | \mathcal{M}_\tau$*. Notice that an equivalent way to express (3.14) is

$$(3.15) \qquad\qquad Q_{\omega'}(\tau(\omega) = \tau(\omega') \quad \text{and} \quad x(s, \omega) = x(s, \omega')$$

$$\text{for} \quad 0 \le s \le \tau(\omega)) = 1.$$

The version of Theorem 1.1.9 which is most suitable for the study of measures on $(\Omega, \mathcal{M})$ is the following:

1.3.5 Theorem. *Let $\{\tau_n: n \ge 0\}$ be a nondecreasing sequence of stopping times and for each n suppose P_n is a probability measure on $(\Omega, \mathcal{M}_{\tau_n})$. Assume that P_{n+1} equals P_n on $\mathcal{M}_{\tau_n}$ for each $n \ge 0$. If $\lim_{n \to \infty} P_n(\tau_n \le t) = 0$ for all $t \ge 0$, then there is a unique probability measure P on $(\Omega, \mathcal{M})$ such that P equals P_n on $\mathcal{M}_{\tau_n}$ for all $n \ge 0$.*

Proof. If P exists, it is obvious that for $A \in \mathcal{M}_t$

$$(3.16) \qquad\qquad P(A) = \lim_{n \to \infty} P_n[A \cap \{\tau_n > t\}].$$

Thus uniqueness of P on $\mathcal{M}_t$ for all $t \ge 0$ is proved; and therefore P, if it exists, is unique on $\mathcal{M}$. To prove existence, first assume that $\tau_n \equiv n$ for each $n \ge 0$. Clearly $(\Omega, \mathcal{M}_n)$ is isomorphic to $(C([0, n]; R^d), \mathcal{B}(C([0, n]; R^d)))$ and therefore we can find a r.c.p.d. $\{Q^n_\omega\}$ of P_n given $\mathcal{M}_{n-1}$ for each $n \ge 1$. Moreover if

$$A_n(\omega) = \cap \{B \in \mathcal{M}_n, \omega \in B\}$$

then

$$A_n(\omega) = \{\omega': x(s, \omega') = x(s, \omega) \quad \text{for} \quad 0 \le s \le n\}.$$

Therefore if $\{\omega_n\}_0^\infty \subseteq \Omega$ has the property

$$\bigcap_0^N A_n(\omega_n) \ne \varnothing$$

for all $N \ge 0$ then the ω determined by

$$x(t, \omega) = x(t, \omega_n), \quad 0 \le t \le n \quad \text{and} \quad n \ge 0$$

is in $\bigcap_0^\infty A_n(\omega_n)$. We can now apply Theorem 1.1.9 to conclude the existence of a P on $(\Omega, \mathcal{M})$ such that P equals P_0 on $\mathcal{M}_0$ and

$$P(A) = \int Q_{\omega'}^n(A)P(d\omega')$$

for all $n \ge 1$ and $A \in \mathcal{M}_n$. By induction on $n \ge 1$, it is easy to see that P equals P_n on $\mathcal{M}_n$ for all $n \ge 0$, and so P is the desired measure.

In general, we first define $\tilde{P}_n$ on $\mathcal{M}_n$ by

$$\tilde{P}_n(A) = \lim_{k \to \infty} P_k(A \cap \{\tau_k > n\}), \quad A \in \mathcal{M}_n.$$

Note that the limit exists, since

$$P_k(A \cap \{\tau_k > n\}) = P_{k+1}(A \cap \{\tau_k > n\}) \le P_{k+1}(A \cap \{\tau_{k+1} > n\}).$$

Also it is clear that $\tilde{P}_n$ is a finitely additive probability distribution on $\mathcal{M}_n$. To see that $\tilde{P}_n$ is countably additive, suppose that $\{A_m\}_1^\infty \subseteq \mathcal{M}_n$ and $A_m \downarrow \varnothing$. Then for all k

$$\tilde{P}_n(A_m) = \tilde{P}_n(A_m \cap \{\tau_k \le n\}) + \tilde{P}_n(A_m \cap \{\tau_k > n\})$$
$$\le P_k(\tau_k \le n) + P_k(A_m \cap \{\tau_k > n\}).$$

Letting m and then $k \to \infty$, we see that $\tilde{P}_n(A_m) \to 0$ as $m \to \infty$. A similar argument shows that $\tilde{P}_{n+1}$ equals $\tilde{P}_n$ on $\mathcal{M}_n$ for $n \ge 0$. Thus, by the preceding paragraph, there is a P on $(\Omega, \mathcal{M})$ such that P equals $\tilde{P}_n$ on $\mathcal{M}_n$ for all $n \ge 0$. Finally we must check that P equals P_k on $\mathcal{M}_{\tau_k}$. Given $A \in \mathcal{M}_{\tau_k} \cap \mathcal{M}_n$, we have that

$$P(A) = \tilde{P}_n(A) = \lim_{l \to \infty} P_l(A \cap \{\tau_l > n\}).$$

But for $l \ge k$

$$|P_k(A) - P_l(A \cap \{\tau_l > n\})| = P_l(A \cap \{\tau_l \le n\}) \to 0$$

as $l \to \infty$. Thus $P(A) = P_k(A)$ for $A \in \mathcal{M}_{\tau_k} \cap \mathcal{M}_n$. But, by Lemma 1.2.2, $\mathcal{M}_{\tau_k}$ is generated by the maps $x(t \wedge \tau_k)$, for $t \geq 0$. Therefore

$$\mathcal{M}_{\tau_k} = \sigma \left(\bigcup_{n=1}^{\infty} (\mathcal{M}_{\tau_k} \cap \mathcal{M}_n) \right).$$

Hence P equals P_k on $\mathcal{M}_{\tau_k}$, and we are done. $\square$

1.4. Martingales and Compactness

In the preceding section we developed necessary and sufficient conditions for the compactness of measures on $(\Omega, \mathcal{M})$. Like most general results, these conditions are not particularly useful when applied to special situations. It is the purpose of this section to develop a useful condition for compactness. The condition that we have in mind is ideally suited to the study of Markov processes and, more generally, processes for which there is a plentiful supply of associated martingales.

Given $\rho > 0$ and $\omega \in \Omega$, define $\tau_0(\omega) = 0$ and for $n \geq 1$:

$$\tau_n(\omega) = \inf\{t \geq \tau_{n-1}(\omega): |x(t, \omega) - x(\tau_{n-1}(\omega), \omega)| \geq \rho/4\}.$$

Here it is understood that $\tau_n(\omega) = \infty$ if either $\tau_{n-1}(\omega) = \infty$ or there fails to exist a $t \geq \tau_{n-1}(\omega)$ such that $|x(t, \omega) - x(\tau_{n-1}(\omega), \omega)| \geq \rho/4$. Since ω is a continuous path, it must always be true that either $\tau_{n-1}(\omega) = \infty$ or $\tau_{n-1}(\omega) < \tau_n(\omega)$ and $\tau_n(\omega) \to \infty$ as $n \to \infty$. Thus for $T > 0$ (this T is arbitrary but fixed throughout), we can define

$$N = N(\omega) = \min\{n: \tau_{n+1}(\omega) > T\}$$

and

(4.1) $$\delta_\omega(\rho) = \min\{\tau_n(\omega) - \tau_{n-1}(\omega): 1 \leq n \leq N(\omega)\}.$$

We need the following lemma.

1.4.1 Lemma. *Let t_1 and t_2 be any pair of points in $[0, T]$ such that $|t_2 - t_1| < \delta_\omega(\rho)$. Then $|x(t_2, \omega) - x(t_1, \omega)| \leq \rho$ and so*

$$\sup\{|x(t_2, \omega) - x(t_1, \omega)|: 0 \leq t_1 < t_2 \leq T \quad \text{and} \quad |t_2 - t_1| < \delta_\omega(\rho)\} \leq \rho.$$

Proof. Consider the partition of $[0, T]$ into the subintervals $[\tau_0(\omega), \tau_1(\omega)), \ldots, [\tau_{N(\omega)-1}(\omega), \tau_{N(\omega)}(\omega))$, and $[\tau_{N(\omega)}(\omega), T]$. All of these subintervals, except possibly the last one, must have length greater than $\delta_\omega(\rho)$. Thus, either both t_1 and t_2 lie in the same subinterval, or they lie in adjacent subintervals. Since over any subinterval the distance of the path from its position at the left hand end never exceeds $\rho/4$, the oscillation of the path over any subinterval must be less than or equal to $\rho/2$. Hence the oscillation over the union of two successive subintervals cannot exceed ρ. In particular, $|x(t_2, \omega) - x(t_1, \omega)| \leq \rho$. $\square$

The preceding lemma shows that the problem of estimating $P(\sup\{|x(t_2) - x(t_1)| : 0 \leq t_1 < t_2 \leq T$ and $t_2 - t_1 < \delta\} \geq \rho)$ reduces to estimating

$$(4.2) \qquad\qquad P(\{\omega : \delta_\omega(\rho) < \delta\}).$$

The method that we are going to use to estimate the latter quantity depends on the following two hypotheses about P:

1.4.2 Hypothesis. *For all non-negative $f \in C_0^\infty(R^d)$ there is a constant $A_f \geq 0$ such that $(f(x(t)) + A_f t, \mathcal{M}_t, P)$ is a non-negative submartingale.*

1.4.3 Hypothesis. *Given a non-negative $f \in C_0^\infty(R^d)$, the choice of A_f in $(1.4.2)$ can be made so that it works for all translates of f.*

Under these hypotheses, we are going to develop an estimate for the quantity in (4.2) which depends only on the constants A_f.

Let $\varepsilon > 0$ be given and choose $f_\varepsilon \in C_0^\infty(R^d)$ so that $f_\varepsilon(0) = 1$, $f_\varepsilon(x) = 0$ for $|x| \geq \varepsilon$, and $0 \leq f_\varepsilon \leq 1$. Given $a \in R^d$, denote by f_ε^a the function defined by $f_\varepsilon^a(x) = f_\varepsilon(x - a)$. Now choose A_ε by Hypotheses 1.4.2 and 1.4.3 so that $(f_\varepsilon^a(x(t)) + A_\varepsilon t, \mathcal{M}_t, P)$ is a non-negative submartingale for all $a \in R^d$. Note that by Theorem 1.2.10. if τ is a stopping time and $\{Q_\omega\}$ is a r.c.p.d. of P given $\mathcal{M}_\tau$, then there is a P-null set $F \in \mathcal{M}_\tau$ such that $(f_\varepsilon^a(x(t)) + A_\varepsilon t)\chi_{[0,\eta]}(\tau)$ is a non-negative Q_ω submartingale for all $\omega' \notin F$. The null set F will of course depend, in general, on $a \in R^d$. However, by taking a countable dense set $D \subseteq R^d$ and choosing F so that $(f_\varepsilon^a(x(t)) + A_\varepsilon t)\chi_{[0,\eta]}(\tau)$ is a Q_ω submartingale for all $a \in D$ and $\omega' \notin F$, we see that it is possible to take F independent of $a \in R^d$. We can now prove the following lemma.

1.4.4 Lemma. *For any $n \geq 0$*

$$P(\tau_{n+1} - \tau_n \leq \delta \,|\, \mathcal{M}_{\tau_n}) \leq \delta A_{\rho/4} \qquad \text{(a.s., } P) \quad \text{on} \quad \{\tau_n < \infty\}.$$

Proof. Let $\varepsilon = \rho/4$ in the preceding discussion and let $\{Q_\omega\}$ be a r.c.p.d. of P given $\mathcal{M}_{\tau_n}$. Then we can choose a P-null set $F \in \mathcal{M}_{\tau_n}$ so that

$$((f_\varepsilon^{\omega'}(x(t)) + A_\varepsilon t)\chi_{[0,\eta]}(\tau_n(\omega')), \mathcal{M}_t, Q_\omega)$$

is a non-negative submartingale for all $\omega' \notin F$, where $f_\varepsilon^{\omega'}(x) = f_\varepsilon(x - x(\tau_n(\omega'), \omega'))$ if $\tau_n(\omega') < \infty$ and $f_\varepsilon^{\omega'}(\cdot) \equiv 1$ otherwise. In particular, by Theorem 1.2.5,

$$E^{Q_\omega}[f_\varepsilon^{\omega'}(x(\tau_{n+1} \wedge (\tau_n(\omega') + \delta))) + A_\varepsilon \delta] \geq 1$$

for $\omega' \notin F$. In other words,

$$E^{Q_{\omega'}}[1 - f_\varepsilon^{\omega'}(x(\tau_{n+1} \wedge (\tau_n(\omega') + \delta)))] \leq A_\varepsilon \delta, \; \omega' \notin F.$$

But $0 \leq 1 - f_\varepsilon^{\omega'} \leq 1$, and $\tau_{n+1} \leq \tau_n(\omega') + \delta$ implies that

$$1 - f_\varepsilon^{\omega'}(x(\tau_{n+1} \wedge (\tau_n(\omega') + \delta))) = 1,$$

if $\tau_n(\omega') < \infty$. Thus

$$Q_{\omega'}(\tau_{n+1} \leq \tau_n(\omega') + \delta) \leq A_\varepsilon \delta$$

for $\omega' \notin F$ such that $\tau_n(\omega') < \infty$. Since

$$P(\tau_{n+1} - \tau_n \leq \delta \,|\, \mathcal{M}_{\tau_n}) = Q.(\tau_{n+1} \leq \tau_n(\cdot) + \delta) \qquad \text{(a.s., } P),$$

this completes the proof. $\quad\square$

1.4.5 Lemma. *Let* $(E, \mathscr{F}, P)$ *be a probability space and* $\{\mathscr{F}_n : n \geq 0\}$ *a non-decreasing sequence of sub σ-algebras of* $\mathscr{F}$. *Let* $\{\xi_n : n \geq 1\}$ *be a non-decreasing sequence of random variables on* $(E, \mathscr{F})$ *taking values in* $[0, \infty) \cup \{\infty\}$, *and assume that* ξ_n *is* $\mathscr{F}_n$*-measurable. Define* $\xi_0 \equiv 0$ *and suppose that for some* $\lambda < 1$ *and all* $n \geq 0$:

$$E[\exp[-(\xi_{n+1} - \xi_n)] \,|\, \mathscr{F}_n] \leq \lambda \text{ a.s.}$$

If for some $T > 0$ *one defines*

$$N(q) = \inf\{n \geq 0 : \xi_{n+1}(q) > T\},$$

then $N < \infty$ *a.s. and in fact*:

$$P(N \geq k) \leq e^T \lambda^k, \qquad k \geq 0.$$

Proof. First note that:

$$E[e^{-\xi_{n+1}} \,|\, \mathscr{F}_n] = e^{-\xi_n} E[\exp[-(\xi_{n+1} - \xi_n)] \,|\, \mathscr{F}_n]$$
$$\leq \lambda e^{-\xi_n} \text{ a.s.}$$

Thus, by induction on n,

$$E[e^{-\xi_n}] \leq \lambda^n.$$

In particular, $\xi_n \to \infty$ a.s., or, in other words, $N < \infty$ a.s. Moreover, we now have for all $k \geq 0$:

$$P(N \geq k) = P(\xi_k \leq T) \leq e^T E[e^{-\xi_k}] \leq e^T \lambda^k. \quad\square$$

We are at last in a position to prove the compactness criterion to which we referred in the introduction to this section.

1.4.6 Theorem. *Let $\mathscr{P}$ be a family of probability measures on $(\Omega, \mathscr{M})$ such that*

$$\lim_{l \nearrow \infty} \sup_{P \in \mathscr{P}} P(|x(0)| \geq l) = 0.$$

Assume that each $P \in \mathscr{P}$ fulfills hypotheses 1.4.2 and 1.4.3 and that the choice of the constants A_f in 1.4.2 and 1.4.3 can be made independent of $P \in \mathscr{P}$. Then $\mathscr{P}$ is precompact.

Proof. In view of Theorem 1.3.1, we need only check that

$$\lim_{\delta \searrow 0} \inf_{P \in \mathscr{P}} P\left(\sup_{\substack{0 \leq s < t \leq T \\ |t-s| < \delta}} |x(t) - x(s)| \leq \rho \right) = 1$$

for every $T > 0$ and $\rho > 0$. Because of Lemma 1.4.1, this will be done once we show that

(4.3) $$\lim_{\delta \searrow 0} \sup_{P \in \mathscr{P}} P(\{\omega : \delta_\omega(\rho) \leq \delta\}) = 0.$$

Note that, from the definition of $\delta_\omega(\rho)$ in (4.1),

$$P(\delta_\cdot(\rho) \leq \delta) \leq P\left(\min_{1 \leq i \leq k} \tau_i - \tau_{i-1} \leq \delta \right) + P(N > k)$$

$$\leq \sum_1^k P(\tau_i - \tau_{i-1} \leq \delta) + P(N > k)$$

$$\leq k\delta A_{\rho/4} + P(N > k),$$

where we have used Lemma 1.4.4 to get the last line. Thus, the proof will be complete once we establish that

(4.4) $$\lim_{k \nearrow \infty} \sup_{P \in \mathscr{P}} P(N > k) = 0.$$

But, by Lemma 1.4.4, we know that for any $t_0 > 0$ and $P \in \mathscr{P}$:

$$E^P[e^{-(\tau_{i+1} - \tau_i)} | \mathscr{M}_{\tau_i}]$$

$$\leq P(\tau_{i+1} - \tau_i \leq t_0 | \mathscr{M}_{\tau_i}) + e^{-t_0} P(\tau_{i+1} - \tau_i > t_0 | \mathscr{M}_{\tau_i})$$

$$\leq e^{-t_0} + (1 - e^{-t_0}) P(\tau_{i+1} - \tau_i \leq t_0 | \mathscr{M}_{\tau_i})$$

$$\leq e^{-t_0} + (1 - e^{-t_0}) t_0 A_{\rho/4} \quad \text{a.s.}$$

Choosing t_0 in a suitable manner, we can make

$$\lambda \equiv e^{-t_0} + (1 - e^{-t_0})t_0 A_{\rho/4} < 1.$$

Thus, by Lemma 1.4.5,

$$\sup_{P \in \mathscr{P}} P(N \geq k) \leq e^T \lambda^k,$$

and this certainly guarantees (4.4). $\square$

Although Theorem 1.4.6 is well-adapted to the study of continuous time Markov processes having continuous paths, it is not suitable, as it now stands, for the approximation of such processes by discrete time parameter processes. We will now make the necessary modifications to get a theorem which covers this situation.

Given an $h > 0$, let Ω_h stand for the subset $\omega \in \Omega$ such that $x(\cdot, \omega)$ is linear over each interval of the form $[jh, (j+1)h]$, $j = 0, 1, \dots$. Given $\omega \in \Omega_h$, define $\tau_0^*(\omega) = 0$ and for $n \geq 1$,

$$\tau_n^*(\omega) = \inf\{t \geq \tau_{n-1}^*(\omega): t = jh \text{ for some } j \geq 0 \text{ and } |x(t, \omega) - x(\tau_{n-1}^*(\omega), \omega| \geq \rho/4\}.$$

We again adopt the convention that $\tau_n^*(\omega) = \infty$ if $\tau_{n-1}^*(\omega) = \infty$ or if $|x(t, \omega) - x(\tau_{n-1}(\omega), \omega)| < \rho/4$ for all $t \geq \tau_{n-1}^*(\omega)$. Once more, either $\tau_{n-1}^*(\omega) = \infty$ or $\tau_n^*(\omega) > \tau_{n-1}^*(\omega)$ for all $n \geq 1$ and

$$\lim_{n \to \infty} \tau_n^*(\omega) = \infty.$$

Define for $T > 0$ (arbitrary but fixed):

$$N^*(\omega) = \inf\{n \geq 0: \tau_{n+1}^*(\omega) > T\},$$

$$\delta_\omega^*(\rho) = \min\{\tau_{n+1}^*(\omega) - \tau_n^*(\omega): 0 \leq n \leq N^*(\omega)\}$$

and

$$\theta_h^*(\omega) = \max\{|x((j+1)h, \omega) - x(jh, \omega)|: 0 \leq jh \leq T\}.$$

In place of Lemma 1.4.1, we now have the following lemma.

1.4.7 Lemma. *If $t_1, t_2 \in [0, T]$ and $0 \leq t_2 - t_1 \leq \delta_\omega^*(\rho)$, then*

$$|x(t_2, \omega) - x(t_1, \omega)| \leq \rho + 2\theta_h^*(\omega).$$

In particular,

$$\sup\{|x(t_2, \omega) - x(t_1, \omega)|: t_1, t_2 \in [0, T] \quad \text{and} \quad 0 \leq t_1 - t_2 \leq \delta_\omega^*(\rho)\}$$
$$\leq \rho + 2\theta_h^*(\omega).$$

Proof. Given t_1 and t_2, let t_1^* and t_2^* be, respectively, the smallest and largest multiple of h such that $t_1^* \geq t_1$ and $t_2^* \leq t_2$. Repeating, word for word, the argument used to prove Lemma 1.4.1, we see that $|x(t_2^*, \omega) - x(t_1^*, \omega)| \leq \rho$. On the other hand, neither $|x(t_1^*, \omega) - x(t_1, \omega)|$ nor $|x(t_2^*, \omega) - x(t_2, \omega)|$ can exceed $\theta_h^*(\omega)$, and so the proof is complete. $\square$

By analogy with hypotheses 1.4.2 and 1.4.3, we now state appropriate hypotheses for a P on $(\Omega, \mathcal{M})$ which is concentrated on Ω_h.

1.4.8 Hypothesis. *For all non-negative $f \in C_0^\infty(R^d)$ there is a constant $A_f \geq 0$ such that $(f(x(jh)) + A_f(jh), \mathcal{M}_{jh}, P)$ is a non-negative submartingale.*

1.4.9 Hypothesis. *Given a non-negative $f \in C_0^\infty(R^d)$, the choice of A_f in 1.4.8 can be made so that it works for all translates of f.*

The analogue of Lemma 1.4.4 in this context is the following lemma, whose proof is identical to the proof of Lemma 1.4.4 when one takes remark 1.2.11 into account. The details are left to the reader.

1.4.10 Lemma. *For any δ which is a multiple of h and any $n \geq 0$,*

$$P(\tau_{n+1}^* - \tau_n^* \leq \delta | \mathcal{M}_{\tau_n^*}) \leq \delta A_{\rho/4} \qquad \text{a.s.}$$

We are now ready to prove the analogue of Theorem 1.4.6.

1.4.11 Theorem. *Let $\{h_n : n \geq 0\}$ be a non-increasing sequence of positive numbers such that $h_n \to 0$ as $n \to \infty$. Let $\{P_n : n \geq 0\}$ be a sequence of probability measures on $(\Omega, \mathcal{M})$ such that P_n is concentrated on Ω_{h_n}. Assume that each P_n satisfies hypotheses 1.4.8 and 1.4.9 (with $h = h_n$) and that the choice of the constants A_f can be made independent of n. If for each $T > 0$ and $\varepsilon > 0$*

$$(4.5) \qquad \lim_{n \to \infty} \sum_{0 \leq jh_n \leq T} P_n(|x((j+1)h_n) - x(jh_n)| \geq \varepsilon) = 0,$$

and

$$(4.6) \qquad \limsup_{l \nearrow \infty} \sup_{n \geq 0} P_n(|x(0)| \geq l) = 0,$$

then $\{P_n : n \geq 0\}$ is precompact.

Proof. The proof here goes by analogy with the proof of Theorem 1.4.6. We must show that

$$(4.7) \qquad \overline{\lim_{\delta \searrow 0}} \ \overline{\lim_{n \to \infty}} \ P_n\left(\sup_{\substack{0 \leq s < t \leq T \\ t - s < \delta}} |x(t) - x(s)| \geq 2\rho \right) = 0$$

for all $T > 0$ and $\rho > 0$.

By Lemma 1.4.7,

$$P_n\left(\sup_{\substack{0 \leq s < t \leq T \\ t-s < \delta}} |x(t) - x(s)| \geq 2\rho\right) \leq P_n(\delta^*(\rho) \leq \delta) + P_n(\theta^*_{h_n} \geq \rho/2).$$

The term $P_n(\delta^*(\rho) \leq \delta)$ is estimated in exactly the same way as $P(\delta.(\rho) \leq \delta)$ was in Theorem 1.4.6, only one now uses Lemma 1.4.10 instead of Lemma 1.4.4. In this way, one proves that

$$\lim_{\delta \searrow 0} \sup_{n \geq 0} P_n(\delta^*(\rho) \leq \delta) = 0.$$

On the other hand, by hypothesis,

$$P_n(\theta^*_{h_n} \geq \rho/2) \leq \sum_{0 \leq jh \leq T} P_n(|x(j+1)h_n) - x(jh_n)| \geq \rho/2)$$

$$\to 0$$

as $n \to \infty$ □

1.5. Exercises

1.5.1. Let E be a non-empty set and $\mathscr{C}$ a collection of subsets of E such that if A and B are in $\mathscr{C}$ then $A \cap B$ is in $\mathscr{C}$. Show that the smallest class $\mathscr{L}$ of subsets of E such that

(i) $\mathscr{C} \subseteq \mathscr{L}$
(ii) $E \in \mathscr{L}$
(iii) $A, B \in \mathscr{L}, A \subset B$ implies $B \backslash A \in \mathscr{L}$
(iv) $A, B \in \mathscr{L}, A \cap B = \varnothing$ implies $A \cup B \in \mathscr{L}$
(v) $\{A_n\}_1^\infty \in \mathscr{L}, A_n \subseteq A_{n+1}$ for $n \geq 1$ implies $\bigcup_1^\infty A_n \in \mathscr{L}$

coincides with $\sigma(\mathscr{C})$, the smallest σ-algebra containing $\mathscr{C}$. The proof of this fact is very much like the monotone class theorem (for details see Dynkin [1960]). As an application, show that if $\mathscr{H}$ is a class of bounded functions on E into R such that

(i) $\chi_C \in \mathscr{H}$ for all $C \in \mathscr{C}$
(ii) $1 \in \mathscr{H}$
(iii) $\mathscr{H}$ is a vector space
(iv) If $\{f_n\}_1^\infty \subseteq \mathscr{H}$ is a nondecreasing sequence of non-negative functions with $\sup_{n,q} f_n(q) < \infty$, then $f = \lim_{n \to \infty} f_n$ is again in $\mathscr{H}$

then $\mathscr{H}$ contains all the bounded measurable functions on $(E, \sigma(\mathscr{C}))$.

1.5.2. Let (X, ρ) be a metric space and denote by $U_\rho(X)$ the class of all bounded, uniformly continuous functions on (X, ρ) into R. Show that the smallest class of functions from X into R which contains $U_\rho(X)$ and is closed under bounded pointwise convergence is $B(X)$, the set of bounded Borel measurable functions on X.

1.5.3. Let $(E, \mathcal{F})$ be a measurable space. Let $q \in E$ be a point. The atom $A(q)$ containing q is defined by

$$A(q) = \cap \, [B\colon B \in \mathcal{F}, q \in B].$$

Give an example to show that $A(q)$ need not be in $\mathcal{F}$. However if $\mathcal{F}$ is countably generated and $\mathcal{A}$ is a countable subalgebra generating $\mathcal{F}$ show that

$$A(q) = \cap \, [B\colon B \in \mathcal{A}, q \in B],$$

and is therefore always an element of $\mathcal{F}$.

1.5.4. Let $(E, \mathcal{F}, P)$ be a probability space and let $f(\cdot)$ and $g(\cdot)$ be two non-negative random variables on it. Assume that g is integrable and that

$$P[f(\cdot) \geq \lambda] \leq \frac{1}{\lambda} E[g(\cdot)\colon f(\cdot) \geq \lambda], \qquad \lambda > 0.$$

Show that if $r > 1$, then $g \in L^r(P)$ implies $f \in L^r(P)$ and that

$$E[f^r]^{1/r} \leq \frac{r}{r-1} E[g^r]^{1/r}.$$

The proof of this inequality turns on a clever application of the formula

$$E[f^r] = \int [f(q)]^r P(dq) = r \int_0^\infty \lambda^{r-1} P[f(\cdot) \geq \lambda] \, d\lambda$$

which is valid for any non-negative f and any $r \geq 1$. See Theorem 3.4 in Chapter 7 of Doob [1952] for details.

1.5.5. Let $(E_1, \mathcal{F}_1)$, $(E_2, \mathcal{F}_2)$ and $(E_3, \mathcal{F}_3)$ be three measurable spaces. Let $F\colon E_1 \times E_3 \to R$ be a measurable map. Let $\mu(q_2, dq_3)$ be a signed measure on $(E_3, \mathcal{F}_3)$ for each $q_2 \in E_2$ and a measurable function of q_2 for each set in $\mathcal{F}_3$. Show that the set $N \subset E_1 \times E_2$ defined by

$$N = \left\{ (q_1, q_2)\colon \int |F(q_1, q_3)| \, |\mu|(q_2, dq_3) < \infty \right\}$$

is a measurable subset of $(E_1 \times E_2, \mathcal{F}_1 \times \mathcal{F}_2)$ and the function $\int F(q_1, q_3)\mu(q_2, dq_3)$ is a measurable function of (q_1, q_2) on the set N. Deduce from this that if $\theta(t, q)$ is a progressively measurable function and $\eta(t, q)$ is a progressively measurable continuous function which is of bounded variation in t over any finite interval $[0, T]$, then $Z(t, q)$ is again a progressively measurable function where

$$Z(t, q) = \int_0^t \theta(s, q)\eta(ds, q) \quad \text{if} \quad \int_0^t |\theta(s, q)| \, |\eta|(ds, q) < \infty$$

$$= 0 \text{ otherwise.}$$

1.5.6. Let E be a non-empty set and η a map of $[0, \infty) \times E$ into a Polish space (X, d). Define on E the σ-field $\mathcal{F} = \sigma[\eta(s): s \geq 0]$. Show that if f is an $\mathcal{F}$ measurable map of E into a Polish space M, then there exists a $\mathcal{B}_{X^{Z^+}}$-measurable map F of X^{Z^+} into $(M, \mathcal{G})$ and a sequence $\{t_n\}_1^\infty \subseteq [0, \infty)$ such that

$$f(q) = F(\eta(t_1, q), \ldots, \eta(t_n, q) \cdots), \qquad q \in E.$$

Next, assume that $\eta(\cdot, q)$ is right continuous for each $q \in E$ and define $\mathcal{F}_t = \sigma[\eta(s): 0 \leq s \leq t]$. Given a measurable function $\theta: [0, \infty) \times E \to M$ such that for each t, $\theta(t, \cdot)$ is $\mathcal{F}_t$-measurable, show that there exists a $\mathcal{B}_{[0,\infty)} \times \mathcal{B}_{X^{Z^+}}$-measurable map F of $[0, \infty) \times X^{Z^+}$ and a sequence $\{t_n\}_1^\infty \subseteq [0, \infty)$ such that for all $t \geq s$ and $q \in E$

$$\theta(s \wedge t, q) = F(s \wedge t, \eta(t_1 \wedge t, q), \ldots, \eta(t_n \wedge t, q), \ldots).$$

In particular conclude that for each fixed t, $\theta(s \wedge t, q)$ is measurable in (s, q) with respect to $\mathcal{B}_{[0, t]} \times \mathcal{F}_t$, and hence $\theta(\cdot, \cdot)$ is progressively measurable.

1.5.7. Let $(E, \mathcal{F}, P)$ be a probability space and $\Sigma \subset \mathcal{F}$ be a sub σ-field. Let $(Y, \mathcal{B})$ be a measurable space and $F: E \times Y \to R$ a measurable function (relative to $\mathcal{F} \times \mathcal{B}$) such that $\sup_{y \in Y} E^P[|F(\cdot, y)|] < \infty$. Show that a version $G(q, y)$ of $E^P[F(\cdot, y)|\Sigma]$ can be chosen so that $G(\cdot, \cdot)$ is $\Sigma \times \mathcal{B}$ measurable. Suppose now that we have a map $f: E \to Y$ which is Σ-measurable. Assuming that $E^P[|F(\cdot, f(\cdot))|] < \infty$, show that

$$E^P[F(\cdot, f(\cdot))|\Sigma] = G(\cdot, f(\cdot)) \qquad \text{a.e.}$$

1.5.8. Suppose $(\theta(t), \mathcal{F}_t, P)$ is a martingale on $(E, \mathcal{F}, P)$. Let $\bar{\mathcal{F}}_{t+0} = \bigcap_{s>t} \bar{\mathcal{F}}_s$ where $\bar{\mathcal{F}}_s$ is the completion of $\mathcal{F}_s$ in $(E, \mathcal{F}, P)$. (That is $\bar{A} \in \bar{\mathcal{F}}_s$ if and only if there is an A in $\mathcal{F}_s$ with $A \triangle \bar{A} \subset B$ where $B \in \mathcal{F}$ and $P(B) = 0$.) Show that $(\theta(t), \bar{\mathcal{F}}_{t+0}, P)$ is a martingale.

1.5.9. Use Theorem 1.2.8 to show that if $(\theta(t), \mathcal{F}_t, P)$ is a continuous real valued martingale which is almost surely of bounded variation, then for almost all q, $\theta(t)$ is a constant in t. Note that this conclusion is definitely false if one drops the assumption of continuity.

1.5.10. Suppose $(\theta(t), \mathcal{F}_t, P)$ is a martingale on $(E, \mathcal{F}, P)$ such that $\sup_{t \geq 0} E(\theta(t))^2 < \infty$. Show that $E(\theta(t))^2$ is an increasing function of t with a finite limit as $t \to \infty$. Use this to show that $\theta(n)$ tends in mean square to a limit $\theta(\infty)$ as $n \to \infty$. Next, use Doob's inequality to prove

$$P\left[\sup_{t \geq s} |\theta(t) - \theta(s)| \geq \varepsilon\right] \leq \frac{1}{\varepsilon^2} E[\theta^2(\infty) - \theta^2(s)].$$

In particular $\theta(t) \to \theta(\infty)$ a.e. as $t \to \infty$. Finally, show that if τ is an extended stopping time relative to $\{\mathcal{F}_t\}$, then $E^P[\theta(\infty)|\mathcal{F}_\tau] = \theta(\tau)$ a.e. P. This is an especially easy case of Doob's Martingale Convergence Theorem.

1.5.11. Let $(E, \mathcal{F})$ be a measurable space and $\mathcal{F}_t, t \geq 0$ a non-decreasing family of σ-fields such that $\mathcal{F} = \sigma(\bigcup_t \mathcal{F}_t)$. Given $A \subseteq [0, \infty) \times E$, we say that A is *progressively measurable* if $\chi_A(\cdot, \cdot)$ is a progressively measurable map from $[0, \infty) \times E$ into R. Show that the class of progressively measurable sets constitute a σ-field and that a function $f: [0, \infty) \times E \to (X, \mathcal{B})$ is progressively measurable if and only if it is a measurable map relative to the σ-field of progressive measurable sets.

1.5.12. With the same notation as above, let $\tau: E \to [0, \infty]$ be an extended non-negative real valued function such that for each $t > 0, \{q : \tau(q) \leq t\} \in \mathcal{F}_t$. If $f: [0, \infty) \times E \to (X, \mathcal{B})$ is progressively measurable show that $\hat{f}(t, q)$ defined by $\hat{f}(t, q) = f(t \wedge \tau(q), q)$ is again a progressively measurable function. In fact show that if Σ is the σ-field of progressively measurable sets in $[0, \infty) \times E$, then the map $\Phi_\tau: [0, \infty) \times E \to [0, \infty) \times E$ defined by $\Phi_\tau(t, q) = (t \wedge \tau(q), q)$ is a measurable map of $([0, \infty) \times E, \Sigma)$ into itself. (Hint: it is enough to verify that for each $t > 0$, the map Φ_τ restricted to $[0, t] \times E$ is a measurable map of $([0, t] \times E, \mathcal{B}_{[0, t]} \times \mathcal{F}_t)$ into itself. Since the second component of Φ_τ is the identity map it is enough to check that $\tau(q) \wedge s: [0, t) \times E \to [0, t]$ is $\mathcal{B}_{[0, t]} \times \mathcal{F}_t$ measurable. Since $\tau(q) \wedge s$ cannot exceed t, we need only check that for each $a < t$ the set $\{(s, q): \tau(q) \wedge s \leq a\}$ is in $\mathcal{B}_{[0, t]} \times \mathcal{F}_t$. Clearly

$$\{\tau(q) \wedge s \leq a\} = ([0, t] \times \{q: \tau(q) \leq a\}) \cup [0, a] \times E.)$$

1.5.13. Let $A(t), t \geq 0$ be a non-increasing family of subsets of E such that $A(t) \in \mathcal{F}_t$ for each $t \geq 0$. Consider the set

$$A = \bigcup_{t \geq 0} (t, A(t)) = \{(t, q): q \in A(t)\}.$$

Define $B(t) = A(t - 0) = \bigcap_{s < t} A(s)$ for $t > 0$ and $B(0) = A(0)$. Show that if A is progressively measurable then

$$B = \bigcup_{t \geq 0} (t, B(t)) = \{(t, q): q \in B(t)\}$$

and

$$B \backslash A$$

are progressively measurable too. (Hint: Consider the function $f(t, q) = \chi_A(t, q)$. From the fact that $f(t, q)$ is progressively measurable show that $\bar{f}$ defined by

$$\bar{f}(t, q) = f(t - 0, q) = \lim_{n \to \infty} f\left(t\left(1 - \frac{1}{n}\right), q\right)$$

is again progressively measurable. Identify $\bar{f}$ as χ_B.)

Markov Processes, Regularity of Their Sample Paths, and the Wiener Measure

2.1. Regularity of Paths

Suppose that for each $n \geq 1$ and $0 \leq t_1 < \cdots < t_n$ we are given a probability distribution $P_{t_1, \ldots, t_n}$ on the Borel subsets of $(R^d)^n$. Assume that the family $\{P_{t_1, \ldots, t_n}\}$ is consistent in the sense that if $\{s_1, \ldots, s_{n-1}\}$ is obtained from $\{t_1, \ldots, t_n\}$ by deleting the kth element t_k, then $P_{s_1, \ldots, s_{n-1}}$ coincides with the marginal distribution of $P_{t_1, \ldots, t_n}$ obtained by removing the kth coördinate. Then it is obvious that the Kolmogorov extension theorem (cf. Theorem 1.1.10) applies and proves the existence of a unique probability measure P on $(R^d)^{[0, \infty)}$ such that the distribution of $(\psi(t_1), \ldots, \psi(t_n))$ under P is $P_{t_1, \ldots, t_n}$. (Here, and throughout this chapter, ψ stands for an element of $(R^d)^{[0, \infty)}$ and $\psi(t)$ is the random variable on $(R^d)^{[0, \infty)}$ giving the position of ψ at time t.)

As easy and elegant as the preceding construction is, it does not accomplish very much. Although it establishes an isomorphism between consistent families of finite dimensional distributions and measures on a function space, the function space is the wrong one because it is too large and the class of measurable subsets is too small. To be precise, no subset of $(R^d)^{[0, \infty)}$ whose description involves an uncountable number of t's (e.g. $\{\psi: \sup_{0 \leq t \leq 1} |\psi(t)| \leq 1\}$) is in the Borel field. A beautiful partial resolution of this objection is provided by Doob's celebrated Separability Theorem. However, we will not discuss Doob's theorem because the processes with which we will be concerned have the property that the measure P gives the set $\Omega = C([0, \infty), R^d)$ outer measure one and therefore, by the following lemma, can be transferred to $(\Omega, \mathcal{M})$.

2.1.1 Lemma. *Let $(E, \mathcal{F}, P)$ be a probability space and $E' \subseteq E$ a subset having P-outer measure one. Then there is a unique probability measure P' on $(E', \mathcal{F}[E'])$ with the property that $P'(A \cap E') = P(A)$ for all $A \in \mathcal{F}$.*

Proof. The uniqueness is trivial. To prove existence it is really enough to make sure that P' is well-defined by $P'(A \cap E') = P(A)$, $A \in \mathcal{F}$. But if $A, B \in \mathcal{F}$ and $A \cap E' = B \cap E'$, then their symmetric difference $A \bigtriangleup B$ must be contained in $E \backslash E'$; and so, since E' has outer measure one, $P(A \bigtriangleup B) = 0$. Thus $P(A) = P(B)$, and so P' is well-defined. $\square$

The question now is: how does one go about determining when the P associated with $\{P_{t_1, \ldots, t_n}\}$ gives Ω outer measure one? The answer is contained in the next lemma.

2.1.2 Lemma. *Let P be a probability measure on $((R^d)^{[0, \infty)}, \mathcal{B}_{(R^d)^{[0, \infty)}})$. Then Ω has P-outer measure one if and only if for every bounded, countable set $S \subseteq [0, \infty)$:*

$$(1.1) \qquad P(\{\psi: \psi|_S \text{ is uniformly continuous}\}) = 1.$$

(Here $\psi|_S$ stands for the restriction of ψ to S.)

Proof. First suppose that Ω has P-outer measure one. Given S, note that $A_S \equiv \{\psi: \psi|_S \text{ is uniformly continuous}\}$ is a Borel set and that $\Omega \subseteq A_S$. Hence $P(A_S) = 1$.

Next assume that (1.1) holds for all S and let A be a Borel set containing Ω. By exercise 1.5.6, there exists a measurable function F on $((R^d)^{Z^+})$ and a countable set $T = \{t_n: n \geq 1\} \subseteq [0, \infty)$ such that

$$\chi_A(\psi) = F(\psi(t_1), \ldots, \psi(t_n), \ldots), \qquad \psi \in (R^d)^{[0, \infty)}.$$

Thus if $S_N = T \cap [0, N]$, $\bigcap_{N=1}^{\infty} \{\psi: \psi|_{S_N} \text{ is uniformly continuous}\} \subseteq A$. But, by (1.3), $P(\{\psi: \psi|_{S_N} \text{ is uniformly continuous}\}) = 1$ for all N; and so, by the monotone convergence theorem, $P(A) = 1$. This shows that Ω has P-outer measure one. $\square$

We are now going to develop a criterion on $\{P_{t_1, \ldots, t_n}\}$ for testing whether the associated P satisfies (1.1). The basic method which we will use goes back to Kolmogorov, although the elegant approach that we employ here is due to Garsia, Rademich, and Rumsey [1970]. It must be admitted that as a tool for studying Markov processes, Kolmogorov's criterion is rather crude by comparison to the machinery which we developed in Section 1.4. On the other hand, it has the important feature that it depends only on the two dimensional marginals $P_{s, t}$, and not on any higher order structural properties of the process. This fact makes it more ubiquitous than more refined results.

2.1.3 Theorem. *Let p and Ψ be continuous, strictly increasing functions on $[0, \infty)$ such that $p(0) = \Psi(0) = 0$ and $\lim_{t \nearrow \infty} \Psi(t) = \infty$. Given $T > 0$ and $\phi \in C([0, T], R^d)$, if*

$$(1.2) \qquad \int_0^T \int_0^T \Psi\left(\frac{|\phi(t) - \phi(s)|}{p(|t - s|)}\right) ds \, dt \leq B,$$

then for $0 \leq s < t \leq T$:

$$(1.3) \qquad |\phi(t) - \phi(s)| \leq 8 \int_0^{(t-s)} \Psi^{-1}\left(\frac{4B}{u^2}\right) p(du).$$

Proof. Define $d_{-1} = T$ and

$$I(t) = \int_0^T \Psi\left(\frac{|\phi(t) - \phi(s)|}{p(|t - s|)}\right) ds.$$

Since $\int_0^T I(t)dt \leq B$, there is a $t_0 \in (0, d_{-1})$ such that $I(t_0) \leq B/T$. We are now going to choose a non-increasing sequence $\{t_n : n \geq 1\} \subseteq [0, t_0]$ as follows. Given t_{n-1}, define d_{n-1} by $p(d_{n-1}) = \frac{1}{2}p(t_{n-1})$. Choose $t_n \in (0, d_{n-1})$ so that

(1.4) $$I(t_n) \leq 2B/d_{n-1}$$

and

(1.5) $$\Psi\left(\frac{|\phi(t_n) - \phi(t_{n-1})|}{p(|t_n - t_{n-1}|)}\right) \leq 2I(t_{n-1})/d_{n-1}.$$

This can be done because the set of $t \in (0, d_{n-1})$ on which either one of these inequalities can fail must have measure less than $d_{n-1}/2$, and so there is a point in $(0, d_{n-1})$ at which they both hold.

Clearly

$$2p(d_{n+1}) = p(t_{n+1}) \leq p(d_n) = \frac{1}{2}p(t_n).$$

Thus $d_{n+1} \leq t_{n+1} \leq d_n \leq t_n$ and $t_n \searrow 0$ as $n \to \infty$. Also,

(1.6) $$p(t_n - t_{n+1}) \leq p(t_n) = 2p(d_n) = 4(p(d_n) - \tfrac{1}{2}p(d_n))$$
$$\leq 4(p(d_n) - p(d_{n+1})).$$

Combining (1.4), (1.5), and (1.6) with the fact that $d_n \leq d_{n-1}$, we see that

$$|\phi(t_n) - \phi(t_{n+1})| \leq \Psi^{-1}(2I(t_n)/d_n)p(t_n - t_{n+1})$$
$$\leq \Psi^{-1}(4B/d_{n-1}d_n)4(p(d_n) - p(d_{n+1}))$$
$$\leq 4\Psi^{-1}(4B/d_n^2)(p(d_n) - p(d_{n+1}))$$
$$\leq 4\int_{d_{n+1}}^{d_n} \Psi^{-1}(4B/u^2)p(du).$$

Summing over $n \geq 0$, we now get

(1.7) $$|\phi(t_0) - \phi(0)| \leq 4\int_0^T \Psi^{-1}(4B/u^2)p(du).$$

Replacing $\phi(t)$ by $\phi(T - t)$ in the preceding argument, we conclude that

(1.8) $$|\phi(T) - \phi(t_0)| \leq 4\int_0^T \Psi^{-1}(4B/u^2)p(du).$$

Adding these, we arrive at

(1.9)
$$|\phi(T) - \phi(0)| \le 8 \int_0^T \Psi^{-1}(4B/u^2)p(du).$$

Note that (1.9) now holds for any Ψ, p, and ϕ such that (1.2) is satisfied. In particular, if $0 \le s < t \le T$ are given and we define

$$\bar{\phi}(u) = \phi\left(s + \frac{t-s}{T}u\right), \qquad u \in [0, T],$$

and

$$\bar{p}(u) = p\left(\frac{t-s}{T}u\right), \qquad u \in [0, T],$$

then

$$\int_0^T \int_0^T \Psi\left(\frac{|\bar{\phi}(u) - \bar{\phi}(v)|}{\bar{p}(|u-v|)}\right) du\, dv = \left(\frac{T}{t-s}\right)^2 \int_s^t \int_s^t \Psi\left(\frac{|\phi(u) - \phi(v)|}{p(|u-v|)}\right) du\, dv$$

$$\le \left(\frac{T}{t-s}\right)^2 B \equiv \bar{B}.$$

Thus, replacing ϕ by $\bar{\phi}$, p by $\bar{p}$, and B by $\bar{B}$ in (1.9), we have

$$|\bar{\phi}(T) - \bar{\phi}(0)| \le 8 \int_0^T \Psi^{-1}(4\bar{B}/u^2)\bar{p}(du),$$

which becomes (1.3) after the obvious change of variables.   □

2.1.4 Corollary. *Let* $(E, \mathcal{F}, P)$ *be a probability space and* $\theta: [0, \infty) \times E \to R^d$ *a* $\mathcal{B}_{[0, \infty)} \times \mathcal{F}$-*measurable function such that* $\theta(\cdot, q)$ *is continuous for all* $q \in E$. *If for each* $T > 0$ *there exist number* $\alpha = \alpha_T > 0$, $r = r_T > 0$, *and* $C = C_T < \infty$ *such that*

(1.10)
$$E[|\theta(t) - \theta(s)|^r] \le C|t - s|^{1+\alpha}, \qquad 0 \le s, t \le T,$$

then for any $\gamma = \gamma_T \in (2, 2 + \alpha_T)$ *and* $\lambda > 0$:

(1.11)
$$P\left(\sup_{0 \le s < t \le T} \frac{|\theta(t) - \theta(s)|}{|t - s|^\beta} \ge \frac{8\gamma}{\gamma - 2}(4\lambda)^{1/r}\right) \le CA/\lambda,$$

where $\beta = \beta_T = (\gamma_T - 2)/r_T$ *and* $A = A_T = \int_0^T \int_0^T |t - s|^{1+\alpha-\gamma}\, ds\, dt$.

Proof. From (1.10) we know that

$$E\left[\int_0^T \int_0^T \left(\frac{|\theta(t) - \theta(s)|}{|t - s|^{\gamma/r}}\right)^r ds\, dt\right] \leq CA.$$

Thus,

$$P\left(\int_0^T \int_0^T \left(\frac{|\theta(t) - \theta(s)|}{|t - s|^{\gamma/r}}\right)^r ds\, dt \geq \lambda\right) \leq CA/\lambda,$$

and by Theorem 2.1.3,

$$|\theta(t) - \theta(s)| \leq 8 \int_0^{|t-s|} \left(\frac{4\lambda}{u^2}\right)^{1/r} du^{\gamma/r}$$

$$= \frac{8\gamma}{\gamma - 2} (4\lambda)^{1/r} |t - s|^\beta, \qquad 0 \leq s \leq t \leq T,$$

if

$$\int_0^T \int_0^T \left(\frac{|\theta(t) - \theta(s)|}{|t - s|^{\gamma/r}}\right)^r ds\, dt \leq \lambda. \qquad \square$$

2.1.5 Corollary. *Let P be a probability measure on* $((R^d)^{[0,\,\infty)}, \mathscr{B}_{(R^d)^{[0,\,\infty)}})$ *and suppose that for each $T > 0$ there exist numbers $\alpha = \alpha_T > 0, r = r_T \geq 1 + \alpha_T$, and $C = C_T < \infty$ such that*

(1.12) $E[\,|\psi(t) - \psi(s)|^r] \leq C\,|t - s|^{1+\alpha}, \qquad 0 \leq s < t \leq T.$

Then Ω has P-outer measure one.

Proof. According to Lemma 2.1.2, we must show that if $T > 0$ and S is a countable subset of $[0, T]$, then

$$P(\{\psi: \psi\,|_S \text{ is uniformly continuous}\}) = 1.$$

Given S, choose for each $N \geq 1$ points $0 = t_{0,\,N} < \cdots < t_{N,\,N} = T$ so that if $S_N = \{t_{i,\,N}: 0 \leq i \leq N\}$, then $S_N \subseteq S_{N+1}$ and $S \subset \bigcup_1^\infty S_N$. Then

(1.13) $P(\{\psi: \psi\,|_S \text{ is uniformly continuous}\})$

$$\geq \lim_{\lambda \nearrow \infty} \lim_{N \nearrow \infty} P\left(\left\{\psi: \sup_{s,\,t \in S_N} \frac{|\psi(t) - \psi(s)|}{|t - s|^\beta} \leq \frac{8\gamma}{\gamma - 2} (4\lambda)^{1/r}\right\}\right),$$

where γ and β are chosen as in Corollary 2.1.4. But if $\psi^{(N)}(\cdot)$ is defined by

$$\psi^N(t) = \frac{(t_{i+1,N} - t)\psi(t_{i,N}) + (t - t_{i,N})\psi(t_{i+1,N})}{t_{i+1,N} - t_{i,N}}, \qquad t_{i,N} \leq t \leq t_{i+1,N},$$

and $\psi^{(N)}(t) = \psi(T)$ for $t \geq T$, then $\psi^{(N)}(\cdot)$ is continuous and an easy calculation shows that (1.12) implies

$$E[|\psi^{(N)}(t) - \psi^{(N)}(s)|^r] \leq 3^r C |t - s|^{1+\alpha}, \qquad 0 \leq s < t \leq T.$$

Thus, Corollary 2.1.4 applied to $\psi^{(N)}$ yields

$$P\left(\left\{\psi: \sup_{s,\, t \in S_N} \frac{|\psi(t) - \psi(s)|}{|t - s|^\beta} \geq \frac{8\gamma}{\gamma - 2}(4\lambda)^{1/r}\right\}\right) \leq 3^r C A / \lambda,$$

and so we see that the right hand side of (1.13) is one. $\quad\square$

We now have all the ingredients necessary for the following theorem of A. N. Kolmogorov.

2.1.6 Theorem. *Let* $\{P_{t_1,\,\ldots,\,t_n}\}$ *be a consistent family of finite dimensional distributions. If for each* $T > 0$ *there exist numbers* $\alpha = \alpha_T > 0, r = r_T \geq 1 + \alpha_T$, *and* $C_T < \infty$ *such that*

$$(1.14) \qquad \iint |y - x|^r P_{s,\,t}(dx \times dy) \leq C_T |t - s|^{1+\alpha}, \qquad 0 \leq s < t \leq T,$$

then there is a unique probability measure P *on* $(\Omega, \mathcal{M})$ *such that*

$$(1.15) \qquad P(x(t_1) \in \Gamma_1, \ldots, x(t_n) \in \Gamma_n) = P_{t_1,\,\ldots,\,t_n}(\Gamma_1 \times \cdots \times \Gamma_n)$$

for all $0 \leq t_1 < \cdots < t_n$ *and* $\Gamma_1, \ldots, \Gamma_n \in \mathcal{B}_{R^d}$.

Proof. Since $\mathcal{M} = \sigma[x(t): t \geq 0]$, the uniqueness is obvious. To prove existence, we proceed as follows. First construct P on $((R^d)^{[0,\,\infty)}, \mathcal{B}_{(R^d)^{[0,\,\infty)}})$, corresponding to $\{P_{t_1,\,\ldots,\,t_n}\}$, as in the first paragraph of this section. Then (1.14) implies (1.12); and therefore, by Corollary 2.1.5 and Lemma 2.1.1, P determines a restriction P' to $(\Omega, \mathcal{B}_{(R^d)^{[0,\,\infty)}}[\Omega])$ given by $P'(A \cap \Omega) = P(A)$, $A \in \mathcal{B}_{(R^d)^{[0,\,\infty)}}$. Since $\mathcal{M} = \mathcal{B}_{(R^d)^{[0,\,\infty)}}[\Omega]$, the proof will be complete once we rename P' as P. $\quad\square$

2.2. Markov Processes and Transition Probabilities

In the theory of Markov processes, the consistent family $\{P_{t_1,\,\ldots,\,t_n}\}$ arises in a special way. To be precise, we define a transition probability function as a function $P(s; x; t, \Gamma)$, $0 \leq s < t$, $x \in R^d$, and $\Gamma \in \mathcal{B}_{R^d}$, satisfying:

> (i) $P(s, x, t, \cdot)$ is a probability measure on $(R^d, \mathcal{B}_{R^d})$ for all $0 \leq s < t$ and $x \in R^d$,

(ii) $P(s, \cdot, t, \Gamma)$ is $\mathscr{B}_{R^d}$-measurable for all $0 \le s < t$ and $\Gamma \in \mathscr{B}_{R^d}$,
(iii) if $0 \le s < t < u$ and $\Gamma \in \mathscr{B}_{R^d}$, then

$$(2.1) \qquad P(s, x; u, \Gamma) = \int P(t, y; u, \Gamma)P(s, x; t, dy).$$

Equation (2.1) is known as the *Chapman–Kolmogorov equation*. A transition probability function should be thought of as giving the conditional distribution of a process at time t given that at time s the process was at x. It turns out that, even if one specifies the initial distribution of the process, a stochastic process is not uniquely determined by insisting that it have the preceding property relative to a given $P(s, x; t, \cdot)$. However, if one goes one step further and demands that $P(s, x; t, \cdot)$ be the conditional distribution of the process at time t given the process before time s and that at time s the process was at x, then the process is uniquely determined as soon as its initial distribution is given. We now have the following formal definition.

2.2.1 Definition. Let $P(s, x; t, \cdot)$ be a transition probability function and μ a probability measure on $(R^d, \mathscr{B}_{R^d})$. A probability measure P on $((R^d)^{[0, \infty)}, \mathscr{B}_{(R^d)^{[0, \infty)}})$ is called the *Markov process with transition function $P(s, x; t, \cdot)$ and initial distribution μ* if

$$(2.2) \qquad P(\psi(0) \in \Gamma) = \mu(\Gamma), \qquad \Gamma \in \mathscr{B}_{R^d}:$$

and for all $0 \le s < t$ and $\Gamma \in \mathscr{B}_{R^d}$:

$$(2.3) \qquad P(\psi(t) \in \Gamma \mid \sigma[\psi(u): 0 \le u \le s]) = P(s, \psi(s); t, \Gamma) \quad (\text{a.s.}, P).$$

If $P(s, x; t, \cdot) = P(t - s, x, \cdot)$, the corresponding Markov process is said to be *time-homogeneous*.

We now turn to the question of the existence of a Markov process with given transition function and initial distribution.

2.2.2 Theorem. *Let $P(s, x; t, \cdot)$ be a transition probability function and μ a probability measure on $(R^d, \mathscr{B}_{R^d})$. Define*

$$(2.4) \qquad P_0(\Gamma) = \mu(\Gamma), \qquad \Gamma \in \mathscr{B}_{R^d},$$

and for $0 \le t_1 < \cdots < t_{n+1}$:

$$(2.5) \qquad P_{t_1, \dots, t_{n+1}}(\Delta) = \int \dots \int_{\Delta} P(t_n, y_n; t_{n+1}, dy_{n+1})P_{t_1, \dots, t_n}(dy_1 \times \dots \times dy_n),$$

where $\Delta \in \mathscr{B}_{(R^d)^{N+1}}$. Then, $\{P_{t_1, \dots, t_n}\}$ is consistent; and a probability measure P on $((R^d)^{[0, \infty)}, \mathscr{B}_{(R^d)^{[0, \infty)}})$ is the Markov process with transition function $P(s, x; t, \cdot)$ and

*initial distribution μ if and only if P is the probability measure on $((R^d)^{[0,\,\infty)},$
$\mathscr{B}_{(R^d)[0,\,\infty)})$ having $\{P_{t_1,\,\ldots,\,t_n}\}$ as finite dimensional distributions. In particular there
exists one and only one Markov process with given transition function and initial
distribution.*

Proof. The consistency of $\{P_{t_1,\,\ldots,\,t_n}\}$ is immediate from the Chapman–Kolmogorov
equation (2.1). Now suppose P on $((R^d)^{[0,\,\infty)},\ \mathscr{B}_{(R^d)[0,\,\infty)})$ has finite dimensional
distributions given by (2.4) and (2.5). Given $t > 0$ and $\Gamma \in \mathscr{B}_{R^d}$, it is clear that

$$P(\psi(0) \in \Delta,\ \psi(t) \in \Gamma) = \int_\Delta P(0,\,y;\,t,\,\Gamma)P_0(dy)$$

$$= E^P[P(0,\,\psi(0);\,t,\,\Gamma),\,\psi(0) \in \Delta],$$

and therefore:

$$P(\psi(t) \in \Gamma \,|\, \sigma[\psi(0)]) = P(0,\,\psi(0);\,t,\,\Gamma) \quad \text{(a.s., } P\text{)}.$$

Next let $0 \le s < t$ and $\Gamma \in \mathscr{B}_{R^d}$ be given and suppose that $0 \le u_1 < \cdots < u_n = s$
and $\Gamma_1,\,\ldots,\,\Gamma_n \in \mathscr{B}_{R^d}$ are chosen. Then

$$P(\psi(u_1) \in \Gamma_1,\,\ldots,\,\psi(u_n) \in \Gamma_n,\,\psi(t) \in \Gamma)$$

$$= P_{u_1,\,\ldots,\,u_n,\,t}(\Gamma_1 \times \cdots \times \Gamma_n \times \Gamma)$$

$$= \iint_{\Gamma_1 \times \cdots \times \Gamma_n} P(s;\,y_n;\,t,\,\Gamma)P_{u_1,\,\ldots,\,u_n}(dy_1 \times \cdots \times dy_n)$$

$$= E^P[P(s,\,\psi(s);\,t,\,\Gamma),\,\psi(u_1) \in \Gamma_1,\,\ldots,\,\psi(u_n) \in \Gamma_n].$$

Thus an easy application Exercise 1.5.1 implies that (2.3) holds.

Finally, assume that P on $((R^d)^{[0,\,\infty)},\ \mathscr{B}_{(R^d)[0,\,\infty)})$ satisfies (2.2) and (2.3). We want
to check that its finite dimensional distributions $\{Q_{t_1,\,\ldots,\,t_n}\}$ are $\{P_{t_1,\,\ldots,\,t_n}\}$. Clearly
(2.2) implies $Q_0 = P_0$. To complete the identification we use induction on n. If
$n = 1$ and $t_1 = 0$, we have already checked $Q_{t_1} = P_{t_1}$. If $n = 1$ and $t_1 > 0$, then for
$\Gamma_1 \in \mathscr{B}_{R^d}$:

$$Q_{t_1}(\Gamma_1) = P(\psi(t_1) \in \Gamma_1) = E^P[P(0,\,\psi(0);\,t_1,\,\Gamma_1)]$$

$$= \int P(0,\,y;\,t_1,\,\Gamma_1)Q_0(dy) = \int P(0,\,y;\,t_1,\,\Gamma_1)P_0(dy)$$

$$= P_{t_1}(\Gamma_1).$$

Now assume that $Q_{t_1,\,\ldots,\,t_n} = P_{t_1,\,\ldots,\,t_n}$. Then for any bounded measurable
$f\colon (R^d)^n \to R$, we have

$$E^P[f(\psi(t_1),\,\ldots,\,\psi(t_n))] = \int \cdots \int f(y_1,\,\ldots,\,y_n)P_{t_1,\,\ldots,\,t_n}(dy_1 \times \cdots \times dy_n).$$

In particular, if $\Gamma_1, \ldots, \Gamma_{n+1} \in \mathscr{B}_{R^d}$ and

$$f(y_1, \ldots, y_n) = \chi_{\Gamma_1}(y_1) \cdots \chi_{\Gamma_n}(y_n) P(t_n, y_n; t_{n+1}, \Gamma_{n+1}),$$

then, by (2.3),

$$Q_{t_1, \ldots, t_{n+1}}(\Gamma_1 \times \cdots \times \Gamma_{n+1}) = E^P[f(\psi(t_1), \ldots, \psi(t_n))]$$

$$= \int \cdots \int f(y_1, \ldots, y_n) P_{t_1, \ldots, t_n}(dy_1 \times \cdots \times dy_n)$$

$$= \iint\limits_{\Gamma_1 \times \cdots \times \Gamma_{n+1}} P(t_n, y_n; t_{n+1}, dy_{n+1}) P_{t_1, \ldots, t_n}(dy_1 \times \cdots \times dy_n)$$

$$= P_{t_1, \ldots, t_{n+1}}(\Gamma_1 \times \cdots \times \Gamma_{n+1}).$$

Thus the proof can be completed by another application of Exercise 1.5.1. $\qquad \square$

Of course, there is no reason why a Markov process should always have to be realized on $((R^d)^{[0, \infty)}, \mathscr{B}_{(R^d)^{[0, \infty)}})$. In fact, we want the following definition.

2.2.3 Definition. Let $(E, \mathscr{F}, P)$ be a probability space and $\{\mathscr{F}_t : t \geq 0\}$ a nondecreasing family of sub σ-algebras of $\mathscr{F}$. Let $P(s, x; t, \cdot)$ be a transition probability function and μ a probability measure on $(R^d, \mathscr{B}_{R^d})$. Given a function $\xi: [0, \infty) \times E \to R^d$, the triple $(\xi(t), \mathscr{F}_t, P)$ is called a *Markov process on* $(E, \mathscr{F})$ *with transition probability function* $P(s, x; t, \cdot)$ *and initial distribution* μ if $\xi(t)$ is $\mathscr{F}_t$-measurable for all $t \geq 0$ and

$$(2.5) \qquad\qquad P(\xi(0) \in \Gamma) = \mu(\Gamma), \qquad \Gamma \in \mathscr{B}_{R^d},$$

and

$$(2.6) \qquad\qquad P(\xi(t) \in \Gamma \,|\, \mathscr{F}_s) = P(s, \xi(s); t, \Gamma) \qquad (\text{a.s.}, P)$$

for all $0 \leq s < t$ and $\Gamma \in \mathscr{B}_{R^d}$.

Notice that if $E = (R^d)^{[0, \infty)}$, $\mathscr{F} = (\mathscr{B}_{R^d})^{[0, \infty)}$, and $\mathscr{F}_t = \sigma[\psi(u): 0 \leq u \leq t]$, then the preceding definition is consistent with the one given in 2.2.1. The case in which we will be most interested is when $E = \Omega$, $\mathscr{F} = \mathscr{M}$, $\mathscr{F}_t = \mathscr{M}_t$, and $\xi(t) = x(t)$. In fact, if $(x(t), \mathscr{M}_t, P)$ is a Markov process on $(\Omega, \mathscr{M})$, we will call it a *continuous Markov process*.

2.2.4 Theorem. *Let* $P(s, x; t, \cdot)$ *be a transition probability function such that for each* $T > 0$ *there exist* $\alpha = \alpha_T > 0, r = r_T \geq 1 + \alpha_T$, *and* $C = C_T$ *for which*

$$(2.7) \qquad \sup_{y_1 \in R^d} \int |y - y_1|^r P(t_1, y_1; t_2, dy) \leq C |t_2 - t_1|^{1 + \alpha}, \quad 0 \leq t_1 < t_2 \leq T.$$

Given a probability measure μ on $(R^d, \mathscr{B}_{R^d})$, there is a unique probability measure $P = P_\mu$ on $(\Omega, \mathscr{M})$ such that $(x(t), \mathscr{M}_t, P)$ is a continuous Markov process with transition function $P(s, x; t, \cdot)$ and initial distribution μ. In particular, for each $s \geq 0$ and $x \in R^d$, there is a unique probability measure $P_{s, x}$ on $(\Omega, \mathscr{M})$ such that

$$(2.8) \qquad P_{s, x}(x(t) \equiv x \;\; for \; all \;\; 0 \leq t \leq s) = 1$$

and

$$(2.9) \qquad P_{s, x}(x(t_2) \in \Gamma \,|\, \mathscr{M}_{t_1}) = P(t, x(t_1); t_2, \Gamma) \qquad (a.s., \; P_{s, x})$$

for all $s \leq t_1 < t_2$ and $\Gamma \in \mathscr{B}_{R^d}$.

Proof. Let Q on $((R^d)^{[0, \infty)}, \mathscr{B}_{(R^d)^{[0, \infty)}})$ be the Markov process with transition function $P(s, x; t, \cdot)$ and initial distribution μ (cf. Theorem 2.2.2). By Theorem 2.1.6, Q admits a restriction P_μ to $(\Omega, \mathscr{M})$ having the same finite dimensional distributions. Since $\mathscr{M}_t = (\sigma[\psi(u): 0 \leq u \leq t])[\Omega]$, it is obvious that P_μ is the desired measure. (The uniqueness of P_μ is clear, since its finite dimensional distribution must be those of Q.)

Finally, suppose $s \geq 0$ and $x \in R^d$ are given. Define

$$P'(t_1, y_1; t_2, \cdot) = \begin{cases} \delta_{y_1} & \text{if } 0 \leq t_1 < t_2 \leq s, \\ P(s, y_1; t_2, \cdot) & \text{if } t_1 \leq s < t_2, \\ P(t_1, y_1; t_2, \cdot) & \text{if } s < t_1 < t_2. \end{cases}$$

Then $P'(t_1, y_1; t_2, \cdot)$ satisfies (2.7). It is easy to check that the desired $P_{s, x}$ is the unique probability measure on $(\Omega, \mathscr{M})$ such that $(x(t), \mathscr{M}_t, P_{s, x})$ is a continuous Markov process with transition function $P'(s, x; t, \cdot)$ and initial distribution δ_x. $\quad\square$

A particularly important application of Theorem 2.2.4 is to the situation in which

$$(2.10) \qquad P(s, x; t, \Gamma) = \int_\Gamma g_d(t - s, y - x) \, dy,$$

where

$$(2.11) \qquad g_d(s, x) = \frac{1}{(2\pi s)^{d/2}} e^{-|x|^2/2s}.$$

It is an elementary exercise to show that (2.7) is satisfied in this case with $\alpha = 1$, $r = 4$, and $C = d$. Thus we have the following famous theorem of N. Wiener.

2.2.5 Theorem. *For each* $s \geq 0$ *and* $x \in R^d$ *there is a unique probability measure* $\mathscr{W}_{s,x}^{(d)}$ *on* $(\Omega, \mathscr{M})$ *such that* $\mathscr{W}_{s,x}^{(d)}(x(t) = x, 0 \leq t \leq s) = 1$ *and*

$$\mathscr{W}_{s,x}^{(d)}(x(t_2) \in \Gamma \mid \mathscr{M}_{t_1}) = \int_{\Gamma} g_d(t_2 - t_1, y - x(t_1)) \, dy \qquad (\text{a.s., } \mathscr{W}_s^{(d)})$$

for all $s \leq t_1 < t_2$ *and* $\Gamma \in \mathscr{B}_{R^d}$.

In the future, we will refer to $\mathscr{W}_{s,x}^{(d)}$ as the *d-dimensional Weiner measure starting at* (s, x). When there is no need to emphasize the dimension, we will neglect the superscript d and simply write $\mathscr{W}_{s,x}$. Also we will use $\mathscr{W}^{(d)}$ (or simply $\mathscr{W}$) to stand for $\mathscr{W}_{0,0}^{(d)}$, and we will call $(\mathscr{W}^{(d)})\mathscr{W}$ simply *(d-dimensional) Wiener measure*. The following theorem is quite elementary (cf. Exercise 2.4.3).

2.2.6 Theorem. *Define* $T_{s,x}: \Omega \to \Omega$ *for* $(s, x) \in [0, \infty) \times R^d$, *by*

$$x(t, T_{s,x}\omega) = x((t - s) \vee 0, \omega) + x.$$

Then $\mathscr{W}_{s,x}^{(d)} = \mathscr{W}^{(d)} \circ T_{s,x}^{-1}$. *Also, if* $\{e_1, \ldots, e_d\}$ *is an orthonormal basis in* R^d *and* $\pi_i : R^d \to R$ *is defined by* $\pi_i x = \langle x, e_i \rangle, 1 \leq i \leq d$, *then* $P = \mathscr{W}^{(d)}$ *if and only if the* σ*-algebras* $\sigma[\pi_i \circ x(s) : s \geq 0]$ *are mutually independent under* P *and* $P \circ \pi_i^{-1} = \mathscr{W}^{(1)}$ *for* $1 \leq i \leq d$. *Finally, the following are equivalent statements about a probability measure* P *on* $(\Omega, \mathscr{M})$:

(i) $P = \mathscr{W}^{(d)}$,
(ii) $E^P[e^{i\langle \theta, x(t) \rangle}] = e^{-(|\theta|^2/2)t}$, *for all* $t \geq 0$ *and* $\theta \in R^d$, *and* $\mathscr{M}_s$ *is independent of* $\sigma[x(t) - x(s): t \geq s]$ *under* P, *for all* $0 \leq s < \infty$,
(iii) *for all* $n \geq 1, 0 \leq t_1 < \cdots < t_n$, *and* $\theta_1, \ldots, \theta_n \in R^d$,

$$E^P\left[\exp\left[i\sum_{k=1}^n \langle \theta_k, x(t_k) \rangle\right]\right] = \exp\left[-\frac{1}{2}\sum_{k,l=1}^n (t_k \wedge t_l)\langle \theta_k, \theta_l \rangle\right].$$

2.3. Wiener Measure

This section is devoted to the development of some of the important properties of Wiener measure. For reasons which will become clear in Chapter 4, we want to couch our discussion in slightly more general terms. Thus we introduce now the next definition.

Definition. Let $(E, \mathscr{F}, P)$ be a probability space and $\{\mathscr{F}_t: t \geq 0\}$ a non-decreasing family of sub σ-algebras of $\mathscr{F}$. Given $s \geq 0$ and a function $\beta: [0, \infty) \times E \to R^d$, we will say that $(\beta(t), \mathscr{F}_t, P)$ is a *d-dimensional s-Brownian motion* (alternatively, when there is no need to emphasize d or $\{\mathscr{F}_t: t \geq 0\}$, $\beta(\cdot)$ is an *s-Brownian motion under* P) if

(i) β is right-continuous and progressively measurable after time s,
(ii) β is P-almost surely continuous,
(iii) $P(\beta(t) = 0$ for $0 \leq t \leq s) = 1$,
(iv) for all $s \leq t_1 < t_2$ and $\Gamma \in \mathcal{B}_{R^d}$.

$$P(\beta(t_2) \in \Gamma \,|\, \mathcal{F}_{t_1}) = \int_\Gamma g_d(t_2 - t_1, y - \beta(t_1)) \, dy \ (\text{a.s., } P),$$

where g_d is given in equation (2.11).

If $s = 0$, we will call $(\beta(t), \mathcal{F}_t, P)$ a *Brownian motion*.

Clearly $(x(t), \mathcal{M}_t, \mathcal{W}^{(d)}_{s,0})$ is an s-Brownian motion. In fact, $(x(t), \mathcal{M}_t, \mathcal{W}^{(d)}_{s,0})$ is the *canonical* s-Brownian motion in that if $(\beta(t), \mathcal{F}_t, P)$ is any s-Brownian motion and $P \circ \beta^{-1}$ is the distribution of $\beta(\cdot)$ under P on Ω (note that by (ii), $q \to \beta(\cdot, q)$ is a map of a set having full P-measure into Ω, and therefore $P \circ \beta^{-1}$ is well-defined on $(\Omega, \mathcal{M})$), then $P \circ \beta^{-1} = \mathcal{W}^{(d)}_{s,0}$. The next lemma gives a partial answer to the question of why one likes to consider other versions of Brownian motion besides the canonical one.

2.3.1 Lemma. *Let $(E, \mathcal{F}, P)$ be a probability space and $(\beta(t), \mathcal{F}_t, P)$ an s-Brownian motion. Denote by $\bar{\mathcal{F}}(\bar{\mathcal{F}}_t)$ the completion of $\mathcal{F}(\mathcal{F}_t)$ under P and use P to denote its own extention to $\bar{\mathcal{F}}$. For $t \geq 0$, set $\bar{\mathcal{F}}_{t+0} = \bigcap_{\delta > 0} \bar{\mathcal{F}}_{t+\delta}$. Then $(\beta(t), \bar{\mathcal{F}}_{t+0}, P)$ is again an s-Brownian motion.*

Proof. Obviously, all we have to check is that $P(\beta(t_2) \in \Gamma \,|\, \bar{\mathcal{F}}_{t_1+0}) = \int_\Gamma g_d(t_2 - t_1, y - \beta(t_1)) \, dy$ (a.s., P) for all $s \leq t_1 < t_2$ and $\Gamma \in \mathcal{B}_{R^d}$. To do this, it is certainly sufficient to show that for all $\phi \in C_0(R^d)$ and $A \in \bar{\mathcal{F}}_{t_1+0}$,

$$(3.1) \qquad E^P[\phi(\beta(t_2)), A] = E^P\left[\int g_d(t_2 - t_1, y - \beta(t_1))\phi(y) \, dy, A\right].$$

Note that (3.1) is obvious if $A \in \bar{\mathcal{F}}_{t_1}$. If $A \in \bar{\mathcal{F}}_{t_1+0}$, then, since $A \in \bar{\mathcal{F}}_{t_1+\varepsilon}$ for all $\varepsilon > 0$,

$$E^P[\phi(\beta(t_2 + \varepsilon)), A] = E^P\left[\int g_d(t_2 - t_1, y - \beta(t_1 + \varepsilon))\phi(y) \, dy, A\right].$$

Since $\beta(\cdot)$ is right-continuous, we can now let $\varepsilon \searrow 0$ and thereby get (3.1). $\square$

We now want to prove one of the basic properties of Brownian motion, namely: "it starts afresh after a stopping time." The first step is the following lemma.

2.3.2 Lemma. *Let* $(\beta(t), \mathcal{F}_t, P)$ *be an s-Brownian motion. Given* $t_0 \geq s$, *define* $\beta_{t_0}(t, q) = \beta(t + t_0, q) - \beta(t_0, q), t \geq 0$ *and* $q \in E$. *If* Φ *is a bounded* $\mathcal{M}$-*measurable function on* Ω, *then*

$$(3.2) \qquad E^P[\Phi \circ \beta_{t_0} | \mathcal{F}_{t_0}] = E^{\mathcal{W}^{(d)}}[\Phi] \qquad (\text{a.s., } P).$$

Proof. We need only prove (3.2) for Φ of the form

$$\Phi(\omega) = \phi_1(x(t_1, \omega)) \cdots \phi_n(x(t_n, \omega)), \qquad \omega \in \Omega,$$

where $0 \leq t_1 < \cdots < t_n$ and $\phi_1, \ldots, \phi_n \in C_b(R^d)$. That is, we must show that if $t_0 \geq s, A \in \mathcal{F}_{t_0}, n \geq 1$, and $\phi_1, \ldots, \phi_n \in C_b(R^d)$, then for $0 < t_1 < \cdots < t_n$:

$$(3.3) \quad E^P[\phi_1(\beta_{t_0}(t_1)) \cdots \phi_n(\beta_{t_0}(t_n)), A] = E^{\mathcal{W}^{(d)}}[\phi_1(x(t_1)) \cdots \phi_n(x(t_n))]P(A).$$

To this end, let $u \geq s$ and $\phi \in C_b(R^d)$ be given. Then for $v > u$ and $A \in \mathcal{F}_u$:

$$E^P[\phi(\beta_u(v)), A] = \int_A E^P[\phi(\beta(v + u) - \beta(u, q)) | \mathcal{F}_u](q)P(dq)$$

$$= \int_A \left(\int \phi(y - \beta(u, q))g_d(v, y - \beta(u, q)) \, dy \right) P(dq)$$

$$= \left(\int \phi(y)g_d(v, y) \, dy \right) P(A).$$

Thus

$$(3.4) \qquad E^P[\phi(\beta_u(v)) | \mathcal{F}_u] = \int \phi(y)g_d(v, y) \, dy \qquad (\text{a.s., } P).$$

In particular, (3.4) proves (3.3) when $n = 1$. Next suppose that (3.3) holds for n. Let $0 < t_1 < \cdots < t_{n+1}$ and $\phi_1, \ldots, \phi_{n+1} \in C_b(R^d)$ be given. Applying (3.4) to $u = t_n + t_0, v = t_{n+1} - t_n$, and $\phi(y) = \phi_{n+1}(y + z)$, we have:

$$E^P[\phi_{n+1}(\beta_{t_n + t_0}(t_{n+1} - t_n) + z) | \mathcal{F}_{t_n + t_0}]$$

$$= \int \phi_{n+1}(y + z)g_d(t_{n+1} - t_n, y) \, dy$$

$$= \int \phi_{n+1}(y)g_d(t_{n+1} - t_n, y - z) \, dy \qquad (\text{a.s., } P).$$

Thus, since (3.3) holds for n:

$$E^P[\phi_1(\beta_{t_0}(t_1)) \cdots \phi_{n+1}(\beta_{t_0}(t_{n+1})), A]$$

$$= \int_A E^P[\phi_{n+1}(\beta_{t_n+t_0}(t_{n+1} - t_n) + \beta_{t_0}(t_n, q)) \,|\, \mathcal{F}_{t_n+t_0}](q)$$

$$\times \phi_1(\beta_{t_0}(t_1, q)) \cdots \phi_n(\beta_{t_0} t_n, q)) P(dq)$$

$$= \int_A \phi_1(\beta_{t_0}(t_1)) \cdots \phi_n(\beta_{t_0}(t_n))$$

$$\times \int \phi_{n+1}(y) g_d(t_{n+1} - t_n, y - \beta_{t_0}(t_n)) \, dy P(dq)$$

$$= E^{W^{(d)}} \left[\phi_1(x(t_1)) \cdots \phi_n(x(t_n)) \right.$$

$$\left. \times \int \phi_{n+1}(y) g_d(t_{n+1} - t_n, y - x(t_n)) \right] P(A)$$

$$= E^{W^{(d)}}[\phi_1(x(t_1)) \cdots \phi_n(x(t_n))\phi_{n+1}(x(t_{n+1}))] P(A)$$

Thus the induction is complete. $\square$

2.3.3 Theorem. *If $(\beta(t), \mathcal{F}_t, P)$ is an s-Brownian motion and τ is a stopping time satisfying $\tau \geq s$ define $\beta_\tau(\cdot)$ by:*

$$\beta_\tau(t, q) = \begin{cases} \beta(t + \tau(q), q) - \beta(\tau(q), q) & \text{if } \tau(q) < \infty \\ 0 & \text{if } \tau(q) = \infty. \end{cases}$$

Then for $A \in \mathcal{F}_\tau$ and Φ a bounded $\mathcal{M}$-measurable function on Ω into R, we have

(3.5) $$E^P[\Phi \circ \beta_\tau, A \cap \{\tau < \infty\}] = E^{W^{(d)}}[\Phi] P(A \cap \{\tau < \infty\}).$$

In particular,

(3.6) $$E^P[\Phi \circ \beta_\tau \,|\, \mathcal{F}_\tau] = E^{W^{(d)}}[\Phi] \quad on \quad \{\tau < \infty\} \quad (a.s., P).$$

Proof. It is certainly enough to check (3.5) when Φ is a bounded continuous function on Ω. But in that case,

$$E^P[\Phi \circ \beta_\tau, A \cap \{\tau < \infty\}] = \lim_{n \to \infty} E^P[\Phi \circ \beta_{\tau_n}, A \cap \{\tau_n < \infty\}],$$

where

$$\tau_n \equiv \begin{cases} s + \dfrac{[n(\tau - s)]}{n} & \textit{if } \ \tau < \infty \\[4mm] \infty & \textit{if } \ \tau = \infty. \end{cases}$$

But $A \in \mathcal{F}_\tau$ implies $A \cap \{\tau_n = s + (k/n)\} \in \mathcal{F}_{s + (k/n)}$ for all $k \geq 1$, and so

$$E^P[\Phi \circ \beta_{\tau_n}, \ A \cap \{\tau_n < \infty\}] = \sum_{k=1}^{\infty} E^P\left[\Phi \circ \beta_{s + (k/n)}, A \cap \left\{\tau_n = s + \frac{k}{n}\right\}\right]$$

$$= \sum_{k=1}^{\infty} E^{\mathcal{W}^{(d)}}[\Phi] P\left(A \cap \left\{\tau_n = s + \frac{k}{n}\right\}\right)$$

$$= E^{\mathcal{W}^{(d)}}[\Phi] P(A \cap \{\tau_n < \infty\}).$$

Since $\{\tau_n < \infty\} = \{\tau < \infty\}$, this completes the proof. $\square$

2.4. Exercises

2.4.1. Extend Theorem 2.1.3 in the following way. Let p and Ψ be strictly increasing continuous functions on $[0, \infty)$ such that $p(0) = \Psi(0) = 0$ and $\lim_{t \uparrow \infty} \Psi(t) = \infty$. Let L be a normed linear space and $f: R^d \rightarrow L$ a function which is strongly continuous on $\overline{B}(a, 2r)$ for some $a \in R^d$ and $r > 0$. Show that

$$\int_{B(a, r)} \int_{B(a, r)} \Psi\left(\frac{\|f(x) - f(y)\|}{p(|x - y|)}\right) dx \, dy \leq B$$

implies that

$$\|f(x) - f(y)\| \leq 8 \int_0^{2|x-y|} \Psi\left(\frac{4^{d+1} B}{\gamma^2 u^{2d}}\right) p(du), \qquad x, y \in B(a, r),$$

where

$$\gamma = \gamma_d \equiv \inf_{x \in B(0, 1)} \ \inf_{0 < \rho \leq 2} \frac{|B(x, \rho) \cap B(0, 1)|}{\rho^d}.$$

The proof mimics that of Theorem 2.1.3. Set

$$I(\cdot) = \int_{B(a, r)} \Psi\left(\frac{\|f(\cdot) - f(y)\|}{p(|\cdot - y|)}\right) dy.$$

Given distinct points $x, y \in B(a, r)$ set $\rho = |x - y|$ and proceed as follows:
  (a) Choose $c \in B((x + y)/2, \rho/2) \cap B(a, r)$ so that

$$I(c) \leq 2^{d+1} \frac{B}{\gamma \rho^d},$$

and set $x_0 = y_0 = c$;

(b) Given x_{n-1} and y_{n-1}, define d_{n-1} and e_{n-1} by:

$$p(d_{n-1}) = \tfrac{1}{2}p(2\,|x_{n-1} - x|) \quad \text{and} \quad p(e_{n-1}) = \tfrac{1}{2}p(2\,|y_{n-1} - y|),$$

respectively, and choose

$$x_n \in B(x, \tfrac{1}{2}d_{n-1}) \cap B(a, r), \qquad y_n \in B(y, \tfrac{1}{2}e_{n-1}) \cap B(a, r)$$

so that

$$I(x_n) \le 2^{d+1}B/\gamma d_{n-1}^d \quad \text{and} \quad \Psi\left(\frac{\|f(x_n) - f(x_{n-1})\|}{p(\,|x_n - x_{n-1}|)}\right) \le 2^{d+1}I(x_{n-1})/\gamma d_{n-1}^d$$

$$I(y_n) \le 2^{d+1}B/\gamma e_{n-1}^d \quad \text{and} \quad \Psi\left(\frac{\|f(y_n) - f(y_{n-1})\|}{p(\,|y_n - y_{n-1}|)}\right) \le 2^{d+1}I(y_{n-1})/\gamma e_{n-1}^d.$$

(c) Conclude that

$$\|f(y) - f(c)\| \le 4\int_0^{2\rho} \Psi^{-1}\left(\frac{4^{d+1}B}{\gamma^2 u^{2d}}\right)p(du)$$

$$\|f(x) - f(c)\| \le 4\int_0^{2\rho} \Psi^{-1}\left(\frac{4^{d+1}B}{\gamma^2 u^{2d}}\right)p(du).$$

2.4.2. Using Corollary (2.14), show that if $\mathscr{P}$ is a set of probability measures on $(\Omega, \mathscr{M})$ such that

$$\lim_{L\uparrow\infty} \sup_{P\in\mathscr{P}} P(\,|x(0)| \ge L) = 0$$

and for each $T > 0$ there exist $\alpha_T > 0$, $r_T \ge 1 + \alpha_T$, and $C_T < \infty$ with

$$\sup_{P\in\mathscr{P}} E^P[\,|x(t) - x(s)|^{r_T}] \le C_T(t - s)^{1+\alpha_T}, \qquad 0 \le s \le t \le T,$$

then $\mathscr{P}$ is precompact. This observation suggests the following derivation of the existence of $\mathscr{W}^{(d)}$. Let $\{X_n : n \ge 1\}$ be independent R^d-valued normal random variables with mean 0 and covariance I on some probability space $(E, \mathscr{F}, P)$. Given $n \ge 1$, define $\Phi_n : E \to \Omega$ by:

$$x(t, \Phi_n(q)) = \frac{1}{n^{1/2}}\left(\sum_1^{[nt]} X_k + n\left(t - \frac{[nt]}{n}\right)X_{[nt]+1}\right),$$

(where $\sum_1^0 X_k \equiv 0$), and set $P_n = P \circ \Phi_n^{-1}$. Check that

$$\sup_{} E^{P_n}[\,|x(t) - x(s)|^4] \le C(t - s)^2, \qquad 0 \le s < t,$$

for some $C < \infty$ and that the finite dimensional distribution of P_n coincide with that of $\mathscr{W}^{(d)}$ at times k_1/n, ..., k_l/n, where $l \geq 1$ and $0 \leq k_1 < \cdots < k_l$. Hence conclude that $P_n \to \mathscr{W}^{(d)}$ as $n \to \infty$.

2.4.3. Prove Theorem 2.2.6.

2.4.4. Prove that if $\beta(\cdot)$ is any Brownian motion, then

$$E[\,|\beta(t) - \beta(s)|^{2r}] = C_r |t - s|^r, \qquad 0 \leq s < t,$$

for any $r \geq 1$. Deduce from this that for any $0 < \alpha < 1/2$, $\beta(\cdot)$ is almost surely Hölder continuous with exponent α on any finite time interval.

2.4.5. Given $\lambda > 0$, define $S_\lambda : \Omega \to \Omega$ by:

$$x(t, S_\lambda \omega) = \lambda^{-1/2} x(\lambda t, \omega).$$

Show that $\mathscr{W}^{(d)}$ is invariant under S_λ (i.e., $\mathscr{W}^{(d)} = \mathscr{W}^{(d)} \circ S_\lambda^{-1}$). Using this fact, prove that a Brownian motion is almost surely not Hölder continuous with exponent $1/2$ even at one time point. With a little more effort one can show that Brownian motion has an exact modulus of continuity: $(2\delta |\log \delta|)^{1/2}$ (cf. McKean [1969]).

2.4.6. Let $a > 0$ be given. Using Theorem 2.3.3, derive the following equality.

$$\mathscr{W}^{(1)}(x(t) > a) = \mathscr{W}^{(1)}(x(t) > a, \tau < t) = \tfrac{1}{2}\mathscr{W}^{(1)}(\tau < t)$$

where $\tau = \inf\{t \geq 0 : x(t) \geq a\}$. Conclude that

$$\mathscr{W}^{(1)}(\tau \leq t) = \frac{a}{(2\pi)^{1/2}} \int_0^t \frac{1}{s^{3/2}} e^{-a^2/2s} \, ds.$$

Next, use this formula to derive the joint distribution under $\mathscr{W}^{(1)}$ of $x(t)$ and τ. Finally, show that

$$\mathscr{W}^{(d)}\left(\sup_{0 \leq s \leq t} |x(s)| \geq a\right) \leq 2d\left(\frac{2t}{\pi}\right)^{1/2} \frac{d^{1/2}}{a} e^{-a^2/2td}.$$

2.4.7. Observe that by the strong law of large numbers,

$$\lim_{n \to \infty} \frac{1}{n} \beta(n) = 0 \qquad \text{(a.s., } P)$$

if $(\beta(t), \mathscr{F}_t, P)$ is a d-dimensional Brownian motion. Combining this with 2.4.6, show that

$$\lim_{t \to 0} t\beta(1/t) = 0 \qquad \text{(a.s., } P).$$

Use this, along with Theorem 2.2.6, to check that if

$$\xi(0) \equiv 0$$

$$\xi(t) = t\beta(1/t), \qquad t > 0,$$

then the distribution of $\xi(\cdot)$ under P is again $\mathcal{W}^{(d)}$. That is, $(\xi(t), \mathcal{F}'_t, P)$ is a Brownian motion, where $\mathcal{F}'_t = \sigma[\xi(u): 0 \le u \le t]$.

2.4.8. Lévy's result on the precise modulus of continuity for Brownian motion asserts that

$$\mathcal{W}^{(1)} \left[\varliminf_{\substack{\delta \to 0 \\ 0 \le s < t \le 1 \\ |t-s| \le \delta}} \sup \frac{|x(t) - x(s)|}{\left(2\delta \log \dfrac{1}{\delta} \right)^{1/2}} = 1 \right] = 1.$$

The difficult part of the result is to prove that

(4.1)
$$\mathcal{W}^{(1)} \left[\varliminf_{\substack{\delta \to 0 \\ 0 \le s < t \le 1 \\ |t-s| \le \delta}} \sup \frac{|x(t) - x(s)|}{\left(2\delta \log \dfrac{1}{\delta} \right)^{1/2}} \le 1 \right] = 1.$$

The other part is a simple estimation of the probability of $\bigcap_{k=1}^{2^n} A_k(\rho)$, where $A_k(\rho)$ are the independent events

$$A_k(\rho) = \left\{ \left| x\left(\frac{k}{2^n} \right) - x\left(\frac{k-1}{2^n} \right) \right| \le (1 - \rho)(2^{-n+1} \log 2^n)^{1/2} \right\}$$

and $0 < \rho < 1$ is arbitrary. (cf. McKean (1969)). It has been pointed out to the authors by R. Wolpert that a slightly weaker version of (4.2), namely

(4.2)
$$\mathcal{W}^{(1)} \left[\varliminf_{\substack{\delta \downarrow 0 \\ 0 \le s < t \le 1 \\ |t-s| \le \delta}} \sup \frac{|x(t) - x(s)|}{\left(2\delta \log \dfrac{1}{\delta} \right)^{1/2}} \le 8 \right] = 1,$$

can be deduced from Theorem 2.1.3.

To this end, take $\psi(u) = e^u - 1$ and for fixed $\alpha > \frac{1}{2}$ take $p(u) = (\alpha u/(\log 1/u))^{1/2}$. Check that

$$E^{\mathcal{W}^{(1)}} \left[\int_0^1 \int_0^1 \psi\left(\frac{|x(t) - x(s)|}{p(|t-s|)} \right) ds \, dt \right] < \infty.$$

Next, apply Theorem 2.1.3 to show that

$$\lim_{B \uparrow \infty} \mathcal{W}^{(1)} \left[|x(t) - x(s)| \le 4\alpha^{1/2} \int_0^{|t-s|} \frac{1}{u^2} \log\left(1 + \frac{4B}{u^2} \right) \frac{1 + \log \dfrac{1}{u}}{\left(\log \dfrac{1}{u} \right)^{3/2}} \, du, \right.$$

$$\left. 0 \le s \le t \le 1 \right] = 1.$$

That is to say, there is a finite random variable B such that

$$|x(t) - x(s)| \leq 4\alpha^{1/2} \int_0^\delta \frac{1}{u^{1/2}} \log\left(1 + \frac{4B}{u^2}\right) \frac{1 + \log\frac{1}{u}}{\left(\log\frac{1}{u}\right)^{3/2}} \, du$$

for all $0 \leq s < t \leq 1$ with $t - s \leq \delta$. By L'Hospital's rule

$$\lim_{\delta \downarrow 0} \frac{\displaystyle\int_0^\delta u^{1/2} \log\left(1 + \frac{4B}{u^2}\right) \frac{1 + \log\frac{1}{u}}{\left(\log\frac{1}{u}\right)^{3/2}} \, du}{\left(\delta \log\frac{1}{\delta}\right)^{1/2}} = 4$$

and therefore

$$\varlimsup_{\substack{\delta \downarrow 0 \\ \substack{0 \leq s < t \leq 1 \\ |t-s| \leq \delta}}} \sup \frac{|x(t) - x(s)|}{\left(2\delta \log\frac{1}{\delta}\right)^{1/2}} \leq \frac{16\alpha^{1/2}}{2^{1/2}}$$

almost surely. Since $\alpha > \frac{1}{2}$ is arbitrary, this implies (4.2). If one is willing to replace 8 in (4.2) by 16, then one can take $\Phi(u) = e^{u^2/2}$ and $p(u) = \alpha u^{1/2}$ for some $\alpha > 1$. The details are somewhat simpler in this case.

Chapter 3

Parabolic Partial Differential Equations

3.1 The Maximum Principle

In Chapter 2, Section 2.2, we showed how one can start with a transition probability function $P(s, x; t, \cdot)$ and end up with a Markov process. The problem is: where does $P(s, x; t, \cdot)$ come from? The example we gave there, namely:

(1.1)
$$P(s, x; t, \Gamma) = \int_\Gamma g_d(t - s, y - x) \, dy$$

is a natural one from the probabilistic point of view because of its connection with independent increments and Gaussian processes. It turns out to be natural from another point of view as well: the theory of second order parabolic partial differential equations. The connection between the $P(s, x; t, \cdot)$ in (1.1) and partial differential equations is well-known and easy to derive. Namely, if $\varphi \in C_b(R^d)$ and

$$f(s, x) = \int g_d(T - s, y - x)\varphi(y) \, dy, \ s < T,$$

then

(1.2)
$$\begin{cases} \dfrac{\partial f}{\partial s} + \dfrac{1}{2}\Delta f = 0, \ s < T, \\ \lim_{s \uparrow T} f(s, \cdot) = \varphi \end{cases}$$

where Δ is Laplace's operator

$$\sum_1^d \frac{\partial^2}{\partial x_i^2}.$$

That is, f solves the backward heat equation with terminal data φ at time T. Besides the kinetic theory of gases and Einstein's famous articles on Brownian motion, there is a purely analytic reason why it is not surprising that the heat equation should be the source of a transition probability function. This reason is the "weak maximum principle" for parabolic equations.

3.1.1 Theorem. *Let a and b be bounded functions on $[0, \infty) \times R^d$ with values in S_d and R^d, respectively. Define*

(1.3) $$L_t = \frac{1}{2} \sum_{i, j=1}^{d} a^{ij}(t, x) \frac{\partial^2}{\partial x_i \partial x_j} + \sum_{i=1}^{d} b^i(t, x) \frac{\partial}{\partial x_i}.$$

If $f \in C^{1, 2}([0, T) \times R^d)$ is bounded below and satisfies:

(1.4) $$\frac{\partial f}{\partial s} + L_s f \leq 0, \, 0 \leq s < T$$

$$\lim_{s \nearrow T} f(s, \cdot) \geq 0,$$

then f is non-negative. In particular, if $f \in C^{1, 2}([0, T) \times R^d) \cap C_b([0, T] \times R^d)$ and $c, g \in C_b([0, T))$ satisfy:

(1.5) $$\frac{\partial f}{\partial s} + L_s f + c(s) f \geq -g(s), \qquad 0 \leq s < T,$$

then for $0 \leq s \leq T$:

(1.6) $$f(s, x) \leq \| f(T, \cdot) \| \exp\left(\int_s^T c(u) \, du \right) + \int_s^T g(t) \exp\left(\int_s^t c(u) \, du \right) dt.$$

Proof. First assume that f satisfies

(1.7) $$\frac{\partial f}{\partial s} + L_s f < 0, \qquad 0 \leq s < T.$$

If $(s_0, x_0) \in [0, T) \times R^d$ is a point with the property that

(1.8) $$f(s_0, x_0) \leq f(s, x), \qquad (s, x) \in [s_0, T) \times R^d,$$

then we would have:

$$\frac{\partial f}{\partial s}(s_0, x_0) \geq 0, \qquad \nabla_x f(s_0, x_0) = 0, \quad \text{and}$$

$$\left(\left(\frac{\partial^2 f}{\partial x_i \partial x_j}(s_0, x_0) \right) \right)_{1 \leq i, j \leq d} \in S_d,$$

where ∇_x denotes gradient in the x-directions. Since

$$L_{s_0} f(s_0, x_0) = \frac{1}{2} \text{Trace}(a(s_0, x_0) H_f(s_0, x_0)) + \langle b(s_0, x_0), \nabla_x f(s_0, x_0) \rangle,$$

where

$$H_f(s, x) \equiv \left(\left(\frac{\partial^2 f}{\partial x_i \, \partial x_j}(s, x) \right) \right)_{1 \leq i, j \leq d}$$

is the Hessian matrix of f, we conclude that:

$$\frac{\partial f}{\partial s}(s_0, x_0) + L_{s_0} f(s_0, x_0) \geq 0,$$

since

$$\text{Trace}(AB) \geq 0 \quad \text{if} \quad A, B \in S_d.$$

But this contradicts (1.7), and therefore there is no $(s_0, x_0) \in [0, T) \times R^d$ for which (1.8) holds. Next suppose that f satisfies (1.4). Given $\delta > 0$ and $\varepsilon > 0$, set

$$f_{\delta, \varepsilon}(s, x) = f(s, x) + \delta(T - s) + \varepsilon e^{-s} |x|^2.$$

Then

$$\frac{\partial f_{\delta, \varepsilon}}{\partial s} = \frac{\partial f}{\partial s} - \delta - \varepsilon e^{-s} |x|^2$$

and

$$L_s f_{\delta, \varepsilon} = L_s f + \varepsilon e^{-s} (\text{Trace } a(s, x) + 2 \langle b(s, x), x \rangle).$$

Hence,

$$\frac{\partial f_{\delta, \varepsilon}}{\partial s} + L_s f_{\delta, \varepsilon} \leq -\delta + \varepsilon e^{-s} (\text{Trace}(a(s, x))$$

$$+ 2 \langle b(s, x), x \rangle - |x|^2).$$

Therefore, for each $\delta > 0$, we can choose $\varepsilon_\delta > 0$ so that $f_{\delta, \varepsilon}$ satisfies (1.7) for $\varepsilon \leq \varepsilon_\delta$. Now suppose that $f_{\delta, \varepsilon}(s, x) < 0$ for some $(s, x) \in [0, T) \times R^d$. Then, since $\lim_{s \nearrow T} f_{\delta, \varepsilon}(s, \cdot) \geq 0$ and $f_{\delta, \varepsilon}(s, x) \to +\infty$ as $|x| \to \infty$, there must be a point $(s_0, x_0) \in [0, T) \times R^d$ at which (1.8) obtains, and this is impossible if (1.7) is to hold. We have therefore proved that for all $\delta > 0$ and all $0 < \varepsilon \leq \varepsilon_\delta, f_{\delta, \varepsilon} \geq 0$ in $[0, T) \times R^d$. From this it is clear that $f \geq 0$ in $[0, T) \times R^d$.

Finally, suppose $f \in C^{1,2}([0, T) \times R^d) \cap C_b([0, T] \times R^d)$ and $c, g \in C_b([0, T])$ satisfy (1.5). Then

$$\| f(T, \cdot) \| - f(s, x) \exp \left(-\int_s^T c(u) du \right) + \int_s^T g(t) \exp \left(-\int_t^T c(u) du \right) dt$$

satisfies (1.4), and so (1.6) follows. $\square$

We will make repeated use of Theorem 3.1.1 throughout the rest of this chapter. However, the reader should notice that Theorem 3.1.1 is a uniqueness theorem, and therefore it is not very powerful except in conjunction with an existence theorem. Combined with an existence result one can then prove the following:

3.1.2 Corollary. *Let L_t be defined as in Theorem 3.1.1. Assume that for each $t > 0$ and $\varphi \in C_0^\infty(R^d)$ there is an $f \in C^{1,2}([0, t) \times R^d) \cap C_b([0, t] \times R^d)$ such that $f(t, \cdot) = \varphi$ and $(\partial f/\partial s) + L_s f = 0, 0 \le s < t$. Then for each $t > 0$ and $\varphi \in C_0^\infty(R^d)$ there is exactly one such f. Moreover, if for $0 \le s < t$ and $x \in R^d$ we define $\Lambda_{s,t}(x)$ on $C_0^\infty(R^d)$ into R^1 by $\Lambda_{s,t}(x)\varphi = f(s, x)$, then $\Lambda_{s,t}(x)$ determines a unique non-negative linear functional on*

$$\hat{C}(R^d) \equiv \left\{ \varphi \in C(R^d) \colon \lim_{|x| \to \infty} \varphi(x) = 0 \right\}$$

such that

$$|\Lambda_{s,t}(x)\varphi| \le \|\varphi\|, \qquad \varphi \in \hat{C}(R^d).$$

Finally, if we define $T_{s,t}, 0 \le s < t$, on $\hat{C}(R^d)$ by

$$T_{s,t}\varphi(x) = \Lambda_{s,t}(x)\varphi,$$

then $T_{s,t}$ is a non-negative contraction on $\hat{C}(R^d)$ into itself, $\lim_{s \uparrow t} T_{s,t}\varphi(x) = \varphi(x)$ for all $x \in R^d$, and

$$T_{t_1,t_3} = T_{t_1,t_2} \circ T_{t_2,t_3}, \qquad 0 \le t_1 < t_2 < t_3.$$

Proof. The uniqueness of f is immediate from Theorem 3.1.1. In fact, from that theorem we see that

$$\min_{x \in R^d} \varphi(\dot{x}) \le f(s, \cdot) \le \max_{x \in R^d} \varphi(x), \qquad 0 \le s < t.$$

This proves that $\Lambda_{s,t}(x)$ is a non-negative linear functional on $C_0^\infty(R^d)$ and that

$$|\Lambda_{s,t}(x)\varphi| \le \|\varphi\|.$$

Since $C_0^\infty(R^d)$ is dense in $\hat{C}(R^d)$, this completes the proof of the assertions about $\Lambda_{s,t}(x)$.

In order to prove that $T_{s,t}$ is a non-negative contraction on $\hat{C}(R^d)$ into itself, all we have to show is that $T_{s,t}\varphi \in \hat{C}(R^d)$ for $\varphi \in C_0^\infty(R^d)$. Since, if $\varphi \in C_0^\infty(R^d)$, $f(s, x) = T_{s,t}\varphi(x)$ is the unique $C^{1,2}([0, t) \times R^d)$-function tending to φ as $s \nearrow t$, it only remains to check that such a function must tend to 0 as $|x| \to \infty$. Thus, the proof reduces to showing that if $M > 0$ and $\varphi_M \in C_0^\infty(R^d)$ satisfies: $0 \le \varphi_M \le 1$, $\varphi_M \equiv 1$ on $\overline{B(0, M)}$, and $\varphi_M \equiv 0$ on $R^d \backslash B(0, M + 1)$, then the unique

$f_M \in C_b^{1,2}([0, t) \times R^d) \cap C_b([0, t] \times R^d)$ satisfying $(\partial f_M/\partial s) + L_s f_M = 0$, $0 \le s < t$, with $f_M(t, \cdot) = \varphi_M$, has the property that $f_M(s, \cdot) \in \hat{C}(R^d)$ for all $0 \le s < t$. To this end, note that we can find positive numbers A and B such that for all $x_0 \in R^d$ the function:

$$\psi_{x_0}(s, x) = A e^{B(t-s)}(t - s) + e^{B(t-s)}|x - x_0|^2, \qquad (s, x) \in [0, t) \times R^d,$$

satisfies $(\partial \psi_{x_0}/\partial s) + L_s \psi_{x_0} \le 0$, $0 \le s < t$. Thus if $\rho \equiv \frac{1}{2}|x_0| \ge M + 1$, then, since $\rho^2 \varphi_M(\cdot) \le \psi_{x_0}(t, \cdot)$, we see from Theorem 3.1.1 that

$$0 \le f_M(s, \cdot) \le \frac{1}{\rho^2} \psi_{x_0}(s, \cdot), \qquad 0 \le s < t.$$

In particular,

$$0 \le f_M(s, x_0) \le \frac{A}{\rho^2} e^{B(t-s)}(t - s) = \frac{2A}{|x_0|^2} e^{B(t-s)}(t - s)$$

for $|x_0| \ge 2M + 2$, and this completes the proof that $T_{s,t} \varphi_M \in \hat{C}(R^d)$.

Finally, we must show that

$$T_{t_1, t_3} = T_{t_1, t_2} \circ T_{t_2, t_3}, \qquad 0 \le t_1 < t_2 < t_3.$$

To this end, let $\varphi \in C_0^\infty(R^d)$ be given and set $f(s, \cdot) = T_{s,t_3}\varphi$, $0 \le s < t_3$. Choose $\{\varphi_n : n \ge 1\} \subseteq C_0^\infty(R^d)$ such that $\varphi_n \to f(t_2, \cdot)$ uniformly, and set $f_n(s, \cdot) = T_{s,t_2}\varphi_n$, $0 \le s < t_2$. Then, $f - f_n \in C_b^{1,2}([0, t_2) \times R^d)$,

$$\frac{\partial(f - f_n)}{\partial s} + L_s(f - f_n) = 0, \qquad 0 \le s < t_2,$$

and

$$\lim_{s \nearrow t_2} (f(s, x) - f_n(s, x)) = f(t_2, x) - \varphi_n(x), \qquad x \in R^d.$$

Thus, by Theorem 3.1.1,

$$\sup_{0 \le s < t_2} \|f(s, \cdot) - f_n(s, \cdot)\| \le \|f(t_2, \cdot) - \varphi_n(\cdot)\| \to 0$$

as $n \to \infty$. That is:

$$T_{t_1, t_3} \varphi = f(t_1, \cdot) = \lim_{n \to \infty} f_n(t_1, \cdot) = \lim_{n \to \infty} T_{t_1, t_2} \varphi_n = T_{t_1, t_2} \circ T_{t_2, t_3} \varphi,$$

which is what we needed to show. $\square$

A two-parameter family $\{T_{s,t}: 0 \le s < t\}$ of operators on $\hat{C}(R^d)$ having the properties given in Corollary 3.1.2 is called a *time-inhomogenous semi-group of non-negative contractions on* $\hat{C}(R^d)$. It is clear that if the linear functional $\Lambda_{s,t}(x)$ given by $\Lambda_{s,t}(x)\varphi = T_{s,t}\varphi(x)$ admits the representation

$$(1.9) \qquad \Lambda_{s,t}(x)\varphi = \int P(s, x; t, dy)\varphi(y), \qquad 0 \le s < t \quad \text{and} \quad x \in R^d,$$

where $P(s, x; t, \cdot)$ is a probability measure, then $P(s, x; t, \cdot)$ is a transition probability function. Our next result shows that the $\{T_{s,t}: 0 \le s < t\}$ constructed in Corollary 3.1.2 admits a representation of the sort in (1.9).

3.1.3 Corollary. *Under the same assumptions on L_t as those in Corollary 3.1.2, there exists a unique transition probability function $P(s, x; t, \cdot)$ such that (1.9) holds. Moreover, there exist numbers A and B, depending only on the bounds on the coefficients of L_t, such that:*

$$(1.10) \qquad \int |y - x|^4 P(s, x; t, dy) \le Ae^{B(t-s)}(t - s)^2, \qquad 0 \le s < t \quad \text{and} \quad x \in R^d.$$

Proof. As we said in the preceding paragraph, the existence of a transition probability function $P(s, x; t, \cdot)$ will be established once we show that for each $0 \le s < t$ and $x \in R^d$ the linear functional $\Lambda_{s,t}(x)$ is given by a probability measure. By the Riesz representation theorem for non-negative linear functionals on $\hat{C}(R^d)$, we know that there is a measure μ on $(R^d, \mathscr{B}_{R^d})$ with total mass less than or equal one such that

$$\Lambda_{s,t}(x_0)\varphi = \int \varphi(y)\mu(dy), \qquad \varphi \in \hat{C}(R^d).$$

We must show that $\mu(R^d) = 1$. To this end, let $\varphi_M \in C_0^\infty(R^d)$ be chosen so that $0 \le \varphi_M \le 1$, $\varphi_M \equiv 1$ on $B(x_0, M)$, and $\varphi_M \equiv 0$ off $B(x_0, M + 1)$. Let $\psi_{x_0}(s, x)$ be defined as in the proof of Corollary 3.1.2. Then $M^2(1 - \varphi_M(\cdot)) \le \psi_{x_0}(t, \cdot)$. Thus, by Theorem 3.1.1, if $f_M(s, \cdot) = T_{s,t}\varphi_M$, then $M^2(1 - f_M(s, \cdot)) \le \psi_{x_0}(s, \cdot)$, $0 \le s \le t$. In particular,

$$\mu(B(x_0, M + 1)) \ge T_{s,t}\varphi_M(x_0) = f_M(s, x_0)$$

$$\ge 1 - \frac{1}{M^2} Ae^{B(t-s)}(t - s),$$

and so $\mu(R^d) = 1$.

It remains to prove (1.10). For this purpose, let $t > 0$ and $x_0 \in R^d$ be given and define

$$\gamma(s, x) = Ae^{B(t-s)}(t - s)^2 + A(t - s)e^{B(t-s)}|x - x_0|^2$$
$$+ e^{B(t-s)}|x - x_0|^4.$$

There is a choice of A and B, depending only on the bounds on the coefficients of L such that

$$\frac{\partial \gamma}{\partial s} + L_s \gamma \le 0, \, 0 \le s < t.$$

Thus, if $\{\varphi_n\}_1^\infty \subseteq C_0^\infty(R^d)$ is a sequence of non-negative functions such that $\varphi_n(x) \nearrow |x - x_0|^4$, $x \in R^d$, then, by Theorem 3.1.1,

$$T_{s,t}\varphi_n \le \gamma(s, \cdot), \qquad 0 \le s < t.$$

In particular,

$$\int \varphi_n(y) P(s, x_0; t, dy) = T_{s,t}\varphi_n(x_0) \le A e^{B(t-s)}(t-s)^2.$$

The estimate (1.10) now follows from the monotone convergence theorem.   □

3.2. Existence Theorems

Theorem 3.1.1, and its corollaries, appears to be a good mill with which to turn out transition probability functions. However, like any mill, it requires grist before it can produce; and the grist in this case comes from the theory of partial differential equations. The following theorem is of just the sort that we need; in fact, it gives us more information than is required. We state it here without proof, because the proof is quite intricate (the parametric method is the one usually employed) and we will not be relying on it for anything outside the present section. A good derivation can be found in the book of A. Friedman [1964].

3.2.1 Theorem. *Let $a: [0, \infty) \times R^d \to S_d$ and $b: [0, \infty) \times R^d \to R^d$ be bounded functions for which there exist numbers $\alpha > 0; 0 < \gamma \le 1$, and $C < \infty$ such that*

(i) $\langle \theta, a(s, x)\theta \rangle \ge \alpha |\theta|^2$ *for all* $(s, x) \in [0, \infty) \times R^d$ *and* $\theta \in R^d$,
(ii) $\|a(s, x) - a(t, y)\| + |b(s, x) - b(t, y)| \le C(|x - y|^\gamma + |t - s|^\gamma)$ *for all* (s, x), $(t, y) \in [0, \infty) \times R^d$.

Define

$$L_s = \frac{1}{2} \sum_{i,j=1}^d a^{ij}(s, x) \frac{\partial^2}{\partial x_i \, \partial x_j} + \sum_{i=1}^d b^i(s, x) \frac{\partial}{\partial x_i}.$$

Then there exists a unique positive function $p(s, x; t, y)$, $0 \le s < t$ and $x, y \in R^d$, which is continuous jointly with respect to all its variables and has the property that if $\varphi \in C_0^\infty(R^d)$ and $g \in C_0^\infty([0, \infty) \times R^d)$, then for each $t > 0$ the function

$$(2.1) \qquad f(s, x) = \int p(s, x; t, y)\varphi(y) \, dy + \int_s^t du \int p(s, x; u, y)g(u, y) \, dy$$

is in $C_b^{1,2}([0, t] \times R^d)$ *and satisfies*

$$\frac{\partial f}{\partial s} + L_s f = -g, \qquad 0 \le s < t,$$

with $f(t, \cdot) = \varphi$. *In particular,* L_t *satisfies the conditions of Corollary 3.1.2.*

An easy consequence of Theorem 3.2.1 is the following.

3.2.2 Corollary. *Let* L_t *be given as in Theorem 3.2.1 and let* $P(s, x; t, \cdot)$ *be the associated transition probability function guaranteed by Corollary 3.1.3. Then for any* $0 \le s < t$ *and* $x \in R^d$,

$$P(s, x; t, \Gamma) = \int_\Gamma p(s, x; t, y) \, dy, \qquad \Gamma \in \mathscr{B}_{R^d},$$

where $p(s, x; t, y)$ *is the function described in Theorem 3.2.1. Also, if* $t > 0$ *and* $f \in C_b^{1,2}([0, t) \times R^d) \cap C_b([0, t] \times R^d)$, *then*

$$(2.2) \quad \int f(t, y)P(s, x; t, dy) - f(s, x)$$

$$= \int_s^t du \int \left(\frac{\partial}{\partial u} + L_u \right) f(u, y)P(s, x; u, dy),$$

Proof. The identification of $p(s, x; t, \cdot)$ as the density of $P(s, x; t, \cdot)$ is immediate from the fact that if $g \equiv 0$ in (2.1), then the corresponding f satisfies $(\partial f/\partial s) + L_s f = 0$, $0 \le s < t$ with $f(t, \cdot) = \varphi$; and therefore $f(s, \cdot) = T_{s,t} \varphi$, $0 \le s < t$. To prove (2.2), it suffices to check it for $f \in C_0^\infty([0, \infty) \times R^d)$. But in that case

$$\psi(s, x) = \int f(t, y)P(s, x; t, dy)$$

$$- \int_s^t du \int \left(\frac{\partial}{\partial u} + L_u \right) f(u, y)P(s, x; u, dy)$$

is in $C^{1,2}([0, t] \times R^d)$ *and satisfies*

$$\frac{\partial \psi}{\partial s} + L_s \psi = \frac{\partial f}{\partial s} + L_s f, \qquad 0 \le s < t,$$

with $\psi(t, \cdot) = f(t, \cdot)$. Thus, by Theorem 3.1.1, $\psi(s, \cdot) = f(s, \cdot)$ for all $0 \le s < t$. $\square$

The rest of this section is devoted to the derivation of a result due to Oleinik. The point of Oleinik's theorem is to get existence theorems when the coefficient a is degenerate (i.e., (i) in Theorem 3.2.1 fails) but there are more stringent smoothness requirements on a and b. We will need the following lemma.

3.2.3 Lemma. *Let* $a: R^1 \to S_d$ *be a function having two continuous derivatives. Assume that*

$$\lambda_0 = \sup\{|D^2 a^{i,j}(x)| : 1 \le i, j \le d \quad \text{and} \quad x \in R^1\}$$

Then for all $1 \le i, j \le d$ *and* $x \in R^1$:

(2.3) $$|(Da^{i,j})(x)| \le (2\lambda_0)^{1/2}(a^{ii}(x) + a^{jj}(x))^{1/2}.$$

Moreover, if u *is any symmetric* $d \times d$-*matrix, then:*

(2.4) $$(\mathrm{Trace}((Da)(x)u))^2 \le 4d^2\lambda_0 \,\mathrm{Trace}(ua(x)u), \qquad x \in R^d.$$

Proof. To begin with, let $\varphi \in C^2(R^1)$ be a non-negative function such that $\alpha = \sup_{x \in R^1}|\varphi''(x)| < \infty$. Then for any $x \in R^1$ and all $y \in R^1$, we have, by Taylor's Theorem

$$0 \le \varphi(x + y) \le \varphi(x) + \varphi'(x)y + \frac{\alpha}{2}y^2.$$

In other words, the quadratic $\alpha/2 \, y^2 + \varphi'(x)y + \varphi(x)$ is negative for no real y. Hence, from the elementary theory of discriminants,

$$(\varphi'(x))^2 - 2\alpha\varphi(x) \le 0,$$

and so

(2.5) $$|\varphi'(x)| \le (2\alpha)^{1/2}(\varphi(x))^{1/2}.$$

We now apply (2.5) to the functions

$$\varphi_\pm(x) = a^{ii}(x) \pm 2a^{ij}(x) + a^{jj}(x).$$

Since $\varphi_\pm(x) = \langle e_i \pm e_j, a(x)(e_i \pm e_j)\rangle$, where $\{e_1, \ldots, e_d\}$ is the standard basis in R^d, $|\varphi''_\pm(x)| \le 4\lambda_0$. Hence

$$|\varphi'_\pm(x)| \le (8\lambda_0)^{1/2}(\varphi_\pm(x))^{1/2}.$$

But $a^{ij}(x) = \tfrac{1}{4}(\varphi_+(x) - \varphi_-(x))$, and so

$$\begin{aligned}|(Da^{ij})(x)| &\le \tfrac{1}{4}(|\varphi'_+(x)| + |\varphi'_-(x)|) \\ &\le (\tfrac{1}{2}\lambda_0)^{1/2}((\varphi_+(x))^{1/2} + (\varphi_-(x))^{1/2}), \\ &\le \lambda_0^{1/2}(\varphi_+(x) + \varphi_-(x))^{1/2} \\ &= (2\lambda_0)^{1/2}(a^{ii}(x) + a^{jj}(x))^{1/2}.\end{aligned}$$

To prove (2.4), we can assume that $a(\cdot)$ is diagonal at the point x in question. We then have, by Schwarz's inequality:

$$
\begin{aligned}
(\text{Trace}((Da)(x)u))^2 &= \left(\sum_{i,j=1}^{d} ((Da)(x))^{ij} u^{ij} \right) \\
&\leq d^2 \sum_{i,j=1}^{d} (((Da)(x))^{ij})^2 (u^{ij})^2 \\
&\leq d^2 2\lambda_0 \sum_{i,j=1}^{d} (a^{ii}(x) + a^{jj}(x))(u^{ij})^2 \\
&= 4d^2 \lambda_0 \sum_{i,j=1}^{d} u^{ij} a^{jj}(x) u^{ji} \\
&= 4d^2 \lambda_0 \, \text{Trace}(ua(x)u). \quad \square
\end{aligned}
$$

Before presenting Oleinik's result, we introduce some standard notation from the theory of partial differential equations. Given φ on $[0, \infty) \times R^d$ into R^1, let

$$
\varphi_t = \frac{\partial}{\partial t} \varphi
$$

and

$$
\varphi_{,i} = \frac{\partial \varphi}{\partial x_i}, \quad \varphi_{,ij} = \frac{\partial^2 \varphi}{\partial x_i \, \partial x_j}, \qquad 1 \leq i, j \leq d.
$$

If $\alpha = (\alpha_1, \ldots, \alpha_d)$ is a multi-index of non-negative integers, we define $|\alpha| = \sum_1^d \alpha_i$ and

$$
D^\alpha \varphi = \frac{\partial^{|\alpha|} \varphi}{\partial x_1^{\alpha_1} \cdots \partial x_d^{\alpha_d}}.
$$

Sometimes $\varphi^{(\alpha)}$ is used in place of $D^\alpha \varphi$. Finally, for $n \geq 0$, define

$$
\| \varphi(s, \cdot) \|^{(n)} = \sum_{|\alpha| \leq n} \| D^\alpha \varphi(s, \cdot) \|.
$$

3.2.4 Theorem. Let $a: [0, \infty) \times R^d \to S_d$, $b: [0, \infty) R^d \to R^d$, and $c: [0, \infty) \times R^d \to R^1$ be bounded continuous functions, and set

$$
L_s = \frac{1}{2} \sum_{i,j=1}^{d} a^{ij}(s, \cdot) \frac{\partial^2}{\partial x_i \, \partial x_j} + \sum_{i=1}^{d} b^i(s, \cdot) \frac{\partial}{\partial x_i}.
$$

Assume that $a \in C_b^{0,m}([0, \infty) \times R^d)$ for some $m \geq 2$ and that $b, c \in C^{0,n}([0, \infty) \times R^d)$ for some $n \geq 1$. Given $T > 0$, $\varphi \in C_b^n(R^d)$, and

$g \in C_b^{0,\,n}([0,\,T] \times R^d)$, suppose that $f \in C_b^{1,\,2}([0,\,T] \times R^d)$ satisfies

$$\frac{\partial f}{\partial s} + L_s f + c(s, \cdot) f = -g, \qquad 0 \leq s < T,$$

with $f(T, \cdot) = \varphi$. If, for some $0 \leq l \leq m \wedge n$,

$$f \in C_b^{0,\,l}([0,\,T] \times R^d) \cap C^{0,\,l+2}([0,\,T) \times R^d),$$

then $f_s \in C^{0,\,l}([0,\,T) \times R^d)$ and there exist numbers A_l and B_l such that

(2.6) $\qquad \|f(s, \cdot)\|^{(l)} \leq A_l(\|\varphi\|^{(l)} + \sup_{s \leq t < T} \|g(t, \cdot)\|^{(l)}) e^{B_l(T-s)}, \qquad 0 \leq s < T,$

Moreover, the constants A_l and B_l in (2.6) can be chosen to depend only on l, d, and the bounds on the spatial derivatives of a up to order $l \vee 2$ and those of b and c up to order l.

Proof. Given α with $|\alpha| \leq l$, we use Leibnitz's rule to derive:

(2.7) $\qquad f_s^{(\alpha)} + L_s f^{(\alpha)} + \frac{1}{2} \sum_k{}' \alpha_k a_{,k}^{ij} f_{,ij}^{(\hat{\alpha}^k)} + \sum_{\beta \leq \alpha} c_{\alpha,\,\beta} f^{(\beta)} = -g^{(\alpha)},$

where $\sum_k'$ means summation over those $1 \leq k \leq d$ such that $\alpha_k \neq 0$, $\hat{\alpha}^k = (\alpha_1, \ldots, \alpha_{k-1}, \alpha_k - 1, \alpha_{k+1}, \ldots, \alpha_d)$, and repeated (ij)-indices are summed (in keeping with the usual tensor theory convention). The coefficients $c_{\alpha,\,\beta}$ are linear combinations of a, b, and c and their spatial derivatives up to order l. Define

$$w = \sum_{|\alpha|=l} (f^{(\alpha)})^2.$$

Then, from (2.7), we obtain:

$$w_s + L_s w + \sum_{|\alpha|=l} \sum_k{}' \alpha_k f^{(\alpha)} a_{,k}^{ij} f_{,ij}^{(\hat{\alpha}^k)} - \sum_{|\alpha|=l} \langle \nabla_x f^{(\alpha)}, a \nabla_x f^{(\alpha)} \rangle$$
$$+ 2 \sum_{|\alpha|=l} \sum_{\beta \leq \alpha} c_{\alpha,\,\beta} f^{(\alpha)} f^{(\beta)} + 2 \sum_{|\alpha|=l} f^{(\alpha)} g^{(\alpha)} = 0$$

We must estimate

$$\sum_{|\alpha|=l} \sum_k{}' \alpha_k f^{(\alpha)} a_{,k}^{ij} f_{,ij}^{(\hat{\alpha}^k)} - \sum_{|\alpha|=l} \langle \nabla_x f^{(\alpha)}, a \nabla_x f^{(\alpha)} \rangle.$$

To this end, note that, by Lemma 3.2.3,

$$(a_{,k}^{ij} f_{,ij}^{(\hat{\alpha}^k)})^2 = \left(\text{Trace}\left(\frac{\partial a}{\partial x_k} H^{(\hat{\alpha}^k)} \right) \right)^2 \leq 4 d^2 \lambda_k \, \text{Trace}(H^{(\hat{\alpha}^k)} a H^{(\hat{\alpha}^k)})$$

where $H^{(\hat{a}k)}$ is the Hessian matrix

$$\left(\left(\frac{\partial^2 f^{(\hat{a}k)}}{\partial x_i\, \partial x_j}\right)\right)_{1 \le i,\, j \le d}$$

of $f^{(\hat{a}k)}$ and $\lambda_k = \sup\{|\langle\theta, \partial^2 a/\partial x_k^2(s, x)\theta\rangle|/|\theta|^2 : (s, x) \in [0, \infty) \times R^d$ and $0 \in R^d\backslash\{0\}\}$. Thus

$$\left(\sum_k' \alpha_k f^{(\alpha)} a^{ij}_{\cdot,k} f^{(\hat{a}k)}_{\cdot,ij}\right)^2 \le (f^{(\alpha)})^2 \left(\sum_k \alpha_k^2\right)\left(\sum_k' (a^{ij}_{\cdot,k} f^{(\hat{a}k)}_{\cdot,ij})^2\right)$$

$$\le (f^{(\alpha)})^2 4d^2 l^2 \lambda_0 \sum' \text{Trace}(H^{(\hat{a}k)} a H^{(\hat{a}k)}),$$

where $\lambda_0 = \max_{1 \le k \le d} \lambda_k$. In particular, there is a C_1 depending on l, d, and $\|a\|^{(2)}$ such that

$$\left(\sum_{|\alpha|=l} \sum_k' \alpha_k f^{(\alpha)} a^{ij}_{\cdot,k} f^{(\hat{a}k)}_{\cdot,ij}\right)^2 \le 4C_1 w\gamma,$$

where

$$\gamma = \sum_{|\alpha|=l} \langle \nabla f^{(\alpha)}, a\, \nabla f^{(\alpha)}\rangle.$$

Hence

$$\sum_{|\alpha|=l} \sum_k' \alpha_k f^{(\alpha)} a^{ij}_{\cdot,k} f^{(\hat{a}k)}_{\cdot,ij} - \sum_{|\alpha|=l} \langle \nabla_x f^{(\alpha)}, a\nabla_x f^{(\alpha)}\rangle$$

$$\le 2C_1^{1/2} w^{1/2}\gamma^{1/2} - \gamma \le C_1 w.$$

since:

$$\sup_{t \ge 0} 2C_1^{1/2} w^{1/2}t - t^2 = C_1 w.$$

We have now arrived at:

(2.9) $$w_s + L_s w + C_1 w + 2 \sum_{|\alpha|=l} \sum_{\beta \le \alpha} c_{\alpha,\beta} f^{(\alpha)} f^{(\beta)} + 2 \sum_{|\alpha|=l} f^{(\alpha)} g^{(\alpha)} \ge 0.$$

Next, observe that

$$2 \sum_{|\alpha|=l} \sum_{\beta \le \alpha} c_{\alpha,\beta} f^{(\alpha)} f^{(\beta)} \le C_2 w + C_3 w^{1/2}\|f(s, \cdot)\|^{(l-1)}$$

$$\le (C_2 + C_3\|f(s, \cdot)\|^{(l-1)})w$$

$$+ C_3\|f(s, \cdot)\|^{(l-1)}$$

where C_2 and C_3 depend only on the bounds on the coefficients $c_{\alpha, \beta}$. Also

$$2 \sum_{|\alpha|=l} f^{(\alpha)} g^{(\alpha)} \leq 2w^{1/2} \|g(s, \cdot)\|^{(l)} \leq \|g(s, \cdot)\|^{(l)}$$
$$+ w \|g(s, \cdot)\|^{(l)}.$$

Using these relations in (2.9), we arrive at:

(2.10) $\qquad w_s + L_s w + (C_1 + C_2 + C_3 \|f(s, \cdot)\|^{(l-1)} + \|g(s, \cdot)\|^{(l)}) w$
$$\geq -(\|g(s, \cdot)\|^{(l)} + C_3 \|f(s, \cdot)\|^{(l-1)}).$$

We can now use induction on l and Theorem 3.1.1 to get (2.6). $\qquad\square$

Theorem 3.2.4 is the basic result of Oleinik on which our existence theorem for degenerate parabolic equations turns. However, before we can apply it, we need the following addendum to Theorem 3.2.1 (cf. Friedman (1969)).

3.2.5 Theorem. *Let* $a: [0, \infty) \times R^d \to S_d$ *and* $b: [0, \infty) \times R^d \to R^d$ *be bounded continuous functions having bounded continuous derivatives of all orders. Set*

$$L_s = \frac{1}{2} \sum_{i, j=1}^{d} a^{ij}(s, \cdot) \frac{\partial^2}{\partial x_i \, \partial x_j} + \sum_{i=1}^{d} b^i(s, \cdot) \frac{\partial}{\partial x_i}.$$

If for some $\alpha > 0$,

$$\langle \theta, a(s, x)\theta \rangle \geq \alpha |\theta|^2, \qquad (s, x) \in [0, \infty) \times R^d \quad and \quad \theta \in R^d,$$

then for each $t > 0$ *and* $\varphi \in C_0^\infty(R^d)$ *there exists an* $f \in C_b^{1, 2}([0, t] \times R^d) \cap C^\infty([0, t) \times R^d)$ *such that* $f(t, \cdot) = \varphi$ *and*

$$\frac{\partial f}{\partial s} + L_s f = 0, \qquad 0 \leq s < t.$$

We are now ready to prove the main result of this section.

3.2.6 Theorem. *Let* $a: [0, \infty) \times R^d \to S_d$ *and* $b: [0, \infty) \times R^d \to R^d$ *be bounded continuous functions having two bounded continuous spatial derivatives. Set*

$$L_s = \frac{1}{2} \sum_{i, j=1}^{d} a^{ij}(s, \cdot) \frac{\partial^2}{\partial x_i \, \partial x_j} + \sum_{i=1}^{d} b^i(s, \cdot) \frac{\partial}{\partial x_i}.$$

Then there exists a unique transition probability function $P(s, x; t, \cdot)$ *such that* $P(s, x; \cdot, \Gamma)$ *is* $\mathcal{B}_{(s, \infty)}$*-measurable, for all* $(s, x) \in [0, \infty) \times R^d$ *and* $\Gamma \in \mathcal{B}_{R^d}$, *and*

(2.11) $\quad \int f(t, y) P(s, x; t, dy) - f(s, x) = \int_s^t du \int \left(\frac{\partial}{\partial u} + L_u \right) f(u, y) P(s, x; u, dy)$

for all $0 \leq s < t$, $x \in R^d$, *and* $f \in C_b^{1,2}([0, \infty) \times R^d)$. *Moreover, for each* $(s, x) \in$ $[0, \infty) \times R^d$, *there is a unique probability measure* $P_{s,x}$ *on* $(\Omega, \mathcal{M})$ *such that*:

$$P_{s,x}(x(t) = x, 0 \leq t \leq s) = 1$$

and

$$P_{s,x}(x(t_2) \in \Gamma \mid \mathcal{M}_{t_1}) = P(t_1, x(t_1); t_2, \Gamma) \qquad (\text{a.s., } P_{s,x})$$

for $s \leq t_1 < t_2$ *and* $\Gamma \in \mathscr{B}_{R^d}$.

Proof. Choose coefficients $\{a_n: n \geq 1\}$ and $\{b_n: n \geq 1\}$ so that

(*i*) a_n and b_n have bounded continuous derivatives of all orders,
(*ii*) for each $n \geq 1$ there is an $\alpha_n > 0$ such that

$$\langle \theta, a_n(s, x)\theta \rangle \geq \alpha_n |\theta|^2$$

for all $(s, x) \in [0, \infty) \times R^d$ and $\theta \in R^d$,
(*iii*) for all $t > 0$

$$\|a_n(s, x) - a(s, x)\| + |b_n(s, x) - b(s, x)| \to 0$$

uniformly on $[0, t] \times R^d$ as $n \to \infty$,
(*iv*) there are bounds on the first two spatial derivatives of a_n and b_n which are independent of n.

Define $\{L_t^n: n \geq 1\}$ accordingly, and let $\{T_{s,t}^n: 0 \leq s < t\}$ be the associated time-inhomogeneous semi-group on $\hat{C}(R^d)$ given in Corollary 3.1.2. Given $t > 0$ and $\varphi \in C_0^\infty(R^d)$, set $f_n(s, \cdot) = T_{s,t}^n \varphi$. Then $f_n \in C^\infty([0, t) \times R^d) \cap C_b^{1,2}([0, t] \times R^d)$, $(\partial f_n / \partial s) + L_s^n f_n = 0$, $0 \leq s < t$, and $f_n(t, \cdot) = \varphi$. By Theorem 3.2.4, $\sup_{0 \leq a < t} \|f_n(s, \cdot)\|^{(2)}$ is bounded independent of n, and therefore:

$$\frac{\partial f_n}{\partial s} + L_s f_n \to 0$$

uniformly on $[0, t) \times R^d$. This proves that for $\varphi \in C_0^\infty(R^d)$, and therefore also for $\varphi \in \hat{C}(R^d)$, $T_{s,t}^n \varphi(\cdot)$ converges uniformly on $[0, t] \times R^d$ to a limit, which we denote by $T_{s,t}\varphi(\cdot)$. Clearly $T_{s,t}^n \to T_{s,t}$ strongly as operators on $\hat{C}(R^d)$. Thus, $\{T_{s,t}: 0 \leq s < t\}$ is again a time-inhomogeneous semigroup of non-negative contractions on $\hat{C}(R^d)$.

We now prove that at most one $P(s, x; t, \cdot)$ satisfying (2.11) exists. Indeed, given $t > 0$ and $\varphi \in C_0^\infty(R^d)$, define $f_n(s, \cdot) = T_{s,t}^n \varphi$ as before. Then (2.11) holds with f_n replacing f. Since $(\partial f_n / \partial s) + L_s f_n \to 0$ uniformly on $[0, t) \times R^d$, we have

$$\int \varphi(y) P(s, x; t, dy) - f_n(s, x) \to 0$$

as $n \to \infty$, and therefore:

$$(2.12) \qquad \int \varphi(y)P(s, x; t, dy) = T_{s,t}\varphi(x).$$

That is, if $P(s, x; t, \cdot)$ exists, (2.12) must hold for all $\varphi \in \hat{C}(R^d)$, and therefore there is at most one such $P(s, x; t, \cdot)$.

Conversely, to show that $P(s, x; t, \cdot)$ exists, let $P_n(s, x; t, \cdot)$ be the transition probability function associated with L^n. By Corollary 3.1.3, there exist A and B, independent of n, such that

$$(2.13) \qquad \int |y - x|^4 P_n(s, x; t, dy) \le Ae^{B(t-s)}(t-s)^2, \qquad 0 \le s < t \quad \text{and} \quad x \in R^d.$$

Thus $\{P_n(s, x; t, \cdot): n \ge 1\}$ is (weakly) compact on $(R^d, \mathscr{B}_{R^d})$ for each $(s, x) \in [0, \infty) \times R^d$ and $t > s$. Since

$$T^n_{s,t}\varphi(x) = \int \varphi(y)P_n(s, x; t, dy), \qquad \varphi \in \hat{C}(R^d),$$

and $T^n_{s,t} \to T_{s,t}$ strongly, it follows that $P_n(s, x; t, \cdot)$ converges weakly to a limit $P(s, x; t, \cdot)$; and for this limit, Equation (2.12) obtains. In particular $P(s, x; t, \cdot)$ is a transition probability function (since $\{T_{s,t}: 0 \le s < t\}$ is a semigroup), $P(s, x; t, \Gamma)$ is jointly measurable in s, x, and t ($s < t$) for each $\Gamma \in \mathscr{B}_{R^d}$, and

$$\int |y - x|^4 P(s, x; t, dy) \le Ae^{B(t-s)}(t-s)^2, \qquad 0 \le s < t \quad \text{and} \quad x \in R^d,$$

where A and B are the same as in (2.13). In view of these remarks, it remains only to prove that $P(s, x; t, \cdot)$ satisfies (2.11). But, by Theorem 3.2.1, (2.11) is satisfied when $P(s, x; t, \cdot)$ is replaced by $P_n(s, x; t, \cdot)$ and L. is replaced by L^n. Since $L^n f \to L. f$ uniformly on $[0, t) \times R^d$, we conclude that (2.11) must be true for the $P(s, x; t, \cdot)$ just constructed.

The existence and uniqueness of $\{P_{s,x}: (s, x) \in [0, \infty) \times R^d\}$ is now an easy consequence of Theorem 2.2.4. $\square$

3.3. Exercises

3.3.1. Let L be a linear operator from $C^\infty(R^d)$ into $C(R^d)$ having the following properties:

(i) L is local (i.e., if $\varphi(x) \equiv 0$ in some neighborhood of a point x^0 in R^d then $(L\varphi)(x^0) = 0$).

(ii) L has the maximum principle, (i.e., if $\varphi \in C^\infty(R^d)$ and φ has a local maximum at the point x^0 then $(L\varphi)(x^0) \le 0$).

Under these conditions show that L must be of the form

$$(L\varphi)(x) = \frac{1}{2} \sum a^{ij}(x) \frac{\partial^2 \varphi}{\partial x_i\, \partial x_j}(x) + \sum b^j(x) \frac{\partial \varphi}{\partial x_j}$$

where $a^{ij}(x)$ and $b^j(x)$ are continuous functions on R^d and $\{a^{ij}(x)\}$ is a non-negative definite matrix for each x.

Proceed as follows for the proof:

(1) First show that for any constant function c, $Lc \equiv 0$.

(2) If $\varphi \in C^2(R^d)$ has the property that for some $x^0 \in R^d$, $|\varphi(x) - \varphi(x^0)| = 0(|x - x^0|^2)$ as $|x - x^0| \to 0$, then (by considering $\varphi(x) + \varepsilon|x - x^0|^2$ for suitably small ε) conclude that $(L\varphi)(x^0) = 0$.

(3) Define $b^j(x) = (L\varphi_j)(x)$ and $a^{ij}(x) = (L\varphi_i \varphi_j)(x) - b^j(x)\varphi_i(x) - b^i(x)\varphi_j(x)$, where $\varphi_i(x) = x_i$, i.e. the ith coordinate of x. Then verify by trying functions of the form $\sum_1^d \theta_i(x_i - x_i^0))^2$ that $\{a^{ij}(x^0)\}$ is positive semidefinite for each x^0.

(4) With the above definition of $a^{ij}(x)$ and $b^j(x)$ (using (2) and Taylor's formula) verify the form of L.

3.3.2. Prove the following extended maximum principle: Let $a: [0, \infty) \times R^d \to S^d$ and $b: [0, \infty) \times R^d \to R^d$ be bounded and continuous. Let $T > 0$ be given. Let $f: [0, T] \times R^d \to R$ be a bounded continuous function such that there exists a sequence of functions $\{f_n\}$ in $C_b^{1,2}([0, T] \times R^d)$ such that

$$\lim_{\substack{n \to \infty \\ 0 \le t \le T \\ x \in R^d}} \sup |f_n(t, x) - f(t, x)| = 0$$

and

$$\overline{\lim}_{\substack{n \to 0 \\ 0 \le t \le T \\ x \in R^d}} \sup \left| \frac{\partial f_n}{\partial t} + L_t f_n(t, x) \right| \le \varepsilon$$

where L_t is the operator associated with (a, b). Prove that

$$\sup_{x \in R^d} |f(t, x)| \le \varepsilon(T - t) + \sup_{x \in R^d} |f(T, x)|$$

(Hint: Apply Theorem 3.1.1 to the functions

$$\{\pm f_n(t, x) + \varepsilon'(t - T)\} \quad \text{with} \quad \varepsilon' > \varepsilon.)$$

3.3.3. Let $a: [0, \infty) \times R^d \to S_d$ and $b: [0, \infty) \times R^d$ be bounded and continuous. Let a have two spatial derivatives which are uniformly bounded. Let b have one spatial derivative which is uniformly bounded. Construct a sequence b_n such that $b_n \to b$ uniformly on $[0, \infty) \times R^d$, the first spatial derivatives of b_n have a uniform bound independent of n, and each b_n has two uniformly bounded spatial derivatives. Denote by $L_t^{(n)}$ the operator corresponding to $[a, b_n]$ and by $P_n(s, x, t, dy)$ the transition probability corresponding to $L_t^{(n)}$ constructed in Theorem 3.2.6. Show that for each $0 \le s \le t < \infty$ and $\varphi \in C(R^d)$

$$\lim_{n \to \infty} (T_{s,t}^{(n)} \varphi)(x) = \lim_{n \to \infty} \int \varphi(y) P_n(s, x, t, dy)$$

$$= (T_{s,t} \varphi)(x) \text{ exists uniformly in } x;$$

and show that $T_{s,t}$ is an inhomogeneous semigroup on $\hat{C}(R^d)$. Prove also that $(T_{s,t} \varphi)(x) = \int \varphi(y) P(s, x, t, dy)$ for some transition probabilities $P(s, x, t, \cdot)$, and the Markov process corresponding to $P(s, x, t, \cdot)$ can again be realized in the space $C([0, \infty): R^d)$ of continuous functions on R^d. (Hint: Define for fixed $\varphi \in C_0(R^d)$ and $0 < t < \infty$

$$u^n(s, x) = (T_{s,t}^{(n)} \varphi)(x) \quad \text{for} \quad 0 \le s \le t.$$

Note that from the construction of $T_{s,t}^{(n)}$ in Theorem 2.3.6, the first order spatial derivatives of $u^{(n)}$ are continuous and are bounded, independent of n. Observe that the maximum principle of 3.3.2 applies to each $u^{(n)}$. The rest of the proof is similar to the proof of Theorem 3.2.6.)

Chapter 4

The Stochastic Calculus of Diffusion Theory

4.1. Brownian Motion

Let $(E, \mathscr{F}, P)$ be a probability space and $(\beta(t), \mathscr{F}_t, P)$ a d-dimensional Brownian motion. The purpose of this section is to point out some of the properties that $(\beta(t), \mathscr{F}_t, P)$ possesses in common with a much larger class of stochastic processes which we will be calling Itô processes. Since we are going to be giving a completely rigorous derivation of these properties in the more general context of Itô processes, our treatment here will be somewhat informal and proofs will not be complete.

The first topic that we want to take up is that of equivalent characterizations of Brownian motion. Given $\theta \in R^d$, define $e_\theta(x) = e^{i\langle \theta, x \rangle}$, $x \in R^d$. Then an easy calculation shows that

$$(1.1) \quad E^P[e_\theta(\beta(t_2)) | \mathscr{F}_{t_1}] = e_\theta(\beta(t_1)) \exp\left[-\frac{t_2 - t_1}{2} |\theta|^2 \right],$$

$$0 \le t_1 < t_2 \text{ and } \theta \in R^d.$$

In fact, (1.1) is equivalent to (iv) in Definition 2.3.1. Another, more concise, statement of (1.1) is that

$$(1.2) \quad (X_{i\theta}(t), \mathscr{F}_t, P) \text{ is a martingale for all } \theta \in R^d,$$

where

$$X_{i\theta}(t) = \exp[i\langle \theta, \beta(t) \rangle + \tfrac{1}{2} |\theta|^2 t].$$

Thus (1.2) is equivalent to property (iv) in Definition 2.3.1. A second way of stating (iv) in 2.3.1 is the following. Starting with (1.1), we have

$$\frac{d}{dt} E^P[e_\theta(\beta(t)) | \mathscr{F}_{t_1}] = -\frac{|\theta|^2}{2} e_\theta(\beta(t_1)) \exp\left[-\frac{t - t_1}{2} |\theta|^2 \right]$$

for $t > t_1$. Thus for $t_2 > t_1$:

$$E^P[e_\theta(\beta(t_2)) - e_\theta(\beta(t_1)) \,|\, \mathscr{F}_{t_1}]$$

$$= -\frac{|\theta|^2}{2} \int_{t_1}^{t_2} e_\theta(\beta(t_1)) \exp\left[-\frac{t-t_1}{2}|\theta|^2\right] dt$$

$$= -\frac{|\theta|^2}{2} \int_{t_1}^{t_2} E^P[e_\theta(\beta(t)) \,|\, \mathscr{F}_{t_1}] \, dt$$

$$= E^P\left[\int_{t_1}^{t_2} \tfrac{1}{2} \Delta e_\theta(\beta(t)) \, dt \,\Big|\, \mathscr{F}_{t_1}\right],$$

where Δ stands for the Laplacian:

$$\sum_1^d \frac{\partial^2}{\partial x_i^2}.$$

That is

(1.3)
$$\left(e_\theta(\beta(t)) - \int_0^t \tfrac{1}{2} \Delta e_\theta(\beta(s)) \, ds, \; \mathscr{F}_t, \; P\right)$$

is a martingale for all $\theta \in R^d$. Since any $f \in C_0^\infty(R^d)$ can be represented by

$$f(x) = \int e^{i\langle\theta, x\rangle} \phi(\theta) \, d\theta$$

for some rapidly decreasing ϕ, (1.3) immediately implies that

(1.4) $(Y_f(t), \; \mathscr{F}_t, \; P)$ is a martingale for all $f \in C_b^2(R^d)$,

where

$$Y_f(t) = f(\beta(t)) - \int_0^t \tfrac{1}{2} \Delta f(\beta(s)) \, ds.$$

Conversely, if (1.4) holds, we can deduce (1.1) and therefore (iv) in Definition 2.3.1. In fact, all we need is (1.3), because from (1.3) we have:

$$\frac{d}{dt} E^P[e_\theta(t) \,|\, \mathscr{F}_{t_1}] = -\frac{|\theta|^2}{2} E^P[e_\theta(t) \,|\, \mathscr{F}_{t_1}], \qquad t > t_1,$$

and so

$$E^P[e_\theta(t_2) \,|\, \mathscr{F}_{t_1}] = e_\theta(t_1) e^{-[(t_2-t_1)/2]|\theta|^2}, \qquad t_2 > t_1.$$

We have now proved the next theorem.

4.1.1 Theorem. *Let $(E, \mathscr{F}, P)$ be a probability space, $\{\mathscr{F}_t : t \geq 0\}$ a non-decreasing family of sub σ-algebras of $\mathscr{F}$, and $\beta: [0, \infty) \times E \to R^d$ a right-continuous, P-almost surely continuous, progressively measurable function such that $P(\beta(0) = 0) = 1$. Then the following are equivalent:*

(i) *$(\beta(t), \mathscr{F}_t, P)$ is a Brownian motion,*
(ii) *$(Y_f(t), \mathscr{F}_t, P)$ is a martingale for all $f \in C_b^2(R^d)$,*
(iii) *$(X_{i\theta}(t), \mathscr{F}_t, P)$ is a martingale for all $\theta \in R^d$.*

Theorem 4.2.1 below shows that Theorem 4.1.1 is a special case of a general result.

Next, assume that $d = 1$. We want to discuss the possibility of defining

(1.5)
$$“\int_0^t \theta(s) \, d\beta(s).”$$

The problem is that $\beta(\cdot, q)$ is a.s. not a function of bounded variation. This fact can be easily seen from Exercise 1.5.6, since $(\beta(t), \mathscr{F}_t, P)$ is a continuous martingale. One can also see it from the following simple computation. Let $t_{k,n} = k/2^n$, $0 \leq k \leq 2^n$ and set

$$W_n = \sum_{k=0}^{2^n - 1} (\beta(t_{k+1,n}) - \beta(t_{k,n}))^2.$$

Then

$$E[W_n] = 1,$$

and a simple argument, using the Borel-Cantelli Lemma, shows that $W_n \to E[W_n]$ a.s. Hence the quadratic variation on $[0, 1]$ of $\beta(\cdot, q)$ over the dyadics is equal to one a.s., and therefore $\beta(\cdot, q)$ is a.s. not of bounded variation. This means that we cannot take an entirely naive approach to the definition of (1.5). To see how one can get around this sticky point, suppose that θ is a smooth function on $[0, \infty)$ and think of (1.5) as being defined through an integration by parts. Then $\int_0^t \theta(s) \, d\beta(s) \equiv \beta(t)\theta(t) - \beta(0)\theta(0) - \int_0^t \beta(s)\theta'(s) \, ds$. An elementary calculation shows that $E[\int_0^t \theta(s) \, d\beta(s)] = 0$ and $E[(\int_0^t \theta(s) \, d\beta(s))^2] = \int_0^t \theta^2(s) \, ds$. Hence $\theta \to \int_0^t \theta(s) \, d\beta(s)$ establishes an isometry from $L^2([0, t], \lambda)$ into $L^2(E, P)$: where λ is Lebesgue measure. One can therefore extend the definition of $\int_0^t \theta(s) \, d\beta(s)$ to cover all $\theta \in L^2([0, t], \lambda)$. This procedure was first carried out, and used with great success, by N. Wiener. The situation is somewhat more complicated when θ depends on q as well as s. To illustrate the sort of phenomenon encountered here, consider $\theta(s, q) = \beta(s, q)$. Suppose first that we attempt to define

(1.6)
$$\int_0^1 \beta(s) \, d\beta(s) = \lim_{n \to \infty} \sum_{k=0}^{2^n - 1} \beta(t_{k,n})(\beta(t_{k+1,n}) - \beta(t_{k,n})).$$

By elementary manipulation, we have that

$$\sum_{k=0}^{2^n-1} \beta(t_{k,n})(\beta(t_{k+1,n}) - \beta(t_{k,n}))$$

$$= \sum_{k=0}^{2^n-1} \frac{(\beta(t_{k+1,n}) + \beta(t_{k,n}))(\beta(t_{k+1,n}) - \beta(t_{k,n}))}{2} - \frac{1}{2}W_n$$

$$= \tfrac{1}{2}\beta^2(1) - \tfrac{1}{2}\beta^2(0) - \tfrac{1}{2}W_n \to \tfrac{1}{2}\beta^2(1) - \tfrac{1}{2}\beta^2(0) - \tfrac{1}{2} \qquad \text{a.s.}$$

as $n \to \infty$. Thus the lack of bounded variation manifests itself here in the appearance of the term $-\tfrac{1}{2}$. We will see later on in this chapter that $-\tfrac{1}{2}$ results from the fact that $d\beta(s) = \text{``}(ds)^{\frac{1}{2}}\text{''}$ and therefore $d\beta^2(s) = 2\beta(s)\,d\beta(s) + (d\beta(s))^2 = 2\beta(s)\,d\beta(s) + ds$ (cf. Itô's formula Theorem 4.4.1). It is important to notice that putting the increments of Brownian motion "in the future" as in (1.4) makes a difference (this is another manifestation of the absence of bounded variation). For instance, one can easily check that

$$(1.7) \qquad \lim_{n\to\infty} \sum_{k=0}^{2^n-1} \beta(t_{k+1,n})(\beta(t_{k+1,n}) - \beta(t_{k,n})) = \tfrac{1}{2}\beta^2(1) - \tfrac{1}{2}\beta^2(0) + \tfrac{1}{2}.$$

Although the difference here seems to be small, it turns out that (1.6) is a far preferable way to define $\int_0^1 \beta(s)\,d\beta(s)$ than the one given by (1.7). Suffice it to say that the advantage with (1.6) is that the following relations hold:

$$E\left[\int_0^1 \beta(s)\,d\beta(s)\right] = 0$$

and

$$E\left[\left(\int_0^1 \beta(s)\,d\beta(s)\right)^2\right] = E\left[\int_0^1 \beta^2(s)\,ds\right].$$

These equations allow one to carry out a completion procedure analogous to that in the case where θ depends only on s.

4.2. Equivalence of Certain Martingales

Let $(E, \mathscr{F}, P)$ be a probability space and $\{\mathscr{F}_t : t \geq 0\}$ a nondecreasing family of sub σ-fields of $\mathscr{F}$. Let $s \geq 0$ be arbitrary and $a : [s, \infty) \times E \to S_d$ and $b : [s, \infty) \times E \to R^d$ be bounded progressively measurable functions. For each $(t, q) \in [s, \infty) \times E$ the components of $a(t, q)$ will be denoted by $\{a^{ij}(t, q)\}$ and those of $b(t, q)$ by $\{b^j(t, q)\}$.

For any function $f(x)$ in the space $C^2(R^d)$ of twice continuously differentiable functions we define $L_t f$ for $t \geq s$ by

$$(L_t(q)f)(x) = \frac{1}{2} \sum_{i,j=1}^{d} a^{ij}(t, q) \frac{\partial^2 f}{\partial x_i \partial x_j}(x) + \sum_{j=1}^{d} b^j(t, q) \frac{\partial f}{\partial x_j}.$$

We note that if $\xi(\cdot, \cdot)$ is any progressively measurable function from $[s, \infty) \times E \to R^d$ then $(L_t(q)f)(\xi(t, q))$ defines another progressively measurable function of t and q for $t \geq s$ and $q \in E$. Usually we will suppress the variable q and write $\xi(t)$, $a(t)$, $b(t)$, L_t for our objects.

We shall suppose that we are given a function $\xi(t, q)$ mapping $[s, \infty) \times E$ into R^d which is progressively measurable, right continuous in t, and almost surely continuous in t. The main result of this section is the following theorem which proves that various types of relations between L_t and $\xi(t)$ are equivalent.

4.2.1 Theorem. For any $\xi(\cdot)$, $a(\cdot)$, $b(\cdot)$ satisfying the conditions described above, the following are equivalent:

(i) $f(\xi(t)) - \int_s^t (L_u f)(\xi(u)) \, du$ is a martingale relative to $(E, \mathscr{F}_t, P)$ for $t \geq s$, for all $f \in C_0^\infty(R^d)$.

(ii) $f(t, \xi(t)) - \int_s^t ((\partial/\partial u) + L_u)f(u, \xi(u)) \, du$ is a martingale relative to $(E, \mathscr{F}_t, P)$ for $t \geq s$, for all $f \in C_b^{1,2}([0, \infty) \times R^d)$.

(iii) $f(t, \xi(t)) \exp[-\int_s^t((\partial/\partial u) + L_u)f/f)(u, \xi(u) \, du]$ is a martingale relative to $(E, \mathscr{F}_t, P)$ for $t \geq s$ and for all $f \in C_b^{1,2}([0, \infty) \times R^d)$ which are uniformly positive.

(iv) If $\theta \in R^d$ and $g \in C_b^{1,2}([0, \infty) \times R^d)$ then

$$X_{\theta,g}(t) = \exp\left[\langle \theta, \xi(t) - \xi(s) - \int_s^t b(u)du \rangle + g(t, \xi(t)) \right.$$

$$- \frac{1}{2} \int_s^t \langle \theta + \nabla g, a(u)(\theta + \nabla g) \rangle(u, \xi(u)) \, du$$

$$\left. - \int_s^t \left(\frac{\partial}{\partial u} + L_u\right)g(u, \xi(u)) \, du\right]$$

is a martingale relative to $(E, \mathscr{F}_t, P)$ for times $t \geq s$.

(v) If $\theta \in R^d$ then

$$X_\theta(t) = \exp\left[\langle \theta, \xi(t) - \xi(s) - \int_s^t b(u) \, du \rangle - \frac{1}{2} \int_s^t \langle \theta, a(u)\theta \rangle \, du\right]$$

is a martingale relative to $(E, \mathscr{F}_t, P)$ for $t \geq s$.

(vi) If $\theta \in R^d$, then

$$X_{i\theta}(t) = \exp\left[i\langle \theta, \xi(t) - \xi(s) - \int_s^t b(u) \, du \rangle + \frac{1}{2} \int_s^t \langle \theta, a(u)\theta \rangle \, du\right]$$

is a martingale relative to $(E, \mathscr{F}_t, P)$ for $t \geq s$.

Moreover if any of the above equivalent relations holds, then for each $t > s$

$$(2.1) \qquad P\left(\sup_{s \leq u \leq t} \left| \xi(u) - \xi(s) - \int_s^u b(v) dv \right| \geq \lambda \right) \leq (2d) \exp\left(-\frac{\lambda^2}{2Ad(t-s)} \right)$$

where $A = \sup_{t \geq s, \, q \in E} \sup_{|\theta|=1} \langle \theta, a(t, q)\theta \rangle$. In particular, for any $r > 0$,

$$(2.2) \qquad E\left[\exp\left[r \sup_{s \leq u \leq t} |\xi(u) - \xi(s)| \right] \right] \leq C$$

and the constant C depends only on $t - s$, r, A and B where

$$B = \sup_{\substack{t \geq s \\ q \in E}} |b(t, q)|.$$

Proof. Assume (i). Let $f \in C_0^\infty([0, \infty) \times R^d)$. Then for $s \leq t_1 \leq t_2$ and $A \in \mathcal{F}_{t_1}$

$$E[f(t_2, \xi(t_2)) - f(t_1, \xi(t_1)); A]$$
$$= E[f(t_2, \xi(t_2)) - f(t_1, \xi(t_2)) + f(t_1, \xi(t_2)) - f(t_1, \xi(t_1)); A]$$
$$= E\left[\int_{t_1}^{t_2} \left(\frac{\partial f}{\partial u} \right)(u, \xi(t_2))\, du; A \right] + E\left[\int_{t_1}^{t_2} (L_v f)(t_1, \xi(v))\, dv; A \right]$$
$$= E\left[\int_{t_1}^{t_2} \left(\frac{\partial f}{\partial u} \right)(u, \xi(u))\, du; A \right]$$
$$\qquad + E\left[\int_{t_1}^{t_2} \left[\left(\frac{\partial f}{\partial u} \right)(u, \xi(t_2)) - \left(\frac{\partial f}{\partial u} \right)(u, \xi(u)) \right] du; A \right]$$
$$\qquad + E\left[\int_{t_1}^{t_2} (L_v f)(v, \xi(v))\, dv; A \right]$$
$$\qquad + E\left[\int_{t_1}^{t_2} (L_v f)(t_1, \xi(v)) - (L_v f)(v, \xi(v))\, dv; A \right]$$
$$= E\left[\int_{t_1}^{t_2} \left(\frac{\partial f}{\partial u} + L_u f \right)(u, \xi(u))\, du; A \right]$$
$$\qquad + E\left[\int_{t_1}^{t_2} du \int_u^{t_2} \left(L_v \frac{\partial f}{\partial u} \right)(u, \xi(v))\, dv; A \right]$$
$$\qquad - E\left[\int_{t_1}^{t_2} dv \int_{t_1}^v \left(\frac{\partial}{\partial u} L_v f \right)(u, \xi(v))\, du; A \right]$$
$$= E\left[\int_{t_1}^{t_2} \left(\frac{\partial f}{\partial u} + L_u f \right)(u, \xi(u))\, du; A \right].$$

The last two terms in the preceding step cancel each other because $L_v(\partial f/\partial u) = (\partial/\partial u) L_v f$ and the two repeated integrals are double integrals over the same region, namely: $t_1 \leq u \leq v \leq t_2$. It is obvious that the validity of the last equality for all $f \in C_0^\infty([0, \infty) \times R^d)$ implies the validity of the same equality for all $f \in C_b^{1,\,2}([0, \infty) \times R^d)$. Hence (i) implies (ii).

Next assume (ii) and let $f \in C_b^{1,\,2}([0, \infty) \times R^d)$ be uniformly positive. Take

$$\alpha(t) = f(t, \xi(t)) - \int_s^t \left(\frac{\partial f}{\partial u} + L_u f\right)(u, \xi(u))\, du$$

and

$$\eta(t) = \exp\left[-\int_s^t \left(\frac{(\partial f/\partial u) + L_u f}{f}\right)(u, \xi(u))\, du\right].$$

Then

$$\alpha(t)\eta(t) - \int_s^t \alpha(s)\, d\eta(s) = f(t, \xi(t)) \exp\left[-\int_s^t \left(\frac{(\partial f/\partial u) + L_u f}{f}\right)(u, \xi(u))\, du\right].$$

Applying Theorem 1.2.8, we see that (ii) implies (iii). Assume (iii) and assume for the moment that $\xi(s) = 0$. Obviously (iv) would follow if we were allowed to take $f(t, x) = \exp[\langle\theta, x\rangle + g(t, x)]$. We cannot do this because f is unbounded and not uniformly positive. To circumvent this problem, we choose for each $M \geq 1$ a uniformly positive function f_M in $C_b^{1,\,2}([0, \infty) \times R^d)$ such that $f_M(t, x) = \exp[\langle\theta, x\rangle + g(t, x)]$ for $x \leq M$ and define

$$\tau_M = (\inf [t \geq s: \sup_{s \leq u \leq t} |\xi(u)| \geq M]) \wedge (M \vee s).$$

Then by (iii) and Corollary 1.2.7, $(X_{\theta,\,g}(t \wedge \tau_M), \mathscr{F}_t, P)$ is a martingale for $t \geq s$. Clearly $X_{\theta,\,g}(t \wedge \tau_M) \to X_{\theta,\,g}(t)$ a.e. P as $M \nearrow \infty$ for each $t \geq s$. Thus we can establish (iv) if we can show that $\{X_{\theta,\,g}(t \wedge \tau_M): M \geq 1\}$ is uniformly integrable for each $t \geq s$. But

$$X_{\theta,\,g}^2(t \wedge \tau_M)$$

$$= X_{2\theta,\,2g}(t \wedge \tau_M) \exp\left[\int_s^{t \wedge \tau_M} \langle\theta + \nabla g, a(u)(\theta + \nabla g)\rangle(u, \xi(u))\, du\right]$$

$$\leq C X_{2\theta,\,2g}(t \wedge \tau_M)$$

and $E[X_{2\theta,\,2g}(t \wedge \tau_M)] = E[X_{2\theta,\,2g}(s)] = \exp[2g(s, 0)]$. This completes the proof when $\xi(s) = 0$. To remove this restriction, we define $\xi'(t) = \xi(t) - \xi(s)$. From (iii) and Exercise 1.5.7 it follows that for uniformly positive functions f in $C_b^{1,2}([0, \infty) \times R^d)$

$$f(t, \xi'(t)) \exp\left[-\int_s^t \left(\frac{(\partial f/\partial u) + L_u f}{f}\right)(u, \xi'(u))\, du\right]$$

is an $(E, \mathscr{F}_t, P)$ martingale for $t \geq s$. From what we proved earlier, it follows that for all $\theta \in R^n$ and $g \in C_b^{1,2}([0, \infty) \times R^d)$, $X'_{\theta, g}(t)$ is an $(E, \mathscr{F}_t, P)$ martingale for $t \geq s$, where $X'_{\theta, g}(t)$ is the expression defined in the same way as $X_{\theta, g}(t)$ with $\zeta'(t)$ replacing $\zeta(t)$. If one defines $g_{x'}(t, x) = g(t, x' + x)$ for $x, x' \in R^d$ we have for $s \leq t_1 \leq t_2$

$$E[X'_{\theta, g_{x'}}(t_2) | \mathscr{F}_{t_1}] = X'_{\theta, g_{x'}}(t_1) \qquad \text{a.e.}$$

Again using Exercise 1.5.7

$$E[X'_{\theta, g_{\zeta(s)}}(t_2) | \mathscr{F}_{t_1}] = X'_{\theta, g_{\zeta(s)}}(t_1) \qquad \text{a.e.}$$

Clearly $X'_{\theta, g_{\zeta(s)}}(t) = X_{\theta, g}(t)$ for all $t \geq s$, and we have therefore proved (iv). Obviously (v) is just the special case of (iv) when $g \equiv 0$. To see that (v) implies (vi) it is enough to show that (v) implies (2.1) and (2.2). Indeed, if the estimates (2.1) and (2.2) hold, then both sides of the equality

$$E[X_\theta(t_2); A] = E[X_\theta(t_1); A]$$

for $s \leq t_1 \leq t_2$ and $A \in \mathscr{F}_{t_1}$, determine entire functions of $\theta \in \mathbb{C}^d$. Therefore the validity of the equation for all $\theta \in R^d$ implies its validity for all $i\theta$ with $\theta \in R^d$. To see that (v) implies (2.1), let $|\theta| = 1$ and $\rho > 0$ be given. Then by Theorem 1.2.3 and (v) we obtain

$$P\left(\sup_{s \leq u \leq t} \langle \theta, \xi(t) - \xi(s) - \int_s^t b(u) \, du \rangle \geq \lambda \right)$$

$$\leq P\left(\sup_{s \leq u \leq t} X_{\rho\theta}(u) \geq \exp\left(-\frac{A(t-s)\rho^2}{2} + \rho\lambda \right) \right)$$

$$\leq \exp\left(-\rho\lambda + \frac{\rho^2 A(t-s)}{2} \right).$$

Taking $\rho = \lambda/A(t-s)$ we get

$$P\left(\sup_{s \leq u \leq t} \langle \theta, \xi(t) - \xi(s) - \int_s^t b(u) \, du \rangle \geq \lambda \right) \leq \exp\left(-\frac{\lambda^2}{2A(t-s)} \right)$$

Replacing θ by $-\theta$ we obtain a similar bound for the infimum. Combining the two, we have

$$P\left(\sup_{s \leq u \leq t} \left| \xi(u) - \xi(s) - \int_s^u b(v) dv \right| \geq \lambda \right) \leq 2d e^{-\lambda^2/2Ad(t-s)}$$

since

$$P\left(\sup_{s \le u \le t} \left| \xi(t) - \xi(s) - \int_s^t b(u)\, ds \right| \ge \lambda \right)$$

$$\le \sum_{j=1}^d P\left(\sup_{s \le u \le t} \left| \xi_j(t) - \xi_j(s) - \int_s^t b^j(u)\, du \right| \ge \lambda/\sqrt{d} \right).$$

This proves (2.1), and (2.2) is readily obtained from (2.1) by estimating the integral in terms of the tail probabilities.

It remains to show that (vi) implies (i). First we observe that, by elementary Fourier analysis, it suffices to prove (i) when $f(x)$ is of the form $f(x) = \exp[i\langle \theta, x \rangle]$ for some $\theta \in R^d$. Set $\alpha(t) = X_{i\theta}(t)$ and

$$\eta(t) = \exp\left[i\left\langle \theta, \xi(s) + \int_s^t b(u)\, du \right\rangle - \frac{1}{2} \int_s^t \langle \theta, a(u)\theta \rangle\, du \right].$$

By Theorem 1.2.8

$$\alpha(t)\eta(t) - \int_s^t \alpha(u)\, d\eta(u)$$

is an $(E, \mathscr{F}_t, P)$ martingale for $t \ge s$. It is easy to check that the last expression reduces to

$$f(\xi(t)) - \int_s^t (L_u f)(\xi(u))\, du$$

with $f(x) = \exp[i\langle \theta, x \rangle]$. This proves (i). $\square$

4.2.2 Corollary. *Let $\xi(t)$ satisfy any one of the equivalent conditions in Theorem 4.2.1 and assume that $E^P[e^{\lambda |\xi(s)|}] < \infty$, for all $\lambda > 0$. Suppose that $f \in C^{1,2}([s, T) \times R^d) \cap C([s, T] \times R^d)$, for some $T > s$, and that there are constants C_1 and C_2 such that f, its first time-derivative, and its first two spatial derivatives are bounded by $C_1 e^{C_2 |x|}$ for all $(t, x) \in [s, T) \times R^d$. Then*

$$f(t \wedge T, \xi(t \wedge T)) - \int_s^{t \wedge T} \left(\frac{\partial}{\partial u} + L_u \right) f(u, \xi(u))\, du$$

is an $(E, \mathscr{F}_t, P)$ martingale after time s. In particular, for all $\theta \in R^d$:

(2.4) $$\left\langle \theta, \xi(t) - \xi(s) - \int_s^t b(u)\, du \right\rangle$$

and

$$(2.5) \qquad \left\langle \theta, \, \xi(t) - \xi(s) - \int_s^t b(u) \, du \right\rangle^2 - \int_s^t \langle \theta, \, a(u)\theta \rangle \, du$$

are $(E, \mathscr{F}_t, P)$ martingales after time s.

Proof. Extend f to $R \times R^d$ so that $f(t, \, \cdot) = f(s, \, \cdot)$ if $t < s$ and $f(t, \, \cdot) = f(T, \, \cdot)$ if $t > T$. Use f to denote this extension. Choose functions ϕ and $\psi \in C_0^\infty((-1, 1))$ so that

$$\int_{R \times R^d} \phi((t^2 + |x|^2)^{1/2}) \, dt \, dx = 1$$

and $\psi \equiv 1$ on $[-\tfrac{1}{2}, \tfrac{1}{2}]$. Define ϕ_n and ψ_n by

$$\phi_n(t, x) = n^{d+1} \phi(n(t^2 + |x|^2)^{1/2})$$

and

$$\psi_n(t, x) = \psi\left(\frac{(t^2 + |x|^2)^{1/2}}{n}\right).$$

Finally, define f_n by

$$f_n(t, x) = \psi_n(t, x)(\phi_n * f)(t, x).$$

Then $f_n \in C_0^\infty(R \times R^d)$. Moreover $f_n(t, x) \to f(t, x)$ for all $(t, x) \in [s, T] \times R^d$, and $f_n, \, \partial f_n/\partial t, \, \partial f_n/\partial x_i,$ and $\partial^2 f_n/\partial x_i \partial x_j$ are all dominated by $C_3 e^{C_2|x|}$. Finally, for $s \le t_1 < t_2 \le T$ and $A \in \mathscr{F}_{t_1}$:

$$E^P[f_n(t_2, \xi(t_2)) - f_n(t_1, \xi(t_1)), \, A] = E^P\left[\int_{t_1}^{t_2} \left(\frac{\partial}{\partial u} + L_u\right) f_n(u, \xi(u)) \, du, \, A\right].$$

Using these facts in conjunction with estimate (2.2) and the hypothesis that $E^P[e^{\lambda|\xi(s)|}] < \infty$ for all $\lambda > 0$, we see that

$$f(t \wedge T, \xi(t \wedge T)) - \int_s^{t \wedge T} \left(\frac{\partial}{\partial u} + L_u\right) f(u, \xi(u)) \, du$$

is indeed a martingale.

The rest of the proof is easy and is left to the reader. $\quad\square$

4.3. Itô Processes and Stochastic Integration

Let $(E, \mathscr{F}, P)$ be a probability space, $\{\mathscr{F}_t : t \geq 0\}$ a non-decreasing family of sub σ-algebras of $\mathscr{F}$, and

$$\xi: [s, \infty) \times E \to R^d, \qquad a: [s, \infty) \times E \to S_d, \qquad b: [s, \infty) \times E \to R^d$$

progressively measurable functions with the properties assumed in Theorem 4.2.1. If ξ, a, and b are related by one of the equivalent conditions given in Theorem 4.2.1, we will say that ξ is an *Itô process (on $(E, \mathscr{F}_t, P)$) with covariance a and drift b after time s*. This sentence will be abbreviated by the notation $\xi \sim \mathscr{I}_d^s(a, b)$ when there is no need to emphasize $(E, \mathscr{F}_t, P)$.

The purpose of the present section is to develop the theory of stochastic integration with respect to an Itô process. Since it is clear that a Brownian motion is an Itô process, the theory which we develop must take into account the pitfalls which we pointed out in Section 4.1. The approach which we will adopt is basically due to K. Itô, and, by modern standards, it is very classical.

We begin our discussion with two simple observations. In the first place, the starting time s of an Itô process plays no essential role. Indeed, if $\xi \sim \mathscr{I}_d^s(a, b)$ on $(E, \mathscr{F}_t, P)$, then it is clear that $\hat{\xi} \sim \mathscr{I}_d^0(\hat{a}, \hat{b})$ on $(E, \hat{\mathscr{F}}_t, P)$, where:

$$\hat{\xi}(t, \cdot) = \xi(t + s, \cdot), \qquad \hat{a}(t, \cdot) = a(t + s, \cdot),$$

$$\hat{b}(t, \cdot) = b(t + s, \cdot),$$

and

$$\hat{\mathscr{F}}_t = \mathscr{F}_{t+s}.$$

As with the theory of martingales in Section 1.2, this observation enables us to restrict our attention to the case $s = 0$ although we will state our theorems for general s. Our second observation is that if $\xi \sim \mathscr{I}_d^s(a, b)$ and we define $\xi'(t) = \xi(t) - \int_s^t b(u)\, du$, then $\xi' \sim \mathscr{I}_d^s(a, 0)$. Thus, if we can assign a meaning to $d\xi(t)$ integrals when $\xi \sim \mathscr{I}_d^s(a, 0)$, then we can assign one to $d\xi(t)$ for $\xi \sim \mathscr{I}_d^s(a, b)$, namely: "$d\xi(t) = d\xi'(t) + b(t)\, dt$." For this reason, until further notice, we will take $b \equiv 0$.

Let $(E, \mathscr{F}, P)$ and $\{\mathscr{F}_t : t \geq 0\}$ be given. A function $\theta: [0, \infty) \times E \to R^d$ is said to be *simple* if θ is bounded, progressively measurable, and there is an integer $n \geq 1$ such that $\theta(t) = \theta([nt]/n)$ for all $t \geq 0$. If $\xi \sim \mathscr{I}_d^0(a, 0)$ on $(E, \mathscr{F}_t, P)$ and θ is a simple function, we define $\int_0^t \langle \theta(u), d\xi(u) \rangle$ by

$$(3.1) \qquad \int_0^t \langle \theta(u), d\xi(u) \rangle \equiv \sum_{k=0}^{\infty} \left\langle \theta\left(\frac{k}{n}\right), \left(\xi\left(\frac{k+1}{n} \wedge t\right) - \xi\left(\frac{k}{n} \wedge t\right) \right) \right\rangle$$

where $n \geq 1$ is any integer for which $\theta(t) = \theta([nt]/n)$, $t \geq 0$. Notice that in keeping with the warning in Section 4.1 the increment appears in the future. (It is easy to

check that the definition does not depend on n so long as $\theta(t) = \theta([nt]/n), t \geq 0$.)
If $0 \leq t_1 \leq t_2$, we define

$$\int_{t_1}^{t_2} \langle \theta(u), d\xi(u) \rangle = \int_{0}^{t_2} \langle \theta(u), d\xi(u) \rangle - \int_{0}^{t_1} \langle \theta(u), d\xi(u) \rangle.$$

4.3.1 Lemma. *If $\xi \sim \mathscr{I}_d^0(a, 0)$ and θ is a simple function, then $\int_0^t \langle \theta(u), d\xi(u) \rangle$ is a right-continuous, almost surely continuous progressively measurable function. Furthermore*

$$\int_0^t \langle \theta(u), d\xi(u) \rangle,$$

$$\left(\int_0^t \langle \theta(u), d\xi(u) \rangle \right)^2 - \int_0^t \langle \theta(u), a(u)\theta(u) \rangle \, du,$$

and

$$X_{\theta(\cdot)}(t) = \exp\left[\int_0^t \langle \theta(u), d\xi(u) \rangle - \frac{1}{2} \int_0^t \langle \theta(u), a(u)\theta(u) \rangle \, du \right]$$

are all $(E, \mathscr{F}_t, P)$ martingales. In particular, if

$$\tilde{\xi}(t) = \int_0^t \langle \theta(u), d\xi(u) \rangle$$

and

$$\tilde{a}(u) = \langle \theta(u), a(u)\theta(u) \rangle,$$

then $\tilde{\xi} \sim \mathscr{I}_1^0(\tilde{a}, 0)$. Finally, if θ_1 and θ_2 are simple functions and λ_1 and λ_2 are real numbers, then $\lambda_1 \theta_1 + \lambda_2 \theta_2$ is a simple function and:

$$\int_0^t \langle \lambda_1 \theta_1(u) + \lambda_2 \theta_2(u), d\xi(u) \rangle = \lambda_1 \int_0^t \langle \theta_1(u), d\xi(u) \rangle + \lambda_2 \int_0^t \langle \theta_2(u), d\xi(u) \rangle.$$

Proof. That

$$\tilde{\xi}(t) = \int_0^t \langle \theta(u), d\xi(u) \rangle$$

is right-continuous, almost surely continuous, and progressively measurable follows immediately from the corresponding facts about ξ. We will now show that

$\xi \sim \mathscr{I}_1^0(\tilde{a}, 0)$. Once this has been done, we can use Corollary 4.2.2 to complete the proof of everything except the final assertion. Since it is clear that if $\lambda \in R$, then

$$\lambda \tilde{\xi}(t) = \int_0^t \langle \lambda \theta(u), d\xi(u) \rangle,$$

we will know, by (v) of Theorem 4.2.1, that $\tilde{\xi} \sim \mathscr{I}_1^0(\tilde{a}, 0)$ once we show that

$$X_{\theta(\cdot)}(t) = \exp\left[\tilde{\xi}(t) - \frac{1}{2} \int_0^t \tilde{a}(u)\, du\right]$$

is an $(E, \mathscr{F}_t, P)$ martingale. To this end, choose n so that $\theta(t) = \theta([nt]/n)$, $t \geq 0$, and let $k/n \leq t_1 \leq t_2 \leq (k + 1)/n$ for some $k \geq 1$. Then

$$\tilde{\xi}(t_2) - \tilde{\xi}(t_1) = \left\langle \theta\left(\frac{k}{n}\right), \xi(t_2) - \xi(t_1) \right\rangle.$$

Since $\theta(k/n)$ is a bounded $\mathscr{F}_{t_1}$-measurable function, we can apply estimate (2.2), Exercise 1.5.7, and (v) of Theorem 4.2.1 to conclude that:

$$E\left[\exp\left[\langle \theta(k/n), \xi(t_2) - \xi(t_1) \rangle \right.\right.$$
$$\left.\left. - \frac{1}{2} \int_{t_1}^{t_2} \langle \theta(k/n), a(u)\theta(k/n) \rangle\, du\right] \bigg| \mathscr{F}_{t_1}\right] = 1 \qquad \text{a.s.},$$

and therefore that

$$E[X_{\theta(\cdot)}(t_2) | \mathscr{F}_{t_1}] = X_{\theta(\cdot)}(t_1) \qquad \text{a.s.}$$

It is now an easy matter to remove the restriction that $k/n \leq t_1 \leq t_2 \leq (k + 1)/n$ and conclude that $X_{\theta(\cdot)}(t)$ is an $(E, \mathscr{F}_t, P)$ martingale.

Finally, suppose θ_1 and θ_2 are simple functions and that $\theta_i(t) = \theta_i([n_i t]/n_i)$, $i = 1, 2$ and $t \geq 0$. Then

$$\lambda_1 \theta_1(t) + \lambda_2 \theta_2(t) = \lambda_1 \theta_1([nt]/n) + \lambda_2 \theta_2([nt]/n), \qquad t \geq 0,$$

when $n = n_1 n_2$. Thus $\lambda_1 \theta_1 + \lambda_2 \theta_2$ is simple and it is easy to check that

$$\int_0^t \langle \lambda_1 \theta_1(u) + \lambda_2 \theta_2(u), d\xi(u) \rangle = \lambda_1 \int_0^t \langle \theta_1(u), d\xi(u) \rangle + \lambda_2 \int_0^t \langle \theta_2(u), d\xi(u) \rangle.$$

One just has to write out $\int_0^t \langle \theta_i(u), d\xi(u) \rangle, i = 1, 2$, in terms of the partition determined by n. $\square$

The next lemma demonstrates that a large class of functions on $[0, \infty) \times E$ into R^d can be approximated by simple ones. To be specific, let $a: [s, \infty) \times E \to S_d$ be a bounded, progressively measurable function after time s and define $(H) \,_d^s(a)$ to be the set of progressively measurable $\theta: [s, \infty) \times E \to R^d$ such that

$$E\left[\int_s^T \langle \theta(u), a(u)\theta(u) \rangle \, du\right] < \infty$$

for all $T > s$. We now prove:

4.3.2 Lemma. *Given $\theta \in (H) \,_d^0(a)$, there exists a sequence of simple functions $\{\theta_n: n \geq 1\}$ such that for every $T > 0$.*

$$(3.2) \qquad \lim_{n \to \infty} E\left[\int_0^T \langle \theta(u) - \theta_n(u), a(u)(\theta(u) - \theta_n(u)) \rangle \, du\right] = 0.$$

Moreover, if

$$C \equiv \sup_{t \geq 0, \, q \in E} |\theta(t, q)| < \infty,$$

then the θ_n's can be chosen so that

$$\sup_{t \geq 0, \, q \in E} |\theta_n(t, q)| \leq C$$

for all $n \geq 1$.

Proof. First assume that

$$C \equiv \sup_{t \geq 0, \, q \in E} |\theta(t, q)| < \infty$$

and that $\theta(\cdot, q)$ is continuous for all $q \in E$. Then we can take $\theta_n(t, q) = \theta([nt]/n, q)$, and clearly $\{\theta_n: n \geq 1\}$ has the desired properties, including the fact that $\sup_{t \geq 0, \, q \in E} |\theta_n(t, q)| \leq C$ for all n. Next, we drop the continuity assumption on $\theta(\cdot, q)$. Choose $\phi \in C_0^\infty(R)$ so that $\phi \geq 0$, $\phi \equiv 0$ off $[0, 1]$, and $\int \phi(t) \, dt = 1$. Extend $\theta(\cdot, q)$ to $(-\infty, \infty)$ by setting $\theta(t, q) = 0$ for $t < 0$, and define

$$\theta_n(t, q) = n \int_{-\infty}^\infty \phi(n(t - u))\theta(u, q) \, du, \qquad n \geq 1 \quad \text{and} \quad t \geq 0.$$

Then, θ_n is progressively measurable, $\theta_n(\cdot, q)$ is continuous, and $\sup_{t \geq 0, \, q \in E} |\theta_n(t, q)| \leq C$. Moreover, by elementary properties of approximate identities,

$$\lim_{n \to \infty} \int_0^T |\theta_n(t, q) - \theta(t, q)|^2 \, dt = 0$$

for all $T > 0$. Combining this with the preceding, we see that the lemma has been completely proved in the case that $\sup_{t \geq 0, q \in E} |\theta(t, q)| < \infty$. In order to remove this restriction, it is enough to show that if $\theta \in \textcircled{H}_d^0(a)$, then there are bounded $\theta_n \in \textcircled{H}_d^0(a)$, $n \geq 1$, such that (3.2) holds. But this is easy. Simply take

$$\theta_n(t, q) = \chi_{[-n, n]}(|\theta(t, q)|)\theta(t, q).$$

Then $\theta_n \in \textcircled{H}_d^0(a)$, θ_n is bounded, $\theta_n(t, \cdot) \to \theta(t, \cdot)$ a.s. for each $t \geq 0$, and

$$\langle \theta_n(t, q), a(t, q)\theta_n(t, q) \rangle \leq \langle \theta(t, q), a(t, q)\theta(t, q) \rangle$$

for all (t, q). Thus, by the Lebesgue Dominated Convergence Theorem, (3.2) holds. $\square$

Aside from one technicality, we now have the basic machinery needed to complete the definition of the stochastic integral. Let $\theta \in \textcircled{H}_d^0(a)$ be given. By the preceding lemma, we can choose simple θ_n's for which (3.2) obtains. Define

$$\tilde{\xi}_n(t) = \int_0^t \langle \theta_n(u), d\xi(u) \rangle, \qquad t \geq 0 \quad \text{and} \quad n \geq 1.$$

By Lemma 4.3.1, for all $m, n \geq 1$:

$$\tilde{\xi}_n(t) - \tilde{\xi}_m(t)$$

and

$$(\tilde{\xi}_n(t) - \tilde{\xi}_m(t))^2 - \int_0^t \langle \theta_n(u) - \theta_m(u), a(u)(\theta_n(u) - \theta_m(u)) \rangle \, du$$

are martingales. Hence, by Doob's inequality (Theorem 1.2.3):

$$E\left[\sup_{0 \leq t \leq T} |\tilde{\xi}_n(t) - \tilde{\xi}_m(t)|^2 \right] \leq 4E\left[\int_0^T \langle \theta_n(u) - \theta_m(u), a(u)(\theta_n(u) - \theta_m(u)) \rangle \, du \right];$$

and so, because of (3.2),

$$\lim_{\substack{m \to \infty \\ n \to \infty}} E\left[\sup_{0 \leq t \leq T} |\tilde{\xi}_n(t) - \tilde{\xi}_m(t)|^2 \right] = 0$$

for all $T > 0$. In particular, there is a sub-sequence $\{\tilde{\xi}_{n'}(\cdot)\}$ such that $\tilde{\xi}_{n'}(\cdot, q)$ converges uniformly on finite intervals for q outside a P-null set N. It is at this point that we encounter the aforementioned technicality. Namely, it would seem reasonable to define

$$\int_0^t \langle \theta(u), d\xi(u) \rangle = \lim_{n' \to \infty} \tilde{\xi}_{n'}(t)$$

off the set N, and let it be defined measurably, but otherwise arbitrarily, on N. We could even guarantee that the resulting function be right-continuous, and it would certainly be almost surely continuous. However, it would not, in general, be progressively measurable relative to the original family $\{\mathscr{F}_t: t \geq 0\}$ because the set N forces one to "anticipate the future." The usual way in which this difficulty is avoided is to complete the σ-algebras $\mathscr{F}_t$ as suggested in Exercise 1.5.8. However, this solution is not entirely suitable for us, since the completion is a function of the underlying measure P and in our applications the measure P changes, whereas the σ-algebras do not. Thus we will work a little harder and obtain a definition of $\int_0^t \langle \theta(u), d\xi(u) \rangle$ which, besides being right-continuous and almost surely continuous, is progressively measurable relative to $\{\mathscr{F}_t: t \geq 0\}$. The essence of the procedure that we have in mind is contained in the following lemma.

4.3.3 Lemma. *Let $\{\eta_n: n \geq 1\}$ be a sequence of right-continuous, almost surely continuous, progressively measurable function on $(E, \mathscr{F}_t, P)$ into R^d. If*

$$\lim_{m \to \infty} \sup_{n \geq m} P\left(\sup_{0 \leq t \leq T} |\eta_n(t) - \eta_m(t)| \geq \varepsilon \right) = 0$$

for all $\varepsilon > 0$ and $T > 0$, then there is a right-continuous, almost surely continuous, progressively measurable η on $(E, \mathscr{F}_t, P)$ into R^d such that

$$\lim_{n \to \infty} P\left(\sup_{0 \leq t \leq T} |\eta_n(t) - \eta(t)| \geq \varepsilon \right) = 0$$

for all $\varepsilon > 0$ and $T > 0$.

Proof. Without loss of generality, we can and will assume that for each $t > 0$,

$$P\left(\lim_{m \to \infty} \sup_{n \geq m} \sup_{0 \leq u \leq t} |\eta_n(u, q) - \eta_m(u, q)| = 0 \right) = 1.$$

Since each η_n is progressively measurable, it is easy to see that A defined by $A = \{(t, q): \lim_{m \to \infty} \sup_{n \geq m} \sup_{0 \leq u \leq t} |\eta_n(u, q) - \eta_m(u, q)| = 0\}$ is progressively measurable (cf. Exercise 1.5.11). Moreover, if $A(t) = \{q: (t, q) \in A\}$, then $P(A(t)) = 1$; and if $J_q = \{t: (t, q) \in A\}$, then J_q is an interval and

$$\eta(\cdot, q) = \lim_{n \to \infty} \eta_n(\cdot, q)$$

uniformly on compact subsets of J_q. It is clear that η is progressively measurable on A. Hence, by Exercise 4.6.8 below, η admits a right-continuous, almost surely continuous, progressively measurable extension to all of $[0, \infty) \times E$. Finally, one easily checks that

$$\lim_{n \to \infty} P\left(\sup_{0 \leq t \leq T} |\eta_n(t) - \eta(t)| \geq \varepsilon \right) = 0$$

for all $\varepsilon > 0$ and $T > 0$. $\square$

In view of the preceding discussion and Lemma 4.3.3, we have now proved the next result.

4.3.4 Lemma. *If* $\theta \in \widehat{H}{}^0_d(a)$, *then there exists a sequence* $\{\theta_n : n \geq 1\}$ *of simple functions and a right-continuous, almost surely continuous, progressively measurable function* $\tilde{\xi}$ *such that*

$$\lim_{n \to \infty} E\left[\int_0^T \langle \theta(u) - \theta_n(u), a(u)(\theta(u) - \theta_n(u)) \rangle \, du\right] = 0$$

and

$$\lim_{n \to \infty} E\left[\sup_{0 \leq t \leq T} \left|\tilde{\xi}(t) - \int_0^t \langle \theta_n(u), d\xi(u) \rangle\right|^2\right] = 0$$

for all $T > 0$.

Obviously, it is our intention to take the function $\tilde{\xi}$ described in the preceding lemma to be our definition of $\int_0^t \langle \theta(u), d\xi(u) \rangle$. However, before we can do that we must check that this definition does not depend in an important way on the choice of the simple approximants

4.3.5 Lemma. *Let* $\theta \in \widehat{H}{}^0_d(a)$. *There exists a right-continuous, almost surely continuous, progressively measurable function* $\tilde{\xi}$ *such that*

$$E\left[\sup_{0 \leq t \leq T} \left|\tilde{\xi}(t) - \int_0^t \langle \theta'(u), d\xi(u) \rangle\right|^2\right] \leq 4E\left[\int_0^T \langle \theta(u) - \theta'(u), a(u)(\theta(u) - \theta'(u)) \rangle \, du\right]$$

for all simple functions θ' *and all* $T > 0$. *In particular, there is, up to a set of P-measure* 0, *exactly one such function and it is the one given in Lemma 4.3.4.*

Proof. Let $\{\theta_n : n \geq 1\}$ and $\tilde{\xi}$ be given as in Lemma 4.3.3. If θ' is a simple function, then

$$\left\{E\left[\sup_{0 \leq t \leq T} \left|\tilde{\xi}(t) - \int_0^t \langle \theta'(u), d\xi(u) \rangle\right|^2\right]\right\}^{\frac{1}{2}}$$

$$\leq \left\{E\left[\sup_{0 \leq t \leq T} \left|\tilde{\xi}(t) - \int_0^t \langle \theta_n(u), d\xi(u) \rangle\right|^2\right]\right\}^{\frac{1}{2}}$$

$$+ \left\{E\left[\sup_{0 \leq t \leq T} \left|\int_0^t \langle \theta'(u) - \theta_n(u), d\xi(u) \rangle\right|^2\right]\right\}^{\frac{1}{2}}$$

$$\leq \left\{E\left[\sup_{0 \leq t \leq T} \left|\tilde{\xi}(t) - \int_0^t \langle \theta_n(u), d\xi(u) \rangle\right|^2\right]\right\}^{\frac{1}{2}}$$

$$+ 2\left\{E\left[\int_0^T \langle \theta'(u) - \theta_n(u), a(u)(\theta'(u) - \theta_n(u)) \rangle \, du\right]\right\}^{\frac{1}{2}}$$

$$\to 2\left\{E\left[\int_0^T \langle \theta'(u) - \theta(u), a(u)(\theta'(u) - \theta(u)) \rangle \, du\right]\right\}^{\frac{1}{2}}. \quad \square$$

As we noted at the beginning of our discussion, anything that we can do when $s = 0$ is immediately extendable to general s. Thus, if $s \geq 0$, we say that $\theta : [s, \infty) \times E \to R^d$ is *simple after time s* if $\theta(\cdot + s)$ is simple relative to $\{\mathscr{F}_{t+s} : t \geq 0\}$ and we define $\int_s^t \langle \theta(u), d\xi(u) \rangle$ accordingly for $\xi \in \mathscr{I}_d^s(a, 0)$. It then follows that if $\xi \sim \mathscr{I}_d^s(a, 0)$ and $\theta \in \textcircled{H}_d^s(a)$, then there is a unique, up to a P-null set, right-continuous, almost surely continuous, progressively measurable function $\tilde{\xi}$ after time s such that

$$
E^P \left[\sup_{s \leq t \leq T} \left| \tilde{\xi}(t) - \int_s^t \langle \theta'(u), d\xi(u) \rangle \right|^2 \right]
$$

$$
\leq 4 E^P \left[\int_s^T \langle \theta(u) - \theta'(u), a(u)(\theta(u) - \theta'(u)) \rangle \, du \right]
$$

for all simple functions θ' after time s and all $T > 0$. This function $\tilde{\xi}$ is what we call the *stochastic integral of θ with respect to* ξ, and we denote it by $\int_s^t \langle \theta(u), d\xi(u) \rangle$. The next theorem is now immediate.

4.3.6 Theorem. *For* $\theta \in \textcircled{H}_d^s(a)$ *and* $\xi \sim \mathscr{I}_d^s(a, 0)$,

$$
\int_s^t \langle \theta(u), d\xi(u) \rangle
$$

and

$$
\left(\int_s^t \langle \theta(u), d\xi(u) \rangle \right)^2 - \int_s^t \langle \theta(u), a(u)\theta(u) \rangle \, du
$$

are $(E, \mathscr{F}_t, P)$ *martingales after time s. Moreover, if* θ_1 *and* θ_2 *are in* $\textcircled{H}_d^s(a)$, *then for* $\lambda_1, \lambda_2 \in R$

$$
\int_s^t \langle \lambda_1 \theta_1(u) + \lambda_2 \theta_2(u), d\xi(u) \rangle
$$

$$
= \lambda_1 \int_s^t \langle \theta_1(u), d\xi(u) \rangle + \lambda_2 \int_s^t \langle \theta_2(u), d\xi(u) \rangle \qquad \text{a.s.,}
$$

and so

$$
E \left[\sup_{s \leq t \leq T} \left| \int_s^t \langle \theta_1(u), d\xi(u) \rangle - \int_s^t \langle \theta_2(u), d\xi(u) \rangle \right|^2 \right]
$$

$$
\leq 4 E \left[\int_s^T \langle \theta_1(u) - \theta_2(u), a(u)(\theta_1(u) - \theta_2(u)) \rangle \, du \right].
$$

In the case that $\theta \in \widehat{(H)}_d^s(a)$ satisfies

(3.3) $$\sup_{t \geq s, q \in E} \langle \theta(t, q), a(t, q)\theta(t, q) \rangle < \infty,$$

we can say more about $\int_s^t \langle \theta(u), d\xi(u) \rangle$.

4.3.7 Theorem. *Suppose* $\xi \sim \mathscr{I}_d^s(a, 0)$ *and* $\theta \in \widehat{(H)}_d^s(a)$ *satisfies* (3.3). *Let* $\tilde{\xi}(t) = \int_s^t \langle \theta(\dot{u}), d\xi(u) \rangle$ *and* $\tilde{a}(t) = \langle \theta(t), a(t)\theta(t) \rangle$. *Then* $\tilde{\xi} \sim \mathscr{I}_1^s(\tilde{a}, 0)$ *on* $(E, \mathscr{F}_t, P)$.

Proof. It is enough to carry out the proof when $s = 0$. We will do so first under the assumption that θ is uniformly bounded. In this case, we can choose simple θ_n's having the same bound as θ such that (3.2) holds. Then

$$\tilde{\xi}_n(t) = \int_0^t \langle \theta_n(u), d\xi(u) \rangle$$

tends in probability to $\tilde{\xi}(t)$ as $n \to \infty$; and, by Lemma 4.3.1,

$$X_\lambda^{(n)}(t) = \exp\left[\lambda \tilde{\xi}_n(t) - \frac{\lambda^2}{2} \int_0^t \tilde{a}_n(u)\, du \right]$$

is an $(E, \mathscr{F}_t, P)$ martingale for all $\lambda \in R$, where

$$\tilde{a}_n(t) = \langle \theta_n(t), a(t)\theta_n(t) \rangle.$$

Clearly $X_\lambda^{(n)}(t) \to X_\lambda(t)$ in probability, where

$$X_\lambda(t) = \exp\left[\lambda \tilde{\xi}(t) - \frac{\lambda^2}{2} \int_0^t \tilde{a}(u)\, du \right].$$

Thus, by (v) of Theorem 4.2.1, the proof will be complete in this case if we can show that $X^{(n)}(t): n \geq 1$ is uniformly integrable. But

$$(X_\lambda^{(n)}(t))^2 = X_{2\lambda}^{(n)}(t)\exp\left[\lambda^2 \int_0^t \tilde{a}_n(u)\, du \right]$$

$$\leq X_{2\lambda}^{(n)}(t)\exp[\lambda^2 t C_n],$$

where $C_n = \sup_{t \geq 0, q \in E} \langle \theta_n(t, q), a(t, q)\theta_n(t, q) \rangle$. Since $\sup_n C_n < \infty$ and $E[X_{2\lambda}^{(n)}(t)] = 1$, it follows that

$$\sup_n E[(X^{(n)}(t))^2] < \infty.$$

To remove the restriction that θ be uniformly bounded, choose bounded θ_n's from $(H)_d^0(a)$ so that (3.2) holds and

$$\sup_{t \geq 0, q \in E} \langle \theta_n(t, q), a(t, q)\theta_n(t, q) \rangle \leq \sup_{t \geq 0, q \in E} \langle \theta(t, q), a(t, q)\theta(t, q) \rangle$$

for all n (cf. the proof of Lemma 4.3.2). One can then repeat the preceding argument to complete the proof. □

There are a few matters that we still have to discuss before closing this section. In the first place, suppose that $\xi \sim \mathscr{I}_d^s(a, b)$, where $b \not\equiv 0$. Let $(H)_d^s(a, b)$ be the class of progressively measurable $\theta: [s, \infty) \times E \to R^d$ such that $\theta \in (H)_d^s(a)$ and

$$E\left[\int_s^T |\langle \theta(u), b(u) \rangle| \, du \right] < \infty$$

for all $T > s$. Given $\theta \in (H)_d^s(a, b)$, we can choose a version (unique up to a P-null set) of $\int_s^t \langle \theta(u), b(u) \rangle \, du$ which is right-continuous, almost surely continuous, progressively measurable and is equal to the ordinary Lebesgue integral $\int_s^t \langle \theta(u, q), b(u, q) \rangle \, du$ if $\int_s^t |\langle \theta(u, q), b(u, q) \rangle| \, du < \infty$ for all $t \geq s$. We can therefore define

$$\int_s^t \langle \theta(u), d\xi(u) \rangle = \int_s^t \langle \theta(u), d\xi'(u) \rangle + \int_0^t \langle \theta(u), b(u) \rangle \, du,$$

where $\xi'(t) = \xi(t) - \int_0^t b(u) \, du$.

The next matter concerns the problem of defining

$$\int_{t_1}^{t_2} \langle \theta(u), d\xi(u) \rangle \quad \text{for} \quad s \leq t_1 < t_2, \qquad \xi \sim \mathscr{I}_d^s(a, b) \quad \text{and} \quad \theta \in (H)_d^s(a, b).$$

There are three possibilities, all of which give the same solution. First, one might note that $\xi \sim \mathscr{I}_d^s(a, b)$ and $\theta \in (H)_d^s(a, b)$ means, in particular, that $\xi \sim \mathscr{I}_d^{t_1}(a, b)$ and $\theta \in (H)_d^{t_1}(a, b)$. This would give one definition of $\int_{t_1}^{t_2} \langle \theta(u), d\xi(u) \rangle$. A second possibility is to define

$$\int_{t_1}^{t_2} \langle \theta(u), d\xi(u) \rangle = \int_s^{t_2} \langle \theta(u), d\xi(u) \rangle - \int_s^{t_1} \langle \theta(u), d\xi(u) \rangle.$$

To see that these two coincide is quite elementary and is left to the reader. Finally, one might adopt

$$\int_s^T \langle \chi_{[t_1, t_2]}(u)\theta(u), d\xi(u) \rangle,$$

for some $T > t_2$, as the definition of $\int_{t_1}^{t_2} \langle \theta(u), d\xi(u) \rangle$. In Exercise 4.6.9 below, the reader is asked to show that this definition is the same as the preceding one.

The final topic to be taken up is the stochastic integration of matrix valued integrands. Let $\xi \sim \mathscr{I}_d^s(a, b)$ and suppose that $\sigma: [0, \infty) \times E \to R^n \otimes R^d$ is a progressively measurable function such that

$$(3.4) \qquad\qquad E\left[\int_s^T \text{Trace}(\sigma a \sigma^*(t))\, dt\right] < \infty$$

and

$$(3.5) \qquad\qquad E\left[\int_s^T |\sigma b(t)|\, dt\right] < \infty$$

for all $T > s$. Then $\sigma^* \theta \in \textcircled{H}_d^s(a, b)$ for all $\theta \in R^n$, and we define $\int_s^t \sigma(u)\, d\xi(u)$ so that

$$(3.6) \qquad\qquad \left\langle \theta, \int_s^t \sigma(u)\, d\xi(u) \right\rangle = \int_s^t \langle \sigma^*(u)\theta,\, d\xi(u)\rangle$$

for all $\theta \in R^n$. It is easy to read off the properties of $\int_s^t \sigma(u)\, d\xi(u)$ from those of $\int_s^t \langle \sigma^*(u)\theta,\, d\xi(u)\rangle$. In particular, Theorem 4.3.7 yields the following result.

4.3.8 Theorem. *Let* $\xi \sim \mathscr{I}_d^s[a, b]$ *and* $\sigma: [s, \infty) \times E \to R^n \otimes R^d$ *be progressively measurable. Assume that*

$$\sup_{\substack{t \geq s \\ q \in E}} \text{Trace}(\sigma a \sigma^*)(t, q) < \infty$$

and

$$\sup_{\substack{t \geq s \\ q \in E}} |(\sigma b)(t, q)| < \infty.$$

If $\tilde{\xi}(t) = \int_s^t \sigma(u)\, d\xi(u)$, *then* $\tilde{\xi}$ *is an Itô process after time* s, *and in fact*

$$\tilde{\xi} \sim \mathscr{I}_n^s[\sigma a \sigma^*,\, \sigma b].$$

Proof. Let $\theta \in R^n$ be arbitrary. From (3.6)

$$\langle \theta, \tilde{\xi}(t)\rangle = \int_s^t \langle \sigma^*(u)\theta,\, d\xi(u)\rangle$$

$$= \int_s^t \langle \sigma^*(u)\theta,\, d\xi'(u)\rangle + \int_s^t \langle \sigma^*(u)\theta,\, b(u)\rangle\, du$$

$$= \int_s^t \langle \sigma^*(u)\theta,\, d\xi'(u)\rangle + \int_s^t \langle \theta,\, (\sigma b)(u)\rangle\, du$$

where $\xi'(t) = \xi(t) - \int_s^t b(u)\, du$ and $\xi' \sim \mathscr{I}_d^s[a, 0]$. It follows from Theorem 4.3.7 that

(3.7) $\langle \theta, \tilde{\xi}(t) \rangle \sim \mathscr{I}_1^s[\theta \sigma a \sigma^* \theta, \theta \sigma b]$.

From Theorem 4.2.1 (using v), the validity of (3.7) for all $\theta \in R^n$ is the same as $\tilde{\xi} \sim \mathscr{I}_n^s[\sigma a \sigma^*, \sigma b]$. $\square$

We close this section with the "chain rule" for stochastic integrals.

4.3.9 Theorem. *Let* $\xi \sim \mathscr{I}_d^s(a, b)$ *and* $\sigma : [s, \infty) \times E \to R^n \otimes R^d$ *be a progressively measurable function with*

$$\sup_{\substack{t \geq s \\ q \in E}} \operatorname{Trace} (\sigma a \sigma^*)(t, q) < \infty$$

and

$$\sup_{\substack{t \geq s \\ q \in E}} |(\sigma b)(t, q)| < \infty.$$

Also, let $\rho : [s, \infty) \times E \to R^m \otimes R^n$ *be a progressively measurable function satisfying*

$$\sup_{\substack{t \geq s \\ q \in E}} \operatorname{Trace} (\rho \sigma a \sigma^* \rho^*)(t, q) < \infty$$

and

$$\sup_{\substack{t \geq s \\ q \in E}} |(\rho \sigma b)(t, q)| < \infty.$$

Then if we take

$$\eta(t) = \int_s^t \sigma(u)\, d\xi(u)$$

and consider the stochastic integral with respect to η, *we have*

$$\int_s^t \rho(u)\, d\eta(u) = \int_s^t \rho(u)\sigma(u)\, d\xi(u).$$

Proof. Consider the $2m$ dimensional process $\alpha(t)$ defined for $t \geq s$ by $(\alpha_1(t), \alpha_2(t))$ where

$$\alpha_1(t) = \int_s^t \rho(u)\sigma(u)\, d\xi(u)$$

and

$$\alpha_2(t) = \int_s^t \rho(u)\, d\eta(u).$$

We define $\gamma(t)$ for $t \geq s$ as a $d + n$ dimensional process with components $\gamma_1(t)$ and $\gamma_2(t)$ where

$$\gamma_1(t) = \xi(t) - \xi(s)$$

and

$$\gamma_2(t) = \eta(t).$$

Clearly

$$\gamma(t) = \int_s^t \binom{I}{\sigma(u)} d\xi(u)$$

and

$$\alpha(t) = \int_s^t \begin{pmatrix} \rho(u)\sigma(u) & 0 \\ 0 & \rho(u) \end{pmatrix} d\gamma(u).$$

We can now take a vector $\theta \in R^m$ and write $\langle \theta, \alpha_1 - \alpha_2 \rangle = \langle \bar{\theta}, \alpha \rangle$ where $\bar{\theta} = (\theta, -\theta) \in R^{2m}$. An elementary computation using Theorem 4.3.8 yields

$$\langle \theta, \alpha_1(t) - \alpha_2(t) \rangle \sim \mathscr{I}_1^s(0, 0).$$

Hence $\alpha_1(t) - \alpha_2(t) \equiv \alpha_1(0) - \alpha_2(0) = 0$.
   This proves the theorem. $\square$

4.4. Itô's Formula

In Section 4.1 we saw that if $\beta(t)$ is a 1-dimensional Brownian motion, then

$$\int_0^1 \beta(s)\, d\beta(s) = \tfrac{1}{2}\beta^2(1) - \tfrac{1}{2}\beta^2(0) - \tfrac{1}{2}.$$

The same argument as we used there yields the more general fact that

$$\int_0^t \beta(s)\, d\beta(s) = \tfrac{1}{2}\beta^2(t) - \tfrac{1}{2}\beta^2(0) - \frac{t}{2}.$$

As we indicated at the time, the appearance of the term $t/2$ on the right hand side results from the fact that "$d\beta(t)$" is not a true differential, in the sense that $(d\beta(t))^2 \neq 0$. One can now ask how this fact manifests itself when one takes the "differential" of a more general function of a Brownian path. That is, suppose f is smooth. What is "$df(\beta(t))$"? When $f(x) = x^2$, we have just seen that

$$df(\beta(t)) = f'(\beta(t))\, d\beta(t) + \tfrac{1}{2}f''(\beta(t))\, dt.$$

In other words, it is necessary to go out two terms in the Taylor's expansion of f in order to arrive at a true infinitesimal. This section is devoted to showing that the preceding is a general fact. A second way in which to view what we are about to do is "the identification of the martingale"

$$f(\beta(t)) - \int_0^t \tfrac{1}{2}\,\Delta f(\beta(s))\, ds$$

as the stochastic integral

$$\int_0^t f'(\beta(s))\, d\beta(s).$$

These preliminary remarks should become clearer as we proceed.

4.4.1 Theorem (Itô's Formula). *Let $\xi \sim \mathscr{I}_d^s(a, b)$ on $(E, \mathscr{F}_t, P)$ with $|\xi(s)|$ bounded. Given a function $f \in C^{1,2}([0, \infty) \times R^d)$ such that $f, \partial f/\partial t, \partial f/\partial x_i,$ and $\partial^2 f/\partial x_i \partial x_j (1 \leq i, j \leq d)$ are bounded pointwise by $A\exp(B|x|)$, for some A and B, one has that $\nabla_x f(\cdot, \xi(\cdot)) \in \textcircled{H}_d^s(a, b)$ and*

$$(4.1) \qquad f(t, \xi(t)) - f(s, \xi(s))$$
$$= \int_s^t \langle \nabla f(u, \xi(u)), d\xi'(u)\rangle + \int_s^t \left(\frac{\partial}{\partial u} + Lu\right)f(u, \xi(u))\,du, \qquad t \geq s,$$

almost surely, where $\xi'(t) = \xi(t) - \int_s^t b(u)\, du$.

Proof. Because of estimate (2.2), it is clear not only that $\nabla_x f(\cdot, \xi(\cdot)) \in \textcircled{H}_d^s(a, b)$ but also that

$$E\left[\int_s^T |\nabla_x f_n(u, \xi(u)) - \nabla_x f(u, \xi(u))|^2\, du\right] \to 0$$

for all $T > s$ as $n \to \infty$, where

$$f_n(t, x) = \phi\left(\frac{t^2 + |x|^2}{n^2}\right)f(t, x)$$

and $\phi \in C_0^\infty(R)$ is a function such that $\phi \equiv 1$ on $[-1, 1]$. Thus we may assume that $f \in C_b^{1, 2}([0, \infty) \times R^d)$.

Given $f \in C_b^{1, 2}([0, \infty) \times R^d)$, define the following progressively measurable functions: $z: [s, \infty) \times E \to R^{d+1}$, $B: [s, \infty) \times E \to R^{d+1}$, and $A: [0, \infty) \times E \to S_{d+1}$ by:

$$z(t) = (\xi(t), f(t, \xi(t))),$$

$$B(t) = \left(b(t), \left(\frac{\partial}{\partial t} + L_t\right) f(t, \xi(t))\right),$$

$$A^{ij}(t) = \begin{cases} \langle \nabla_x f(t, \xi(t)), a(t)\nabla_x f(t, \xi(t))\rangle & \text{if } i = j = d + 1 \\ \sum_{l=1}^{d} a^{il}(t) \frac{\partial f}{\partial x_l}(t, \xi(t)) & \text{if } 1 \leq i < j = d + 1 \\ a^{ij}(t) & \text{if } 1 \leq i \leq j \leq d. \end{cases}$$

Using (iv) of Theorem 4.2.1, one sees immediately that $z \sim \mathscr{I}_{d+1}^s(A, B)$. In particular, if $\theta(t) = (\nabla f(t, \xi(t)), -1)$, then θ is bounded; and so

$$\exp\left[\lambda \int_s^t \langle \theta(u), dz'(u)\rangle - \frac{\lambda^2}{2} \int_s^t \langle \theta(u), A(u)\theta(u)\rangle \, du\right]$$

is a martingale for all $\lambda \in R$, where $z'(t) = z(t) - \int_s^t B(u) \, du$. But $\langle \theta(t), A(t)\theta(t)\rangle \equiv 0$, and therefore

$$\int_s^t \langle \theta(u), dz'(u)\rangle \equiv 0, \qquad t \geq s,$$

almost surely. When this is written out, it is easily seen to be just equation (4.1). $\square$

Itô's formula in the form given by (4.1) above "identifies the martingale" in the definition of an Itô process. However, from a differential calculus point of view, (4.1) is not the most intuitive expression of Itô's formula. Indeed, if one wishes to emphasize the differential aspect of the formula, one should write:

$$(4.2) \quad f(t, \xi(t)) - f(s, \xi(s))$$

$$= \int_s^t \langle \nabla_x f(u, \xi(u)), d\xi(u)\rangle + \int_s^t \left(\frac{\partial}{\partial u} + L_u^0\right) f(u, \xi(u)) \, du,$$

where $L_t^0 = \frac{1}{2} \sum_{i, j=1}^{d} a^{ij}(t) \, \partial^2/\partial x_i \, \partial x_j$ is the second order part of L_t. Of course, (4.1) is equivalent to (4.2). The reason that (4.2) is more pleasing from a differential point of view is that it lends itself to the intuitively appealing differential formula:

$$(4.2') \qquad df(t, \xi(t)) = \frac{\partial f}{\partial t}(t, \xi(t)) \, dt + \langle \nabla_x f(t, \xi(t)), \, d\xi(t) \rangle$$

$$+ \frac{1}{2} \sum_{i, j = 1}^{d} \frac{\partial^2 f}{\partial x_i \, \partial x_j}(t, \xi(t)) \, d\xi_i(t) \, d\xi_j(t),$$

where

$$(4.3) \qquad d\xi_i(t) \, d\xi_j(t) = a^{ij}(t) \, dt.$$

Equation (4.2') is, perhaps, the best and most concise statement of the "second order nature" of Itô processes. The reader may wish to compare this purely probabilistic statement with the analytic one given in Exercise 3.3.1.

One more general comment about Itô's formula is in order. Equation (4.2') emphasizes the local nature of this formula and makes it seem unnatural that we have had to impose a global growth condition on f (cf. Theorem 4.4.1). The fact is that the growth condition is, in some sense, totally unnecessary from a path-by-path point of view; it is only there for reasons of integrability. The interested reader should consult the book of H.P. McKean Jr. [1969] or Exercise 4.6.10 below for a more expanded treatment of this point.

4.5. Itô Processes as Stochastic Integrals

Let $(\beta(t), \mathscr{F}_t, P)$ be a d-dimensional s-Brownian motion and suppose that $\sigma : [s, \infty) \times E \to R^d \otimes R^d$ and $b : [s, \infty) \times E \to R^d$ are bounded progressively measurable functions. An immediate consequence of Theorem 4.3.8 is the fact that any process $\xi(t)$ satisfying

$$(5.1) \qquad \xi(t) - \xi(s) = \int_s^t \sigma(u) \, d\beta(u) + \int_s^t b(u) \, du, \qquad t \geq s,$$

is an Itô process having covariance $a(\cdot) = \sigma\sigma^*(\cdot)$ and drift $b(\cdot)$. In this section it is our purpose to prove the converse theorem. That is, suppose that $\xi \sim \mathscr{I}_d^s(a, b)$ on $(E, \mathscr{F}_t, P)$. What we want to show is that there exists an s-Brownian motion $\beta(\cdot)$ for which (5.1) holds. Obviously, this will not be possible in general if we insist that $\beta(\cdot)$ live on $(E, \mathscr{F}_t, P)$. For instance, suppose that E consists of exactly one point q, $\mathscr{F} = \mathscr{F}_t = \{\phi, \{q\}\}$, and $P = \delta_q$. If $\xi(t, q) = \sin t$, then $\xi \sim \mathscr{I}_1^0(0, b)$, where $b(t, q) = \cos t$. On the other hand, there is no way in which one could put a 1-dimensional Brownian motion on $(E, \mathscr{F}_t, P)$. Although this example may appear to be extreme, it demonstrates the necessity of distinguishing the situation in which $a(\cdot)$ may degenerate. For this reason, we are going to prove two versions of the converse theorem: one when $a(\cdot)$ is non-degenerate, and the other when it may degenerate.

4.5.1 Theorem. *Suppose that $a: [s, \infty) \times E \to S_d^+$ and $b: [s, \infty) \times E \to R^d$ are bounded progressively measurable functions after time s on $(E, \mathscr{F})$ relative to $\{\mathscr{F}_t: t \geq 0\}$. (Remember that S_d^+ is the set of strictly positive definite $d \times d$-symmetric matrices.) Let $\sigma: [s, \infty) \times E \to R^d \times R^d$ be a progressively measurable function for which $a(\cdot) = \sigma\sigma^*(\cdot)$. If $\xi: [s, \infty) \times E \to R^d$ is a progressively measurable function and P is a probability measure on $(E, \mathscr{F})$ such that $\xi \sim \mathscr{I}_d^s(a, b)$ on $(E, \mathscr{F}_t, P)$, then there exists a d-dimensional s-Brownian motion $\beta(\cdot)$ on $(E, \mathscr{F}_t, P)$ for which (5.1) holds.*

Proof. Note that $\sigma(t, q)$ is non-singular for all $(t, q) \in [s, \infty) \times E$, and therefore $(\sigma(t, q))^{-1}$ exists and determines a progressively measurable function $\sigma^{-1}: [s, \infty) \times E \to R^d \otimes R^d$. Moreover, $\sigma^{-1}a(\sigma^{-1})^* \equiv I$; and so, by Theorem 4.3.8, if

$$\beta(t) = \int_s^t \sigma^{-1}(u) \, d\xi'(u), \qquad t \geq s,$$

where $\xi'(t) = \xi(t) - \int_s^t b(u) \, du$, then $\beta \sim \mathscr{I}_d^s(I, 0)$. Thus, if $\beta(t) \equiv 0$ for $t < s$, then $\beta(\cdot)$ is an s-Brownian motion on $(E, \mathscr{F}_t, P)$ after time s. Finally, by Theorem 4.3.9,

$$\int_s^t \sigma(u) \, d\beta(u) = \int_s^t \sigma(u)\sigma^{-1}(u) \, d\xi'(u) = \int_s^t d\xi'(u)$$

$$= \xi(t) - \xi(s) - \int_s^t b(u) \, du \qquad \text{a.s.,}$$

and this completes the proof. $\square$

We now turn to the case in which $a(\cdot)$ is allowed to degenerate. The idea is Doob's (cf. [1952]). Intuitively speaking, what is going on here is that we first have to ramify the original sample space and thereby create a space on which we can fit a Brownian motion which is independent of the original Itô process. We then hold this new independent Brownian in reserve and only call on it to fill in the gaps created by lapses in "randomness" of the original Itô process (lack of randomness corresponds to degeneracy in $a(\cdot)$). In this way we get a full Brownian motion for which (5.1) holds, although we will, in general, have had to introduce an external source of randomness in order to do so.

4.5.2 Theorem. *Suppose $\xi \sim \mathscr{I}_d^s(a, b)$ on $(E, \mathscr{F}_t, P)$, and suppose that $\sigma: [s, \infty) \times E \to R^d \otimes R^d$ is a progressively measurable function such that $a = \sigma\sigma^*$. Then there is a probability space $(\tilde{E}, \tilde{\mathscr{E}}, \tilde{P})$, a non-decreasing family of sub σ-algebras $\{\tilde{\mathscr{F}}_t: t \geq 0\}$, and progressively measurable functions:*

$$\tilde{a}: [s, \infty) \times \tilde{E} \to S^d$$

$$\tilde{\sigma}: [s, \infty) \times \tilde{E} \to R^d \otimes R^d$$

$$\tilde{b}: [s, \infty) \times \tilde{E} \to R^d$$

$$\tilde{\xi}: [s, \infty) \times \tilde{E} \to R^d$$

and

$$\tilde{\beta}: [s, \infty) \times \tilde{E} \to R^d$$

such that $\tilde{a}, \tilde{\sigma}, \tilde{b},$ and $\tilde{\xi}$ are jointly distributed under $\tilde{P}$ in the same way as $a, \sigma, b,$ and ξ are under P, $\tilde{\beta}(\cdot)$ is a d-dimensional s-Brownian motion on $(\tilde{E}, \tilde{\mathscr{F}}_t, \tilde{P})$, and

$$\tilde{\xi}(t) - \tilde{\xi}(s) = \int_s^t \tilde{\sigma}(u) \, d\tilde{\beta}(u) + \int_s^t \tilde{b}(u) \, du, \qquad t \geq s,$$

$\tilde{P}$-almost surely.

Proof. Let $\Omega = \Omega_d$ as in Chapter 1, define $x(\cdot)$, $\mathcal{M}$ and $\{\mathcal{M}_t: t \geq 0\}$ accordingly, and let $\mathscr{W}^{(d)}_{s,0}$ be d-dimensional Wiener measure starting at $(s, 0)$. Set $\tilde{E} = E \times \Omega$, $\tilde{\mathscr{F}} = \mathscr{F} \times \mathcal{M}, \tilde{\mathscr{F}}_t = \mathscr{F}_t \times \mathcal{M}_t$, and $\tilde{P} = P \times \mathscr{W}^{(d)}_{s,0}$. Extend $a, \sigma, b,$ and ξ from E to $\tilde{E}$ and x from Ω to $\tilde{E}$ in the obvious way, and denote these extensions by $\tilde{a}, \tilde{\sigma}, \tilde{b}, \tilde{\xi}$, and $\tilde{x}$, respectively. It is easy to check that all these extensions are progressively measurable after time s with respect to $\{\tilde{\mathscr{F}}_t: t \geq 0\}$ and that

$$\tilde{\Xi} \equiv \begin{pmatrix} \tilde{\xi} \\ \tilde{x} \end{pmatrix} \sim \mathscr{I}^s_{2d} \left(\begin{pmatrix} \tilde{a} & 0 \\ 0 & I \end{pmatrix}, \begin{pmatrix} \tilde{b} \\ 0 \end{pmatrix} \right) \quad \text{on} \quad (\tilde{E}, \tilde{\mathscr{F}}_t, \tilde{P})$$

(cf. Exercise 4.6.4 below).

Let $\Pi(t, \tilde{q})$ be the orthogonal projection onto the range of $\tilde{a}(t, \tilde{q})$ and $\hat{\Pi}(t, \tilde{q})$ the orthogonal projection onto the range of $\tilde{\sigma}^*(t, \tilde{q})\tilde{\sigma}(t, \tilde{q})$. Note that

$$\Pi(t, \tilde{q}) = \lim_{\varepsilon \searrow 0} \tilde{a}(t, \tilde{q})(\varepsilon I + \tilde{a}(t, \tilde{q}))^{-1}$$

and Π is therefore progressively measurable. Similarly, $\hat{\Pi}$ is progressively measurable. Define $\tilde{\alpha}$ by

$$\tilde{\alpha}(t, \tilde{q}) = \lim_{\varepsilon \searrow 0} (\varepsilon I + \tilde{a}(t, \tilde{q}))^{-1} \Pi(t, \tilde{q})$$

and set $\tilde{\tau} = \tilde{\sigma}^* \tilde{\alpha}$. The following relations are easily deduced from elementary linear algebra:

(5.3) $$\tilde{\tau}\tilde{\sigma} = \hat{\Pi}$$

and

(5.4) $$\tilde{\sigma}\tilde{\tau} = \Pi.$$

We now define $\tilde{\Sigma}(u)$ to be the $d \times 2d$-matrix $(\tilde{\tau}(u), I - \hat{\Pi}(u))$ and

$$\tilde{\beta}(t) = \int_s^t \tilde{\Sigma}(u)\, d\tilde{\Xi}'(u),$$

where

$$\tilde{\Xi}'(t) = \tilde{\Xi}(t) - \int_s^t \begin{pmatrix} b(u) \\ 0 \end{pmatrix} du.$$

Note that this integral is defined, since, in fact,

$$\tilde{\Sigma}\begin{pmatrix} \tilde{a} & 0 \\ 0 & I \end{pmatrix}\tilde{\Sigma}^* = \tilde{\tau}\tilde{a}\tilde{\tau}^* + I - \hat{\Pi} = (\tilde{\tau}\tilde{\sigma})(\tilde{\tau}\tilde{\sigma})^* + I - \hat{\Pi} = I$$

by (5.3). In particular, $\tilde{\beta} \sim \mathscr{I}_d^s(I, 0)$ and is therefore a d-dimensional s-Brownian motion. Finally,

$$\int_s^t \tilde{\sigma}(u)\, d\tilde{\beta}(u) = \int_s^t \tilde{\sigma}(u)\tilde{\Sigma}(u)\, d\tilde{\Xi}'(u) = \int_s^t \tilde{\sigma}\tilde{\tau}(u)\, d\tilde{\xi}'(u) + \int_s^t \tilde{\sigma}(I - \hat{\Pi})(u)\, d\tilde{x}(u)$$

where $\tilde{\xi}'(t) = \tilde{\xi}(t) - \int_s^t \tilde{b}(u)\, du$. But, by (5.4), $\tilde{\sigma}\tilde{\tau} = \Pi$, and therefore

$$E\left[\left|\tilde{\xi}'(t) - \tilde{\xi}'(s) - \int_s^t \tilde{\sigma}\tilde{\tau}(u)\, d\tilde{\xi}'(u)\right|^2\right]$$

$$= \tilde{E}\left[\int_s^t \text{Trace}[(I - \Pi)\tilde{a}(I - \Pi)(u)]\, du\right] = 0.$$

On the other hand, $\tilde{\sigma}(I - \hat{\Pi}) = 0$ since the null space of σ coincides with that of $\sigma\sigma^*$. Thus

$$\tilde{E}\left[\left|\int_s^t \tilde{\sigma}(I - \hat{\Pi})(u)\, d\tilde{x}(u)\right|^2\right] = 0.$$

We can therefore conclude that

$$\int_s^t \tilde{\sigma}(u)\, d\tilde{\beta}(u) = \tilde{\xi}'(t) - \tilde{\xi}'(s) = \tilde{\xi}(t) - \tilde{\xi}(s) - \int_s^t \tilde{b}(u)\, du, \qquad t \geq s,$$

$\tilde{P}$-almost surely, and this completes the proof. □

4.6. Exercises

4.6.1. Let $\xi \sim \mathscr{I}_d^s(0, b)$ on $(E, \mathscr{F}_t, P)$. A special case of Theorem 4.5.2 is the fact that

$$\xi(t) - \xi(s) = \int_s^t b(u)\, du, \qquad t \ge s,$$

P-almost surely. However, one doesn't really need any of the machinery developed in this chapter to prove it. In fact, it is instructive to derive it directly from Theorem 1.2.8 and Leibnitz's rule.

4.6.2. Let $\xi \sim \mathscr{I}_d^s(a, b)$ on $(E, \mathscr{F}_t, P)$, where P is defined on $(E, \mathscr{F})$. For each $t \ge 0$, define $\bar{\mathscr{F}}_t$ to be the set of $A \in \mathscr{F}$ such that there exists a $B \in \mathscr{F}_t$ satisfying $P(A \Delta B) = 0$. Next, take $\bar{\mathscr{F}}_{t+0} = \bigcap_{\delta > 0} \bar{\mathscr{F}}_{t+\delta}$. Show that $\xi \sim \mathscr{I}_d^s(a, b)$ on $(E, \bar{\mathscr{F}}_{t+0}, P)$ (cf. Exercise 1.5.8).

4.6.3. Show that the condition that an Itô process be almost surely continuous is superfluous in the following sense. Let $(E, \mathscr{F}, P)$ be a probability space, $\{\mathscr{F}_t : t \ge 0\}$ a non-decreasing family of sub σ-algebras of $\mathscr{F}$, and $a \colon [s, \infty) \times E \to S_d$ and $b \colon [s, \infty) \times E \to R^d$ bounded progressively measurable function after time s. Suppose $\xi \colon [s, \infty) \times E \to R^d$ is a right-continuous progressively measurable function after time s which satisfies any one of the conditions of Theorem 4.2.1. Note that $C = \{q \colon \xi(\cdot, q)$ is continuous on $[s, \infty)\}$ is an element of $\mathscr{F}$, and show that $P(C) = 1$. Perhaps the simplest approach is to first note that almost sure continuity was never used in the proof of Theorem 4.2.1, and therefore that ξ must satisfy (ii) of that theorem. Conclude that $E[\,|\xi(t_2) - \xi(t_1)|^4] \le A(t_2 - t_1)^2 e^{B(t_2 - t_1)}, s \le t_1 < t_2$.

4.6.4. Let $(E, \mathscr{F}, P)$ be a probability space and $\{\mathscr{F}_t : t \ge 0\}$ a non-decreasing family of sub σ-algebras of $\mathscr{F}$. Suppose $\xi \sim \mathscr{I}_d^s(a, b)$ on $(E, \mathscr{F}_t, P)$. Given a second probability space $(E', \mathscr{F}', P')$, extend ξ, a, and b in the natural way (i.e., $\xi(t, (q, q')) = \xi(t, q)$ etc.) to $E'' = E \times E'$ and call these extensions $\tilde{\xi}$, $\tilde{a}$, and $\tilde{b}$. Define $P'' = P \times P'$ on $(E'', \mathscr{F}'')$, where $\mathscr{F}'' = \mathscr{F} \times \mathscr{F}'$. Show that $\tilde{\xi} \sim \mathscr{I}_d^s(\tilde{a}, \tilde{b})$ on $(E'', \mathscr{F}_t \times \mathscr{F}', P'')$. Use this result to show that if $\xi' \sim \mathscr{I}_{d'}^s(a', b')$ on $(E', \mathscr{F}_t', P')$, where $\{\mathscr{F}_t' : t \ge 0\}$ is a non-decreasing family of sub σ-algebras of $\mathscr{F}'$, and if

$$\xi''(t, q'') = \begin{pmatrix} \xi(t, q) \\ \xi'(t, q') \end{pmatrix}$$

$$a''(t, q'') = a(t, q) \oplus a'(t, q')$$

$$b''(t, q'') = \begin{pmatrix} b(t, q) \\ b(t, q') \end{pmatrix}$$

for $t \ge s$ and $q'' = (q, q') \in E''$, then $\xi'' \sim \mathscr{I}_{d''}^s(a'', b'')$ on $(E'', \mathscr{F}_t'', P'')$, where $d'' = d + d'$ and $\mathscr{F}_t'' = \mathscr{F}_t \times \mathscr{F}_t'$.

4.6.5. The following is an interesting, and sometimes useful, representation theorem for one-dimensional Itô processes. Let $(E, \mathscr{F}, P)$ and $\{\mathscr{F}_t: t \geq 0\}$ be as in the preceding and suppose that $\xi \sim \mathscr{I}_d^s(a, 0)$. Assume, for the moment, that there is an $\alpha > 0$ such that $a(t, q) \geq \alpha$ for all $(t, q) \in [s, \infty) \times E$. Then for each $q \in E$, $t \to s + \int_s^t a(u, q)\, du$ is a strictly increasing function from $[s, \infty)$ onto $[s, \infty)$. Let $\tau(\cdot, q)$ be the inverse function. Check that $\{\tau(t, \cdot): t \geq s\}$ is a non-decreasing family of bounded stopping times and that $\tau(\cdot, q)$ is continuous on $[s, \infty)$ onto $[s, \infty)$ for all $q \in E$. Apply Doob's stopping time theorem (Corollary 1.2.6) to show that if $\beta(t, q) = \xi(\tau(t, q), q) - \xi(s, q)$, $t \geq s$, then $\beta(\cdot)$ is a 1-dimensional s-Brownian motion on $(E, \mathscr{F}_{\tau(t)}, P)$. In particular, one will have shown that

$$(6.1) \qquad \xi(t) = \xi(s) + \beta\left(\int_s^t a(u)\, du\right), \qquad t \geq s,$$

where the distribution of $\beta(\cdot)$ under P is $\mathscr{W}_{s,0}^{(1)}$. In other words, the paths of such a 1-dimension Itô process are the same as those of a 1-dimensional Brownian motion; the difference being entirely in the rate at which the process follows these paths. When one puts it this way, it is clear that there should be no essential reason for the insistence on the uniform positivity of a. We will now outline how to proceed when this condition is dropped. As in Theorem 4.5.2, the difficulty caused by allowing a to be zero is that there may not be sufficient "randomness" to support a full-blown Brownian motion. The development outlined below is based on the ideas in H. P. McKean's book [1969].

Suppose that $\xi \sim \mathscr{I}_1^s(a, 0)$ on $(E, \mathscr{F}_t, P)$. Because of Exercise 4.6.22, we can and will assume that $\mathscr{F}_t = \mathscr{F}_{t+0}$. Define $\zeta = s + \int_s^\infty a(u)\, du$. The first step is to prove that $\lim_{t \nearrow \infty} \xi(t)$ exists almost surely on $\{\zeta < \infty\}$. To see this, let $\tau(t) = \sup\{u \geq s: s + \int_s^u a(v)\, dv \leq t\}$ if $\zeta > t$, and set $\tau(t) = \infty$ if $\zeta \leq t$. Note that $\{\tau(t): t \geq s\}$ is a non-decreasing, right continuous family of stopping times. Also, $\xi(\cdot) = \xi(\cdot \wedge \tau(T))$ on $\{\zeta \leq T\}$, and $E[|\xi(t \wedge \tau(T)) - \xi(s)|^2] \leq T$ for all $t \geq s$. Conclude from this, and Exercise 1.5.10, that $\lim_{t \nearrow \infty} \xi(t \wedge \tau(T))$ exists (a.s., P) for each $T \geq s$. Thus $\lim_{t \nearrow \infty} \xi(t)$ exists (a.s., P) on $\{\zeta < \infty\}$. Let $\xi_\infty: E \to \mathbb{R}$ be an $\mathscr{F}$-measurable function such that $\xi_\infty = \lim_{t \nearrow \infty} \xi(t)$ (a.s., P) on $\{\zeta < \infty\}$, and define $\eta: [s, \infty) \times E \to \mathbb{R}$ by

$$\eta(t, q) = \begin{cases} \xi(\tau(t, q), q) - \xi(s, q) & \text{if } \zeta(q) > t \\ \xi_\infty(q) - \xi(s, q) & \text{if } \zeta(q) \leq t. \end{cases}$$

Note that $\eta(\cdot, q)$ is right-continuous for all $q \in E$. Using Corollary 1.2.6 and Exercise 1.5.10, prove that for all $\lambda \in \mathbb{R}$:

$$\left(\exp\left[\lambda \eta(t) - \frac{\lambda^2}{2}((t \wedge \zeta) - s)\right], \mathscr{F}_{\tau(t)}, P\right)$$

is a martingale after time s. With Exercise 4.6.3, conclude from this that $\eta \sim \mathscr{I}_1^s(\chi_{(\cdot, \infty)}(\zeta), 0)$ on $(E, \mathscr{F}_{\tau(t)}, P)$.

To complete the proof, let $\mathcal{W}_{s,0}^{(1)}$ be 1-dimensional Wiener measure on $(\Omega, \mathcal{M})$ starting at $(s, 0)$. Let $\tilde{E} = E \times \Omega$, $\tilde{\mathcal{F}} = \mathcal{F} \times \mathcal{M}$, $\tilde{\mathcal{F}}_t = \mathcal{F}_{\tau(t)} \times \mathcal{M}_t$, and $\tilde{P} = P \times \mathcal{W}_{s,0}^{(1)}$. For $\tilde{q} = (q, \omega) \in \tilde{E}$ and $t \geq s$, define

$$\tilde{\beta}(t, \tilde{q}) = \eta(t, q) + x(t, \omega) - x(t \wedge \zeta(q), \omega).$$

Using Exercise 4.6.4, show that $\tilde{\beta}$ is a 1-dimensional s-Brownian motion on $(\tilde{E}, \tilde{\mathcal{F}}_t, \tilde{P})$. Finally, check that if $\tilde{\zeta}$ and $\tilde{a}$ are the natural extensions of ζ and a, respectively, to $[s, \infty) \times \tilde{E}$, then

$$\tilde{\zeta}(t) = \tilde{\zeta}(s) + \tilde{\beta}\left(s + \int_s^t \tilde{a}(u) \, du\right), \qquad t \geq s,$$

$\tilde{P}$-almost surely. This boils down to proving that $\zeta(t) = \zeta(\tau(s + \int_s^t a(u) \, du))$ (a.s., P), which can be done most easily by calculating the second moment of the difference.

4.6.6. In Exercise 4.6.3, we pointed out that almost sure continuity is an unnecessary assumption in our definition of an Itô process. That is because we have adopted a very strong characterization of an Itô process. If, instead, we had chosen a characterization more in keeping with Lévy's description of 1-dimensional Brownian motion as the only almost surely continuous martingale $\beta(t)$ such that $\beta(0) \equiv 0$ and $\beta^2(t) - t$ is a martingale, then the assumption of continuity would have been essential. Indeed, if $\eta(t)$ is a Poisson process with constant intensity 1, then $\eta(t) - t$ and $(\eta(t) - t)^2 - t$ are martingales, but $\eta(t) - t$ is certainly not a Brownian motion. We will now outline how one can characterize Itô processes in a way which is consistent with Lévy's idea. The procedure which we outline below is adopted from Kunita and Watenabe [1967].

Let $(E, \mathcal{F}, P)$ be a probability space, $\{\mathcal{F}_t : t \geq 0\}$ a nondecreasing family of sub σ-algebras of $\mathcal{F}$, and $a: [s, \infty) \times E \to S_d$ and $b: [s, \infty) \times E \to R^d$ bounded progressively measurable functions. Let $\zeta: [s, \infty) \times E \to R^d$ be a right-continuous, almost surely continuous progressively measurable function. Then $\zeta \sim \mathcal{I}_d^s(a, b)$ if and only if for each $\theta \in R^d$:

$$\left\langle \theta, \xi(t) - \xi(s) - \int_s^t b(u) \, du \right\rangle$$

and

$$\left\langle \theta, \xi(t) - \xi(s) - \int_s^t b(u) \, du \right\rangle^2 - \int_s^t \langle \theta, a(u)\theta \rangle \, du$$

are $(E, \mathcal{F}_t, P)$ martingales. The "only if" statement is immediate from Corollary 4.2.2. To prove "if", it is certainly sufficient to treat the case when $b \equiv 0$, since otherwise we can simply replace $\xi(\cdot)$ by $\xi(\cdot) - \int_s^\cdot b(u)\,du$. Given $f \in C_0^\infty(R^d)$, $s \le t_1 < t_2$, and $A \in \mathcal{M}_{t_1}$, define, for $n \ge 1$, $\tau_0^{(n)} \equiv t_1$ and

$$\tau_k^{(n)} = \left(\inf\left\{t \ge \tau_{k-1}^{(n)} : \sup_{\tau_{k-1}^{(n)} \le u \le t} |\xi(u) - \xi(\tau_{k-1}^{(n)})| \vee |t - \tau_{k-1}^{(n)}| \ge \frac{1}{n}\right\}\right) \wedge t_2.$$

Since $\xi(\cdot)$ is almost surely continuous, for almost all $q \in E$ there is a $k(q)$ such that $\tau_k^{(n)}(q) = t_2$ for $k \ge k_n(q)$. Thus

$$E[f(\xi(t_2)) - f(\xi(t_1)), A] = \lim_{K \nearrow \infty} \sum_{k=1}^K E[f(\xi(\tau_k^{(n)})) - f(\xi(\tau_{k-1}^{(n)})), A].$$

Assuming that $b \equiv 0$, show that

$$E[f(\xi(\tau_k^{(n)})) - f(\xi(\tau_{k-1}^{(n)})), A]$$

$$= \frac{1}{2} \sum_{i,j=1}^d E\left[\frac{\partial^2 f}{\partial x_i \, \partial x_j}(\xi(\tau_{k-1}^{(n)})) \int_{\tau_{k-1}^{(n)}}^{\tau_k^{(n)}} a^{ij}(u)\,du, A\right]$$

$$+ 0\left(\frac{1}{n^{1/2}}\right) E[\tau_k^{(n)} - \tau_{k-1}^{(n)}].$$

From this it is easy to conclude that:

$$E[f(\xi(t_2)) - f(\xi(t_1)), A] = E\left[\int_{t_1}^{t_2} L_u f(\xi(u))\,du, A\right].$$

4.6.7. Let $\xi \sim \mathcal{I}_d^s(a, b)$ relative to $(E, \mathcal{F}_t, P)$. Given a progressively measurable function $V: [s, \infty) \times E \to R$ which is bounded above, show that for any $f \in C_b^{1,2}([s, \infty) \times R^d)$

$$\exp\left(\int_s^t V(u)du\right) f(t, \xi(t)) - \int_s^t \exp\left(\int_s^u V(\sigma)d\sigma\right)\left(\frac{\partial f}{\partial u} + L_u f + V f\right)(u, \xi(u))du$$

is a martingale after time s. The proof is a simple application of Theorem 1.2.8. As simple as this fact is to prove, it is nevertheless extremely powerful and leads immediately to a general form of the Feynman-Kac formula.

4.6.8. Let $(E, \mathcal{E})$ be a measurable space and $\{\mathcal{F}_t : t \ge 0\}$ a non-decreasing family of sub σ-algebras of $\mathcal{F}$. Suppose that $A \subseteq [s, \infty) \times E$ is a progressively measurable set after time s (cf. Exercises 1.5.11–1.5.13) such that $(t_1, q) \in A$ whenever $t_1 \ge s$ and there is a $t_2 \ge t_1$ such that $(t_2, q) \in A$. Define $A(t) = \{q : (t, q) \in A\}$, $J(q) = \{t : (t, q) \in A\}$, and $\tau(q) = \sup\{t \ge s : (t, q) \in A\}$. Show that $\{\tau < t\} \in \mathcal{F}_t$ for all $t \ge s$. Next, let $\bar{J}(q) = [s, \tau(q)]$, the closure of the interval $J(q)$, and show that $\bar{A} \equiv \{(t, q) : t \in \bar{J}(q)\}$ is again progressively measurable. Now suppose that f is a progressively measurable function on A into a metric space X. Choose $x_0 \in X$ and define

$\bar{f}(t, q) = x_0$ for $(t, q) \in \bar{A} A$ and $\bar{f} = f$ on A. Show that $\bar{f}$ is progressively measurable on $\bar{A}$. Finally, define $\hat{f}$ on $[s, \infty) \times E$ by $\hat{f}(t, q) = \bar{f}(t \wedge \tau(q), q)$ and check that $\hat{f}$ is progressively measurable after time s. Observe that if $f(\cdot, q)$ is right-continuous on $J(q)$ for all q, then $\hat{f}(\cdot, q)$ is right-continuous on $[s, \infty)$ for all q.

4.6.9. Let $\xi \sim \mathscr{I}_d^s(a, 0)$ on $(E, \mathscr{F}_t, P)$ and suppose that $\theta \in \widehat{(H)}_d^s(a)$. Given stopping times τ_1 and τ_2 such that $s \leq \tau_1 \leq \tau_2 \leq T$, for some $T > s$, show that $\chi_{[\tau_1, \tau_2)}(\cdot)\theta(\cdot) \in \widehat{(H)}_d^s(a)$ and that

$$\int_{\tau_1}^{\tau_2} \langle \theta(u), d\xi(u) \rangle \equiv \left(\int_s^{\tau_2} \langle \theta(u), d\xi(u) \rangle - \int_s^{\tau_1} \langle \theta(u), d\xi(u) \rangle \right)$$

$$= \int_s^T \chi_{[\tau_1, \tau_2)}(u)\langle \theta(u), d\xi(u) \rangle \qquad \text{(a.s., } P),$$

where $\int_s^{\tau_i} \langle \theta(u), d\xi(u) \rangle = \tilde{\xi}(\tau_i)$ when $\tilde{\xi}(\cdot) \equiv \int_s^\cdot \langle \theta(u), d\xi(u) \rangle$. This can be done without ever resorting to simple approximations. The idea is to use nothing but the basic properties of stochastic integrals plus Doob's stopping time theorem. By the same sort of reasoning, show that if τ is a stopping time and $\theta: E \to R^d$ is an $\mathscr{F}_\tau$-measurable function satisfying:

$$E \left[\int_s^T \langle \theta; a(u)\theta \rangle \, du \right] < \infty$$

for all $T > s$, then $\chi_{[\cdot, \infty)}(\tau)\theta \in \widehat{(H)}_d^s(a, 0)$ and

$$\int_s^t \langle \chi_{[u, \infty)}(\tau)\theta, d\xi(u) \rangle = \langle \theta, \xi(t) - \xi(\tau \wedge t) \rangle.$$

4.6.10. Kunita and Watanabe [1967] gave an elegant and easy derivation of Itô's formula based on the results of Exercise 4.6.9. Define the stopping time $\tau_k^{(n)}$ as in Exercise 4.6.6 with $t_1 = s$ and $t_2 = t$. Then, almost surely,

$$f(t, \xi(t)) - f(s, \xi(s)) = \sum_{t=1}^{\infty} \left(f(\tau_k^{(n)}, \xi(\tau_k^{(n)})) - f(\tau_{k-1}^{(n)}, \xi(\tau_{k-1}^{(n)})) \right).$$

One now expands f in a Taylor's expansion, up to first order in time and second order in space, around $(\tau_{k-1}^{(n)}, \xi(\tau_{k-1}^{(n)}))$. By Exercise 4.6.9, one can see that the terms corresponding to the first order spatial derivatives tends, as $n \to \infty$, to

$$\int_s^t \langle \nabla_s f(u, \xi(u)), d\xi(u) \rangle.$$

The other terms are easily recognized to be what they should be in the limit as $n \to \infty$. The form of Itô's formula at which one arrives in this way is the one given in (3.2). This is entirely natural, since the approach is based on a differential technique.

4.6.11. Suppose that $\xi \sim \mathscr{I}_d^s(a, 0)$ on $(E, \mathscr{F}_t, P)$ and $\theta: [s, \infty) \times E \to R^d$ is progressively measurable. Does one really need the condition

$$E\left[\int_s^T \langle \theta(u), a(u)\theta(u) \rangle \, du \right\rangle\right] < \infty, \qquad T > s,$$

in order to define $\int_s^t \langle \theta(u), d\xi(u) \rangle$? The answer is no, as we will now see. In fact, suppose that instead

(6.1) $P\left(\int_s^T \langle \theta(u), a(u)\theta(u) \rangle \, du < \infty \qquad \text{for all} \qquad T > s\right) = 1.$

Define, for $l \geq 1$,

$$\theta_l(t, q) = \chi_{[0, l]}\left(\int_s^t \langle \theta(u, q), a(u, q)\theta(u, q) \rangle \, du\right)\theta(t, q)$$

and check that $\theta_l \in (H)_d^s(a)$. Set

$$\tilde{\xi}_l(t) = \int_s^t \langle \theta_l(u), d\xi(u) \rangle.$$

Prove that $\tilde{\xi}_{l+1}(t) = \tilde{\xi}_l(t)$ a.s. on $\{\int_s^t \langle \theta(u), a(u)\theta(u) \rangle \, du \leq l\}$. It then follows that there is a right-continuous, almost surely continuous progressively measurable $\tilde{\xi}$ such that $\tilde{\xi} = \lim_{l \to \infty} \tilde{\xi}_l$ a.s. Of course $\tilde{\xi}$ will not, in general, be a martingale, since $\tilde{\xi}(t)$ need not even be integrable. Thus, the most suitable context in which to discuss this extension of Itô's integral is that of "local martingales". To demonstrate that this extension is meaningful, the reader should use it to prove that Itô's formula holds for all $f \in C^{1, 2}([0, \infty) \times R^d)$ without any growth conditions of f. Finally, it is important to notice that, in some sense, the condition (6.1) is the optimal one. In fact, one can show that the result proved in Exercise 4.6.4 can be extended to any stochastic integral $\tilde{\xi}(\cdot) = \int_s^{\cdot} \langle \theta(u), d\xi(u) \rangle$, where $\theta(\cdot)$ satisfies (6.1). That is, $\tilde{\xi}(\cdot)$ looks like a 1-dimensional s-Brownian motion run at the clock $\int_s^{\cdot} \langle \theta(u), a(u)\theta(u) \rangle \, du$. Since a 1-dimensional Brownian motion has limit superior equal $+\infty$ and limit inferior equal to $-\infty$ as time tends to ∞, it is clear that our notion of stochastic integral breaks down completely if (6.1) doesn't hold. The interested reader is referred to the book of H. P. McKean [1969] for more details on this point, and also on the version of Itô's formula mentioned above.

4.6.12. Let $\xi \sim \mathscr{I}_d^s(a, 0)$ on $(E, \mathscr{F}_t, P)$ and assume that $\xi(s) \equiv 0$. Show that for each $2 \leq p < \infty$ there is a constant $C_p < \infty$, depending only on p and d, such that

(6.2) $E\left[\sup_{s \leq t \leq T} |\xi(t)|^p\right] \leq C_p E\left[\left(\int_s^T \text{Trace } a(u) \, du\right)^{p/2}\right]$

for all $T > s$. Estimate (6.2) is one fourth of a family of beautiful inequalities discovered originally by D. Burkholder [1966] (See Ex A.3.3). The easy proof outlined below is based on ideas of A. Garsia. A more analytic approach can be found in the book of H. P. McKean [1969]. Here is Garsia's proof:

$$E\left[\sup_{s \le t \le T} |\xi(t)|^p\right] \le \left(\frac{p}{p-1}\right)^p E[|\xi(T)|^p]$$

$$\le \left(\frac{p}{p-1}\right)^p \frac{p(p-1)}{2} E\left[\int_s^T \text{Trace } a(t)|\xi(t)|^{p-2}\, dt\right]$$

$$\le \left(\frac{p}{p-1}\right)^p \frac{p(p-1)}{2} E\left[\sup_{s \le t \le T} |\xi(t)|^{p-2} \int_s^T \text{Trace } a(t)\, dt\right]$$

$$\le \left(\frac{p}{p-1}\right)^p \frac{p(p-1)}{2} E\left[\sup_{s \le t \le T} |\xi(t)|^p\right]^{1-2/p} E\left[\left(\int_s^T \text{Trace } a(t)\, dt\right)^{p/2}\right]^{2/p}.$$

We have used here Doob's inequality, the fact that $f(\xi(t)) - \int_s^t L_u f(\xi(u))\, du$ is a martingale with $f(x) = |x|^p$, and finally Hölder's inequality.

4.6.13. Let $\xi \sim \mathscr{I}_1^0(a, 0)$ on $(E, \mathscr{F}_t, P)$ and assume that $\xi(0) \equiv 0$. We are going to outline a proof of the existence of a function $l_x(t, q)$ for $t \ge 0$, $x \in R$ and $q \in E$ such that:

(i) l_x is progressively measurable for each $x \in R$,
(ii) $l_x(\cdot, q)$ is right-continuous, non-decreasing, and $l_x(0, q) = 0$ for all $x \in R$ and $q \in E$,
(iii) for almost all $q \in E$, $(x, t) \to l_x(t, q)$ is jointly continuous,
(iv) for almost all $q \in E$,

$$(6.3) \qquad \int_\Gamma l_x(t, q)\, dx = \frac{1}{2} \int_0^t a(u)\chi_\Gamma(\xi(u))\, du$$

for all $t \ge 0$ and $\Gamma \in \mathscr{B}_R$.

In the case $a(\cdot) \equiv 1$, (6.3) shows that $2l.(t, q)$ is identifiable as the density, with respect to Lebesgue measure, of the occupation time functional, up until time t, of the path $\xi(\cdot, q)$. For this reason, l_x is called *the local time at x*. The existence of a local time was first observed by P. Lévy [1948], but it was H. Trotter who first provided a rigorous proof. The tack which we adopt here is due to H. Tanaka. (See McKean [1975].) Although it is somewhat obscure in the present proof, what underlies the existence of a local time for a 1-dimensional Itô process is the fact that points have positive capacity in one dimension. The way in which this fact is used here is hidden in the boundedness at its pole of the fundamental solution for $\frac{1}{2}\, d^2/dx^2$. To be precise, the functional for which we are looking is given, at least formally, by

$$``\frac{1}{2}\int_0^t a(u)\, \delta(\xi(u) - x)\, du,"$$

where $\delta(\cdot)$ stands for the Dirac δ-function. If $f(x) = x \vee 0$ for $x \in R$ then $f''(\cdot) = \delta(\cdot)$, and then Itô's formula predicts that

$$\frac{1}{2} \int_0^t a(u)\, \delta(\xi(u) - x)\, du = f(\xi(t) - x) - f(-x) - \int_0^t \chi_{[x,\,\infty)}(\xi(u))\, d\xi(u).$$

The terms on the right of this expression make sense and therefore indicate how we should go about realizing the left hand side. This observation is the basis of Tanaka's approach. We now outline some of the details.

For $n \geq 1$, let f_n be defined on R by:

$$f_n(x) = \begin{cases} 0 & \text{if} \quad x < -\dfrac{1}{n} \\[2mm] n\left(x + \dfrac{1}{n}\right)^2 \Big/ 4 & \text{if} \quad -\dfrac{1}{n} \leq x \leq \dfrac{1}{n} \\[2mm] x & \text{if} \quad x \geq \dfrac{1}{n}. \end{cases}$$

Of course f_n is not in $C^2(R)$, but an obvious mollification argument shows that Itô's formula can be extended to yield:

$$f_n(\xi(t) - x) - f_n(-x) - \int_0^t f'_n(\xi(u) - x)\, d\xi(u)$$

$$= \frac{n}{4} \int_0^t a(u)\chi_{[-1/n,\,1/n)}(\xi(u) - x)\, du$$

almost surely. It is easy to check that for each $x \in R$ and $T > 0$:

$$P\left(\sup_{0 \leq t \leq T} \left| \left(f_n(\xi(t) - x) - f_n(-x) - \int_0^t f'_n(\xi(u) - x)\, d\xi(u) \right) \right. \right.$$

$$\left. \left. - \left(f(\xi(t) - x) - f(-x) - \int_0^t \chi_{[x,\,\infty)}(\xi(u))\, d\xi(u) \right) \right| \geq \varepsilon \right)$$

tends to zero as $n \to \infty$ for each $\varepsilon > 0$. Here $f(x) = x \vee 0$. With this, together with the method used to construct stochastic integrals, one can conclude the existence, for each $x \in R$, of a right-continuous, non-decreasing, progressively measurable function $\overline{l}_x$ such that $\overline{l}_x(0) \equiv 0$ and almost surely

$$(6.4) \qquad \overline{l}_x(t) = \xi(t) \vee x - 0 \vee x + \int_0^t \chi_{[x,\,\infty)}(\xi(u))\, d\xi(u), \qquad t \geq 0.$$

The next step is to show that we can modify $\bar{l}_x$ so that the resulting function satisfies (i), (ii), and (iii) above. To this end, show that for each $T > 0$ and $L > 0$ there is a $C = C(T, L)$ such that

$$E\left[\sup_{0 \leq t \leq T} |\bar{l}_x(t) - \bar{l}_y(t)|^4\right] \leq C|y - x|^2$$

for all $x, y \in R$ such that $0 \leq y - x \leq L$. This can be done most easily by observing that, by Exercise 4.6.12,

$$E\left[\sup_{0 \leq t \leq T} \left(\int_0^t \chi_{[x,\, y]}(\xi(u))\, d\xi(u)\right)^4\right] \leq 36(4/3)^8 E\left[\left(\int_0^T a(u)\chi_{[x,\, y]}(\xi(u))\, du\right)^2\right].$$

One can now use Exercise 4.6.5 to estimate the right hand side in terms of a similar quantity involving a 1-dimensional Brownian motion, and the desired estimate then follows easily. We now define

$$l_x^{(n)}(t, q) = \bar{l}_{[2^n x]/2^n}(t, q) + 2^n(x - [2^n x]/2^n)\bar{l}_{([2^n x] + 1)/2^n}(t, q).$$

It follows immediately from the preceding estimate that there exists $C' = C'(T, L)$ such that

$$\sup_n E\left[\sup_{0 \leq t \leq T} |l_y^{(n)}(t) - l_x^{(n)}(t)|^4\right] \leq C'|y - x|^2$$

for all $0 \leq y - x \leq L$. We can therefore apply Exercise 2.4.1 and the method used to construct stochastic integrals to conclude the existence of an $l_x(t, q)$ satisfying (i), (ii), and (iii) as well as the property that $l_x(\cdot) = \bar{l}_x(\cdot)$ almost surely.

Finally, note that

$$\xi(t) \vee x - 0 \vee x - l_x(t)$$

is an $(E, \mathscr{F}_t, P)$ martingale for all $x \in R$. Thus, if $\varphi \in C_0^\infty(R)$ and F is given by $F(x) = \int (x \vee y)\varphi(y)\, dy$, then

$$F(\xi(t)) - F(0) - \int l_x(t)\varphi(x)\, dx$$

is an $(E, \mathscr{F}_t, P)$ martingale. On the other hand, $F'' = \varphi$, and so

$$F(\xi(t)) - F(0) - \frac{1}{2}\int_0^t a(u)\varphi(\xi(u))\, du$$

is a $(E, \mathscr{F}_t, P)$ martingale. It follows from Exercise 1.5.9 that

(6.5) $$\frac{1}{2}\int_0^t a(u)\varphi(\xi(u))\, du = \int l_x(t)\varphi(x)\, dx, \qquad t \geq 0,$$

almost surely. It is now easy to choose one P-null set N so that (6.5) holds off of N for all $\varphi \in C_0^\infty(R)$, and (iv) is immediate from this.

4.6.14. Let $\mathscr{W}$ be the one-dimensional Wiener measure. Wiener [1930] discovered a remarkable decomposition of the Hilbert space $L^2(\mathscr{W})$ into mutually orthogonal subspaces of "homogeneous chaos." This means that every Φ in $L^2(\mathscr{W})$ can be written as $\Sigma_0^\infty \Phi^{(m)}$, where $\Phi^{(m)}$ comes from the subspace of "chaos of mth order."

For each $n \geq 1$, let Δ_n be the set $\Delta_n = \{\tilde{s} \in R^n; 0 \leq s_1 \leq s_2 \cdots \leq s_n < \infty\}$. Given $f \in L_2(\Delta_n)$ (with respect to the Lebesgue measure) we define

$$(6.6) \qquad \int_{\Delta_n} f(\tilde{s}) \, d^{(n)}x(\tilde{s}) = \int_0^\infty dx(s_n) \int_0^{s_n} dx(s_{n-1}) \cdots \int_0^{s_2} f(\tilde{s}) \, dx(s_1)$$

and show that

$$(6.7) \qquad E\left[\left(\int_{\Delta_n} f(\tilde{s}) \, d^{(n)}x(\tilde{s})\right)\left(\int_{\Delta_m} g(\tilde{s}) \, d^{(m)}x(\tilde{s})\right)\right]$$

$$= \begin{cases} 0 & \text{if } n \neq m, \\[2ex] \int_{\Delta_n} f(\tilde{s})g(\tilde{s}) \, d\tilde{s} & \text{if } n = m. \end{cases}$$

To do this, we first define

$$(6.8) \qquad \int_{\Delta_n(t)} f(\tilde{s}) \, d^{(n)}x(\tilde{s}) = \int_0^t dx(s_n) \int_0^{s_n} dx(s_{n-1}) \cdots \int_0^{s_2} f(\tilde{s}) \, dx(s_1)$$

by repeated stochastic integration, moving from right to left. Here $\Delta_n(t) = \{\tilde{s} \in R^n : 0 \leq s_1 \leq s_2 \ldots \leq s_n \leq t\}$. By using Itô's formula, we show, by induction, the analog of (6.7) with $\Delta_n(t)$ replacing Δ_n. We then use exercise 1.5.10 and let $t \to \infty$. We can in this manner see that (6.6) makes sense and these multiple stochastic integrals satisfy (6.7). We can also check along the way that the left hand side of (6.8) can be defined as a progressively measurable, right-continuous, almost surely continuous process on Wiener space.

We now define $\mathscr{H}^{(0)}$ to be constants and $\mathscr{H}^{(n)}$ for $n \geq 1$ to be $\{\int_{\Delta_n} f(\tilde{s}) d^{(n)}x(\tilde{s}) : f \in L_2(\Delta_n)\}$. Using (6.7), one easily verifies that $\mathscr{H}^{(n)}$ are for $n \geq 0$ mutually orthogonal subspaces of $L^2(\mathscr{W})$. Wiener's theorem is that $L^2(\mathscr{W}) = \bigoplus_0^\infty \mathscr{H}^{(n)}$. To prove this fact, first define

$$X_\phi(t) = \exp\left[\int_0^t \phi(s) \, dx(s) - \frac{1}{2}\int_0^t \phi^2(s) \, ds\right]$$

for $\phi \in L_2[0, \infty)$. Show by Itô's formula that

$$X_\phi(t) = 1 + \int_0^t \phi(s) X_\phi(s) \, dx(s).$$

By induction, show that

$$X_\phi(t) = 1 + \sum_{j=1}^m \int_{\Delta_j(t)} \phi(s_1) \cdots \phi(s_j) \, d^{(j)}x(\tilde{s}) + R_m(t),$$

where

$$R_m(t) = \int_0^t \phi(s_{m+1}) \, dx(s_{m+1}) \int_0^{s_{m+1}} \phi(s_m) \, dx(s_m) \dots \int_0^{s_2} \phi(s_1) X_\phi(s_1) dx(s_1)$$

Then estimate $E[R_m^2(t)]$ and show that $E[R_m^2(t)] \to 0$ as $m \to \infty$. Conclude from this that $X_\phi(t) \in \bigoplus_0^\infty \mathscr{H}^{(m)}$ for all t. Finally, prove Wiener's theorem by observing that the only function in $L^2(\mathscr{W})$ that is orthogonal to $X_\phi(t)$ for all ϕ and t is the zero function.

Now if $X(t)$ is a martingale on the Wiener space with $\sup_t E[X^2(t)] < \infty$, then by Exercise 1.5.10, $X(t) = E[X | \mathscr{M}_t]$ and use the representation $X = \Sigma_0^\infty \Phi^{(n)}$ in terms of homogeneous chaos to conclude that $X(t)$ is given by a stochastic integral

$$X(t) = \int_0^t \theta(s) \, dx(s),$$

where

$$E \int_0^\infty \theta^2(s) \, ds < \infty.$$

In particular, $X(t)$ is almost surely continuous. This result has been generalized by Kunita and Watanabe [1967] and Jacod [1976]. See McKean [1973] for a more detailed discussion of Wiener's ideas.

Chapter 5

Stochastic Differential Equations

5.0. Introduction

Starting with coefficients $a(t, x) = ((a^{ij}(t, x)))_{1 \le i, j \le d}$ and $b(t, x) = (b^i(t, x))_{1 \le i \le d}$, we saw in Chapter 3 how the associated parabolic equation

(0.1)
$$\frac{\partial u}{\partial t} + L_t u = 0$$

can be a source of a transition probability function on which to base a continuous Markov process. Although this is a perfectly legitimate way to pass from given coefficients to a diffusion process, it has the unfortunate feature that it involves an intermediate state (namely, the construction of the fundamental solution to (0.1)) in which the connection between the coefficients and the resulting process is somewhat obscured. Aside from the loss of intuition caused by this rather circuitous route, as probabilists our principal objection to the method given in Chapter 3 is that everything it can say about the Markov process from a knowledge of its coefficients must be learned by studying the associated parabolic equation. The result is that probabilistic methods take a back seat to analytic ones, and the probabilist ends up doing little more than translating the hard work of analysts into his own jargon.

In this chapter we present K. Itô's recipe for constructing a diffusion with given coefficients. Itô's method has the advantage for us, that it does not rely on the theory of partial differential equations; and, at the same time, provides probabilistic insight into the structure of diffusions with non-constant coefficients. The reader may be interested, if somewhat disappointed, to note that the class of coefficients to which Itô's method applies is (at least in the degenerate case) negligibly different from the class that we handled by the methods of Chapter 3.

The origin of Itô's method is the following intuitive idea. Let coefficients $a(t, x)$ and $b(t, x)$ be given. The diffusion with these coefficients should be, essentially, a Brownian motion $\xi(t)$ which has been altered, at each instant, by changing its instantaneous covariance to $a(t, \xi(t))$ and its mean to $b(t, \xi(t))$. That is, for small $\delta > 0$:

$$\xi(t + \delta) - \xi(t) \sim a^{\frac{1}{2}}(t, \xi(t))(\beta(t + \delta) - \beta(t)) + b(t, \xi(t)) \, \delta$$

where $\beta(t)$ is a Brownian motion with the property that $\{\beta(t + \delta) - \beta(t): \delta > 0\}$ is independent of $\xi(t)$. In other words, $\xi(\cdot)$ ought to solve the *stochastic integral equation*:

$$(0.2) \qquad \xi(t) - \xi(s) = \int_s^t a^{\frac{1}{2}}(u, \xi(u))d\beta(u) + \int_s^t b(u, \xi(u))du.$$

There is a second, more formally acceptable but less intuitively appealing, way to arrive at (0.2). Namely, suppose that a and b are amenable to the techniques of Chapter 3, and let $\{P_{s,x}: (s, x) \in [0, \infty) \times R^d\}$ be the associated Markov family of measures on $(\Omega, \mathcal{M})$. Then, for $f \in C_0^\infty(R^d)$ and $s \leq t_1 < t_2$ and $x \in R^d$:

$$E^{P_{s,x}}[f(x(t_2)) - f(x(t_1))] = E^{P_{s,x}}\left[\int_{t_1}^{t_2} L_u f(x(u))\, du\right],$$

where L_t is the operator having $a(t, \cdot)$ and $b(t, \cdot)$ as its coefficients. By the Markov property, one therefore has:

$$E^{P_{s,x}}[f(x(t_2)) - f(x(t_1))\,|\,\mathcal{M}_{t_1}] = E^{P_{t_1,x(t_1)}}[f(x(t_2)) - f(x(t_1))]$$

$$= E^{P_{t_1,x(t_1)}}\left[\int_{t_1}^{t_2} L_u f(x(u))\, du\right]$$

$$= E^{P_{s,x}}\left[\int_{t_1}^{t_2} L_u f(x(u))\, du\,|\,\mathcal{M}_{t_1}\right].$$

But this means that

$$f(x(t)) - \int_s^t L_u f(x(u))\, du$$

is an $(\Omega, \mathcal{M}_t, P_{s,x})$ martingale after time s. Because this is true for any $f \in C_0^\infty(R^d)$, we conclude that $x(\cdot) \sim \mathscr{I}_d^s(a(\cdot, x(\cdot)), b(\cdot, x(\cdot)))$ on $(\Omega, \mathcal{M}_t, P)$. If one sets $\xi(\cdot) = x(\cdot)$, Equation (0.2) can therefore be seen as a consequence of Theorem 4.5.2. In this connection, note that, by Theorem 4.3.8, if ξ satisfies (0.2), then conversely $\xi \sim \mathscr{I}_d^s(a(\cdot, \xi(\cdot)), b(\cdot, \xi(\cdot)))$.

Regardless of the manner in which one arrives at (0.2), the question remains as to how one can exploit this equation to construct a diffusion process. Following Itô, what we are going to do in this chapter is to treat (0.2) as a equation for the trajectories of the diffusion in much the same way as one would approach an ordinary differential equation. In fact, it has been recently discovered by H. Sussman [1977] that, in special cases, Itô's method can be interpreted as a natural extension of the theory of ordinary differential equations.

5.1. Existence and Uniqueness

Let $(E, \mathscr{F}, P)$ be a probability space, $\{\mathscr{F}_t: t \geq 0\}$ a non-decreasing family of sub σ-algebras of $\mathscr{F}$, and $\beta(\cdot)$ is d-dimensional Brownian motion on $(E, \mathscr{F}_t, P)$. Suppose that $\sigma: [0, \infty) \times R^d \to R^d \otimes R^d$ are measurable functions which satisfy the conditions:

$$(1.1) \qquad \sup_{\substack{t \geq 0 \\ x \in R^d}} \|\sigma(t, x)\| \leq A, \qquad \sup_{\substack{t \geq 0 \\ x \in R^d}} |b(t, x)| \leq A$$

and for all $x, y \in R^d$ and $t \geq 0$:

$$(1.2) \qquad \|\sigma(t, x) - \sigma(t, y)\| + |b(t, x) - b(t, y)| \leq A |x - y|,$$

where $A < \infty$. Given $s \geq 0$ and an $\mathscr{F}_s$-measurable $\eta: E \to R^d$ satisfying $E[|\eta|^2] < \infty$, we want to prove the existence of a right-continuous, almost surely continuous, progressively measurable function $\xi: [s, \infty) \times E \to R^d$ such that almost surely:

$$(1.3) \qquad \xi(t) = \eta + \int_s^t \sigma(u, \xi(u)) \, d\beta(u) + \int_s^t b(u, \xi(u)) \, du, \ t \geq s.$$

Moreover, it will be shown that $\xi(\cdot)$ is uniquely determined by these properties up to a P-null set.

5.1.1 Theorem. *Assume that σ and b satisfy (1.1) and (1.2). Given $s \geq 0$ and an $\mathscr{F}_s$-measurable $\eta: E \to R^d$ such that $E[|\eta|^2] < \infty$, define $\xi_n(t) = \xi_n(t; s, \eta)$ for $t \geq s$ by $\xi_0(\cdot) \equiv \eta$ and*

$$\xi_n(t; s, \eta) = \eta + \int_s^t \sigma(u, \xi_{n-1}(u)) \, d\beta(u) + \int_s^t b(u, \xi_{n-1}(u)) \, du.$$

Then there is a right-continuous, almost surely continuous progressively measurable function $\xi(\cdot) = \xi(\cdot; s, \eta)$ on $[s, \infty) \times E$ into R^d such that

$$(1.4) \qquad \lim_{n \to \infty} E\left[\sup_{s \leq t \leq T} |\xi(t) - \xi_n(t)|^2 \right] = 0$$

for all $T > s$. In particular, $\xi(\cdot)$ solves (1.3).

Proof. We will first show that there is a progressively measurable function $\tilde{\xi}: [s, \infty) \times E \to R^d$ such that

$$(1.5) \qquad \lim_{n \to \infty} E\left[\int_s^T |\tilde{\xi}(t) - \xi_n(t)|^2 \, dt \right] = 0$$

for all $T > s$. To this end, set

$$\Delta_n(t) = E[\,|\xi_n(t) - \xi_{n-1}(t)|^2], \qquad t \geq s.$$

Then:

$$\Delta_{n+1}(t) \leq 2E\left[\left|\int_s^t (\sigma(u, \xi_n(u)) - \sigma(u, \xi_{n-1}(u)))\, d\beta(u)\right|^2\right]$$

$$+ 2E\left[\left|\int_s^t (b(u, \xi_n(u)) - b(u, \xi_{n-1}(u)))\, du\right|^2\right]$$

$$\leq 2A^2 \int_s^t \Delta_n(u)\, du + 2A^2(t - s) \int_s^t \Delta_n(u)\, du$$

$$= 2A^2(1 + (t - s)) \int_s^t \Delta_n(u)\, du,$$

and

$$\Delta_1(t) \leq 2E\left[\left|\int_s^t \sigma(u, \eta)\, d\beta(u)\right|^2\right]$$

$$+ 2E\left[\left|\int_s^t b(u, \eta)\, du\right|^2\right] \leq 2A^2(1 + (t - s))(t - s)$$

Working by induction on $n \geq 1$, we conclude that

$$\sup_{s \leq t \leq T} \Delta_n(t) \leq \frac{(2A^2(1 + (T - s)))^{n+1}}{n + 1!}$$

It follows immediately from this that for each $T > s$:

$$\lim_{\substack{m \to \infty \\ n \geq m}} \sup E\left[\int_s^T |\xi_n(t) - \xi_m(t)|^2\, dt\right] = 0.$$

By the Riesz–Fischer Theorem and Exercise 1.5.11, we conclude that there is a progressively measurable $\tilde{\xi}\colon [s, \infty) \times E \to R^d$ for which (1.5) holds.

We next set

(1.6) $\qquad \xi(t) = \eta + \int_s^t \sigma(u, \tilde{\xi}(u))\, d\beta(u) + \int_s^t b(u, \tilde{\xi}(u))\, du, \qquad t \geq s.$

Certainly $\xi(\cdot)$ is right-continuous, almost surely continuous, and progressively measurable. If we can show that (1.4) holds for all $T > s$ for this $\xi(\cdot)$, then the proof will be complete since we will then know that $\tilde{\xi}(\cdot)$ can be replaced by $\xi(\cdot)$ on the right hand side of (1.6). To prove (1.4), note that

$$E\left[\sup_{s\leq t\leq T}|\xi(t)-\xi_n(t)|^2\right]$$

$$\leq 2E\left[\sup_{s\leq t\leq T}\left|\int_s^t(\sigma(u,\tilde{\xi}(u))-\sigma(u,\xi_{n-1}(u)))\,d\beta(u)\right|^2\right]$$

$$+2(T-s)E\left[\int_s^T|b(u,\tilde{\xi}(u))-b(u,\xi_{n-1}(u))|^2\,du\right]$$

$$\leq 2A^2(4+(T-s))E\left[\int_s^T|\tilde{\xi}(u)-\xi_{n-1}(u)|^2\,du\right]\to 0$$

as $n\to\infty$ by (1.5).   □

5.1.2 Corollary. *Let σ, b, and η be as in the preceding. Then there is exactly one solution to (1.3); namely, the solution $\xi(\cdot\,;s,\eta)$ constructed in Theorem 5.1.1. Moreover, for $T>s$:*

(1.7) $$E\left[\sup_{s\leq t\leq T}|\xi(t;s,\eta)-\xi(t;s,\eta')|^2\right]$$

$$\leq 3E[|\eta-\eta'|^2]\exp[3A^2(4+(T-s))(T-s)]$$

for all square integrable $\mathscr{F}_s$-measurable η and η' on E into R^d.

Proof. Let η and η' be given and suppose $\xi(\cdot)$ and $\zeta'(\cdot)$ are solutions to (1.3) for η and η', respectively. Then:

$$E\left[\sup_{s\leq t\leq T}|\xi(t)-\zeta'(t)|^2\right]\leq 3E[|\eta-\eta'|^2]$$

$$+3E\left[\sup_{s\leq t\leq T}\left|\int_s^t(\sigma(u,\xi(u))-\sigma(u,\zeta'(u)))\,d\beta(u)\right|^2\right]$$

$$+3E\left[\sup_{s\leq t\leq T}\left|\int_s^t(b(u,\xi(u))-b(u,\zeta'(u)))\,du\right|^2\right]$$

$$\leq 3E[|\eta-\eta'|^2]+12E\left[\left|\int_s^T(\sigma(u,\xi(u))-\sigma(u,\zeta'(u)))\,d\beta(u)\right|^2\right]$$

$$+3(T-s)E\left[\int_s^T|b(u,\xi(u))-b(u,\zeta'(u))|^2\,du\right]$$

$$\leq 3E[|\eta-\eta'|^2]+3A^2(4+(T-s))\int_s^T E[|\xi(t)-\zeta'(t)|^2]\,dt.$$

Putting $\Delta(T) = E[\sup_{s \leq t \leq T} |\xi(t) - \xi'(t)|^2]$, we have:

$$\Delta(T) \leq 3E[|\eta - \eta'|^2] + 3A^2(4 + (T - s)) \int_s^T \Delta(t) \, dt.$$

An application of Gromwall's inequality now yields:

$$\Delta(T) \leq 3E[|\eta - \eta'|^2]\exp[3A^2(4 + (T - s)(T - s)].$$

This proves, in particular, that $\xi(\cdot) = \xi'(\cdot)$ almost surely if $\eta = \eta'$. Obviously, since we now know that $\xi(\cdot)$ and $\xi'(\cdot)$ must be $\xi(\cdot; s, \eta)$ and $\xi(\cdot; s, \eta')$ respectively, it also proves (1.7). $\square$

5.1.3 Corollary. *Let σ and b be as before. For $(s, x) \in [0, \infty) \times R^d$, let $\zeta_{s,x}(\cdot) \equiv \xi(\cdot \vee s; s, \eta)$, where $\eta \equiv x$. Then $\zeta_{s,x}(\cdot)$ can be chosen to be $\sigma(\beta(t) - \beta(s): t \geq s)$-measurable. Moreover, if $\beta'(\cdot)$ is a second Brownian motion on a second space $(E', \mathscr{F}'_t, P')$ and if $\zeta'_{s,x}(\cdot)$ is defined accordingly for the primed system, then the distribution of $\zeta_{s,x}(\cdot)$ under P coincides with the distribution of $\zeta'_{s,x}(\cdot)$ under P'.*

Proof. To prove the first part, simply note that if $\eta \equiv x$ in Theorem 5.1.1, then, by induction, $\xi_n(\cdot; s, \eta)$ can be chosen to be $\sigma(\beta(t) - \beta(s): t \geq s)$-measurable for each $n \geq 0$. Thus the assertion follows from (1.4). To prove the second part, define $\xi'_n(\cdot; s, \eta)$ with $\eta \equiv x$ for the primed systems as in Theorem 5.1.1. By induction, the distribution of $\xi_n(\cdot; s, \eta)$ under P and $\xi'_n(\cdot; s, \eta)$ under P' coincides for all $n \geq 0$. Therefore the same must be true of the distributions of $\zeta_{s,x}(\cdot)$ and $\zeta'_{s,x}(\cdot)$. $\square$

Corollaries 5.1.2 and 5.1.3 tell us that when σ and b satisfy (1.1) and (1.2), we can unequivocally talk about the distribution of the solution to (1.3) when η is constant. Since $\zeta_{s,x}(\cdot)$ is P-almost surely continuous, we think of its distribution as a probability measure $P_{s,x}$ on $(\Omega, \mathscr{M})$. We urge the reader to bear in mind that, because of Corollary 5.1.3, the measure $P_{s,x}$ is a function of σ and b alone; and not of the underlying Brownian motion $\beta(\cdot)$ or the space $(E, \mathscr{F}_t, P)$.

5.1.4 Theorem. *Let $\{P_{s,x}: (s, x) \in [0, \infty) \times R^d\}$ be the family of measures on $(\Omega, \mathscr{M})$ associated with functions σ and b satisfying (1.1) and (1.2). Then for bounded continuous $\Phi: \Omega \to \mathbb{C}$, $(s, x) \to E^{P_{s,x}}[\Phi]$ is a continuous function. In particular, $\{P_{s,x}: (s, x) \in [0, \infty) \times R^d\}$ is measurable in the sense that $(s, x) \to E^{P_{s,x}}[\Phi]$ is measurable for all bounded $\mathscr{M}$-measurable $\Phi: \Omega \to \mathbb{C}$.*

Proof. Clearly it suffices to prove that

$$\lim_{\delta \searrow 0} \sup_{0 \leq s - s' < \delta} \sup_{|x - x'| < \delta} E\left[\sup_{0 \leq t \leq T} |\zeta_{s,x}(t) - \zeta_{s',x'}(t)|^2\right] = 0$$

for each $T > 0$. Let $0 \le s' \le s < T$ and $x', x \in R^d$ be given. For $t \ge s$, $\xi_{s', x'}(t) = \xi(t; s, \eta')$ almost surely, where

$$\eta' = x' + \int_{s'}^{s} \sigma(u, \xi_{s', x'}(u)) \, d\beta(u) + \int_{s'}^{s} b(u, \xi_{s', x'}(u)) \, du.$$

Thus, by (1.7),

$$E\left[\sup_{s \le t \le T} |\xi_{s, x}(t) - \xi_{s', x'}(t)|^2 \right] \le 3E[|\eta' - x|^2]\exp[3A^2(4 + (T - s))(T - s)],$$

and

$$E[|\eta' - x'|^2] \le 3|x - x'|^2 + 3A^2(s - s') + 3A^2(s - s')^2.$$

If $s' \le t \le s$, then

$$\xi_{s', x'}(t) - \xi_{s, x}(t) = x' - x + \int_s^t \sigma(u, \xi_{s', x'}(u)) \, d\beta(u) + \int_s^t b(u, \xi_{s', x'}(u)) \, du,$$

and so

$$E\left[\sup_{s' \le t \le s} |\xi_{s, x}(t) - \xi_{s', x'}(t)|^2 \right] \le 3|x - x'|^2 + 3A^2(s - s') + 3A^2(s - s')^2.$$

Since $\xi_{s, x}(t) - \xi_{s', x'}(t) = x - x'$ for $0 \le t \le s'$, this completes the proof. $\quad\square$

We are now ready to prove the main result of this section.

5.1.5 Theorem. *Let σ and b satisfy* (1.1) *and* (1.2) *and define* $\{P_{s, x}: (s, x) \in [0, \infty) \times R^d\}$ *accordingly. For $0 \le s < t$, $x \in R^d$, and $\Gamma \in \mathcal{B}_{R^d}$, define*

$$P(s, x; t, \Gamma) = P_{s, x}(x(t) \in \Gamma).$$

Then $(s, x) \to P(s, x; t, \Gamma)$ is measurable on $[0, t) \times R^d$; and for $s \le t_1 < t_2$ and $\Gamma \in \mathcal{B}_{R^d}$:

$$P_{s, x}(x(t_2) \in \Gamma \,|\, \mathcal{M}_{t_1}) = P(t_1, x(t_1); t_2, \Gamma) \qquad (\text{a.s., } P_{s, x}).$$

In particular, $\{P_{s, x}: (s, x) \in [0, \infty) \times R^d\}$ is a continuous Markov family and $P(s, x; t, \cdot)$ is its transition probability function. Finally, for $f \in C_b^{1, 2}([0, \infty) \times R^d)$, $0 \le s < t$, and $x \in R^d$:

$$\int f(t, y)P(s, x; t, dy) - f(s, x) = \int_s^t du \int \left(\frac{\partial}{\partial u} + L_u \right) f(u, y)P(s, x; u, dy),$$

where

$$L_t = \frac{1}{2} \sum_{i,\,j=1}^{d} (\sigma\sigma^*)^{ij}(t,\,x) \frac{\partial^2}{\partial x_i\,\partial x_j} + \sum_{i=1}^{d} b^i(t,\,x) \frac{\partial}{\partial x_i}.$$

Proof. To prove the Markov property, let $f \in C_b(R^d)$ and $s \le t_1 < t_2$ be given. Note that $\xi_{s,x}(t) = \xi(t; t_1, \xi_{s,x}(t_1))$, $t \ge t_1$, almost surely. Thus if $0 \le s_1 < \cdots < s_n \le t_1$ and $\Gamma_1, \dots, \Gamma_n \in \mathscr{B}_{R^d}$, then

$$E^{P_{s,x}}[f(x(t_2)), \{x(s_1) \in \Gamma_1, \dots, x(s_n) \in \Gamma_n\}]$$
$$= E[f(\xi(t_2; t_1, \xi_{s,x}(t_1))), \{\xi_{s,x}(s_1) \in \Gamma_1, \dots, \xi_{s,x}(s_n) \in \Gamma_n\}].$$

Hence, if we can show that

$$E[f(\xi(t_2; t_1, \xi_{s,x}(t_1))) \mid \mathscr{F}_{t_1}] = \int f(y)P(t_1, \xi_{s,x}(t_1); t_2, dy) \qquad \text{a.s.,}$$

then the proof will be complete. To this end, note that for any $z \in R^d$, $\xi_{t_1,z}(t_2)$ is $\sigma(\beta(t) - \beta(t_1): t \ge t_1)$-measurable and is therefore independent of $\mathscr{F}_{t_1}$. Hence

$$E[f(\xi_{t_1,z}(t_2)) \mid \mathscr{F}_{t_1}] = E[f(\xi_{t_1,z}(t_2))]$$
$$= \int f(y)P(t_1, z; t_2, dy) \qquad \text{a.s.}$$

Next, suppose that $\eta: E \to R^d$ is $\mathscr{F}_{t_1}$-measurable and takes on only a finite number of values $z_1, \dots, z_N$. Then it is easy to see, from uniqueness and Exercise 4.6.8, that

$$\xi(t_2; t_1, \eta) = \sum_1^N \chi_{\{\eta = z_j\}} \xi_{t_1, z_j}(t_2) \qquad \text{a.s.}$$

Thus, for such an η, we have:

$$E[f(\xi(t_2; t_1, \eta)) \mid \mathscr{F}_{t_1}] = \int f(y)P(t_1, \eta; t_2, dy) \qquad \text{a.s.}$$

Finally, choose a sequence $\{\eta_n\}_1^\infty$ of such η's so that $E[|\eta_n - \xi_{s,x}(t_1)|^2] \to 0$ as $n \to \infty$. Then $\xi(t_2; t_1, \eta_n) \to \xi(t_2; t_1, \xi_{s,x}(t_1))$ in probability, and so

$$E[f(\xi(t_2; t_1, \xi_{s,x}(t_1))) \mid \mathscr{F}_{t_1}] = \lim_{n \to \infty} E[f(\xi(t_2; t_1, \eta_n)) \mid \mathscr{F}_{t_1}]$$
$$= \lim_{n \to \infty} \int f(y)P(t_1, \eta_n; t_2, dy) = \int f(y)P(t_1, \xi_{s,x}(t_1); t_2, dy)$$

almost surely. We have used here the fact that

$$\int f(y)P(t_1, \cdot; t_2, dy) = E^{P_{t_1}}[f(x(t_2))]$$

and is therefore a bounded continuous function.

The final assertion is an immediate consequence of the fact that $x(\cdot) \sim \mathscr{I}_d^s(\sigma\sigma^*(\cdot, x(\cdot)), b(\cdot, x(\cdot)))$ on $(\Omega, \mathscr{M}_t, P_{s,x})$. $\quad\square$

5.1.6 Remark. If σ and b satisfy (1.1) and (1.2), then we have seen that the unique solution to

$$(1.8) \qquad \xi(t) = x + \int_0^t \sigma(u, \xi(u)) \, d\beta(u) + \int_0^t b(u, \xi(u)) \, du$$

is a measurable functional of $\beta(\cdot)$. This brings up a rather subtle point which we will have occasion to discuss in greater detail in Chapter 8. Suffice it to say now that when σ and b fail to satisfy (1.1) and (1.2), there are situations in which $\xi(\cdot)$ can satisfy (1.8) without it being true that $\xi(\cdot)$ is a measurable functional of $\beta(\cdot)$. See Exercise 5.4.2.

5.1.7 Remark. We have already observed that if σ and b satisfy (1.1) and (1.2), then the measures $P_{s,x}$ depend only on σ and b. On the other hand, σ has no intrinsic meaning, since it is $a \equiv \sigma\sigma^*$ which is important and there may be a continuum of choices of σ such that $a = \sigma\sigma^*$. Thus we would like to know that $P_{s,x}$ really only depends on $\sigma\sigma^*$ and b. That this is indeed the case will be shown in Section 5.3.

5.1.8 Remark. The condition (1.1) can be considerably weakened. A good discussion of what one can say when (1.1) is abandoned can be found in the book of H. P. McKean [1969].

5.1.9 Remark. Suppose that σ and b satisfy (1.1) and (1.2) and that, in addition, they are independent of time. Let $\xi_{s,x}(\cdot)$, $(s, x) \in [0, \infty) \times R^d$, be defined accordingly relative to $\beta(\cdot)$ on $(E, \mathscr{F}_t, P)$. Given $s \geq 0$, set $\mathscr{F}_t' = \mathscr{F}_{t+s}$, $t \geq s$, and $\beta'(t) = \beta(t + s) - \beta(s)$. Then $\beta'(\cdot)$ is a Brownian motion on $(E, \mathscr{F}_t', P)$. Moreover, if $\xi(\cdot) = \xi_{s,x}(\cdot + s)$, then

$$\xi(t) = x + \int_0^t \sigma(\xi(u)) \, d\beta'(u) + \int_0^t b(\xi(u)) \, du, \qquad t \geq s.$$

Hence $\xi(\cdot) = \xi_{0,x}'(\cdot)$, where $\xi_{0,x}'(\cdot)$ is defined relative to $\beta'(\cdot)$ on $(E, \mathscr{F}_t', P)$. From this we see that the distribution of $\xi_{s,x}(\cdot + s)$ coincides with that of $\xi_{0,x}(\cdot)$. In other words, if $\theta_s: \Omega \to \Omega$ is the map given by $x(t, \theta_s\omega) = x(t + s, \omega)$, $t \geq 0$, then $P_{0,x} = P_{s,x} \circ \theta_s^{-1}$. In particular, for $0 \leq s < t$:

$$P(s, x; t, \Gamma) = P_{s,x}(x(t) \in \Gamma) = P_{s,x}(\theta_s^{-1}(\{x(t - s) \in \Gamma\}))$$

$$= P_{0,x}(x(t - s) \in \Gamma) = P(0, x; t - s, \Gamma).$$

Hence, if $P(t, x, \cdot) \equiv P(0, x, t, \cdot)$, then

$$P(s, x; t, \cdot) = P(t - s, x, \cdot), \qquad 0 \le s < t.$$

The Markov process is therefore *homogeneous* in time.

5.2. On the Lipschitz Condition

We saw in Section 5.1 that if we are given coefficients a and b such that b satisfies (1.2) and a can be written as $a = \sigma\sigma^*$ with σ satisfying (1.2) then we can construct a diffusion corresponding to a and b. Unfortunately the assumptions on σ are not directly transferable into assumptions on a. However we have the following theorems giving a wide class of coefficients a for which σ exists with $\sigma\sigma^* = a$ and σ satisfying (1.2).

5.2.1 Lemma. *Let $0 < \alpha < A$ be given, and suppose that $a \in S_d^+$ has all its eigenvalues in $[\alpha, A]$. Then the non-negative definite symmetric square root $a^{\frac{1}{2}}$ of a is given by the absolutely convergent series:*

$$A^{\frac{1}{2}} \sum_0^{\infty} c_n (I - a/A)^n$$

where c_n is the nth coefficient in the power series expansion of $(1 - x)^{\frac{1}{2}}$ around $x = 0$. In particular, the map $a \to a^{\frac{1}{2}}$ is analytic on S_d^+ into S_d^+.

Proof. Simply write $a = A(I - (I - a/A))$ and note that

$$0 \le \langle \theta, (I - a/A)\theta \rangle = |\theta|^2 - \frac{d}{A} \langle \theta, a\theta \rangle \le \left(1 - \frac{\alpha}{A}\right) |\theta|^2, \qquad \theta \in R^d. \quad \square$$

5.2.2 Theorem. *Let $a: [0, \infty) \times R^d \to S_d$ be a function such that $\langle \theta, a(s, x)\theta \rangle \ge \alpha|\theta|^2$ for all $(s, x) \in [0, \infty) \times R^d$, $\theta \in R^d$, and some $\alpha > 0$. If there is a $C < \infty$ such that*

$$\|a(s, x) - a(s, y)\| \le C|x - y|, \qquad s \ge 0 \quad and \quad x, y \in R^d,$$

then

$$\|a^{\frac{1}{2}}(s, x) - a^{\frac{1}{2}}(s, y)\| \le C/2\alpha^{\frac{1}{2}}|x - y|, \qquad s \ge 0 \quad and \quad x, y \in R^d.$$

Proof. Clearly it is enough to show that if $x \to a(x)$ is a C^1-function on R^1 into S_d such that $\langle \theta, a(x)\theta \rangle \ge \alpha|\theta|^2$, $x \in R^1$ and $\theta \in R^d$, then

$$\|(a^{\frac{1}{2}}(x))'\| \le \frac{1}{2\alpha^{\frac{1}{2}}} \|(a(x))'\|.$$

Moreover, we can assume that $a(\cdot)$ is diagonal at the point x in question. But

$$(a(x))' = a^{\frac{1}{2}}(x)(a^{\frac{1}{2}}(x))' + (a^{\frac{1}{2}}(x))'a^{\frac{1}{2}}(x);$$

and therefore if $a(\cdot)$ is diagonal at x,

(2.1) $$((a^{\frac{1}{2}}(x))')^{ij} = \frac{(a^{ij}(x))'}{(a^{ii}(x))^{\frac{1}{2}} + (a^{jj}(x))^{\frac{1}{2}}},$$

and this completes the proof. $\square$

We now want to look at the situation when a is allowed to degenerate.

5.2.3 Theorem. *Let* $a: [0, \infty) \times R^d \to S_d$ *be a function having two continuous spatial derivatives. If*

$$\max_{1 \le i \le d} \left| \left\langle \theta, \frac{\partial^2 a}{\partial x_i^2}(s, x)\theta \right\rangle \right| \le \lambda_0 |\theta|^2, \qquad \theta \in R^d,$$

for all $(s, x) \in [0, \infty) \times R^d$, *then*

$$\|a^{\frac{1}{2}}(s, x) - a^{\frac{1}{2}}(s, y)\| \le d(2\lambda_0)^{\frac{1}{2}} |x - y|, s \ge 0 \quad and \quad x, y \in R^d.$$

Proof. It is sufficient for us to show that if $x \to a(x)$ is a C^2-function on R^1 into S_d^+, then

$$\sup_{x \in R^d} |\langle \theta, (a(x))''\theta \rangle| \le \lambda_0 |\theta|^2, \qquad \theta \in R^d,$$

implies

$$\|(a^{\frac{1}{2}}(x))'\| \le d(2\lambda_0)^{\frac{1}{2}}$$

for each $x \in R^1$. Moreover, we may assume that $a(\cdot)$ is diagonal at the x in question. But in that case, we can use (2.1), and so:

$$|(a^{\frac{1}{2}}(x)')^{ij}| \le \frac{|a^{ij}(x)'|}{(a^{ii}(x) + a^{jj}(x))^{\frac{1}{2}}}$$

The desired estimate now follows from Lemma 3.2.3. $\square$

5.3. Equivalence of Different Choices of the Square Root

Suppose $\beta(\cdot)$ is a d-dimensional Brownian motion on some space $(E, \mathscr{F}_t, P)$. Given bounded measurable functions $\sigma: [0, \infty) \otimes R^d \to R^d \times R^d$ and $b: [0, \infty) \times R^d \to R^d$, suppose that $\xi(\cdot)$ is a right-continuous, almost surely continuous, progressively measurable solution to

$$(3.1) \qquad \xi(t) = x + \int_s^t \sigma(u, \xi(u)) \, d\beta(u) + \int_s^t b(u, \xi(u)) \, du, \qquad t \geq s.$$

Then, by Theorem 4.3.8, $\xi(\cdot)$ is an Itô process after time s with covariance $\sigma(\cdot, \xi(\cdot))\sigma^*(\cdot, \xi(\cdot))$ and drift $b(\cdot, \xi(\cdot))$. Thus we have the next result.

5.3.1 Theorem. *Let* $a: [0, \infty) \times R^d \to S_d$ *and* $b: [0, \infty) \times R^d \to R^d$ *be bounded measurable functions and suppose that* $\sigma: [0, \infty) \times R^d \to R^d \otimes R^d$ *is a measurable function satisfying* $a = \sigma\sigma^*$. *If* $\beta(\cdot)$ *is a d-dimensional Brownian motion on some space* $(E, \mathcal{F}_t, P)$ *and* $\xi(\cdot)$ *is a right-continuous, almost surely continuous, progressively measurable solution to* (3.1), *then* $\xi \sim \mathcal{I}_d^s(a(\cdot, \xi(\cdot)), b(\cdot, \xi(\cdot)))$ *on* $(E, \mathcal{F}_t, P)$.

We are now ready to resolve the question raised in Remark 5.1.2. Indeed, let $\sigma: [0, \infty) \times R^d \to R^d \times R^d$ and $b: [0, \infty) \times R^d \to R^d$ be bounded measurable functions satisfying (1.1) and (1.2), and define $\{P_{s, x}: (s, x) \in [0, \infty) \times R^d\}$ accordingly. Let $\rho: [0, \beta) \times R^d \to R^d \otimes R^d$ be a measurable function for which $\rho\rho^* = \sigma\sigma^*$. Suppose that there is a probability space $(E, \mathcal{F}, P)$ and a non-decreasing family $\{\mathcal{F}_t: t \geq 0\}$ of sub σ-algebras of $\mathcal{F}$ relative to which there is a d-dimensional Brownian motion β and a right-continuous, almost surely continuous, progressively measurable function $\xi: [s, \infty) \times E \to R^d$ such that:

$$\xi(t) = x + \int_s^t \rho(u, \xi(u)) \, d\beta(u) + \int_s^t b(u, \xi(u)) \, du, \qquad t \geq s.$$

Set $\xi_s(\cdot) = \xi(\cdot \vee s)$, and note that, by Theorem 5.3.1, $\xi_s \sim \mathcal{I}_d^0(\chi_{[s, \infty)}(\cdot)a(\cdot, \xi_s(\cdot)), \chi_{[s, \infty)}(\cdot)b(\cdot, \xi_s(\cdot)))$. Hence, by Theorem 4.5.2, there is another space $(E', \mathcal{F}', P')$ and a non-decreasing family $\{\mathcal{F}_t': t \geq 0\}$ of sub σ-algebras of $\mathcal{F}'$ on which there live a Brownian motion $\beta'(\cdot)$ and a right-continuous, almost surely continuous progressively measurable function $\xi'(\cdot)$ such that $\xi'(\cdot)$ is distributed under P' in the same way as $\xi_s(\cdot)$ is under P and

$$\xi'(t) = x + \int_0^t \chi_{[s, \infty)}(u)\sigma(u, \xi'(u)) \, d\beta'(u)$$

$$+ \int_0^t \chi_{[s, \infty)}(u)b(u, \xi'(u)) \, du.$$

But this means that $P_{s, x}$ is the distribution of $\xi'(\cdot)$ under P', and therefore it is also the distribution of $\xi(\cdot)$ under P. This argument not only proves the assertion which we set out to check, it also proves the next theorem.

5.3.2 Theorem. *Let* $\sigma: [0, \infty) \times R^d \to R^d \otimes R^d$ *and* $b: [0, \infty) \times R^d \to R^d$ *be bounded measurable functions satisfying* (1.1) *and* (1.2). *Define* $\{P_{s, x}: (s, x) \in [0, \infty) \times R^d\}$ *accordingly, and set* $a = \sigma\sigma^*$. *Let* $(E, \mathcal{F}, P)$ *be a probability space,* $\{\mathcal{F}_t: t \geq 0\}$ *a* non-decreasing family of sub σ-algebras of $\mathcal{F}$, and $\xi: [s,\infty) \times E \to R^d$ a right-continuous, almost surely continuous, progressively measurable function such that $P(\xi(s) = x) = 1$ and $\xi \sim \mathcal{I}_d^s(a(\cdot, \xi(\cdot)), b(\cdot, \xi(\cdot)))$ on $(E, \mathcal{F}_t, P)$. Then the distribution of $\xi(\cdot \vee s)$ under P is $P_{s,x}$.

5.4. Exercises

5.4.1. Let $\sigma: [0, \infty) \times R^d \to R^d \otimes R^d$ and $b: [0, \infty) \times R^d \to R^d$ be bounded measurable functions satisfying (1.1) and (1.2). Given a space $(E, \mathcal{F}, P)$, a non-decreasing family $\{\mathcal{F}_t : t \geq 0\}$ of sub σ-algebras of $\mathcal{F}$, and a d-dimensional Brownian motion $\beta(\cdot)$ on $(E, \mathcal{F}_t, P)$, define $\{\xi_{s,x}(\cdot): (s, x) \in [0, \infty) \times R^d\}$ as in Section 5.1. Using Exercise 4.6.12 show that for each $2 \leq p < \infty$ and $T > 0$, there is a constant $A_{p,T} < \infty$ such that

$$E\left[\sup_{0 \leq t \leq T} |\xi_{s_1, x_1}(t) - \xi_{s_2, x_2}(t)|^p\right] \leq A_{p,T}(|s_1 - s_2| + |x_1 - x_2|^2)^{p/2}$$

for all $s_1, s_2 \in [0, T]$ and $x_1, x_2 \in R^d$. Next, apply Exercise 2.4.1, and the technique used in choosing a nice version of a local time (cf. Exercise 4.6.13) to find a new family $\{\bar{\xi}_{s,x}(\cdot): (s, x) \in [0, \infty) \times R^d\}$ of right-continuous progressively measurable functions $\bar{\xi}_{s,x}(\cdot)$ such that $P(\xi_{s,x}(\cdot) = \bar{\xi}_{s,x}(\cdot)) = 1$ for all $(s, x) \in [0, \infty) \times R^d$ and for P-almost all $q \in E$ the map $(s, x, t) \to \bar{\xi}_{s,x}(t, q)$ is continuous.

5.4.2. There is a distinction between the type of "path-wise" uniqueness proved in Corollary 5.1.2 and the "distribution" uniqueness proved in Theorem 5.3.2. In fact, this distinction will be the topic of Chapter 8. To prove that there really is a difference, let $\beta(\cdot)$ be a 1-dimensional Brownian motion on some space $(E, \mathcal{F}_t, P)$. Define

$$\sigma(x) = \begin{cases} 1 & \text{if } x \geq 0 \\ -1 & \text{if } x < 0 \end{cases}$$

and

$$\bar{\beta}(t) = \int_0^t \sigma(\beta(s)) \, d\beta(s), \qquad t \geq 0.$$

Note that $\bar{\beta}(\cdot)$ is again a Brownian motion on $(E, \mathcal{F}_t, P)$. Next observe that

$$\beta(t) = \int_0^t \sigma(\beta(u)) \, d\bar{\beta}(u), \qquad t \geq 0,$$

and

$$-\beta(t) = \int_0^t \sigma(-\beta(u)) \, d\bar{\beta}(u), \qquad t \geq 0.$$

(The latter equation turns on the fact that

$$P\left(\int_0^t \chi_{\{0\}}(\beta(u))\, du = 0\right) = 1$$

for all $t \geq 0$.) Thus we have here a σ with the property that there is more than one solution to the corresponding Itô stochastic integral equation; and yet it is clear that all solutions have the same distribution, namely $\mathcal{W}_{0,0}^{(1)}$. This example was discovered by H. Tanaka (cf. Yamada-Watanabe [1971]).

Next observe that $\bar{\beta}(t) = |\beta(t)| - 2l_0(t)$, where $l_0(\cdot)$ is the local time of $\beta(\cdot)$ at 0 (cf. Exercise 4.6.13). Thus $\bar{\beta}(\cdot)$ is $\sigma(|\beta(t)|: t \geq 0)$-measurable. On the other hand, $\beta(\cdot)$ certainly is not $\sigma(|\beta(t)|: t \geq 0)$-measurable. Thus, in spite of the fact that

$$\beta(t) = \int_0^t \sigma(\beta(u))\, d\bar{\beta}(u), \qquad t \geq 0$$

$\beta(\cdot)$ is not $\sigma(\bar{\beta}(t): t \geq 0)$-measurable. Here is the example promised in Remark 5.1.1.

Chapter 6

The Martingale Formulation

6.0. Introduction

We have now seen two approaches to the construction of diffusion processes corresponding to given coefficients a and b. Let us quickly recapitulate the essential steps involved in these different methods and see if we cannot extract their common features.

The approach presented in Chapter 3 involved solving certain differential equations in which a and b appear as coefficients. If enough of these equations can be solved, then we are able to interpret the "fundamental solution" as a transition probability function $P(s, x; t, \cdot)$. Moreover, $P(s, x; t, \cdot)$ automatically satisfies estimates which permit us to apply the results of Chapter 2 and thereby construct, for each $(s, x) \in [0, \infty) \times R^d$, a probability measure $P_{s,x}$ on $\Omega = C([0, \infty), R^d)$ with the properties that:

$$(0.1) \qquad\qquad P_{s,\,x}(x(t) = x, \, 0 \le t \le s) = 1$$

and

$$(0.2) \qquad P_{s,\,x}(x(t_2) \in \Gamma \,|\, \mathscr{M}_{t_1}) = P(t_1, x(t_1); t_2, \Gamma) \qquad (\text{a.s., } P_{s,\,x})$$

for all $s \le t_1 < t_2$ and $\Gamma \in \mathscr{B}_{R^d}$. The relationship between $P_{s,\,x}$ and the coefficients a and b is via $P(s, x; t, \cdot)$ and can be most concisely summarized by the equation:

$$(0.3) \qquad \int_{R^d} P(s, x; t, dy) f(y) - f(x) = \int_s^t du \int_{R^d} P(s, x; u, dy) L_u f(y),$$

where

$$(0.4) \qquad\qquad L_t = \frac{1}{2} \sum_{i,\,j=1}^d a^{ij}(t, \cdot) \frac{\partial^2}{\partial x_i \, \partial x_j} + \sum_{i=1}^d b^i(t, \cdot) \frac{\partial}{\partial x_i}$$

and Equation (0.3) holds for $f \in C_0^\infty(R^d)$. Combining (0.2) and (0.4), we can eliminate $P(s, x; t, \cdot)$ and state the connection between $P_{s,x}$ and the coefficients a and b by the equation:

$$E^{P_{s,x}}[f(x(t_2))\,|\,\mathcal{M}_{t_1}] - f(x) = E^{P_{s,x}}\left[\int_s^t L_u f(x(u))\,du\,\Big|\,\mathcal{M}_{t_1}\right];$$

or, more succinctly,

$$(0.5) \qquad\qquad f(x(t)) - \int_s^t L_u f(x(u))\,du$$

is a $P_{s,x}$-martingale for all $f \in C_0^\infty(R^d)$.

The second approach that we have discussed for constructing diffusions is Itô's method of stochastic integral equations. Here again there are intervening quantities between the coefficients a and b and the measures $P_{s,x}$ (described in the paragraph preceding Theorem 5.1.4). Namely, there is the underlying Brownian motion with respect to which the stochastic integral equation is defined and there is the choice of σ satisfying $a = \sigma\sigma^*$. That neither of these quantities is canonical can be seen most dramatically when one tries to carry out Itô's method on a differentiable manifold (cf. McKean [1969]). But, if one focuses on the direct relationship between $P_{s,x}$ and the coefficients a and b, matters simplify; and, in spite of the apparent differences between Itô's method and the construction via partial differential equations, one is led once again to Equations (0.1) and (0.5) as being the most concise description of the connection between $P_{s,x}$ and a and b (cf. Theorem 5.3.1).

With the preceding discussion in mind, it seems only natural to ask if, in fact, (0.1) and (0.5) do not characterize what we should call "the diffusion with coefficients a and b starting from x at time s." Indeed, any measure with a legitimate claim to that title satisfies (0.1) and (0.5). Thus the real question is whether (0.1) and (0.5) are sufficient to uniquely determine a measure. That this may not be too much to hope for is already indicated in Theorem 4.1.1, from which it is easy to show that Wiener measure is characterized by (0.1) and (0.5) with $a \equiv I$ and $b \equiv 0$. On the other hand, one must not allow oneself to be over-optimistic because it is easy to find coefficients a and b for which (0.1) and (0.5) do not uniquely determine a measure. For instance, let $a \equiv 0$ and suppose that b is a continuous vector field which admits several integral curves through a given point. Then it is obvious that each integral curve determines a different measure satisfying (0.1) and (0.5). Thus we should guess that there will be cases in which (0.1) and (0.5) together determine a unique measure, but that it cannot be entirely trivial to recognize such cases. Obviously, an interesting class of questions come out of these and related considerations. Before investigating them, as we will be throughout the remainder of this book, we will formalize the statement of the problem around which these questions revolve.

Given bounded measurable functions $a: [0, \infty) \times R^d \to S_d$ and $b: [0, \infty) \times R^d \to R^d$, define L_t by (0.4). Given $(s, x) \in [0, \infty) \times R^d$, *a solution to the martingale problem for L_t (or a and b) starting from (s, x)* is a probability measure P on $(\Omega, \mathcal{M})$ satisfying

$$P(x(t) = x, \; 0 \le t \le s) = 1$$

such that

$$f(x(t)) - \int_s^t L_u f(x(u)) \, du$$

is a P-martingale after time s for all $f \in C_0^\infty(R^d)$. The basic questions which arise in connection with the martingale problem for given coefficients are the following:

(i) Does there exist at least one solution starting from a specified point (s, x)?

(ii) Is there at most one solution starting from (s, x)?

(iii) What conclusions can be drawn if there is exactly one solution for each (s, x)?

Questions (i) and (ii) are, respectively, those of *existence* and *uniqueness*; while the answers to (iii) presumably contain the justification for our interest in (i) and (ii). By analogy with Hadamard's codification of problems in the theory of partial differential equations, we will say that the martingale problem for a and b is *well-posed* if, for each (s, x), there is exactly one solution to that martingale problem starting from (s, x).

To the student familiar with the modern theory of Markov processes via semi-groups and resolvents the preceding formulation may seem somewhat odd. In particular when the coefficients do not depend explicitly on t, the modern theories describe the diffusion process generated by L as a Markov family of measures $\{P_x : x \in R^d\}$ whose transition probability function comes from a semigroup which is generated by an extension of L. On the other hand in our formulation each measure P_x or, in the time dependent case, each measure $P_{s,x}$ is described separately, and no mention is made about the connection between one $P_{s,x}$ and the others. There is cause for concern here, because unless one proves uniqueness, there will in general be no nice relation among the solutions starting from various points. (Although, as we will see in Chapter 12, under quite general conditions one can make a "Markov Selection" from among the multiplicity of solutions available for each starting point.) Thus, it should be clear that uniqueness is very vital for our formulation to yield a completely satisfactory theory. Once established, uniqueness will lead to many important consequences. For these reasons, after quickly establishing a reasonably general existence result, we are going to devote much of this chapter to the development of some techniques that will play an important role in establishing uniqueness results.

6.1. Existence

We open this section with an easy construction which will serve us well in the sequel.

6.1.1 Lemma. *Let $s \geq 0$ be given and suppose that P is a probability measure on $(\Omega, \mathcal{M}^s)$, where $\mathcal{M}^s \equiv \sigma(x(t): t \geq s)$. If $\eta \in C([0, s], R^d)$ and $P(x(s) = \eta(s)) = 1$, then there is a unique probability measure $\delta_\eta \otimes_s P$ on $(\Omega, \mathcal{M})$ such that $\delta_\eta \otimes_s P(x(t) = \eta(t), 0 \leq t \leq s) = 1$ and $\delta_\eta \otimes_s P(A) = P(A)$ for all $A \in \mathcal{M}^s$.*

Proof. The uniqueness is obvious. As for existence, let δ_η be the point-mass on $C([0, s], R^d)$ at η (i.e., $\delta_\eta(\{\alpha \in C([0, s], R^d): \alpha(t) = \eta(t), 0 \leq t \leq s\}) = 1$) and let $\Phi: \Omega \to C([s, \infty), R^d)$ be the map defined by $\Phi(\omega)(t) = x(t, \omega), t \geq s$. Clearly Φ is measurable on $(\Omega, \mathcal{M}^s)$, and therefore $P \circ \Phi^{-1}$ is well defined. Define $\tilde{P} = \delta_\eta \times (P \circ \Phi^{-1})$ on $\tilde{X} \equiv C([0, s], R^d) \times C([s, \infty), R^d)$ and set $X = \{(\alpha, \beta) \in C([0, s], R^d) \times C([s, \infty), R^d): \alpha(s) = \beta(s)\}$. Then X is obviously a measurable subset of $\tilde{X}$ and $\tilde{P}(X) \geq \delta_\eta(\{\alpha \in C([0, s], R^d): \alpha(s) = \eta(s)\})P \circ \Phi^{-1}(\{\beta \in C([s, \infty), R^d): \beta(s) = \eta(s)\}) = 1$. Thus $\tilde{P}$ can be restricted to X. Finally, $\Psi: X \to C([0, \infty), R^d)$ defined by

$$\Psi((\alpha, \beta))(t) = \begin{cases} \alpha(t) & \text{if } 0 \leq t < s \\ \beta(t) & \text{if } t > s \end{cases}$$

is clearly a measurable map of X onto $C([0, \infty), R^d)$, and therefore the restriction of $\tilde{P}$ to X determines, via Ψ, a probability measure on $(\Omega, \mathcal{M})$. It is easy to check that this is the desired measure $\delta_\eta \otimes_s P$. $\square$

6.1.2 Theorem. *Let τ be a finite stopping time on Ω. Suppose that $\omega \to Q_\omega$ is a mapping of Ω into probability measures on $(\Omega, \mathcal{M})$ such that*

(i) $\omega \to Q_\omega(A)$ is $\mathcal{M}_\tau$-measurable for all $A \in \mathcal{M}$,
(ii) $Q_\omega(x(\tau(\omega), \cdot) = x(\tau(\omega), \omega)) = 1$ for all $\omega \in \Omega$.

Given a probability measure P on $(\Omega, \mathcal{M})$, there is a unique probability measure $P \otimes_{\tau(\cdot)} Q.$ on $(\Omega, \mathcal{M})$ such that $P \otimes_{\tau(\cdot)} Q.$ equals P on $(\Omega, \mathcal{M}_\tau)$ and $\{\delta_\omega \otimes_{\tau(\omega)} Q_\omega\}$ is a r.c.p.d. of $P \otimes_{\tau(\cdot)} Q. | \mathcal{M}_\tau$. In particular, suppose that $\tau \geq s$ and that $\theta: [s, \infty) \times \Omega \to C$ is a right-continuous, progressively measurable function after time s such that $\theta(t)$ is $P \otimes_{\tau(\cdot)} Q.$-integrable for all $t \geq s$, $(\theta(t \wedge \tau), \mathcal{M}_t, P)$ is a martingale after time s, and $(\theta(t) - \theta(t \wedge \tau(\omega)), \mathcal{M}_t, Q_\omega)$ is a martingale after time s for each $\omega \in \Omega$. Then $(\theta(t), \mathcal{M}_t, P \otimes_{\tau(\cdot)} Q.)$ is a martingale after time s.

Proof. The second assertion is an immediate consequence of Theorem 1.2.10 once we have proved the first part. Moreover, the uniqueness assertion of the first part is obvious. Finally, to prove the existence of $P. \otimes_{\tau(\cdot)} Q.$, it is enough to check that $\omega \to \delta_\omega \otimes_{\tau(\omega)} Q_\omega(A)$ is $\mathcal{M}_\tau$-measurable for all $A \in \mathcal{M}$ and then set

$$R(A) = E^P[\delta. \otimes_{\tau(\cdot)} Q.(A)], \qquad A \in \mathcal{M}.$$

Once it is known that $\omega \to \delta_\omega \otimes_{\tau(\omega)} Q_\omega(A)$ is $\mathcal{M}_\tau$-measurable, the proof that R has the desired properties of $P . \otimes_{\tau(\cdot)} Q .$ is easy. But if $A = \{x(t_1) \in \Gamma_1, \ldots, x(t_n) \in \Gamma_n\}$, where $n \geq 1$, $0 \leq t_1 < \cdots < t_n$, and $\Gamma_1, \ldots, \Gamma_n \in \mathcal{B}_{R^d}$, then

$$\delta_\omega \otimes_{\tau(\omega)} Q_\omega(A) = \chi_{[0, t_1)}(\tau(\omega)) Q_\omega(A)$$

$$+ \sum_{k=1}^{n-1} \chi_{[t_k, t_{k+1})}(\tau(\omega)) \chi_{\Gamma_1}(x(t_1, \omega)) \cdots \chi_{\Gamma_k}(x(t_k, \omega))$$

$$\times Q_\omega(x(t_{k+1}) \in \Gamma_{k+1}, \ldots, x(t_n) \in \Gamma_n)$$

$$+ \chi_{[t_n, \infty)}(\tau(\omega)) \chi_{\Gamma_1}(x(t_1, \omega)) \cdots \chi_{\Gamma_n}(x(t_n, \omega)),$$

and this is clearly $\mathcal{M}_\tau$-measurable. $\square$

We will soon use Theorem 6.1.2 to prove a quite general existence result. But before we do, we want to establish a very useful theorem which complements Theorem 6.1.2.

6.1.3 Theorem. *Let* $a: [s, \infty) \times \Omega \to S_d$ *and* $b: [s, \infty) \times \Omega \to R^d$ *be bounded progressively measurable functions. Suppose that* P *is a probability measure on* $(\Omega, \mathcal{M})$ *and that* $\xi: [s, \infty) \times \Omega \to R^d$ *is a progressively measurable, right continuous P-almost surely continuous function for which* $\xi(\cdot) \sim \mathcal{I}_d^s(a(\cdot), b(\cdot))$ *with respect to* $(\Omega, \mathcal{M}_t, P)$. *Given a stopping time* $\tau \geq s$ *and a r.c.p.d.* $\{Q_\omega\}$ *of* $P \mid \mathcal{M}_\tau$, *there exists a P-null set* $N \in \mathcal{M}_\tau$ *such that* $\xi(\cdot) \sim \mathcal{I}_d^{\tau(\omega)}(a(\cdot), b(\cdot))$ *with respect to* $(\Omega, \mathcal{M}_t, Q_\omega)$ *for each* $\omega \notin N$.

Proof. Let $\textcircled{H} \subseteq C_0^\infty(R^d)$ be a countable, dense subset. For each $f \in \textcircled{H}$, we can apply Theorem 1.2.10 to find a P-null set $N_f \in \mathcal{M}_\tau$ such that

$$X_f^\omega(t) = f(\xi(t, \omega)) - f(\xi(\tau(\omega), \omega)) - \int_{\tau(\omega)}^t (L_u f)(\xi(u, \omega)) \, du$$

is a martingale after $\tau(\omega)$ with respect to $(\Omega, \mathcal{M}_t, Q_\omega)$ for each $\omega \notin N_f$. Define $N = \bigcup_{f \in \textcircled{H}} N_f$. Then N is a P-null set and $N \in \mathcal{M}_\tau$. Moreover, for any $f \in C_0^\infty(R^d)$ and $\omega \notin N$, we can find $\{f_n\}_1^\infty \subseteq \textcircled{H}$ such that $f_n \to f$ in $C_0^\infty(R^d)$ and therefore $X_{f_n}^\omega(t) \to X_f^\omega(t)$ boundedly and point-wise. Hence $(X_f^\omega(t), \mathcal{M}_t, Q_\omega)$ is a martingale after $\tau(\omega)$ for all $f \in C_0^\infty(R^d)$ and $\omega \notin N$. By Theorem 4.2.1, this completes the proof. $\square$

6.1.4 Lemma. *Let* $c: [0, \infty) \to S_d$ *and* $b: [0, \infty) \to R^d$ *be bounded measurable functions and set*

$$L_t = \frac{1}{2} \sum_{i, j=1}^d c^{ij}(t) \frac{\partial^2}{\partial x_i \, \partial x_j} + \sum_{i=1}^d b^i(t) \frac{\partial}{\partial x_i} .$$

Then the martingale problem for L_t is well-posed. Moreover, if $\{P_{s,x}: (s, x) \in [0, \infty) \times R^d\}$ denotes the family of solutions determined by L_t, then $(s, x) \to P_{s,x}(A)$ is measurable for all $A \in \mathcal{M}$.

Proof. To prove the existence of a measurable family of solutions, we could invoke Theorems 5.1.4 and 5.3.2. Moreover, uniqueness could be proved by an application of Theorem 4.5.2 and Corollary 5.1.3. However, we prefer to take a different route.

Define $C(s, t) = \int_s^t c(u) \, du$ and $B(s, t) = \int_s^t b(u) \, du$ for $0 \le s \le t$. Then for all $0 \le s \le t$ and $x \in R^d$ the function

$$\varphi(s, x; t, \theta) = \exp[i\langle \theta, x + B(s, t)\rangle - \tfrac{1}{2}\langle \theta, C(s, t)\theta\rangle]$$

is, as a function of $\theta \in R^d$, the characteristic function of a probability measure $P(s, x; t, \cdot)$. Clearly $(s, x; t) \to P(s, x; t, \cdot)$ is weakly continuous, and so $(s, x; t) \to P(s, x; t, \Gamma)$ is measurable for all $\Gamma \in \mathcal{B}_{R^d}$. Moreover, it is easy to check that

$$P(s, x; t_2, \Gamma) = \int P(t_1, y; t_2, \Gamma)P(s, x; t_1, dy)$$

for $s \le t_1 < t_2$, since

$$\varphi(s, x; t_2, \theta) = \varphi(s, x; t_1, \theta)\varphi(t_1, 0; t_2, \theta).$$

We can now prove the uniqueness assertion. Given $(s, x) \in [0, \infty) \times R^d$ and a solution P starting from (s, x), we have, from Theorem 4.2.1, that

$$E^P[\exp(i\langle \theta, x(t_2)\rangle)|\mathcal{M}_{t_1}] = \varphi(t_1, x(t_1); t_2, \theta) \qquad \text{(a.s., } P)$$

for all $s \le t_1 \le t$ and $\theta \in R^d$. Hence, if $f \in B(R^d)$, then

$$E^P[f(x(t_2))|\mathcal{M}_{t_1}] = \int f(y)P(t_1, x(t_1); t_2, dy) \qquad \text{(a.s., } P)$$

for $s \le t_1 < t_2$. Working by induction on n, one can easily conclude that if $0 \le u_1 < \cdots < u_m = s < t_1 < \cdots < t_n$ and $g_1, \ldots, g_m, f_1, \ldots, f_n \in B(R^d)$, then

$$E^P[g_1(x(u_1)) \cdots g_m(x(u_m))f_1(x(t_1)) \cdots f_n(x(t_n))]$$

$$= g(x) \cdots g_m(x) \int_{\Gamma_1 \times \cdots \times \Gamma_m} \cdots \int f_1(y_1)$$

$$\cdots f_n(y_n)P(s, x; t_1, dy_1) \cdots P(t_{n-1}, y_{n-1}; t_n, dy_n).$$

This equation not only demonstrates that P is unique, it also guarantees measurability as a function of (s, x). Thus it only remains to prove existence. But this is easy, since $P(s, x; t, \cdot)$ certainly satisfies the conditions of Theorem 2.2.4; and if $\{P_{s, x} : (s, x) \in [0, \infty) \times R^d\}$ is the corresponding Markov family constructed in that theorem, then

$$E^{P_{s, x}}\left[\exp\left[i\left\langle \theta, x(t_2)\right\rangle - \int_s^{t_2} b(u)\, du\right\rangle + \frac{1}{2}\int_s^{t_2} \langle \theta, c(u)\theta\rangle\, du\right]\Big| \mathcal{M}_{t_1}\right]$$

$$= \varphi(t_1, x(t_1); t_2, \theta)\exp[-i\langle\theta, B(s, t_2)\rangle + \tfrac{1}{2}\langle\theta, C(s, t_2)\theta\rangle]$$

$$= \exp\left[i\left\langle \theta, x(t_1)\right\rangle - \int_s^{t_1} b(u)\, du\right\rangle + \frac{1}{2}\int_s^{t_1} \langle\theta, c(u)\theta\rangle\, du\right]$$

$$(\text{a.s., } P_{s, x})$$

for all $s \leq t_1 \leq t_2$ and $\theta \in R^d$. Hence, by Theorem 4.2.1, $P_{s, x}$ is a solution starting from (s, x). $\square$

We will use $\mathcal{W}_{s, x}^{(c, b)}$ to denote the unique solution for $c(t)$ and $b(t)$ starting from (s, x). In the case when $b \equiv 0$, we will use $\mathcal{W}_{s, x}^{(c)}$ instead of $\mathcal{W}_{s, x}^{(c, 0)}$.

6.1.5 Lemma. *Let $a: [0, \infty) \times \Omega \to S_d$ and $b: [0, \infty) \times \Omega \to R^d$ be bounded progressively measurable function such that there exists an $n \geq 1$ for which $a(t)$ and $b(t)$ are $\mathcal{M}_{[nt]/n}$-measurable for all $t \geq 0$. Then for each probability measure μ on R^d there is a unique probability measure P on $(\Omega, \mathcal{M})$ such that $P(x(0) \in \Gamma) = \mu(\Gamma)$ for $\Gamma \in \mathcal{B}_{R^d}$ and $x(\cdot) \sim \mathcal{I}_d^0(a(\cdot), b(\cdot))$ with respect to $(\Omega, \mathcal{M}_t, P)$. Moreover, if for $k \geq 0$ and $\omega \in \Omega$ the element ω_k of Ω is determined by $x(t, \omega_k) = x(t \wedge k/n, \omega)$, $t \geq 0$, and if $\{P_\omega^k\}$ is a r.c.p.d. of $P|\mathcal{M}_{k/n}$, then for P-almost every $\omega \in \Omega$, P_ω^k equals $\delta_\omega \otimes_{k/n} \mathcal{W}_{k/n, x(k/n, \omega)}^{(a(\cdot, \omega_k), b(\cdot, \omega_k))}$ on $\mathcal{M}_{k+1/n}$.*

Proof. Define $P^{(0)}$ on $(\Omega, \mathcal{M})$ by

$$P^{(0)}(A) = \int \delta_x(A)\mu(dx), \qquad A \in \mathcal{M},$$

where δ_x is the measure on $(\Omega, \mathcal{M})$ such that $\delta_x(x(t) = x, t \geq 0) = 1$. For $k \geq 0$, define $a_k(\cdot, \omega) = a(\cdot, \omega_k)$, $b_k(\cdot, \omega) = b(\cdot, \omega_k)$, and

$$P^{(k+1)} = P^{(k)} \otimes_{k/n} \mathcal{W}_{k/n, x(k/n, \cdot)}^{(a_k(t, \cdot), b_k(t, \cdot))}.$$

By Theorem 1.3.5, there is a unique P on $(\Omega, \mathcal{M})$ such that P equals $P^{(k)}$ on $\mathcal{M}_{k/n}$ for all $k \geq 0$. Given $f \in C_0^\infty(R^d)$, set

$$\theta(t) = f(x(t)) - \int_0^t L_u f(x(u))\, du$$

where

$$L_t = \frac{1}{2} \sum_{i,j=1}^{d} a^{ij}(t, \cdot) \frac{\partial^2}{\partial x_i \, \partial x_j} + \sum_{i=1}^{d} b^i(t, \cdot) \frac{\partial}{\partial x_i}.$$

Certainly $(\theta(t \wedge 0), \mathcal{M}_t, P^{(0)})$ is a martingale. Assume that $(\theta(t \wedge k/n), \mathcal{M}_t, P^{(k)})$ is a martingale. Since

$$\left(\theta\left(t \wedge \frac{k+1}{n} \right) - \theta\left(t \wedge \frac{k}{n} \right), \mathcal{M}_t, \mathcal{W}_{k/n, \, x(k/n, \, \omega)}^{(a_k(t, \, \omega), \, b_k(t, \, \omega))} \right)$$

is a martingale for all ω, it follows from Theorem 6.1.2 that $(\theta(t \wedge [(k+1)/n]), \mathcal{M}_t,$ $P^{(k+1)})$ is a martingale. Thus, by induction, $(\theta(t), \mathcal{M}_t, P)$ is a martingale, and so we have completed the proof of existence.

Clearly the uniqueness will follow if we can prove the last assertion. Given P with the prescribed properties and $k \geq 0$, define

$$Q = P \otimes_{(k+1)/n} \mathcal{W}_{(k+1)/n, \, x((k+1)/n, \, \cdot)}^{(a_k(t, \, \cdot), \, b_k(t, \, \cdot))}.$$

By Theorem 6.1.2, we see that $x(\cdot) \sim \mathscr{I}_d^0(a_k(\cdot), b_k(\cdot))$ with respect to $(\Omega, \mathcal{M}_t, Q)$. Thus by Theorem 6.1.3, if $\{Q_\omega\}$ is a r.c.p.d. of $Q \mid \mathcal{M}_{k/n}$, then, for Q-almost all $\omega, x(\cdot) \sim \mathscr{I}_d^{k/n}(a_k(\cdot, \omega), b_k(\cdot, \omega))$. Hence, by Theorem 6.1.3, for Q-almost all ω,

$$Q_\omega = \delta_\omega \otimes_{k/n} \mathcal{W}_{k/n, \, x(k/n, \, \omega)}^{(a_k(t, \, \omega), \, b_k(t, \, \omega))}.$$

Since P equals Q on $\mathcal{M}_{k+1/n}$, this finishes the proof. $\quad\square$

6.1.6 Theorem. *Let* $a: [0, \infty) \times \Omega \to S_d$ *and* $b: [0, \infty) \times \Omega \to R^d$ *be bounded progressively measurable functions. If for each* $t \geq 0$, $a(t, \cdot)$ *and* $b(t, \cdot)$ *are continuous, then for each probability measures* μ *on* R^d *there is a* P *on* $(\Omega, \mathcal{M})$ *such that* $P(x(0) \in \Gamma) = \mu(\Gamma)$, $\Gamma \in \mathscr{B}_{R^d}$, *and* $x(\cdot) \sim \mathscr{I}_d^0(a(\cdot), b(\cdot))$ *with respect to* $(\Omega, \mathcal{M}_t, P)$.

Proof. For each $n \geq 1$, define $a_n: [0, \infty) \times \Omega \to S_d$ and $b_n: [0, \infty) \times \Omega \to R^d$ by

$$a_n(t, \omega) = a(t, \tilde{\omega}) \quad \text{and} \quad b_n(t, \omega) = b(t, \tilde{\omega})$$

where $\tilde{\omega} \in \Omega$ is determined by $x(u, \tilde{\omega}) = x(u \wedge [nt]/n, \omega)$, $u \geq 0$. Then a_n and b_n satisfy the conditions of Lemma 6.1.4, and so we can find a P_n on $(\Omega, \mathcal{M})$ such that $P_n(x(0) \in \Gamma) = \mu(\Gamma)$, $\Gamma \in \mathscr{B}_{R^d}$, and $x(\cdot) \sim \mathscr{I}_d^0(a_n(\cdot), b_n(\cdot))$ with respect to $(\Omega, \mathcal{M}_t, P_n)$. Moreover, by Theorem 1.4.6, $\{P_n: n \geq 1\}$ is relatively weakly compact on $(\Omega, \mathcal{M})$. Let P be any limit point of $\{P_n: n \geq 1\}$ and let $\{P_{n'}\}$ be any subsequence such that $P_{n'} \to P$. Clearly

$$\int f(y)\mu(dy) = \lim_{n' \to \infty} E^{P_{n'}}[f(x(0))] = E^P[f(x(0))]$$

for all $f \in C_b(R^d)$, and so P satisfies the correct initial condition. Moreover, if $f \in C_0^\infty(R^d)$, then

$$f(x(t)) - \int_0^t L_u^{(n)} f(x(u))\, du$$

is a P_n-martingale, where

$$L_t^{(n)} = \frac{1}{2} \sum_{i,\,j=1}^d a_n^{ij}(t, \cdot) \frac{\partial^2}{\partial x_i\, \partial x_j} + \sum_{i=1}^d b_n^i(t, \cdot) \frac{\partial}{\partial x_i}.$$

Since for each $t \geq 0$, the functions

$$\int_0^t L_u^{(n)} f(x(u))\, du$$

are equicontinuous at every point of Ω, bounded independently of n, and tend to $\int_0^t L_u f(x(u))\, du$ as $n \to \infty$, we can apply Corollary 1.1.5 to prove that

$$E^P\left[\left(f(x(t)) - \int_0^t L_u f(x(u))\, du\right)\Phi\right]$$

$$= \lim_{n' \to \infty} E^{P_{n'}}\left[\left(f(x(t)) - \int_0^t L_u^{(n')} f(x(u))\, du\right)\Phi\right].$$

for all $t \geq 0$ and bounded continuous Φ. In particular, if $0 \leq t_1 < t_2$ and Φ is a bounded continuous $\mathcal{M}_{t_1}$-measurable function, then we have:

$$E^P\left[\left(f(x(t_2)) - \int_0^{t_2} L_u f(x(u))\, du\right)\Phi\right]$$

$$= E^P\left[\left(f(x(t_1)) - \int_0^{t_1} L_u f(x(u))\, du\right)\Phi\right],$$

since this equality certainly holds when P and L_u are replaced by $P_{n'}$ and $L_u^{(n')}$, respectively. It is now easy to see that this equality persists when Φ is simply bounded and $\mathcal{M}_{t_1}$-measurable, and so we have shown that $x(\cdot) \sim \mathscr{I}_d^0(a(\cdot), b(\cdot))$ with respect to $(\Omega, \mathcal{M}_t, P)$. $\square$

6.1.7 Theorem. *Let* $a: [0, \infty) \times R^d \to S_d$ *and* $b: [0, \infty) \times R^d \to R^d$ *be bounded measurable functions such that* $a(t, \cdot)$ *and* $b(t, \cdot)$ *are continuous for all* $t \geq 0$. *For each* $s \geq 0$ *and each probability measure* μ *on* R^d, *there is a probability measure* P *on* $(\Omega, \mathcal{M})$ *such that*

$$P(x(t) \in \Gamma, 0 \leq t \leq s) = \mu(\Gamma), \qquad \Gamma \in \mathscr{B}_{R^d},$$

and $x(\cdot) \sim \mathscr{I}_d^s(a(\cdot, x(\cdot)), b(\cdot, x(\cdot)))$ *with respect to* $(\Omega, \mathscr{M}_t, P)$. *In particular, the martingale problem for a and b has at least one solution starting from each* $(s, x) \in [0, \infty) \times R^d$.

Proof. Clearly we need only prove the first assertion. To this end, let s and μ be given and define a_s and b_s by $a_s(t, x) = a(t + s, x)$ and $b_s(t, x) = b(t + s, x)$. By Theorem 6.1.6, we can find a Q on $(\Omega, \mathscr{M})$ with the properties that

$$Q(x(0) \in \Gamma) = \mu(\Gamma), \qquad \Gamma \in \mathscr{B}_{R^d},$$

and $x(\cdot) \sim \mathscr{I}_d^0(a_s(\cdot, x(\cdot)), b_s(\cdot, x(\cdot)))$ with respect to $(\Omega, \mathscr{M}_t, Q)$. Therefore, if we now define $\Phi_s: \Omega \to \Omega$ by $x(t, \Phi_s\omega) = x((t - s) \vee 0, \omega)$, $t \geq 0$, and set $P = Q \circ \Phi_s^{-1}$, then it is easily seen that P has the desired properties. $\square$

6.2. Uniqueness: Markov Property

We now have our basic existence result. Unfortunately, the corresponding uniqueness theorem is not so easy to prove; in fact, it is not even true. The purpose of the present section is to start laying the groundwork for a reduction procedure with which we will eventually be able to obtain a reasonably satisfactory uniqueness theorem. At the same time, the preparations that we make here will serve us well when it comes time to discuss the consequences of uniqueness.

In order to appreciate what is going on in this section, it is helpful to think about how one might try proving uniqueness. Given two probability measures P and Q on $(\Omega, \mathscr{M})$, one knows that they are equal if and only if

$$P(x(t_1) \in \Gamma_1, \ldots, x(t_n) \in \Gamma_n) = Q(x(t_1) \in \Gamma_1, \ldots, x(t_n) \in \Gamma_n)$$

for all $n \geq 1$, $0 \leq t_1 < \cdots < t_n$, and $\Gamma_1, \ldots, \Gamma_n \in \mathscr{B}_{R^d}$. That is, if the *finite dimensional marginal distributions* determined by P and Q agree. If one knows *a priori* that P and Q are Markovian in the sense that for any $0 \leq t_1 < t_2$ the conditional distribution of $x(t_2)$ given $\mathscr{M}_{t_1}$ is a function of t_1, t_2, and $x(t_1)$ alone and if this function is the same for P and Q, then $P = Q$ if $x(0)$ has the same distribution under P and Q. Of course this line of reasoning cannot be applied without the *a priori* knowledge that P and Q are Markovian. Nonetheless, a variant of it is applicable to the study of martingale problems; and it is this variant which we now want to develop.

6.2.1 Theorem. *Let* $a: [0, \infty) \times R^d \to S_d$ *and* $b: [0, \infty) \times R^d \to R^d$ *be bounded measurable functions and suppose that* P *on* $(\Omega, \mathscr{M})$ *has the properties that* $P(x(\sigma) = y) = 1$ *and* $x(\cdot) \sim \mathscr{I}_d^\sigma(a(\cdot, x(\cdot)), b(\cdot, x(\cdot)))$ *under* P *for some* $\sigma \geq 0$ *and* $y \in R^d$. *Then for any* $\eta \in C([0, \infty); R^d)$ *satisfying* $\eta(\sigma) = y, x(\cdot) \sim \mathscr{I}_d^\sigma(a(\cdot, x(\cdot)), b(\cdot, x(\cdot)))$ *under* $\delta_\eta \otimes_\sigma P$. *In particular, if* $x(\cdot) \sim \mathscr{I}_d^s(a(\cdot, x(\cdot)), b(\cdot, x(\cdot)))$ *under* P *and if* $\{P_\omega\}$ *is a r.c.p.d. of* $P|\mathscr{M}_\tau$ *where* $\tau \geq s$ *is a finite stopping time, then there is a P-null set*

$N \in \mathcal{M}_\tau$ such that $\delta_{x(\tau(\omega)),\,\omega} \otimes_{\tau(\omega)} P_\omega$ solves the martingale problem for a and b starting from $(\tau(\omega), x(\tau(\omega), \omega))$ whenever $\omega \notin N$. (Here, and in the similar contexts in the
future, δ_x is the point-mass on Ω concentrated on the path which stays at x for all time).

Proof. In view of Theorem 6.1.3, the second part is an immediate consequence of
the first part. To prove the first assertion, let $f \in C_0^\infty(R^d)$ and set

$$\theta(t) = f(x(t)) - f(x(t \wedge \sigma)) - \int_{t \wedge \sigma}^t L_u f(x(u))\, du,$$

where L_u is defined in terms of a and b by (0.4). Certainly $(\theta(t \wedge \sigma), \mathcal{M}_t, \delta_\eta)$ is a
martingale, and, by assumption $(\theta(t) - \theta(t \wedge \sigma), \mathcal{M}_t, P)$ is a martingale. Hence, by
Theorem 6.1.2, $(\theta(t), \mathcal{M}_t, \delta_\eta \otimes_\sigma P)$ is a martingale. But this means that

$$f(x(t)) - \int_\sigma^t L_u f(x(u))\, du$$

is a $\delta_\eta \otimes_\sigma P$-martingale after time σ; and so the proof is complete. $\square$

One immediate application of Theorem 6.2.1 is the following.

6.2.2 Theorem. *Let* $a: [0, \infty) \times R^d \to S_d$ *and* $b: [0, \infty) \times R^d \to R^d$ *be bounded measurable functions for which the martingale problem is well-posed, and assume that the
corresponding family of solutions* $\{P_{s,\,x}: (s, x) \in [0, \infty) \times R^d\}$ *is measurable (i.e.,*
$(s, x) \to P_{s,\,x}(A)$ *is measurable for all* $A \in \mathcal{M}$). *Then* $\{P_{s,\,x}: (s, x) \in [0, \infty) \times R^d\}$ *is
strong Markov in the sense that if* $\tau \geq s$ *is a finite stopping time then*
$\{\delta_\omega \otimes_{\tau(\omega)} P_{\tau(\omega),\,x(\tau(\omega),\,\omega)}\}$ *is a r.c.p.d. of* $P_{s,\,x} | \mathcal{M}_\tau$.

Proof. By Theorem 6.2.1 and uniqueness, we know that if $\{P_\omega\}$ is a r.c.p.d. of
$P_{s,\,x} | \mathcal{M}_\tau$, then $\delta_{x(\tau(\omega)),\,\omega} \otimes_{\tau(\omega)} P_\omega = P_{\tau(\omega),\,x(\tau(\omega),\,\omega)}$ for all ω outside of $P_{s,\,x}$-null set
$N \in \mathcal{M}_\tau$. Thus

$$P_\omega = \delta_\omega \otimes_{\tau(\omega)} (\delta_{x(\tau(\omega)),\,\omega} \otimes_{\tau(\omega)} P_\omega)$$

$$= \delta_\omega \otimes_{\tau(\omega)} P_{\tau(\omega),\,x(\tau(\omega),\,\omega)}$$

for all $\omega \notin N$. $\square$

It is worth remarking that the measurability condition on $\{P_{s,\,x}: (s, x) \in
[0, \infty) \times R^d\}$ in Theorem 6.2.2 is not really necessary. Indeed, the above proof
shows that there is a $P_{s,\,x}$-null set $N \in \mathcal{M}_\tau$ such that $P_\omega = \delta_\omega \otimes_{\tau(\omega)} P_{\tau(\omega),\,x(\tau(\omega),\,\omega)}$ off
of N; and this is all that is important. Moreover, it is shown in Exercise 6.7.4 below
that $\{P_{s,x}: (s, x) \in [0, \infty) \times R^d\}$ is necessarily measurable if the martingale problem
is well-posed. However, the proof of this fact requires the application of a rather
hard theorem about Polish spaces and there is no circumstance in which we have
been able to prove uniqueness without verifying measurability at the same time.
We will therefore avoid using Exercise 6.7.4 whenever possible.

We now turn to the result, alluded to earlier, which will help us to prove uniqueness criteria. Before stating this result, we recall the notion of a *determining class of functions*. Namely, $\Phi \subseteq B(R^d)$ is said to be determining if whenever two probability measure μ_1 and μ_2 on R^d satisfy

$$\int \varphi \, d\mu_1 = \int \varphi \, d\mu_2$$

for all $\varphi \in \Phi$ the measures μ_1 and μ_2 must be equal.

6.2.3 Theorem. *Let* $a: [0, \infty) \times R^d \to S_d$ *and* $b: [0, \infty) \times R^d \to R^d$ *be bounded measurable functions. Suppose that for every* $(s, x) \in [0, \infty) \times R^d$, $t \geq s$, *and* φ *from a given determining class* Φ,

$$E^{P^1}[\varphi(x(t))] = E^{P^2}[\varphi(x(t))]$$

whenever P^1 *and* P^2 *both solve the martingale problem for* a *and* b *starting from* (s, x). *Then starting from each* (s, x) *there is at most one solution to the martingale problem for* a *and* b.

Proof. Let P^1 and P^2 both be solutions starting from (s, x). Certainly P^1 and P^2 agree on $\mathcal{M}_s$; and therefore, since $\mathcal{M}_s$ is independent of $\sigma(x(t): t \geq s)$ under both P^1 and P^2, it suffices to check that

(2.1) $$E^{P^1}[f_1(x(t_1)) \cdots f_n(x(t_n))] = E^{P^2}[f_1(x(t_1)) \cdots f_n(x(t_n))]$$

for all $n \geq 1$, $s \leq t_1 < \cdots < t_n$, and $f_1, \ldots, f_n \in B(R^d)$.

Certainly (2.1) is true if $n = 1$. Next, assume it is true for n. Set $\tilde{\mathcal{M}} = \sigma(x(t_j)$: $1 \leq j \leq n)$ and let $\{\tilde{P}^i\}$ be a r.c.p.d. of $P^i | \tilde{\mathcal{M}}$, $i = 1, 2$. We now show that there is an $\tilde{N} \in \tilde{\mathcal{M}}$ such that $P^1(\tilde{N}) = P^2(\tilde{N}) = 0$ and, for $\omega \notin \tilde{N}$, $\delta_{x(t_n, \omega)} \otimes_{t_n} \tilde{P}^i_\omega$ solves the martingale problem for a and b starting from $(t_n, x(t_n, \omega))$, $i = 1, 2$. To this end, let $\{P^i_\omega\}$ be a r.c.p.d. of $P^i | \mathcal{M}_{t_n}$. Then it is easy to check that $\int P^i(\cdot)\tilde{P}^i_\omega(d\omega')$ is a r.c.p.d. of $P^i | \tilde{\mathcal{M}}$, and so

(2.2) $$\tilde{P}^i_\omega = \int P^i_{\omega'}(\cdot)\tilde{P}^i_\omega(d\omega')$$

for ω not in a P^i-null set $\tilde{A}_i \in \tilde{\mathcal{M}}$. Next, choose a P^i-null set $N_i \in \mathcal{M}_{t_n}$ so that $x(\cdot) \sim \mathcal{S}^{t_n}_d(a(\cdot, x(\cdot)), b(\cdot, x(\cdot)))$ under P^i_ω for all $\omega \notin N_i$. Then there is a P^i-null set $\tilde{B}_i \in \tilde{\mathcal{M}}$ such that $\tilde{P}^i_\omega(N_i) = 0$ for $\omega \notin \tilde{B}_i$. It follows, from (2.2), that if $\omega \notin \tilde{N}_i = \tilde{A}_i \cup \tilde{B}_i$, then $x(\cdot) \sim \mathcal{S}^{t_n}_d(a(\cdot, x(\cdot)), b(\cdot, x(\cdot)))$ under $\tilde{P}^i_\omega$. But, by induction hypothesis, P^1 equals P^2 on $\tilde{\mathcal{M}}$, and therefore $P^1(\tilde{N}_2) = P^2(\tilde{N}_2) = 0$ and $P^2(\tilde{N}_1) = P^1(\tilde{N}_1) = 0$. Thus, if $\tilde{N} = \tilde{N}_1 \cup \tilde{N}_2$, then $P^i(\tilde{N}) = 0$ and, by Theorem

6.2.1, $\delta_{x(t_n, \omega)} \otimes_{t_n} \tilde{P}^i_\omega$ solves the martingale problem for a and b starting from $(t_n, x(t_n, \omega))$ for all $\omega \notin \tilde{N}$. We therefore have:

$$E^{P^1_\omega}[f_{n+1}(x(t_{n+1}))] = E^{\delta_{x(t_n, \omega)} \otimes_{t_n} P^1_\omega}[f_{n+1}(x(t_{n+1}))]$$

$$= E^{\delta_{x(t_n, \omega)} \otimes_{t_n} P^2_\omega}[f_{n+1}(x(t_{n+1}))]$$

$$= E^{P^2_\omega}[f_{n+1}(x(t_{n+1}))]$$

for all $\omega \notin \tilde{N}$; and so there is a bounded $\tilde{\mathcal{M}}$-measurable function H such that

$$H = E^P[f_{n+1}(x(t_{n+1})) | \tilde{\mathcal{M}}] \qquad \text{(a.s., } P^i\text{)}$$

for $i = 1$ and 2. Since P^1 equals P^2 on $\tilde{\mathcal{M}}$, this completes the induction. $\square$

6.2.4 Corollary. *Let* $a: [0, \infty) \times R^d \to S_d$ *and* $b: [0, \infty) \times R^d \to R^d$ *be bounded measurable functions. Suppose that for all* $(s, x) \in [0, \infty) \times R^d$ *and any pair of solutions* P^1 *and* P^2 *to the martingale problem for* a *and* b *starting from* (s, x), *one of the following holds:*

(i) $E^{P^1}[\int_s^T \varphi(t, x(t)) \, dt] = E^{P^2}[\int_s^T \varphi(t, x(t)) \, dt]$ *for all* $T > s$ *and* φ *in some determining class on* $[s, T) \times R^d$,

(ii) $E^{P^1}[\int_s^\infty e^{-\lambda t}\varphi(x(t)) \, dt] = E^{P^2}[\int_s^\infty e^{-\lambda t}\varphi(x(t)) \, dt]$ *for all* $\lambda > 0$ *and* φ *in some determining class on* R^d.

Then the solution is unique for each starting point (s, x).

Proof. From either (i) or (ii) it is easy to see that the conditions of Theorem 6.2.3 are met. $\square$

6.2.5 Corollary. *Let* $a: [0, \infty) \times R^d \to S_d$ *and* $b: [0, \infty) \times R^d \to R^d$ *be bounded measurable functions and assume that the martingale problem for* a *and* b *has at least one solution starting from each* $(s, x) \in [0, \infty) \times R^d$. *Suppose that* $\Phi \subseteq B(R^d)$ *has the property that it generates* $B(R^d)$ *under bounded pointwise convergence; and assume that for each* $T > 0$ *and* $\varphi \in \Phi$ *there is an* $f_{T, \varphi} \in B([0, T] \times R^d)$ *such that*

$$f_{T, \varphi}(s, x) = E^P[\varphi(x(T))]$$

whenever P *solves the martingale problem for* a *and* b *starting from* $(s, x) \in [0, T) \times R^d$. *Then the martingale problem for* a *and* b *is well-posed, and the associated family of solutions is measurable.*

Proof. Clearly uniqueness follows from Theorem 6.2.3. Let $\{P_{s, x}: (s, x) \in [0, \infty) \times R^d\}$ denote the family of solutions determined by a and b. We first note that $(s, x) \to E^{P_{s,x}}[\psi(x(T))]$ is measurable on $[0, T) \times R^d$ for all $T > 0$ and $\psi \in B(R^d)$. Indeed, this is true by hypothesis if $\psi \in \Phi$, and the class of $\psi \in B(R^d)$ for which it is true is closed under bounded point-wise convergence. Thus we can, and will, assume that $\Phi = B(R^d)$.

The rest of the proof is easy. Let $n \geq 1$, $0 = t_0 < \cdots < t_n \leq s$, and $\varphi_0, \ldots,$ $\varphi_n \in B(R^d)$ be given. Certainly

$$(s, x) \to E^{P_{s,x}}[\varphi_0(x(t_0)) \cdots \varphi_n(x(t_n))] = \varphi_0(x) \cdots \varphi_n(x)$$

is measurable on $[t_n, \infty) \times R^d$. Next suppose that $t_m \leq s < t_{m+1}$ for some $0 \leq m < n$. Define

$$\psi_n(\cdot) = \varphi_n(\cdot)$$

and

$$\psi_k(\cdot) = \varphi_k(\cdot) f_{t_{k+1}, \psi_{k+1}}(t_k, \cdot)$$

for $0 \leq k < n$. Then by repeated applications of Theorem 6.2.1,

$$E^{P_{s,x}}[\varphi_0(x(t_0)) \cdots \varphi_n(x(t_n))] = \varphi_0(x) \cdots \varphi_m(x) f_{t_{m+1}, \psi_{m+1}}(s, x)$$

which explicitly gives

$$(s, x) \to E^{P_{s,x}}[\varphi_0(x(t_0)) \cdots \varphi_n(x(t_n))]$$

as a measurable function on $[t_m, t_{m+1}) \times R^d$. $\square$

The following variant of Corollary 6.2.5 is the form in which we will find it to be most useful.

6.2.6 Corollary. *Let* $a: [0, \infty) \times R^d \to S_d$ *and* $b: [0, \infty) \times R^d \to R^d$ *be bounded measurable functions and assume that the martingale problem for* a *and* b *has at least one solution starting at each* $(s, x) \in [0, \infty) \times R^d$. *Further, assume that for each* $0 \leq s < T$ *and* $\varphi \in C_0^\infty([s, T] \times R^d)$ *there is an* $f_{T, \varphi} \in B([0, T) \times R^d)$ *such that*

$$f_{T, \varphi}(s, x) = E^P\left[\int_s^T \varphi(t, x(t)) \, dt\right]$$

whenever P *solves the martingale problem for* a *and* b *starting from* $(s, x) \in R^d$. *Then the martingale problem for* a *and* b *is well-posed and the associated family of solutions is measurable.*

6.3. Uniqueness: Some Examples

Before continuing with the preparations for our main uniqueness theorem, it is important to recognize that the contents of Chapters 3 and 5 provide us with some highly non-trivial examples of well-posed martingale problems. Indeed, when dealing with degenerate coefficients (i.e. if the matrix $a(s, x)$ is allowed to be singular at some point (s, x)), then the uniqueness theorems obtained by these methods are not covered by the one toward which we are working. On the other hand, it must be admitted that when Chapters 3 or 5 are applicable, the martingale formulation does not add greatly to our understanding, although even here it does provide a unifying principle.

6.3.1 Lemma. *Let* $a\colon [s, \infty) \times \Omega \to S_d$ *and* $b\colon [s, \infty) \times \Omega \to R^d$ *be bounded progressively measurable functions and suppose that* P *is a probability measure on* $(\Omega, \mathcal{M})$ *such that* $x(\cdot) \sim \mathcal{I}_a^s(a(\cdot), b(\cdot))$ *under* P. *Let* $\mathcal{G} \subseteq [s, \infty) \times R^d$ *be an open set and define* $\tau = \inf\{t \geq s\colon (t, x(t)) \notin \mathcal{G}\}$. *Then for any* $f \in C_b^{1, 2}(\mathcal{G}) \cap C_b(\mathcal{G})$

$$f(t \wedge \tau, x(t \wedge \tau)) - \int_s^{t\wedge\tau} \left(\frac{\partial}{\partial u} + L_u\right) f(u, x(u)) du$$

is a P-*martingale after time* s, *where* L_u *is defined in terms of* $a(\cdot)$ *and* $b(\cdot)$ *as in Chapter 4.*

Proof. Let $\{\mathcal{H}_n\}_1^\infty$ be a sequence of bounded open subsets of $\mathcal{G}$ such that $\bar{\mathcal{H}}_n \subseteq \mathcal{H}_{n+1}, n \geq 1$, and $\mathcal{G} = \bigcup_1^\infty \mathcal{H}_n$. By standard real-variable techniques, we can find $\{\varphi_n\}_1^\infty \subseteq C_0^\infty([0, \infty) \times R^d)$ such that $\chi_{\mathcal{H}_n} \leq \varphi_n \leq \chi_{\mathcal{H}_{n+1}}$. Define $f_n = \varphi_n \cdot f$ and $\tau_n = \inf\{t \geq s\colon (t, x(t)) \notin \mathcal{H}_n\}$. Then

$$f_n(t, x(t)) - \int_s^t \left(\frac{\partial}{\partial u} + L_u\right) f_n(u, x(u)) du$$

is a P-martingale after time s, and so is

$$f_n(t \wedge \tau_n, x(t \wedge \tau_n)) - \int_s^{t \wedge \tau_n} L_u f_n(u, x(u))\, du.$$

But

$$f_n(t \wedge \tau_n, x(t \wedge \tau_n)) = f(t \wedge \tau_n, x(t \wedge \tau_n))$$

and

$$\left(\frac{\partial}{\partial u} + L_u\right) f_n(u, x(u)) = \left(\frac{\partial}{\partial u} + L_u\right) f(u, x(u)), \quad s \leq u \leq \tau_n.$$

Thus

$$f(t \wedge \tau_n, x(t \wedge \tau_n)) - \int_s^{t\wedge\tau_n} \left(\frac{\partial}{\partial u} + L_u\right) f(u, x(u)) du$$

is a P-martingale. Finally, $\tau_n \nearrow \tau$, and so the proof of the lemma is completed by an obvious limit argument. $\square$

6.3.2 Theorem. *Let $a: [0, \infty) \times R^d \to S_d$ and $b: [0, \infty) \times R^d \to R^d$ be bounded measurable functions and define L_t accordingly. Suppose that one of the following conditions holds:*

(i) *for each $T > 0$ and $\varphi \in C_0^\infty(R^d)$ there exists a function $f \in B([0, T] \times R^d)$ with $f(T, \cdot) = \varphi(\cdot)$ and a sequence $\{f_n\} \subseteq C_b^{1,2}([0, T] \times R^d) \cap C_b([0, T] \times R^d)$ such that $f_n \to f$ on $[0, T] \times R^d$ and $(\partial f_n/\partial t) + L_t f_n \to 0$ on $[0, T] \times R^d$ boundedly and point-wise;*

(ii) *for each $T > 0$ and $\varphi \in C_0^\infty([0, T) \times R^d)$ there exists a function $f \in B([0, T] \times R^d)$ with $f(T, \cdot) = 0$ and a sequence $\{f_n\} \subseteq C_b^{1,2}([0, T) \times R^d) \cap C_b([0, T] \times R^d)$ such that $f_n \to f$ on $[0, T] \times R^d$ and $(\partial f_n/\partial t) + L_t f_n \to \varphi$ on $[0, T) \times R^d$ boundedly and point-wise.*

Then for each $(s, x) \in [0, \infty) \times R^d$ there is at most one solution to the martingale problem for a and b starting from (s, x).

In fact, if P solves starting from (s, x) and (i) obtains, then

$$(3.1) \qquad\qquad f(s, x) = E^P[\varphi(x(T))];$$

and if (ii) obtains

$$(3.2) \qquad\qquad f(s, x) = E^P\left[\int_s^T \varphi(t, x(t))\, dt\right].$$

In particular, if (i) or (ii) holds and, in addition, there is a solution starting from each $(s, x) \in [0, T] \times R^d$, then the martingale problem for a and b is well-posed and the associated family of solutions is measurable.

Proof. In view of Theorem 6.2.3 and Corollaries 6.2.5 and 6.2.6, we need only show that (i) implies (3.1) and that (ii) implies (3.2). Since the proofs are so similar, we will only show that (i) implies (3.1).

Let $T > 0$ and $\varphi \in C_0(R^d)$ be given. Choose f and $\{f_n\}_1^\infty$ accordingly. By Lemma 6.3.1, with $\mathscr{G} = [0, T) \times R^d$,

$$f_n(t \wedge T, x(t \wedge T)) - \int_s^{t \wedge T} \left(\frac{\partial}{\partial u} + L_u\right) f_n(u, x(u))\, du$$

is a P-martingale after time s. Thus

$$E^P[f_n(T, x(T))] - f_n(s, x) = E^P\left[\int_s^T \left(\frac{\partial f_n}{\partial u} + L_u f_n\right)(u, x(u))\, du\right].$$

Upon letting $n \to \infty$, we get (3.1). $\square$

6.3.3 Corollary. *Let $a: [0, \infty) \times R^d \to S_d$ and $b: [0, \infty) \times R^d \to R^d$ be bounded continuous functions having two bounded continuous spatial derivatives and define L_t accordingly. Then the martingale problem for a and b is well-posed and the asso-*

ciated family of solutions $\{P_{s,x}: (s, x) \in [0, \infty) \times R^d\}$ *is Feller continuous (i.e.,* $P_{s_n, x_n} \to P_{s, x}$ *weakly if* $(s_n, x_n) \to (s, x)$*). Moreover,* $\{P_{s, x}: (s, x) \in [0, \infty) \times R^d\}$ *coincides with the family constructed in Theorem 3.2.6.*

Proof. We can either use Theorem 3.2.6 or Theorem 6.1.7 to produce solutions starting from a given (s, x). To prove uniqueness, we will use (i) of Theorem 6.3.2. Indeed, in the proof of Theorem 3.2.6, we produced of time-inhomogeneous semigroup $\{T_{s,t}: 0 \leq s < t\}$ on $C_b(R^d)$ with the property that if $\varphi \in C_0^\infty(R^d)$ and $f(s, \cdot) = T_{s,T} \varphi(\cdot), 0 \leq s \leq T$, then $f(T, \cdot) = \varphi(\cdot)$ and there is a sequence $\{f_n\}_1^\infty \subseteq C_b^{1,2}([0, T] \times R^d) \cap C_b([0, T] \times R^d)$ such that $f_n \to f$ on $[0, T] \times R^d$ and $(\partial f_n/\partial t) + L_t f_n \to 0$ on $[0, T) \times R^d$ uniformly. Thus the proof that the martingale problem for a and b is well-posed is now complete. To prove that $P_{s_n, x_n} \to P_{s, x}$ if $(s_n, x_n) \to (s, x)$, note that $\{P_{s_n, x_n}: n \geq 1\}$ is relatively weakly compact (cf. Theorem 1.4.6) and that any limit point is a solution to the martingale problem for a and b starting from (s, x) (cf. the proof of Theorem 6.1.7). Since there is only one such solution, namely $P_{s, x}$, the proof is complete. $\square$

We could equally well have started with the result of Exercise 3.3.3 and thereby have proved the preceding corollary under the weaker assumptions on a and b given there. Similarly, we could have taken the assumptions on a and b given in Theorem 3.2.1, and, using that theorem, proved the analogue of Corollary 6.3.3 for such coefficients. Indeed, Theorem 6.3.2 is just waiting to be fed results from analysis. The preceding examples should provide ample evidence of its usefulness.

We close this section with a discussion of the application of Itô's method to the study of the martingale problem. Actually, all the work has already been done, and all that remains for us to do is make a translation into the martingale formulation of results proved in Chapter 5.

6.3.4 Theorem. *Let* $a: [0, \infty) \times R^d \to S_d$ *and* $b: [0, \infty) \times R^d \to R^d$ *be bounded measurable functions and suppose that* $\sigma: [0, \infty) \times R^d \to R^d \times R^d$ *is a bounded measurable function such that* $a = \sigma\sigma^*$*. Assume that there is an* A *such that*

$$\|\sigma(t, x) - \sigma(t, y)\| + |b(t, x) - b(t, y)| \leq A|x - y|$$

for all $x, y \in R^d$*. Then the martingale problem for* a *and* b *is well-posed and the corresponding family of solutions* $\{P_{s, x}: (s, x) \in [0, \infty) \times R^d\}$ *is Feller continuous.*

Proof. Existence follows from Theorem 5.3.1 or Theorem 6.1.7. As for uniqueness, Theorem 5.3.2 clearly suffices. Finally, the Feller continuity is proved in exactly the same way as in Theorem 6.3.3. $\square$

6.4. Cameron–Martin–Girsanov Formula

In this section we present the Cameron–Martin–Girsanov formula. Our reason for doing so at this point is that it will play an important role in our study of existence and uniqueness for the martingale problem; particularly when the coefficient a is non-degenerate. The essence of the matter is contained in the following lemma.

6.4.1 Lemma. *Let* $a : [0, \infty) \times \Omega \to S_d$, $b : [0, \infty) \times \Omega \to R^d$, *and* $c : [0, \infty) \times \Omega \to R^d$ *be progressively measurable functions such that* a, b *and* $\langle c, a c \rangle$ *are bounded. Let* P *be a probability measure on* $(\Omega, \mathcal{M})$ *such that* $x(\cdot) \sim \mathscr{I}_d^s(a(\cdot), b(\cdot))$ *and define* $b(\cdot))$ *and define*

$$R(t) = \exp \left[\int_s^{t \vee s} \langle c(u), d\bar{x}(u) \rangle - \frac{1}{2} \int_s^{t \vee s} \langle c(u), a(u)c(u) \rangle \, du \right]$$

where $\bar{x}(\cdot) = x(\cdot) - \int_s^{t \vee s} b(u) \, du$. *Then there is a unique probability measure* Q *on* $(\Omega, \mathcal{M})$ *such that* $Q \ll P$ *on* $\mathcal{M}_t$ *for all* $t \geq 0$ *and* dQ/dP *equals* $R(t)$ *on* $\mathcal{M}_t$, $t \geq 0$. *Moreover,* $x(\cdot) \sim \mathscr{I}_d^s(a(\cdot), (b + ac)(\cdot))$ *under* Q.

Proof. Since $(R(t), \mathcal{M}_t, P)$ is a non-negative martingale (by Theorem 4.3.7) with mean 1, the measures Q_t, $t \geq 0$, defined by

$$Q_t(A) = E^P[R(t), A], \quad A \in \mathcal{M}_t,$$

are consistently defined probability measures. Thus, by Theorem 1.3.5, there is a unique Q on $(\Omega, \mathcal{M})$ which coincides with Q_t on $\mathcal{M}_t$ for all $t \geq 0$. Finally, if $\theta \in R^d$, $s \leq t_1 < t_2$, and $A \in \mathcal{M}_{t_1}$, then, by Theorem 4.3.7,

$$E^Q \left[\exp \left[\left\langle \theta, x(t_2) - x(s) - \int_s^{t_2} (b + ac)(u) du \right\rangle - \frac{1}{2} \int_s^{t_2} \langle \theta, a(u)\theta \rangle du \right], A \right]$$

$$= E^P \left[\exp \left[\int_s^{t_2} \langle \theta + c(u), d\bar{x}(u) \rangle \right.\right.$$
$$\left.\left. - \frac{1}{2} \int_s^{t_2} \langle \theta + c(u), a(u)(\theta + c(u)) \rangle \, du \right], A \right]$$

$$= E^P \left[\exp \left[\int_s^{t_1} \langle \theta + c(u), d\bar{x}(u) \rangle \right.\right.$$
$$\left.\left. - \frac{1}{2} \int_s^{t_1} \langle \theta + c(u), a(u)(\theta + c(u)) \rangle \, du \right], A \right]$$

$$= E^Q \left[\exp \left[\langle \theta, x(t_1) - x(s) - \int_s^{t_1} (b + ac)(u) du \rangle \right.\right.$$
$$\left.\left. - \frac{1}{2} \int_s^{t_1} \langle \theta, a(u)\theta \rangle du \right], A \right].$$

Hence, by Theorem 4.2.1, $x(\cdot) \sim \mathscr{I}_d^s(a(\cdot), (b + ac)(\cdot))$ under Q. □

6.4.2 Theorem. *Let* $a: [0, \infty) \times R^d \to S_d$, $b: [0, \infty) \times R^d \to R^d$, *and* $c:$ $[0, \infty) \times R^d \to R^d$ *be measurable functions such that* a, b, *and* $\langle c, ac \rangle$ *are bounded. Then for each* $(s, x) \in [0, \infty) \times R^d$ *there is a one to one correspondence between solutions* P *to the martingale problem for* a *and* b *starting from* (s, x) *and solutions* Q *for* a *and* $b + ac$ *starting from* (s, x). *The correspondence is such that* $P \to Q$ *where* $Q \ll P$ *on* $\mathscr{M}_t$, $t \geq 0$, *and* dQ/dP *on* $\mathscr{M}_t$ *equals*

$$(4.1) \quad \exp \left[\int_s^{t \vee s} \langle c(u, x(u)), d\bar{x}(u) \rangle - \frac{1}{2} \int_s^{t \vee s} \langle c(u, x(u)), a(u, x(u))c(u, x(u)) \rangle \, du \right]$$

with $\bar{x}(t) = x(t) - \int_s^{t \vee s} b(u, x(u)) \, du$, $t \geq 0$.

Proof. In view of Lemma 6.4.1, it suffices for us to check that if Q is a solution for a and $b + ac$ starting from (s, x), then there is a solution P for a and b starting from (s, x) such that $Q \ll P$ on $\mathscr{M}_t$ and dQ/dP on $\mathscr{M}_t$ is given by (4.1), $t \geq 0$. To this end, let

$$\tilde{x}(t) = x(t) - \int_s^{t \vee s} (b + ac)(u, x(u)) \, du, \qquad t \geq 0,$$

and define, relative to Q,

$$\xi(t) = - \int_s^{t \vee s} \langle c(u, x(u)), d\tilde{x}(u) \rangle, \qquad t \geq 0.$$

Then, by Lemma 6.4.1, there is a unique P such that

$$\frac{dP}{dQ} = Y(t) \equiv \exp \left[\xi(t) - \frac{1}{2} \int_s^{t \vee s} \langle c(u, x(u)), a(u, x(u))c(u, x(u)) \rangle \, du \right]$$

on $\mathscr{M}_t$, $t \geq 0$. Moreover, by that same lemma, P solves for a and b starting from (s, x). Finally, since $Y(t) > 0$, $Q \ll P$ on $\mathscr{M}_t$ and $dQ/dP = (Y(t))^{-1}$ on $\mathscr{M}_t$, $t \geq 0$. Thus, we will be done if we can identify $(Y(t))^{-1}$ as the expression given in (4.1). Formally, there seems to be nothing to such an identification, since we would like to simply write:

$$\xi(t) - \frac{1}{2} \int_s^{t \vee s} \langle c(u, x(u), a(u, x(u))c(u, x(u)) \rangle \, du$$

$$= - \int_s^{t \vee s} \langle c(u, x(u)), d\tilde{x}(u) \rangle$$

$$+ \int_s^{t \vee s} \langle c(u, x(u)), a(u, x(u))c(u, x(u)) \rangle \, du$$

$$- \frac{1}{2} \int_s^{t \vee s} \langle c(u, x(u)), a(u, x(u))c(u, x(u)) \rangle \, du$$

$$= - \left(\int_s^{t \vee s} \langle c(u, x(u)), d\tilde{x}(u) \rangle \right.$$

$$\left. - \frac{1}{2} \int_s^{t \vee s} \langle c(u, x(u)), a(u, x(u))c(u, x(u)) \rangle \, du \right)$$

and thereby finish the proof. However, this calculation is valid only if we can interpret all the stochastic integrals involved relative to Q. In order to use it to complete our proof, we must therefore show that

$$\int_s^{t \vee s} \langle c(u, x(u)), d\bar{x}(u) \rangle$$

is the same no matter whether we interpret the stochastic integral relative to P or to Q. But, as in the proof of Lemma 4.3.2, we can find simple functions θ_n: $[0, \infty) \times \Omega \to R^d$ such that

$$\sup_n \sup_{t, \omega} \langle \theta_n(t, \omega), a(t, x(t, \omega)) \theta_n(t, \omega) \rangle < \infty$$

and

(4.2) $$\int_0^T \langle \theta_n(t) - c(t, x(t)), a(t, x(t))(\theta_n(t) - c(t, x(t))) \rangle \, dt \to 0$$

in Q-probability for each $T > 0$. Since P and Q are equivalent on $\mathcal{M}_t$ for all $t \geq 0$, this means that (4.2) holds in P-probability as well. Hence, for a given $t \geq 0$, there is a set $N \in \mathcal{M}_t$ such that $P(N) = Q(N) = 0$ and

(4.3) $$\int_s^{t \vee s} \langle c(u, x(u)), d\bar{x}(u) \rangle = \lim_{n \to \infty} \int_s^{t \vee s} \langle \theta_n(u), d\bar{x}(u) \rangle$$

off N, whether the left hand side is interpreted relative to P or Q. Because the right hand side of (4.3) is defined independent of the measure involved, this completes the proof. $\square$

The importance of Theorem 6.4.2 to the study of the martingale problem should be obvious. Namely, it allows us to relate the questions of existence, uniqueness, and measurability for one martingale problem to the same questions about a second martingale problem. Typical of the sort of use to which we will be putting Theorem 6.4.2 is the following.

6.4.3 Theorem. *Let $a: [0, \infty) \times R^d \to S_d^+$ and $b: [0, \infty) \times R^d \to R^d$ be bounded measurable functions and assume that there is a $\lambda > 0$ such that*

(4.4) $$\langle \theta, a(s, x) \theta \rangle \geq \lambda |\theta|^2, \qquad (s, x) \in [0, \infty) \times R^d \quad \text{and} \quad \theta \in R^d.$$

Then, for each $(s, x) \in [0, \infty) \times R^d$, the martingale problem for a and b starting from (s, x) has at least one (at most one) solution if and only if the martingale problem for a and 0 starting from (s, x) has at least one (at most one) solution; and if there is exactly one solution $P_{s, x}$ for a and 0, then the measure $Q_{s, x}$ given by

(4.5) $$\left. \frac{dQ_{s, x}}{dP_{s, x}} \right|_{\mathcal{M}_t} = \exp \left[\int_s^{t \vee s} \langle a^{-1} b(u, x(u)), dx(u) \rangle \right.$$

$$\left. - \frac{1}{2} \int_s^{t \vee s} \langle b, a^{-1} b \rangle (u, x(u)) \, du \right]$$

for all $t \geq 0$ is the only solution for a and b. Finally, if the martingale problem for a and 0 (and therefore also for a and b) is well-posed, then the family of solutions associated with a and b is measurable if the one associated with a and 0 is.

Proof. Only the final assertion requires comment. Suppose that the martingale problem for a and 0 is well-posed and that the associated family $\{P_{s,x}: (s, x) \in [0, \infty) \times R^d\}$ is measurable. Then the family $\{Q_{s,x}: (s, x) \in [0, \infty) \times R^d\}$ for a and b is given by (4.5). In particular, if $t \geq 0$ and $A \in \mathcal{M}_t$, then

$$Q_{s,x} = E^{P_{s,x}}[R_c^s(t), A]$$

where $c = a^{-1}b$ and

$$R_c^s(t) = \exp\left[\int_s^{t \vee s} \langle c(u, x(u)), dx(u)\rangle - \frac{1}{2}\int_s^{t \vee s} \langle c, ac\rangle(u, x(u))\, du\right].$$

Thus all that we have to show is that

$$(s, x) \to E^{P_{s,x}}[R_c^s(t), A]$$

is measurable for all bounded measurable $c: [0, \infty) \times R^d \to R^d$. Since the class of c for which this will be true is closed under bounded point-wise convergence, we need only prove it for bounded continuous c's. But if c is bounded and continuous, then for all (s, x): $R_c^s(t) = \lim_{n \to \infty} \widehat{H}_n^s(t)$ in $P_{s,x}$-probability, where

$$\widehat{H}_n^s(t) = \exp\left[\int_s^{t \vee s} \langle \theta_n(u), dx(u)\rangle - \frac{1}{2}\int_2^{t \vee s} \langle \theta_n(u), a(u, x(u))\theta_n(u)\rangle\, du\right]$$

and θ_n is the simple function given by $\theta_n(t) = c(([nt]/n), x([nt]/n))$. Since $\widehat{H}_n^s(t, \omega)$ is jointly measurable in s and ω, $(s, x) \to E^{P_{s,x}}[\widehat{H}_n^s(t), A]$ is measurable for each $n \geq 1$. Thus, it only remains to check that $\widehat{H}_n^s(t) \to R_c^s(t)$ in $L^1(P_{s,x})$ for all (s, x). But

$$E^{P_{s,x}}[|\widehat{H}_n^s(t)|^2] \leq e^{B(t \vee s - s)} E^{P_{s,x}}\left[\exp\left[\int_s^{t \vee s} \langle 2\theta_n(u), dx(u)\rangle\right.\right.$$

$$\left.\left. - \frac{1}{2}\int_s^{t \vee s} \langle 2\theta_n(u), a(u, x(u))2\theta_n(u)\rangle\, du\right]\right]$$

$$= e^{B(t \vee s - s)}$$

where $B = \sup_{t,y} \|a(t, y)\| \sup_{t,y} |c(t, y)|^2$; and therefore the $L^1(P_{s,x})$-convergence follows from the convergence in $P_{s,x}$-probability. $\square$

One immediate application of Theorem 4.6.3 to the martingale problem is that the martingale problem is well-posed and has a measurable family of solutions when a satisfies the conditions of Theorem 3.2.1 and b is merely bounded and measurable. A second application is the next corollary.

6.4.4 Corollary. *Let* a: $[0, \infty) \times R^d \to S_d^+$ *and* b: $[0, \infty) \times R^d \to R^d$ *be bounded measurable functions such that* a *satisfies* (4.4) *for some* $\lambda > 0$ *and* $a(s, \cdot)$ *is continuous for all* $s \geq 0$. *Then the martingale problem for* a *and* b *has at least one solution starting from each* $(s, x) \in [0, \infty) \times R^d$.

Proof. Simply use Theorems 6.4.3 in conjunction with Theorem 6.1.7 applied to a and 0. $\square$

It should be emphasized that the Cameron–Martin–Girsanov formula (4.5) has many important applications to questions other than those coming from the martingale problem. Indeed, it provides the most explicit known expression of the analytic fact that the first order terms of L_t can be thought of as a compact perturbation of the second order terms. We will have ample occasion to exploit this explicitness later on.

6.5. Uniqueness: Random Time Change

Before proceeding with the main theme, we want to indulge in another brief digression

Suppose that a: $[0, \infty) \times R^1 \to (0, \infty)$ is a bounded, uniformly positive, measurable function and let P on $(\Omega, \mathcal{M})$ solve the martingale problem for a and 0 starting from (s, x). Then $x(\cdot) \sim \mathcal{I}_1^s(a(\cdot, x(\cdot)), 0)$ under P. Following Exercise 4.6.5, define $\{\tau(t): t \geq 0\}$ by

$$
(5.1) \qquad\qquad s + \int_s^{\tau(t)} a(u, x(u))\, du = t, \qquad t \geq s
$$

and set $\beta(t) = x(\tau(t)) - x$, $t \geq s$. Then $(\beta(t), \mathcal{M}_{\tau(t)}, P)$ is a Brownian motion after time s and

$$
(5.2) \qquad\qquad x(t) = x + \beta\left(s + \int_s^t a(u, x(u))du\right), \qquad t \geq s.
$$

What we want to do in this section is explore the possibility of using (5.2) to study the martingale problem. In particular, if we can show that $\int_s^t a(u, x(u))\, du$ must be a measurable functional of $\beta(\cdot)$, then (5.2) gives $x(\cdot)$ as a measurable functional of $\beta(\cdot)$; and therefore we will know that the distribution of $x(\cdot)$ is uniquely determined; that is, the martingale problem for a and 0 can have at most one solution. But $\int_s^t a(u, x(u))\, du$ is just the inverse of $\tau(t)$, as a function of $t \in [s, \infty)$; and so this question of measurability can be transferred to $\tau(t)$. Moreover,

$$
(5.3) \qquad\qquad \tau'(t) = \frac{1}{a(\tau(t), \beta(t))}, \quad t \geq s, \qquad \tau(s) = s.
$$

From (5.3), it is clear that $\tau(t)$ must be a measurable function of $\beta(\cdot)$ if a, as a function of t, satisfies sufficiently strong regularity conditions. (For instance, if a is Lipschitz continuous with respect to t with a Lipschitz constant which is independent of x.) In particular, if a is independent of t, then

$$\tau(t) = s + \int_s^t \frac{1}{a(\beta(u))}\, du, \qquad t \geq s,$$

and is therefore obvious $\beta(\cdot)$-measurable. In order to make the presentation as simple as possible, we will restrict our attention to the case when a is independent of t.

6.5.1 Lemma. *Let $a: R^d \to S_d$ and $b: R^d \to R^d$ be bounded measurable functions. Then for each $(s, x) \in [0, \infty) \times R^d$ there is a one-to-one correspondence between solutions to the martingale problem for a and b starting from $(0, x)$ and those starting from (s, x). In particular, existence (uniqueness) holds for solutions starting from (s, x) if existence (uniqueness) obtains for solutions starting from $(0, x)$.*

Proof. Given $s \geq 0$, define $\Phi_s: \Omega \to \Omega$ so that $x(t, \Phi_s(\omega)) = x((t-s) \vee 0, \omega)$, $t \geq 0$. For $t \geq 0$, note that Φ_s^{-1} determines a mapping of $\mathcal{M}_{t+s}$ onto $\mathcal{M}_t$. Moreover, if P solves the martingale problem for a and b starting from $(0, x)$, then

$$E^{P \cdot \Phi_s^{-1}}\left[f(x(t_2)) - \int_s^{t_2} Lf(x(u))\, du, A \right]$$

$$= E^P\left[f(x(t_2 - s)) - \int_0^{t_2 - s} Lf(x(u))\, du, \Phi_s^{-1}A \right]$$

$$= E^P\left[f(x(t_1 - s)) - \int_0^{t_1 - s} Lf(x(u))\, du, \Phi_s^{-1}A \right]$$

$$= E^{P \cdot \Phi_s^{-1}}\left[f(x(t_1)) - \int_s^{t_1} Lf(x(u))\, du, A \right]$$

for $f \in C_0^\infty(R^d)$, $s \leq t_1 < t_2$, and $A \in \mathcal{M}_{t_1}$.

$$\text{(In this computation } L = \frac{1}{2} \sum_{i,j=1}^d a^{ij}(\cdot)\frac{\partial^2}{\partial x_i\, \partial x_j} + \sum_{i=1}^d b^i(\cdot)\frac{\partial}{\partial x_i}.)$$

Also $P \circ \Phi_s^{-1}(x(t) = x, \ 0 \leq t \leq s) = P(x(0) = x) = 1$. Thus $P \to P \circ \Phi_s^{-1}$ establishes a mapping from solutions starting from $(0, x)$ into solutions starting from (s, x). It is easy to see, by essentially the same reasoning, that if $\Psi_s: \Omega \to \Omega$ is defined by $x(t, \Psi_s\omega) = x(t + s, \omega)$, $t \geq 0$, then $P \to P \circ \Psi_s^{-1}$ establishes the inverse mapping from solutions starting from (s, x) into those starting from $(0, x)$. $\square$

Because of Lemma 6.5.1, we restrict our attention to solutions starting at $(0, x)$ when studying the martingale problem for time-independent coefficients.

6.5.2 Theorem. *Let Φ be the class of all bounded, measurable, uniformly positive functions $\varphi: R^d \to (0, \infty)$. Given $\varphi \in \Phi$, define $\tau_\varphi: [0, \infty) \times \Omega \to [0, \infty)$ by the relation:*

$$\int_0^{\tau_\varphi(t, \, \omega)} \frac{1}{\varphi(x(u, \omega))} \, du = t,$$

and let $S_\varphi: \Omega \to \Omega$ be the map determined by $x(t, S_\varphi \omega) = x(\tau_\varphi(t, \omega), \omega), t \geq 0$. Then, for each ω, $\tau_\varphi(\cdot, \omega)$ is a continuous, strictly increasing map of $[0, \infty)$ onto itself; and for each $t \geq 0$, $\tau_\varphi(t, \cdot)$ is a bounded stopping time. Furthermore, S_φ is measurable from $(\Omega, \mathcal{M}_{\tau(t)})$ onto $(\Omega, \mathcal{M}_t)$ for all $t \geq 0$. Finally, if $a: R^d \to S_d$ and $b: R^d \to R^d$ are bounded measurable functions and if P solves the martingale problem for a and b starting from $(0, x)$, then $P \circ S_\varphi^{-1}$ solves the martingale problem for $\varphi(\cdot)a(\cdot)$ and $\varphi(\cdot)b(\cdot)$ starting from $(0, x)$.

Proof. Everything except the last assertion is left to the reader. To prove the last assertion, let $f \in C_0^\infty(R^d)$, $0 \leq t_1 < t_2$, and $A \in \mathcal{M}_{t_1}$ be given. Then

$$E^{P \, \circ \, S_\varphi^{-1}}\left[f(x(t_2)) - \int_0^{t_2} \varphi(x(u))Lf(x(u)) \, du, \, A \right]$$

$$= E^P\left[f(x(\tau_\varphi(t_2))) - \int_0^{t_2} \varphi(x(\tau_\varphi(u)))Lf(x(\tau_\varphi(u))) \, du, \, S_\varphi^{-1}A \right]$$

where L is defined in terms of a and b. Note that, by a change of variables,

$$\int_0^t \varphi(x(\tau_\varphi(u)))Lf(x(\tau_\varphi(u))) \, du = \int_0^{\tau_\varphi(t)} Lf(x(u)) \, du, \qquad t \geq 0.$$

Thus, since

$$\left(f(x(\tau_\varphi(t))) - \int_0^{\tau_\varphi(t)} Lf(x(u)) \, du, \, \mathcal{M}_{\tau_\varphi(t)}, \, P \right)$$

is a martingale and $S_\varphi^{-1}A \in \mathcal{M}_{\tau_\varphi(t_1)}$,

$$E^{P \, \circ \, S_\varphi^{-1}}\left[f(x(t_2)) - \int_0^{t_2} \varphi(x(u))Lf(x(u)) \, du, \, A \right]$$

$$= E^P\left[f(x(\tau_\varphi(t_1))) - \int_0^{\tau_\varphi(t_1)} Lf(x(u)) \, du, \, S_\varphi^{-1}A \right]$$

$$= E^{P \, \circ \, S_\varphi^{-1}}\left[f(x(t_1)) - \int_0^{t_1} \varphi(x(u))Lf(x(u)) \, du, \, A \right].$$

In other words, $x(\cdot) \sim \mathscr{I}_d^0(\varphi(x(\cdot))a(x(\cdot)), \varphi(x(\cdot))b(x(\cdot)))$ under $P \circ S_\varphi^{-1}$. Since $P \circ S_\varphi^{-1}(x(0) = x) = P(x(\tau_\varphi(0)) = 0) = P(x(0) = x) = 1$, this completes the proof. $\square$

6.5.3 Lemma. *Let Φ and S_φ, $\varphi \in \Phi$, be defined as in the preceding. Then*

$$S_{\varphi \cdot \psi} = S_\varphi \circ S_\psi = S_\psi \circ S_\varphi$$

for all $\varphi, \psi \in \Phi$. In particular, S_φ is invertible and its inverse is $S_{1/\varphi}$.

Proof. Clearly all that we have to show is that $\tau_{\varphi \cdot \psi}(t, \omega) = \tau_\psi(\tau_\varphi(t, S_\psi \omega), \omega)$. Since, for any $\eta \in \Phi$, $\tau_\eta(\cdot, \omega)$ is characterized by the condition that $\tau_\eta(0, \omega) = 0$ and $\tau'_\eta(t, \omega) = \eta(x(\tau_\eta(t, \omega), \omega))$, all that has to be checked is that if $\sigma(t, \omega) = \tau_\psi(\tau_\varphi(t, S_\psi \omega), \omega)$, then $\sigma'(t, \omega) = \varphi \cdot \psi(x(\sigma(t, \omega), \omega))$. But

$$
\begin{aligned}
\sigma'(t, \omega) &= \psi(x(\tau_\psi(\tau_\varphi(t, S_\psi \omega), \omega), \omega))\tau'_\varphi(t, S_\psi \omega) \\
&= \psi(x(\sigma(t, \omega), \omega))\varphi(x(\tau_\varphi(t, S_\psi \omega), S_\psi \omega)) \\
&= \psi(x(\sigma(t, \omega), \omega))\varphi(x(\tau_\psi(\tau_\varphi(t, S_\psi \omega), \omega), \omega)) \\
&= \varphi \cdot \psi(x(\sigma(t, \omega), \omega)),
\end{aligned}
$$

and so the proof is complete. $\square$

6.5.4 Theorem. *Let $a: R^d \to S_d$ and $b: R^d \to R^d$ be bounded measurable functions and let $\varphi: R^d \to (0, \infty)$ be a bounded, measurable, uniformly positive function. For each $x \in R^d$, there is a one to one correspondence between solutions to the martingale problem for a and b starting from $(0, x)$ and those for $\varphi(\cdot)a(\cdot)$ and $\varphi(\cdot)b(\cdot)$ starting from the same point. The correspondence sends a solution P for a and b into $P \circ S_\varphi^{-1}$, where S_φ is described in Theorem 6.4.5.*

Theorem 6.5.4 is particularly satisfactory in the one-dimensional case. Indeed, combining Theorems 6.5.4 and 6.4.3 together with Lemma 6.5.1, we arrive at the following corollary.

6.5.5 Corollary. *Let $a: R^1 \to (0, \infty)$ be a bounded, measurable, uniformly positive function and $b: R^1 \to R^1$ a bounded measurable function. Then the martingale problem for a and b is well-posed and the associated family of solutions is measurable.*

Proof. Define $S_a: \Omega \to \Omega$ as in Theorem 6.5.2 and $\Phi_s: \Omega \to \Omega$ by $x(t, \Phi_s \omega) = x((t - s) \vee 0, \omega)$, $t \geq 0$. According to Lemma 6.5.1 and Theorem 6.5.4, $P_{s,x} \equiv \mathscr{W}^{(1)}_{0,x} \circ S_a^{-1} \circ \Phi_s^{-1}$ is the only solution to the martingale problem for a and 0 starting from (s, x). Clearly $(s, x) \to P_{s,x}$ is measurable. The rest of the proof is accomplished by an application of Theorem 6.4.3. $\square$

With the possible exception of Itô's method, the preceding result is, from a purely probabilistic point of view, the most satisfactory resolution of a martingale problem which we will present. It is not so much that we are surprised that one can handle the situation dealt with in Corollary 6.5.5 (indeed, considering Feller's magnificent success in understanding one-dimensional, time-homogeneous diffusions, it would have been very disappointing if we had not been able to), but that we are able to do so without having to invoke anything from the theory of partial differential equations. Unfortunately, we will not get off so lightly when we insist on being more ambitious.

6.6. Uniqueness: Localization

In this section we will show that the basic questions about the martingale problem reduce to local considerations. To be precise, we will prove the following theorem, which will play an essential part in Chapter 7.

6.6.1 Theorem. *Let* $a: [0, \infty) \times R^d \to S_d$ *and* $b: [0, \infty) \times R^d \to R^d$ *be bounded measurable functions. Suppose that for each* $(s, x) \in [0, \infty) \times R^d$ *there is an open set* $\mathcal{G} \ni (s, x)$ *and bounded measurable coefficients* $\tilde{a}$ *and* $\tilde{b}$ *such that:*

(i) *$\tilde{a}$ equals a and $\tilde{b}$ equals b on $\mathcal{G}$,*
(ii) *the martingale problem for $\tilde{a}$ and $\tilde{b}$ is well-posed and the associated family of solutions is measurable.*

Then the martingale problem for a and b is well-posed and the associated family of solutions is measurable.

The proof of this theorem will be broken into several steps, each one being a lemma. Thus in the statement of the following lemmas, we will assume that we are dealing with the situation described in Theorem 6.6.1 and will not mention it each time.

6.6.2 Lemma. *Given* $R > 0$, *set* $\Gamma_R = [0, R] \times \overline{B(0, R)}$. *There exist open subsets* $V_1, \dots, V_M$ *of* $[0, \infty) \times R^d$ *covering* Γ_R *and bounded measurable coefficients* $\tilde{a}_1, \dots, \tilde{a}_M$ *and* $\tilde{b}_1, \dots, \tilde{b}_M$ *satisfying* (ii) *of Theorem 6.6.1 such that* $\tilde{a}_m = a$ *and* $\tilde{b}_m = b$ *on* V_m *for* $1 \le m \le M$. *Moreover, the number*

$$\rho \equiv \inf_{(s, x) \in \Gamma_R} \max_{1 \le m \le M} \{(t - s) + |y - x| : t > s \text{ and } (t, y) \notin V_m\}$$

is strictly positive.

Proof. This lemma is an easy application of the Heine-Borel property for Γ_R. $\square$

Now suppose that $R > 0$ is given. Let $\Gamma_R^0 = [0, R) \times B(0, R)$ and choose $V_1, \dots, V_M$ as in Lemma 6.6.2 for $\Gamma_R = \overline{\Gamma_R^0}$. For each $(s, x) \in \Gamma_R$, let $m(s, x)$ be the smallest $1 \le m \le M$ for which $(t - s) + |y - x| \ge \rho$ whenever $t > s$ and $(t, y) \notin V_m$. We define the following stopping times on Ω. First

$$\zeta_R = \inf\{t \ge 0: (t, x(t)) \notin \Gamma_R^0\}.$$

Next, for $0 \leq s \leq R$, set $\tau_0^s = s$ and

$$
\tau_n^s =
\begin{cases}
\tau_{n-1}^s + 1 & \text{if } \tau_{n-1}^s \geq \zeta_R \\
\left(\inf\{t \geq \tau_{n-1}^s : (t, x(t)) \notin V_{m(\tau_{n-1}^s, x(\tau_{n-1}^s))}\}\right) \wedge \zeta_R \\
& \text{if } \tau_{n-1}^s < \zeta_R.
\end{cases}
$$

Finally, if $(s, x) \notin \Gamma_R^0$, let $Q_{s,x}^{(R)}$ be the measure on $(\Omega, \mathcal{M})$ satisfying $Q_{s,x}^{(R)}(x(t) = x, t \geq 0, = 1)$; and if $(s, x) \in \Gamma_R^0$, let $Q_{s,x}^{(R)}$ be the solution to the martingale problem for $\bar{a}_{m(s, x)}$ and $\bar{b}_{m(s, x)}$ starting from (s, x). Note that $\{Q_{s,x}^{(R)} : (s, x) \in [0, \infty) \times R^d\}$ is measurable.

6.6.3 Lemma. *For each $(s, x) \in [0, \infty) \times R^d$ there is a unique $P_{s,x}^{(R)}$ on $(\Omega, \mathcal{M})$ with the properties that $P_{s,x}^{(R)}$ equals $Q_{s,x}^{(R)}$ on $\mathcal{M}_{\tau_0^s}$; and, for each $n \geq 0$, $P_{s,x}^{(R)}$ coincides on $\mathcal{M}_{\tau_{n+1}^s}$ with $P_{s,x}^{(R)} \otimes_{\tau_n^s} Q_{\tau_n^s, x(\tau_n^s)}^{(R)}$. Moreover, the family $\{P_{s,x}^{(R)} : (s, x) \in [0, \infty) \times R^d\}$ is measurable.*

Proof. Clearly, since $\tau_n^s(\omega) \nearrow \infty$ as $n \to \infty$ for all $\omega \in \Omega$, if such measures $P_{s,x}^{(R)}$ exist, they are unique and $\{P_{s,x}^{(R)} : (s, x) \in [0, \infty) \times R^d\}$ is measurable. To prove their existence define $P_{s,x}^0 = Q_{s,x}$ and, for $n \geq 1$,

$$
P_{s,x}^n = P_{s,x}^{n-1} \otimes_{\tau_n^s} Q_{\tau_n^s, x(\tau_n^s)}^{(R)}.
$$

If we can show that

$$
(6.1) \qquad \lim_{n \to \infty} P_{s,x}^n(\tau_{n+1}^s \leq T) = 0, \qquad T \geq 0
$$

then there is a unique $P_{s,x}^{(R)}$ on $(\Omega, \mathcal{M})$ such that $P_{s,x}^{(R)}$ equals $P_{s,x}^n$ on $\mathcal{M}_{\tau_{n+1}^s}$, $n \geq 0$, and clearly this is the measure we are looking for.

The proof of (6.1) reduces to finding a $\delta > 0$ for which

$$
(6.2) \qquad \lim_{n \to \infty} \sup_{(s, x)} P_{s,x}^n(\tau_{n+1}^s \leq s + \delta) = 0.
$$

Indeed, from (6.2), we can use induction on k to show that

$$
(6.3) \qquad \lim_{n \to \infty} \sup_{(s, x)} P_{s,x}^n(\tau_{n+1}^s \leq s + k\delta) = 0.
$$

To see this, assume (6.2) and (6.3). Then

$$
P_{s,x}^n(\tau_{n+1}^s \leq s + (k+1)\delta) \leq P_{s,x}^m(\tau_{m+1}^s \leq s + k\delta) + P_{s,x}^n(\tau_{n+1}^s - \tau_{m+1}^s \leq \delta)
$$

for any $0 \le m < n$. The first term on the right goes to 0 as $m \to \infty$ at a rate independent of (s, x). As for the second term, note that

$$P_{s,x}^n(\tau_{n+1}^s - \tau_{m+1}^s \le \delta) = E^{P_{s,x}^m}[P_{\tau_{m+1}^s, x(\tau_{m+1}^s)}^{n-m}(\tau_{n+1}^s \le \tau_{m+1}^s + \delta)]$$

$$\le \sup_{t,y} P_{t,y}^{n-m}(\tau_{n-m}^t \le t + \delta) \to 0$$

as $n - m \to \infty$; which completes the proof of (6.3) with $(k + 1)$ replacing k.

To prove (6.2), let A and B be the bounds of the coefficients a and b, respectively; and choose $0 < \delta < 1$ so that $B\delta < \rho$ and

$$2d \, \exp[-(\rho - B\delta)^2/2 \, dA\delta] \le \tfrac{1}{2}.$$

Note that for any $(s, x) \in [0, \infty) \times R^d$

(6.4) $$P_{s,x}^1(\tau_2^s \le s + \delta) \le \tfrac{1}{2}.$$

Indeed, if $(s, x) \notin \Gamma_R^0$, then $\tau_2^s = s + 1$ (a.s., $Q_{s,x}$), and so (6.4) is trivial. On the other hand, if $(s, x) \in \Gamma_R^0$, then

$$P_{s,x}^1(\tau_2^s \le s + \delta) = P_{s,x}^1(\tau_2^s \le s + \delta, \tau_1^s < \zeta_R)$$

since $\tau_2^s = s + 1$ if $\tau_1^s \ge \zeta_R$. But

$$P_{s,x}^1(\tau_2^s \le s + \delta, \tau_1^s < \zeta_R) \le Q_{s,x}(\tau_1^s \le s + \delta, \tau_1^s < \zeta_R)$$

$$\le Q_{s,x}\left(\sup_{s \le t \le s+\delta} |x(t) - x| \ge \rho\right) \le \tfrac{1}{2},$$

by the estimate in Theorem 4.2.1 applied to $x(t \wedge \tau_1^s) - \int_s^{t \wedge \tau_1^s} b(u, x(u)) \, du$. Thus (6.4) is established. Starting with (6.4), we have:

$$P_{s,x}^{2n+1}(\tau_{2n+2}^s \le s + \delta) = E^{P_{s,x}^{2n-1}}[P_{\tau_{2n}^s, x(\tau_{2n}^s)}^1(\tau_{2n}^{\tau_{2n}^s} \le s + \delta), \tau_{2n}^s - s + \delta]$$

$$\le \tfrac{1}{2} P_{s,x}^{2n-1}(\tau_{2n}^s \le s + \delta),$$

and so

$$P_{s,x}^{2n-1}(\tau_{2n}^s \le s + \delta) \le (\tfrac{1}{2})^n,$$

which certainly proves (6.2). $\square$

6.6.4 Lemma. *For each $R > 0$ and $(s, x) \in [0, \infty) \times R^d$, the measure $P_{s,x}^{(R)}$ described in Lemma 6.6.3 is uniquely characterized by the facts that $P_{s,x}^{(R)}(x(t) = x, 0 \le t \le s) = 1$ and that $x(\cdot) \sim \mathscr{I}_d^s(\chi_{[0,\zeta_R)}(\cdot)a(\cdot, x(\cdot)), \chi_{[0,\zeta_R)}(\cdot)b(\cdot, x(\cdot)))$ under $P_{s,x}^{(R)}$.*

Proof. We first show that $x(\cdot) \sim \mathscr{I}_d^s(\chi_{[0,\,\zeta_R)}(\cdot)a(\cdot,\,x(\cdot)),\,\chi_{[0,\,\zeta_R)}(\cdot)b(\cdot,\,x(\cdot)))$ under $P_{s,\,x}^{(R)}$. To this end, set $a_R(\cdot) = \chi_{[0,\,\zeta_R)}(\cdot)a(\cdot,x(\cdot))$ and $b_R(\cdot) = \chi_{[0,\,\zeta_R)}(\cdot,x(\cdot))\,b(\cdot,x(\cdot))$, and let

$$L_t^{(R)} = \frac{1}{2} \sum_{i,\,j=1}^{d} a_R^{ij}(\cdot) \frac{\partial^2}{\partial x_i\,\partial x_j} + \sum_{i=1}^{d} b_R^i(\cdot) \frac{\partial}{\partial x_i}.$$

Given $f \in C_0(R^d)$, we must show that

$$\theta(t) \equiv f(x(t)) - \int_s^t L_u^{(R)}f(x(u))\,du$$

is a $P_{s,\,x}^{(R)}$-martingale after time s. This will be done by showing that, for all $n \geq 0$, $\theta(t \wedge \tau_n^s)$ is a $P_{s,\,x}^{(R)}$-martingale after time s. There is nothing to check when $n = 0$. Assume now that $\theta(t \wedge \tau_n^s)$ is a $P_{s,\,x}^{(R)}$-martingale after time s. Observe that if $\tau_n^s(\omega) < \zeta_R(\omega)$, then

$$f(x(t)) - \int_{\tau_n^s(\omega)}^t \tilde{L}_u^{(\omega)}f(x(u))\,du$$

is a $Q_{\tau_n^s(\omega),\,x(\tau_n^s(\omega)),\,\omega}^{(R)}$-martingale after time $\tau_n^s(\omega)$, where $\tilde{L}_t^{(\omega)}$ is the operator corresponding to the coefficients $\tilde{a}_{m(\tau_n^s(\omega),\,x(\tau_n^s(\omega)),\,\omega)}$ $\tilde{b}_{m(\tau_n^s(\omega),\,x(\tau_n^s(\omega)),\,\omega)}$. Since $\tilde{a}_{m(\tau_n^s(\omega),\,x(\tau_n^s(\omega)),\,\omega)}(t,\,x(t)) = a(t,\,x(t))$ and $\tilde{b}_{m(\tau_n^s(\omega),\,x(\tau_n^s(\omega)),\,\omega)}(t,\,x(t)) = b(t,\,x(t))$ for $\tau_n^s(\omega) \leq t < \tau_1^{\tilde{n}(\omega)}$ and because $\tau_1^{\tilde{n}(\omega)} = \tau_{n+1}^s$ (a.s., $\delta_\omega \otimes_{\tau_n^s(\omega)} Q_{\tau_n^s(\omega),\,x(\tau_n^s(\omega)),\,\omega}^{(R)}$), it follows that

$$\theta((t \wedge \tau_{n+1}^s) \wedge \tau_n^s) - \theta(\tau_n^s)$$

is a $\delta_\omega \otimes_{\tau_n^s(\omega)} Q_{\tau_n^s(\omega),\,x(\tau_n^s(\omega)),\,\omega}^{(R)}$-martingale after time s if $\tau_n^s(\omega) < \zeta_R(\omega)$. On the other hand, if $\tau_n^s(\omega) \geq \zeta_R(\omega)$, then $Q_{\tau_n^s(\omega),\,x(\tau_n^s(\omega)),\,\omega}^{(R)}(x(t) = x(\tau_n^s(\omega)),\,\omega),\,t \geq 0) = 1$, and so it is trivial to check that

$$\theta((t \wedge \tau_{n+1}^s) \wedge \tau_n^s) - \theta(\tau_n^s)$$

is a $\delta_\omega \otimes_{\tau_n^s(\omega)} Q_{\tau_n^s(\omega),\,x(\tau_n^s(\omega)),\,\omega}^{(R)}$-martingale after time s. Combining these remarks with the induction hypothesis and applying Theorem 6.1.2, we arrive at the conclusion that $\theta(t \wedge \tau_{n+1}^s)$ is a $P_{s,\,x}^{(R)} \otimes_{\tau_n^s} Q_{\tau_n^s,\,x(\tau_n^s)}^{(R)}$-martingale after time s. Since $P_{s,\,x}^{(R)}$ equals $P_{s,\,x}^{(R)} \otimes_{\tau_n^s} Q_{\tau_n^s,\,x(\tau_n^s)}^{(R)}$ on $\mathscr{M}_{\tau_{n+1}^s}$, it is now easy to see that $\theta(t \wedge \tau_{n+1}^s)$ is a $P_{s,\,x}^{(R)}$-martingale after time s. We have therefore proved that $x(\cdot) \sim \mathscr{I}_d^s(a_R(\cdot),\,b_R(\cdot))$ under $P_{s,\,x}^{(R)}$.

It remains for us to show that any other P on $(\Omega,\,\mathscr{M})$ satisfying $P(x(t) = x,\,0 \leq t \leq s) = 1$ and $x(\cdot) \sim \mathscr{I}_d^s(a_R(\cdot),\,b_R(\cdot))$ must be equal to $P_{s,\,x}^{(R)}$. In other words, we must show that such a P equals $Q_{s,\,x}^{(R)}$ on $\mathscr{M}_{\tau_0^s}$ and equals $P \otimes_{\tau_n^s} Q_{\tau_n^s,\,x(\tau_n^s)}^{(R)}$ on $\mathscr{M}_{\tau_{n+1}^s}$, $n \geq 0$. The first part is trivial. As for the second, let $\{P_\omega\}$ be a r.c.p.d. of $P|\mathscr{M}_{\tau_n^s}$. Then there is a P-null set $N \in \mathscr{M}_{\tau_n^s}$ such that $x(\cdot) \sim \mathscr{I}_d^{\tau_n^s(\omega)}(a_R(\cdot),\,b_R(\cdot))$ under P_ω for all $\omega \notin N$. Define $\tau^\omega = \tau_1^{\tilde{n}(\omega)}$; and for $\omega \notin N$ satisfying $\tau_n(\omega) < \zeta_R(\omega)$

set $m(\omega) = m(\tau_n^s(\omega), x(\tau_n^s(\omega), \omega))$ and $\tilde{P}_\omega = P_\omega \otimes_{\tau\omega} \tilde{P}_{\tau\omega, x(\tau\omega)}^{m(\omega)}$ where $\tilde{P}_{t, y}^m$ is the unique solution to the martingale problem for $\tilde{a}_m$ and $\tilde{b}_m$ starting from (t, y). Then, if $\omega \notin N$ and $\tau_n^s(\omega) < \zeta_R(\omega)$, $a_R(t) = \tilde{a}_{m(\omega)}(t, x(t))$ and $b_R(t) = \tilde{b}_{m(\omega)}(t, x(t))$, $\tau_n^s(\omega) \le t < \tau^\omega$, and so $x(\cdot) \sim \mathscr{I}_d^{\tau_n^s(\omega)}(\tilde{a}_{m(\omega)}(\cdot), \tilde{b}_{m(\omega)}(\cdot))$ under $\tilde{P}_\omega$. Using the uniqueness of solutions to the martingale problem for $\tilde{a}_{m(\omega)}$ and $\tilde{b}_{m(\omega)}$, we now see that $\tilde{P}_\omega = \delta_\omega \otimes_{\tau_n^s(\omega)} \tilde{P}_{\tau_n^s(\omega), x(\tau_n^s(\omega), \omega)}^{m(\omega)}$ for all $\omega \notin N$ satisfying $\tau_n^s(\omega) < \zeta_R(\omega)$; and, therefore, for such ω we have that P equals $\delta_\omega \otimes_{\tau_n^s} Q_{\tau_n^s(\omega), x(\tau_n^s(\omega), \omega)}^{(R)}$ on $\mathscr{M}_{\tau^\omega}$. On the other hand, if $\omega \notin N$ and $\tau_n^s(\omega) \ge \zeta_R(\omega)$, then $P_\omega(x(t) = x(\tau_n^s(\omega), \omega)$, $t \ge \tau_n^s(\omega)) = 1$; and so P_ω certainly equals $\delta_\omega \otimes_{\tau_n^s(\omega)} Q_{\tau_n^s(\omega), x(\tau_n^s(\omega), \omega)}^{(R)}$ on $\mathscr{M}$ in this case. Combining these remarks with the fact that $\tau^\omega = \tau_{n+1}^s$ (a.s., P_ω) for all ω, we conclude that P does indeed equal $P \otimes_{\tau_n^s} Q_{\tau_n^s, x(\tau_n^s)}^{(R)}$ on $\mathscr{M}_{\tau_{n+1}^s}$. $\square$

It is now a relatively easy task to complete the proof of Theorem 6.6.1. Let us first establish the existence and measurability assertions. Let $0 < R_1 < \cdots < R_n \cdots$ increase to infinity as $n \to \infty$ and use $P_{s, x}^{(n)}$ to denote $P_{s, x}^{(R_n)}$ and ζ_n to denote ζ_{R_n}. For each $n \ge 1$, $\{P_{s, x}^{(n)} : (s, x) \in [0, \cdot\infty) \times R^d\}$ is measurable. Moreover, one can easily check that $P_{s, x}^{(n+1)} \otimes_{\zeta_n} \delta_{x(\zeta_n)}$ satisfies the criteria characterizing $P_{s, x}^{(n)}$ and so $P_{s, x}^{(n+1)}$ equals $P_{s, x}^{(n)}$ on $\mathscr{M}_{\zeta_n}$. Finally, by the estimate in Theorem 4.2.1,

$$P_{s, x}^{(n)}(\zeta_n \le T) \le 2d \exp[-(n - B(T - s))^2 / 2Ad(T - s)]$$

for $T > s$ such that $n > B(T - s)$. (Here A and B stand for the bounds on a and b, respectively). Thus (cf. Theorem 1.1.10) for each (s, x) there is a unique $P_{s,x}$ on $(\Omega, \mathscr{M})$ such that $P_{s,x}$ equals $P_{s,x}^{(n)}$ on $\mathscr{M}_{\zeta_n}$ for all $n \ge 1$. Clearly $\{P_{s,x} : (s, x) \in [0, \infty) \times R^d\}$ is measurable, and $P_{s,x}$ solves the martingale problem for a and b starting from (s, x).

Finally, uniqueness is proved as follows. Suppose P is a solution starting at (s, x). Given $n \ge 1$, note that $P \otimes_{\zeta_n} \delta_{x(\zeta_n)}$ equals $P_{s, x}^{(n)}$ for all $n \ge 1$; and therefore P equals $P_{s, x}^{(n)}$ on $\mathscr{M}_{\zeta_n}$, $n \ge 1$. But this means that P equals the $P_{s, x}$ just constructed; and so we are done. $\square$

6.7. Exercises

6.7.1. It is interesting to see what the martingale formulation looks like when dealing with discrete time processes. For this purpose consider the following set-up.

Let $\mathscr{N} = \{0, 1, \ldots, n, \ldots\}$, $\tilde{\Omega} = (R^d)^{\mathscr{N}}$, and for $n \in \mathscr{N}$ and $\tilde{\omega} \in \tilde{\Omega}$ set $x(n, \tilde{\omega})$ equal to the nth coordinate of $\tilde{\omega}$ (i.e., "the position of $\tilde{\omega}$ at time n"). Let $\tilde{\mathscr{M}}_n = \sigma(x(k) : 0 \le k \le n)$ and $\tilde{\mathscr{M}} = \sigma(x(k) : k \ge 0)$.

Given a sequence of transition functions $\Pi_n(x, \cdot)$, $n \in \mathscr{N}$, on R^d into itself, define

$$A_n f(x) = \int (f(y) - f(x)) \Pi_n(x, dy)$$

for $f \in C_b(R^d)$. We will say that a probability measure P on $(\tilde{\Omega}, \tilde{\mathscr{M}})$ solves the martingale problem for A_n starting from (m, x) if $P(x(n) = x, 0 \le n \le m) = 1$ and

$$\left(f(x(n)) - \sum_{m}^{n-1} A_k f(x(k)), \tilde{\mathscr{M}}_k, P \right)$$

is a (discrete-parameter) martingale after time $(m + 1)$ for all $f \in C_b(R^d)$. Show that this martingale problem is well-posed. In fact, show that P solves it starting from (m, x) if and only if

$$P(x(0) \in \Gamma_0, \ldots, x(n) \in \Gamma_n)$$

$$= \chi_{\Gamma_0}(x) \cdots \chi_{\Gamma_m}(x) \int_{\Gamma_{m+1}} \cdots \int_{\Gamma_n} \Pi_m(x, dy_1) \cdots \Pi_{n-1}(y_{n-1}, dy_n)$$

for all $n > m$ and $\Gamma_1, \ldots, \Gamma_n \in \mathscr{B}_{R^d}$.

6.7.2. For some technical purposes, the following is sometimes a useful observation. Suppose that $a: [0, \infty) \times R^d \to S_d$ and $b: [0, \infty) \times R^d \to S_d$ are bounded measurable functions and L_t on $C_0^\infty(R^d)$ is defined accordingly. Next define $\tilde{a}: R^{d+1} \to S_{d+1}$ and $\tilde{b}: R^{d+1} \to R^{d+1}$ by

$$\tilde{a}^{i, j}(\tilde{x}) = \begin{cases} 0 & \text{if } i \text{ or } j = 0 \\ a^{ij}(x_0 \vee 0, x) & \text{otherwise} \end{cases}$$

$$\tilde{b}^i(\tilde{x}) = \begin{cases} 1 & \text{if } i = 0 \\ b^i(x_0 \vee 0, x) & \text{otherwise} \end{cases}$$

for $\tilde{x} = (x_0, x) \in R^1 \times R^d = R^{d+1}$. Let $\tilde{L}$ be the time independent operator on $C_0^\infty(R^{d+1})$ with coefficients $\tilde{a}$ and $\tilde{b}$. Given $(s, x) \in [0, \infty) \times R^d$, show that there is a one to one correspondence between solutions to the martingale problem for L_t starting from x at time s and solutions to the martingale problem for $\tilde{L}$ starting from $\tilde{x} = (s, x)$ at time 0. In fact, show that if $\Psi_s: \Omega_d \to \Omega_{d+1}$ is given by

$$\tilde{x}(t, \Psi_s(\omega)) = (t + s, x(t + s, \omega))$$

then P solves the martingale problem for L_t starting at (s, x) if and only if $P \circ \Psi_s^{-1}$ solves the martingale problem for $\tilde{L}$ starting from $\tilde{x} = (s, x)$ at time 0. Conversely check that the mapping $\Phi_s: \Omega_{d+1} \to \Omega_d$ given by:

$$x(t, \Phi_s(\tilde{\omega})) = (x_1((t - s) \vee 0, \omega), \ldots, x_d((t - s) \vee 0, \omega))$$

has the property that $\tilde{P}$ solves the martingale problem for $\tilde{L}$ starting from $\tilde{x} = (s, x)$ at time 0 if and only if $\tilde{P} \circ \Phi_s^{-1}$ solves for L_t starting from x at time s. Finally show that $(s, x) \to \tilde{P}_{(s, x)} \circ \Phi_s^{-1}$ is measurable (continuous) from $[0, \infty) \times R^d \to M(\Omega_d)$ if $\tilde{x} \to \tilde{P}_{\tilde{x}}$ is measurable (continuous) from $R^{d+1} \to M(\Omega_{d+1})$.

6.7.3. Anyone who is familiar with Meyer's celebrated version of Doob's decomposition theorem might be wondering if the following formulation could not be used to advantage.

Let $a: [0, \infty) \times R^d \to S_d$ and $b: [0, \infty) \times R^d \to R^d$ be bounded measurable functions and define L_t accordingly. Say that a P on $(\Omega, \mathcal{M})$ *solves the submartingale problem for* L_t *starting from* (s, x) if $P(x(t) = x, \ 0 \le t \le s) = 1$ and $(f(t, x(t)), \mathcal{M}_t, P)$ is a submartingale after time s for all $f \in C_b^{1,2}([0, \infty) \times R^d)$ satisfying $(\partial f/\partial t) + L_t f \ge 0$. Clearly, a solution to the martingale problem for L_t starting from (s, x) is a solution to the submartingale problem since $f(t, x(t))$ is then the sum of a martingale and the non-decreasing function $\int_0^t ((\partial f/\partial u) + L_u f) \times (u, x(u)) \, du$ when $(\partial f/\partial t) + L_t f \ge 0$. What Meyer's theorem guarantees is a partial converse. Namely, if P solves the submartingale problem and $(\partial f/\partial t) + L_t f \ge 0$, then there is a unique continuous, progressively measurable, non-decreasing $\xi_f: [s, \infty) \times \Omega \to [0, \infty)$ such that $\xi_f(s) = 0$ and $(f(t, x(t)) - \xi_f(t), \mathcal{M}_t, P)$ is a martingale after time s. Under the assumption that a and b are continuous on $[0, \infty) \times R^d$, identify $\xi_f(t)$, $t \ge s$, as $\int_s^t ((\partial f/\partial u) + L_u f)(u, x(u)) \, du$; and thereby show that, in this case, any solution to the submartingale problem solves the martingale problem.

It is unfortunate that the continuity assumption seems to be essential in going from the submartingale to the martingale problem. Prove that a solution to the submartingale problem always exists even if a and b are merely bounded and measurable. However, the following example demonstrates that not every solution of every submartingale problem is a solution of the corresponding martingale problem. Namely, let $a: R \to [0, \infty)$ and $b: R \to R$ be given by

$$a(x) = \begin{cases} 2 & \text{if} \ \ x \text{ is rational} \\ 1 & \text{if} \ \ x \text{ is irrational} \end{cases}$$

$$b(x) = 0 \quad \text{for all} \ \ x \in R.$$

Check that the only solution to the martingale problem for a and b starting from $(0, 0)$ is 1-dimensional Wiener measure; whereas the solution to the martingale problem for $L = \partial^2/\partial x^2$ starting from $(0, 0)$ is a second solution to the submartingale problem for a and b.

6.7.4. Following Theorem 6.2.2, we mentioned that the measurability of $\{P_{s,x}: (s, x) \in [0, \infty) \times R^d\}$ is a consequence of the associated martingale problem being well posed. We outline the proof of this claim. First observe that in view of Exercise 6.7.2, it suffices to treat the case when the coefficients are time independent and to prove in this case that if the martingale problem is well posed then the family $\{P_x: x \in R^d\}$ of solutions starting from time 0 is measurable. To this end let $\mathcal{A}$ stand for the set of $P \in M(\Omega)$ such that $P(x(0) = x) = 1$ for some $x \in R^d$ and

$$\left(f(x(t)) - \int_0^t (Lf)(x(s)) \, ds, \mathcal{M}_t, P\right)$$

is a martingale for all $f \in C_0^\infty(R^d)$. Check that $\mathscr{A}$ is a measurable subset of $M(\Omega)$. This is easily done by writing down a countable number of relations for P to be a member of $\mathscr{A}$, each of which is the vanishing of a Borel functional on $M(\Omega)$. Next define a map $F: \mathscr{A} \to R^d$ by insisting that $P[x(0) = F(P)] = 1$. Note that F is a one to one map of $\mathscr{A}$ onto R^d by the assumption that the martingale problem is well posed. Check that the mapping F is continuous and therefore measurable from $\mathscr{A}$ onto R^d. By a theorem of Kuratowski [1948] a one to one measurable map (F) from any Borel subset $(\mathscr{A})$ of a Polish space $(M(\Omega))$ onto another Polish space (R^d) has a measurable inverse (P_x). That is to say P_x is a Borel map of R^d into $M(\Omega)$.

6.7.5. The Cameron-Martin formula has many important applications; one of which we now outline. More details can be found in Stroock and Varadhan [1970].

Let $a: [0, \infty) \times \Omega \to S_d^+$ and $b: [0, \infty) \times \Omega \to R^d$ be bounded progressively measurable functions such that $\langle \theta, a(t, \omega)\theta \rangle \geq \lambda |\theta|^2$ for all $t \geq 0$, $\omega \in \Omega$, $\theta \in R^d$, and some $\lambda > 0$. Assume that P is a probability measure on $(\Omega, \mathscr{M})$ such that $P(x(t) = x, 0 \leq t \leq s) = 1$ and $x(\cdot) \sim \mathscr{I}_d^s(a(\cdot), b(\cdot))$ under P. Show that for all $T > s$ and $\varepsilon > 0$:

$$(7.1) \qquad P\left(\sup_{s \leq t \leq T} |x(t) - x| > \varepsilon \right) < 1.$$

The idea behind the proof of (7.1) is the following. First, assume that $s = 0$ and $x = 0$. Next, note that, because of Theorem 6.4.2, (7.1) holds for the given choice of a and b if and only if it holds for a and any other choice of bounded progressively measurable $\tilde{b}: [0, \infty) \times \Omega \to R^d$. The trick is to make a judicious choice of $\tilde{b}$. Namely, define

$$(7.2) \qquad \tilde{b}(t) = -\left(\frac{\text{Trace } a(t)}{|x(t)|^2} \chi_{[\varepsilon/2, \infty)}(|x(t)|) \right) x(t).$$

We now have to show that (7.1) holds when $s = 0$, $x = 0$, $P(x(0) = 0)$, and $x(\cdot) \sim \mathscr{I}_d^0(a(\cdot), \tilde{b}(\cdot))$. To this end, choose $\varphi \in C_0^\infty(R^d)$ so that $\varphi(x) = |x|^2$ for $|x| \leq 2\varepsilon$, and set $\eta(t) = \varphi(x(t))$. Then, by Itô's formula, $\eta(\cdot) \sim \mathscr{I}_1^0(\langle \nabla\varphi(x(\cdot)), a(\cdot)\nabla\varphi(x(\cdot)), \gamma(\cdot))$ under P, where $\gamma: [0, \infty) \times \Omega \to R$ is a bounded progressively measurable function such that $\gamma(t, \omega) = 0$ if $\varepsilon/2 \leq |x(t, \omega)| \leq 2\varepsilon$. But this means that so long as $|x(\cdot)| \in [\varepsilon/2, 2\varepsilon]$, $\eta(\cdot)$ is distributed, under P, like a 1-dimensional Brownian motion run with the clock $\int_0^t \langle \nabla\varphi(x(u)), a(u)\nabla\varphi(x(u)) \rangle \, du$. Using this observation, one can easily deduce that (7.1) (with this choice of s, x, and $\tilde{b}$) could fail only if it were possible for a 1-dimensional Brownian to escape from a neighborhood of 0 before time $(\sup_{(t, \omega)} \langle \nabla\varphi(x(t, \omega)), a(t, \omega)\nabla\varphi(x(t, \omega)) \rangle)T$ with probability one. Since it is well known that a Brownian motion cannot make such an escape, the proof of (7.1) is complete.

Our interest in (7.1) is the following application. Let $a: [0, \infty) \times R^d \to S_d^+$ and $b: [0, \infty) \times R^d \to R^d$ be bounded measurable functions such that $a(t, \cdot)$ is continuous for all $t \geq 0$ and $\langle \theta, a(t, y)\theta \rangle \geq \lambda |\theta|^2$ for all $t \geq 0$, $y \in R^d$, $\theta \in R^d$, and some

$\lambda > 0$. Let L_t be the associated operator. For each $(s, x) \in [0, \infty) \times R^d$, let $P_{s, x}$ be a solution to the martingale problem for a and b starting from (s, x). Using (7.1) and the Cameron-Martin formula, show that if $\psi \in C([0, \infty), R^d)$, then for all $0 \le s < T$ and $\varepsilon > 0$:

$$P_{s, x}\left(\sup_{s \le t \le T} |x(t) - x - \int_s^t \psi(u) \, du| > \varepsilon \right) < 1.$$

In this way, conclude that the support of $P_{s, x}$ in Ω coincides with the class $C(s, x)$ of continuous $\varphi: [0, \infty) \to R^d$ such that $\varphi(t) = x$ for $0 \le t \le s$. In particular, show that if $\mathscr{G}$ is a bounded open subset of $[0, \infty) \times R^d$ containing (s, x) and if $\zeta = \inf\{t \ge s: (t, x(t)) \notin \mathscr{G}\}$, then

$$P_{s, x}((t, x(t \wedge \zeta)) \in U) > 0$$

so long as U is an open set in $[0, \infty) \times R^d$ for which there is a $\varphi \in C(s, x)$ with the property that $(t, \varphi(t)) \in U$ and $(u, \varphi(u)) \in \mathscr{G}$ for all u between s and t. Finally, apply this fact to derive Nirenberg's strong maximum principle: if $f \in C_b^{1, \, 2}(\mathscr{G})$ and $(\partial f/\partial t) + L_t f \ge 0$ on $\mathscr{G}$, then $f(s, x) = \max_{(t, \, y) \in \mathscr{G}} f(t, y)$ for some $(s, x) \in \mathscr{G}$ implies that $f(t, y) = f(s, x)$ for any $(t, y) \in \mathscr{G}$ with the property that there is a $\varphi \in C(s, x)$ satisfying $(u, \varphi(u)) \in \mathscr{G}$, $s \le u \le t$, and $\varphi(t) = y$. In view of the preceding considerations, the proof is an easy consequence of the relation:

$$f(s, x) \le E^{P_{s, x}}[f(t \wedge \zeta, x(t \wedge \zeta))].$$

For more information about these and related matters, consult Stroock and Varadhan [1970] and [1972].

6.7.6. The following simple example demonstrates that there exist bounded coefficients for which the associated martingale problem possesses *no* solutions. Consider, for instance, the time independent coefficients on R^1 given by $a(x) \equiv 0$ and $b(x) = -1$ for $x > 0$ and 1 for $x \le 0$. If P solves the martingale problem for a and b, starting from 0 at time 0, then P must be concentrated on paths ω satisfying

$$x(t, \omega) = \int_0^t b(x(s, \omega)) \, ds, \qquad \text{for } t \ge 0.$$

Show that there are no such paths and conclude thereby that no such P exists.

6.7.7. We know by Theorem 6.1.7 that if a and b are bounded and continuous then there is at least one solution to the martingale problem for any given starting point. It is easy to construct examples to show that uniqueness does not hold in that generality. For example, on R^1, define $a(x) \equiv 0$ and $b(x) = |x|^\alpha \wedge 1$ for some $0 < \alpha < 1$. Show that the ordinary differential equation $\dot{x}(t) = b(x(t))$ for $t \ge 0$ with $x(0) = 0$ has more than one solution. Conclude that the martingale problem for a and b starting from 0 at time 0 has more than one solution.

A more disturbing example of non-uniqueness is the following. On R^1 define $b(x) \equiv 0$ and let $a(x)$ be any bounded continuous function which is positive for $x \neq 0$ and 0 for $x = 0$ with $\int_{-1}^{1} dx/a(x) < \infty$. (Take for instance $a(x) = 1 \wedge |x|^{\alpha}$ for some $0 < \alpha < 1$). It is trivial to check that one solution to the martingale problem for a and b starting from 0 at time 0 is the process P_0 with $P_0[x(t) \equiv 0$ for $t \geq 0] = 1$. We now outline the construction of another solution. Let $\mathscr{W}_0$ be the one-dimensional Wiener measure and define $\sigma(t, \omega)$, for $t \geq 0$, by

$$\int_0^{\sigma(t, \omega)} \frac{ds}{a(x(s, \omega))} = t.$$

Show that $\sigma(t, \omega)$ is well defined and $\mathscr{W}_0[\sigma(t, \omega) > 0] = 1$ for $t > 0$ and use the technique of Section 6.5 to check that $P_\infty = \mathscr{W}_0 \circ \Phi^{-1}$ is a second solution where $x(t, \Phi(\omega)) = x(\sigma(t, \omega), \omega)$. Even worse show that if $l_0(t)$ is the local time at 0 of $x(t, \omega)$ (cf. Exercise 4.6.12) and $0 < \lambda < \infty$, then $P_\lambda = \mathscr{W}_0 \circ \Phi_\lambda^{-1}$ is also a solution where $x(t, \Phi_\lambda(\omega)) = x(\sigma_\lambda(t, \omega), \omega)$ and

$$\int_0^{\sigma_\lambda(t, \omega)} \frac{1}{a(x(s, \omega))}\left(ds + \frac{1}{\lambda} l_0(ds)\right) = t.$$

Each of these P_λ is a different solution and each is the member starting at 0 of a different Feller-continuous time homogeneous strong Markov family.

Chapter 7

Uniqueness

7.0. Introduction

We are going to show in this chapter that if $a: [0, \infty) \times R^d \to S_d$ and $b: [0, \infty) \times R^d \to R^d$ are bounded measurable functions and if a is uniformly positive definite on compacts and satisfies:

$$(0.1) \qquad \lim_{y \to x} \sup_{0 \leq s \leq T} \|a(s, y) - a(s, x)\| = 0$$

for all $T > 0$ and $x \in R^d$, then the martingale problem for a and b is well-posed and the associated family of solutions $\{P_{s,x}: (s, x) \in [0, \infty) \times R^d\}$ is (strongly) Feller continuous. Most of the work involved is devoted to proving this result for the situation in which $b \equiv 0$ and a is very nearly independent of x. Once this case has been thoroughly understood, the general case follows quite easily with the aid of the reduction procedures developed in Chapter 6. Since the details may obscure the idea, we now outline the reasoning behind our analysis.

Let $a: [0, \infty) \times R^d \to S_d^+$ be measurable and have the property that

$$(0.2) \qquad \lambda |\theta|^2 \leq \langle \theta, a(s, x)\theta \rangle \leq \Lambda |\theta|^2, \qquad \theta \in R^d \quad \text{and} \quad (s, x) \in [0, \infty) \times R^d,$$

for some $0 < \lambda < \Lambda < \infty$. Assume, in addition, that there is an $x^0 \in R^d$ such that

$$(0.3) \qquad \varepsilon \equiv \sup_{s \geq 0} \sup_{x \in R} \|a(s, x) - a(s, x^0)\|$$

is small. Set $c(s) = a(s, x^0)$, $s \geq 0$, and define

$$(0.4) \qquad g_c(s, x; t, y) = \frac{1}{(2\pi)^{d/2}(\det C(s, t))^{1/2}} \exp[-\tfrac{1}{2}\langle y - x, C(s, t)^{-1}(y - x)\rangle]$$

for $0 \leq s < t$, where

$$C(s, t) = \int_s^t c(u)\, du.$$

For $0 \leq s < T$ and $f \in C_0^\infty([0, T] \times R^d)$, define

$$(0.5) \qquad G_c^T f(s, x) = \int_s^T dt \int_{R^d} f(t, y) g_c(s, x; t, y) \, dy.$$

Then $G_c^T f$ solves

$$\frac{\partial u}{\partial s} + L_s^{(c)} u = -f$$

in $[0, T]$ with $\lim_{t \nearrow T} u(t, \cdot) = 0$, where

$$L_t^{(c)} = \frac{1}{2} \sum_{i, j = 1}^{} c^{ij}(t) \frac{\partial^2}{\partial x_i \, \partial x_j}.$$

Thus

$$\frac{\partial G_c^T f}{\partial s} + L_s G_c^T f = -f + K^T f, \qquad 0 \leq s < T,$$

where

$$K^T f(s, x) = (L_s - L_s^{(c)}) G_c^T f(s, x).$$

Note that

$$|K^T f(s, x)| \leq \frac{\varepsilon d^2}{2} \max_{1 \leq i, j \leq d} \left| \frac{\partial^2 G_c^T f}{\partial x_i \, \partial x_j}(s, x) \right|;$$

and therefore, if $1 < p < \infty$, then

$$\|K^T f\|_{L^p([0, T] \times R^d)} \leq \frac{\varepsilon}{2} d^2 C_d(p, \lambda, \Lambda) \|f\|_{L^p([0, T] \times R^d)},$$

where $C_d(p, \lambda, \Lambda)$ is the constant in equation (0.4) of the appendix. In particular, if $1 < p < \infty$ is fixed and $\varepsilon > 0$ is chosen so that

$$\varepsilon d^2 C_d(P, \lambda, \Lambda) \leq 1,$$

then $I - K^T$ admits a bounded inverse on $L^p([0, T] \times R^d)$; and, at least formally, we have:

$$(0.6) \qquad \frac{\partial G^T f}{\partial s} + L_s G^T f = -f$$

in $[0, T)$ with

(0.7) $$\lim_{t \nearrow T} G^T f(t, \cdot) = 0,$$

where

$$G^T = G_c^T \circ (I - K^T)^{-1}.$$

It is therefore reasonable to hope that if P solves the martingale problem for a and 0 starting from some $(s, x) \in [0, T) \times R^d$, then

$$G^T f(t \wedge T, x(t \wedge T)) + \int_s^{t \wedge T} f(u, x(u)) \, du$$

will be a P-martingale and, therefore,

(0.8) $$G^T f(s, x) = E^P \left[\int_s^T f(u, x(u)) \, du \right].$$

We should then apply Corollary 6.2.4 and conclude that there is only one such P. Obviously, the flaw in this line of reasoning is that we have not been careful about the sense in which (0.6) holds. In fact, we have not even discussed in what sense $G^T f$ is defined and (0.7) holds. This latter point is easily dispensed with, since by assuming that p is large enough we can guarantee that for $f \in C_0^\infty([0, T) \times R^d)$, $G^T f$ is well defined as a bounded continuous function on $[0, T] \times R^d$ with $G^T f(T, \cdot) = 0$. On the other hand, the problem of interpreting (0.6) will not go away so easily. The fact is that the best that can be said about the second spatial derivatives of $G^T f$ is that they exist as elements of $L^p([0, T) \times R^d)$. In other words, we can find functions $\varphi_n \in C_b^{1,2}([0, \infty) \times R^d)$ such that $\varphi_n \to G^T f$ uniformly on $[0, T] \times R^d$ and

$$\frac{\partial \varphi_n}{\partial t} + L_t \varphi_n \to -f$$

in $L^p([0, T) \times R^d)$. Thus, in order to prove (0.8), we must show that, as a function of f, the right-hand side of (0.8) is continuous with respect to $L^p([0, T) \times R^d)$-convergence. Put another way, if we expect to extract a uniqueness proof out of this procedure, we must show a priori that for any P solving the martingale problem the measure μ on $[0, T) \times R^d$ defined by:

$$\int f \, d\mu = E^P \left[\int_s^T f(u, x(u)) \, du \right]$$

admits a density which is in $L^{p'}([0, T) \times R^d)$ where p' satisfies $(1/p) + (1/p') = 1$. This is the crux of our argument.

7.1. Uniqueness: Local Case

In this section we will make a detailed study of martingale problems in which the diffusion coefficients are nearly independent of the space variables. The main focus of our discussion will be the question of uniqueness. However, starting with Lemma 7.1.8, we will be concerned with some regularity properties of the family of solutions to such martingale problems. Although these regularity properties are not needed to establish uniqueness, they will be used to answer some of the more refined questions of the type that will be taken up in the next section following Theorem 7.2.2.

Let $c: [0, \infty) \to S_d$ be a measurable function for which there exist $0 < \lambda \leq \Lambda < \infty$ satisfying:

$$(1.1) \qquad \lambda |\theta|^2 \leq \langle \theta, c(s)\theta \rangle \leq \Lambda |\theta|^2, \qquad s \geq 0 \quad \text{and} \quad \theta \in R^d.$$

Define $g_c(s, x; t, y)$ for $0 \leq s < t$ as in (0.4) and G_c^T as in (0.5).

7.1.1   Lemma.   *If* $f \in C_0^\infty([0, T] \times R^d)$ *and* $(d + 2)/2 < p \leq \infty$,   *then* $G_c^T f \in C_b^{0, \infty}([0, T] \times R^d)$ *and*

$$(1.2) \qquad |G_c^T f(s, x)| \leq A_d(p, \lambda, \Lambda) T^{1 - (d + 2)/2p} \|f\|_{L^p([s, T) \times R^d)},$$

where $A_d(p, \lambda, \Lambda)$ depends only on d, p, λ, and Λ. Moreover, if $(E, \mathcal{F}, \mu)$ is a probability space, $\{\mathcal{F}_t: t \geq 0\}$ a non-decreasing family of sub σ-algebras of $\mathcal{F}$, and ξ: $[s, \infty)E \to R^d$ a progressively measurable, right-continuous, μ-almost surely continuous function such that $\xi \sim \mathcal{I}_d^s(a, 0)$ for some bounded progressively measurable function $a: [0, \infty) \times E \to S_d$, then

$$G_c^T f(t \wedge T, \xi(t \wedge T)) + \int_s^{t \wedge T} f(u, \xi(u)) \, du - \int_s^{t \wedge T} D_u^T f(u, \xi(u)) \, du$$

is a martingale after time s relative to $(E, \mathcal{F}_t, \mu)$, where

$$D_u^T f(u, \xi(u)) = \frac{1}{2} \sum_{i, j = 1}^d (a^{ij}(u) - c^{ij}(u)) \frac{\partial^2 G_c^T}{\partial x_i \, \partial x_j} (u, \xi(u)).$$

Proof. That $G_c^T f \in C_b^{0, \infty}([0, T] \times R^d)$ is obvious from the fact that

$$\frac{\partial^{|\alpha|} G_c^T f}{\partial x_1^{\alpha_1} \cdots \partial x_d^{\alpha_d}} = G_c^T \frac{\partial^{|\alpha|} f}{\partial x_1^{\alpha_1} \cdots \partial x_d^{\alpha_d}}.$$

Moreover, the estimate in (1.2) follows easily from the inequality

$$|g_c(s, x; t, y)| \leq \frac{1}{(2\pi\lambda(t - s))^{d/2}} \exp[-|y - x|^2/2\Lambda(t - s)]$$

and Hölder's inequality. Finally, to prove the last assertion, let $\{c_n\}_1^\infty$ be a sequence of continuous functions on $[0, \infty)$ into S_d, each of which satisfies (1.1), such that

$$\int_0^T |c_n(u) - c(u)| \, du \to 0.$$

Given $f \in C_0^\infty([0, T] \times R^d)$, note that $G_{c_n}^T f \in C_b^{1, 2}([0, T] \times R^d)$ and that $G_{c_n}^T f \to G_c^T f$ uniformly in $[0, T] \times R^d$. Hence all spatial derivatives of $G_{c_n}^T f$ tend uniformly to the corresponding spatial derivatives of $G_c^T f$. Thus, since

$$\frac{\partial G_{c_n}^T f}{\partial t} + \frac{1}{2} \sum_{i, j=1}^d c_n^{ij}(t) \frac{\partial^2 G_{c_n}^T f}{\partial x_i \, \partial x_j} = -f,$$

we see that

$$\int_0^T \left| \frac{\partial G_{c_n}^T f}{\partial t} (t, \cdot) + \frac{1}{2} \sum_{i, j=1}^d c^{ij}(t) \frac{\partial^2 G_{c_n}^T f}{\partial x_i \, \partial x_j} (t, \cdot) + f(t, \cdot) \right| dt$$

tends uniformly to zero as $n \to \infty$. It is now clear that

$$G_c^T f(t \wedge T, \xi(t \wedge T)) + \int_s^{t \wedge T} f(u, \xi(u)) \, du - \int_s^{t \wedge T} D_u^T f(u, \xi(u)) \, du$$

is a μ-martingale after time s, since for each $n \geq 1$,

$$G_{c_n}^T f(t \wedge T, \xi(t \wedge T)) - \int_s^{t \wedge T} \left(\frac{\partial G_{c_n}^T f}{\partial u} + \frac{1}{2} \sum_{i, j=1}^d a^{ij}(u) \frac{\partial^2 G_{c_n}^T f}{\partial x_i \, \partial x_j} \right) (u, \xi(u)) \, du$$

has the same property. $\square$

7.1.2 Lemma. *Let* $a: [0, \infty) \times \Omega \to S_d$ *be a progressively measurable function which satisfies*

$$\lambda |\theta|^2 \leq \langle \theta, a(t, \omega)\theta \rangle \leq \Lambda |\theta|^2, \qquad t \geq 0, \quad \omega \in \Omega, \quad \text{and} \quad \theta \in R^d,$$

for some $0 < \lambda \leq \Lambda < \infty$. *Assume further that for some* $N \geq 1$, $a(t)$ *is* $\mathcal{M}_{[Nt]/N}$-*measurable for all* $t \geq 0$. *Let* P *be a probability measure on* $(\Omega, \mathcal{M})$ *and* ξ: $[s, \infty) \times \Omega \to R^d$ *a right continuous, P-almost surely continuous progressively measurable function such that* $\xi \sim \mathcal{I}_a^s(a(\cdot), 0)$ *relative to* $(\Omega, \mathcal{M}_t, P)$. *Then for each* $(d + 2)/2 < p \leq \infty$ *there is a constant* $C < \infty$ *depending only on* $d, p, \lambda, \Lambda, T,$ *and* N *such that*

$$\left| E^P \left[\int_s^T f(t, \xi(t)) \, dt \right] \right| \leq CT^{1 - (d+2)/2p} \|f\|_{L^p([s, t] \times R^d)}$$

for all $T > s$ *and* $f \in C_0^\infty([s, T] \times R^d)$.

Proof. Set $k = [Ns]$, $K = [NT]$, and $t_l = (l/N \vee s) \wedge T$ for $k \leq l \leq K + 1$. Clearly:

$$\left| E^P \left[\int_s^T f(t, \xi(t)) \, dt \right] \right| \leq \sum_{l=k}^K E^P \left[\left| E^{P^l} \cdot \left[\int_{t_l}^{t_{l+1}} f(t, \xi(t)) \, dt \right] \right| \right],$$

where $\{P_\omega^l\}$ is a r.c.p.d. of $P \mid \mathcal{M}_{t_l}$. Note that if ω_l denotes the element of Ω such that $x(t, \omega_l) = x(t \wedge t_l, \omega)$, $t \geq 0$, then $a(t, \cdot) = a(t, \omega_l)$, $0 \leq t < t_{l+1}$, P_ω^l-almost surely. Hence, by Theorem 6.1.3 and Lemma 7.1.1,

$$E^{P_\omega^l} \left[\int_{t_l}^{t_{l+1}} f(t, \xi(t)) dt \right] = E^{P_\omega^l} \left[G_{a(\cdot, \omega_l)}^{t_{l+1}} f(t_l, \omega) \right]$$

for P-almost all ω. Thus, by (1.2):

$$\left| E^{P_\omega^l} \left[\int_{t_l}^{t_{l+1}} f(t, \xi(t)) \, dt \right] \right| \leq A_d(p, \lambda, \Lambda) T^{1 - (d+2)/2p} \| f \|_{L^p([s, T) \times R^d)};$$

and so

$$\left| E^P \left[\int_s^T f(t, \xi(t)) \, dt \right] \right| \leq (K - k + 1) A_d(p, \lambda, \Lambda) T^{1 - (d+2)/2p} \| f \|_{L^p([s, T) \times R^d)}.$$

Since $K - k + 1 \leq (T - s + 1)N$, this completes the proof. $\square$

7.1.3 Lemma. *Let* $a : [0, \infty) \times \Omega \to S_d$, $\xi : [s, \infty) \times \Omega \to R^d$, *and* P *on* $(\Omega, \mathcal{M})$ *be as in Lemma 7.1.2. Assume, in addition, that there exists a measurable* $c : [0, \infty) \to S_d$ *satisfying* (1.1) *such that*

$$\sup_{(t, \omega)} \| a(t, \omega) - c(t) \| \leq (d^2 C_d(p, \lambda, \Lambda))^{-1}$$

for some $(d + 2)/2 < p < \infty$, *where* $C_d(p, \lambda, \Lambda)$ *is given in* (0.4) *of the appendix. Then for any* $T > s$ *and* $f \in C_0^\infty([s, T] \times R^d)$:

$$\left| E^P \left[\int_s^T f(t, \xi(t)) \, dt \right] \right| \leq 2 A_d(p, \lambda, \Lambda) T^{1 - (d+2)/2p} \| f \|_{L^p([s, T) \times R^d)}.$$

Proof. Given $T > s$, define μ on $([s, T) \times R^d, \mathcal{B}_{[s, T) \times R^d})$ by $\mu(\Delta \times \Gamma) = E^P[\int_\Delta \chi_\Gamma(\xi(t)) \, dt]$ for $\Delta \in \mathcal{B}_{[s, t]}$ and $\Gamma \in \mathcal{B}_{R^d}$. By Lemma 7.1.2, we know that μ is absolutely continuous and that its density φ is an element of $L^{p'}([s, T) \times R^d)$ (here $(1/p) + (1/p') = 1$). We want to estimate $\| \varphi \|_{L^{p'}([s, T] \times R^d)}$. To this end, let $f \in C_0^\infty([s, T) \times R^d)$ be given. By Lemma 7.1.1,

$$E^P[G_c^T f(s, \xi(s))] = E^P \left[\int_s^T f(u, \xi(u)) \, du \right] - E^P \left[\int_s^T D_u^T f(u, \xi(u)) \, du \right].$$

Note that

$$|E^P[G_c^T f(s, \xi(s))]| \leq A_d(p, \lambda, \Lambda) T^{1-(d+2)/2p} \|f\|_{L^p([s, T) \times R^d)}$$

and

$$|D_u^T f(u, \xi(u))| \leq \frac{1}{2d^2 C_d(p, \lambda, \Lambda)} \left(\sum_{i, j=1}^d \left| \frac{\partial^2 G_c f}{\partial x_i \partial x_j} (u, \xi(u)) \right|^2 \right)^{\frac{1}{2}}.$$

Thus if

$$H(t, y) = \left(\sum_{i, j=1}^d \left| \frac{\partial^2 G_c f}{\partial x_i \partial x_j} (t, y) \right|^2 \right)^{\frac{1}{2}},$$

then:

$$A_d(p, \lambda, \Lambda) T^{1-(d+2)/2p} \|f\|_{L^p([s, T) \times R^d)}$$

$$\geq \left| \int_s^T dt \int_{R^d} f(t, y)\varphi(t, y) \, dy \right|$$

$$- \frac{1}{2d^2 C_d(p, \lambda, \Lambda)} \left| \int_s^T dt \int_{R^d} H(t, y)\varphi(t, y) \, dy \right|.$$

But this means that

$$A_d(p, \lambda, \Lambda) T^{1-(d+2)/2p} \|f\|_{L^p([s, T) \times R^d)}$$

$$+ \frac{1}{2} \|\varphi\|_{L^{p'}([s, T) \times R^d)} \|f\|_{L^p([s, T) \times R^d)}$$

$$\geq \left| \int_s^T dt \int_{R^d} f(t, y)\varphi(t, y) \, dy \right|,$$

since

$$\|H\|_{L^p([s, T) \times R^d} \leq d^2 C_d(p, \lambda, \Lambda) \|f\|_{L^p([s, T) \times R^d)}.$$

Because this holds for all $f \in C_0^\infty([s, T] \times R^d)$, we conclude that

$$A_d(p, \lambda, \Lambda) T^{1-(d+2)/2p} + \frac{1}{2} \|\varphi\|_{L^{p'}([s, T) \times R^d)} \geq \|\varphi\|_{L^{p'}([s, T) \times R^d)},$$

which is what we wanted to prove. □

7.1.4 Theorem. *Let* $a: [0, \infty) \times \Omega \to S_d$ *be a progressively measurable function satisfying:*

$$\lambda|\theta|^2 \leq \langle \theta, a(t, \omega)\theta \rangle \leq \Lambda|\theta|^2, \quad t \geq 0, \omega \in \Omega, \quad \text{and} \quad \theta \in R^d,$$

for some $0 < \lambda \leq \Lambda < \infty$. *Assume that there is a* $(d + 2)/2 < p < \infty$ *and a measurable* $c: [0, \infty) \to S_d$ *satisfying* (1.1) *such that*

$$\sup_{(t, \omega)} \|a(t, \omega) - c(t)\| \leq \frac{1}{d^2 C_d(p, \lambda, \Lambda)},$$

where $C_d(p, \lambda, \Lambda)$ *is the constant in* (0.4) *of the appendix. Suppose that* P *is a probability measure on* $(\Omega, \mathcal{M})$ *for which* $x(\cdot) \sim \mathcal{S}_d^s(a(\cdot), 0)$. *Then for all* $T > s$ *and* $f \in C_0^\infty([s, T) \times R^d)$:

$$(1.3) \qquad \left| E^P\left[\int_s^T f(t, x(t))\, dt\right] \right| \leq 2A_d(p, \lambda, \Lambda) T^{1 - (d + 2/2p)} \|f\|_{L^p([s, T] \times R^d)}.$$

Proof. Let $\rho \in C_0^\infty(R^1)$ be non-negative, supported on the right half-line and satisfy $\int \rho(t)\, dt = 1$. For $\varepsilon > 0$, set $\rho_\varepsilon(t) = 1/\varepsilon\rho(t/\varepsilon)$, $t \in R^1$ and define

$$a_\varepsilon(t) = \int_0^t \rho_\varepsilon(t - s)a(s)\, ds, \qquad t \geq 0,$$

$$c_\varepsilon(t) = \int_0^t \rho_\varepsilon(t - s)c(s)\, ds, \qquad t \geq 0.$$

For $n \geq 1$, define $a_{\varepsilon, n}(t) = a_\varepsilon([nt]/n)$, $t > 0$, and $c_{\varepsilon, n}(t) = c_\varepsilon([nt]/n)$. It is easy to check that for each $\varepsilon > 0$ and $n \geq 1$: $a_{\varepsilon, n}: [0, \infty) \times \Omega \to S_d$ is progressively measurable; $a_{\varepsilon, n}(t)$ is $\mathcal{M}_{[nt]/n}$-measurable for $t \geq 0$;

$$\lambda|\theta|^2 \leq \langle \theta, a_{\varepsilon, n}(t, \omega)\theta \rangle \leq \Lambda|\theta|^2 \quad \text{and} \quad \lambda|\theta|^2 \leq \langle \theta, c_{\varepsilon, n}(t)\theta \rangle \leq \Lambda|\theta|^2$$

for all $t \geq 0$, $\omega \in \Omega$, and $\theta \in R^d$; and $\sup_{(t, \omega)} \|a_{\varepsilon, n}(t, \omega) - c_{\varepsilon, n}(t)\| \leq 1/d^2 C_d(p, \lambda, \Lambda)$. Moreover, for each $T > s$:

$$E^P\left[\int_0^T \|a_{\varepsilon, n}(t) - a(t)\|^2\, dt\right] \to 0$$

as $n \to \infty$ and then $\varepsilon \searrow 0$.

Now define $\beta(t) = \int_s^t a^{-\frac{1}{2}}(u)\, dx(u)$, $t \geq s$. Then $(\beta(t), \mathcal{M}_t, P)$ is a d-dimensional Brownian motion after time s and

$$x(t) = x(s) + \int_s^t a^{\frac{1}{2}}(u)\, d\beta(u), \qquad t \geq s,$$

P-almost surely. Next set

$$\xi_{\varepsilon, n}(t) = x(s) + \int_s^t a_{\varepsilon, n}^{\frac{1}{2}}(u)\, d\beta(u), \qquad t \geq s.$$

For each $\varepsilon > 0$ and $n \geq 1$, $\xi_{\varepsilon, n}(\cdot) \sim \mathscr{I}_d^s(a_{\varepsilon, n}(\cdot), 0)$; and therefore, by Lemma 7.1.3,

$$\left| E^P \left[\int_s^T f(t, \xi_{\varepsilon, n}(t)) \, dt \right] \right| \leq 2 A_d(p, \lambda, \Lambda) T^{1 - (d + 2)/2p} \| f \|_{L^p([s, T) \times R^d)}$$

for all $T > s$ and $f \in C_0^\infty([s, T] \times R^d)$. But

$$E^P \left[\sup_{s \leq t \leq T} |\xi_{\varepsilon, n}(t) - x(t)|^2 \right] \leq 4 E^P \left[\int_s^T \| a_{\varepsilon, n}(t) - a(t) \|^2 \, dt \right]$$

and therefore tends to zero as $n \to \infty$ and then $\varepsilon \searrow 0$. In particular,

$$\left| E^P \left[\int_s^T f(t, x(t)) \, dt \right] \right| = \lim_{\varepsilon \searrow 0} \lim_{n \to \infty} \left| E^P \left[\int_s^T f(t, \xi_{\varepsilon, n}(t)) \, dt \right] \right|$$

$$\leq 2 A_d(p, \lambda, \Lambda) T^{1 - (d + 2)/2p} \| f \|_{L^p([s, T) \times R^d)};$$

and this completes the proof. $\square$

7.1.5 Lemma. *Let* $c: [0, \infty) \to S_d$ *be a measurable function which satisfies* (1.1) *for some* $0 < \lambda \leq \Lambda$. *Suppose that* $a: [0, \infty) \times R^d \to S_d$ *is a measurable function such that*

$$(1.4) \qquad \lambda |\theta|^2 \leq \langle \theta, a(t, y)\theta \rangle \leq \Lambda |\theta|^2, \quad t \geq 0, \quad y \in R^d, \quad \text{and} \quad \theta \in R^d,$$

and define

$$(1.5) \qquad K^T f(s, x) = \frac{1}{2} \sum_{i, j = 1}^d (a^{ij}(s, x) - c^{ij}(s)) \frac{\partial^2 G_c^T f}{\partial x_i \partial x_j}(s, x)$$

for $f \in C_0^\infty([0, T) \times R^d)$. *Then for each* $1 < p < \infty$, K^T *admits a unique extension (again denoted by* K^T) *as a continuous operator on* $L^p([0, T) \times R^d)$ *into itself and the extensions corresponding to different p's in* $(1, \infty)$ *are consistent. Finally, if* $1 < p_1 < p_2 < \infty$ *and*

$$(1.6) \qquad \sup_{(t, y)} \| a(t, y) - c(t) \| \leq 1/d^2 C_d(p, \lambda, \Lambda), \qquad p_1 \leq p \leq p_2,$$

then $I - K^T$ *admits a bounded inverse* $(I - K^T)^{-1}$ *on* $L^p([s, T) \times R^d)$ *for* $p_1 \leq p \leq p_2$, *the bound on* $(I - K^T)^{-1}$ *is less than or equal to two, and* $(I - K^T)^{-1}$ *is consistently defined for different p's in* $[p_1, p_2]$.

Proof. Clearly:

$$K^T f(s, x) \leq \frac{1}{2} \| a(s, x) - c(s) \| \left(\sum_{i, j = 1}^d \left| \frac{\partial^2 G_c^T f}{\partial x_i \partial x_j}(s, x) \right|^2 \right)^{\frac{1}{2}}$$

$$\leq \frac{1}{2} \sup_{(t, y)} \| a(t, y) - c(t) \| \sum_{i, j = 1}^d \left| \frac{\partial^2 G_c^T}{\partial x_i \partial x_j}(s, x) \right|;$$

for $f \in C_0^\infty([0, T] \times R^d)$. This proves that K^T admits a unique extension to $L^p([0, T) \times R^d)$ and that the norm of the extension is no greater than $d^2/2C_d(p, \lambda, \Lambda) \sup_{(t, y)} \|a(t, y) - c(t)\|$. Moreover, since if f is in $L^p([0, T) \times R^d) \cap L^q([0, T) \times R^d)$, where $1 < p, q < \infty$, we can find $\{f_n\}_1^\infty \subseteq C_0^\infty([0, T] \times R^d)$ such that $f_n \to f$ in both $L^p([0, T] \times R^d)$ and $L^q([0, T] \times R^d)$, we see that the extensions of K^T are consistently defined for different p's. Finally, if (1.6) holds, then the norm of K^T on $L^p([0, T) \times R^d)$, $p_1 \le p \le p_2$, does not exceed $\frac{1}{2}$; and so $(I - K^T)^{-1}$ exists and is given by

$$\sum_0^\infty (K^T)^n.$$

It follows that for $p \in [p_1, p_2]$, $(I - K^T)^{-1}$ has norm no greater than 2 and that $(I - K^T)^{-1}$ is consistently defined for p's in this range.   □

7.1.6 Theorem. *Let* $a: [0, \infty) \times R^d \to S_d$ *be a bounded measurable function satisfying* (1.4) *for some* $0 < \lambda \le \Lambda < \infty$. *Suppose that there exists a measurable* $c: [0, \infty) \to S_d$ *satisfying* (1.1) *and* $(d + 2)/2 < p < \infty$ *such that*

$$(1.7) \qquad \sup_{(t, y)} \|a(t, y) - c(t)\| \le 1/d^2 C_d(p, \lambda, \Lambda);$$

define K^T *and* $(I - K^T)^{-1}$ *on* $L^p([0, T) \times R^d)$ *as in Lemma 7.1.5 for each* $T > 0$. *Then the martingale problem for* a *and 0 is well-posed; and if* $\{P_{s, x}: (s, x) \in [0, \infty) \times R^d\}$ *is the associated family of solutions, then for* $0 \le s < T$, $x \in R^d$, *and* $f \in C_0^\infty([s, T] \times R^d)$:

$$(1.8) \qquad E^{P_{s, x}}\left[\int_s^T f(t, x(t)) \, dt\right] = G_c^T \circ (I - K^T)^{-1}f(s, x).$$

In particular, $\{P_{s, x}: (s, x) \in [0, \infty) \times R^d\}$ *is measurable.*

Proof. We will first prove the existence of solutions. To this end, let $\rho \in C_0^\infty(R^d)$ be a non-negative function with $\int_{R^d} \rho(x) \, dx = 1$, and define $\rho_n(x) = n^d \rho(nx)$ for $n \ge 1$ and $x \in R^d$. Set $a_n(t, \cdot) = \rho_n * a(t, \cdot)$. Clearly $a_n(t, \cdot)$ is continuous for all $t \ge 0$, (1.4) and (1.7) are satisfied with a_n replacing a, and for each compact $K \subseteq [0, \infty) \times R^d$ and $1 \le q < \infty$

$$\lim_{n \to \infty} \int_K \int \|a_n(t, y) - a(t, y)\|^q \, dt \, dy = 0.$$

Given $(s, x) \in [0, \infty) \times R^d$, let $P^{(n)}$ be a solution to the martingale problem for a_n and 0 starting from (s, x). Because the a_n's are bounded independent of n, $\{P^{(n)}: n \ge 1\}$ is weakly compact on Ω. Let $\{P^{(n')}\}$ be a convergent subsequence of $\{P^{(n)}\}$ and let P be its limit. We want to show that P solves for a and 0 starting from (s, x). This is easily seen to reduce to showing that if $s \le t_1 < t_2$ and $\Phi: \Omega \to R^1$ is a bounded $\mathcal{M}_{t_1}$-measurable function, then for $f \in C_0^\infty(R^d)$:

$$(1.9) \qquad \lim_{n' \to \infty} E^{P^{(n')}}\left[\Phi \int_{t_1}^{t_2} L_t^{(n')} f(x(t))\, dt\right] = E^P\left[\Phi \int_{t_1}^{t_2} L_t f(x(t))\, dt\right],$$

where

$$L_t = \frac{1}{2} \sum_{i,j=1}^d a^{ij}(t, \cdot)\frac{\partial^2}{\partial x_i\, \partial x_j}$$

and

$$L_t^{(n)} = \frac{1}{2} \sum_{i,j=1}^d a_n^{ij}(t, \cdot)\frac{\partial^2}{\partial x_i\, \partial x_j}.$$

To prove (1.9), let $\varepsilon > 0$ be given and choose N so that

$$(1.10) \qquad \left(\int_s^{t_2} dt \int_{\text{supp}(f)} \|a_{n'}(t, y) - a_N(t, y)\|^p\, dy\right)^{1/p} \le \varepsilon$$

for all $n' \ge N$. Note that:

$$\limsup_{n' \to \infty}\left|E^{P^{(n')}}\left[\Phi \int_{t_1}^{t_2} L_t^{(n')} f(x(t))\, dt\right] - E^P\left[\Phi \int_{t_1}^{t_2} L_t f(x(t))\, dt\right]\right|$$

$$\le \limsup_{n' \to \infty}\left|E^{P^{(n')}}\left[\Phi \int_{t_1}^{t_2} (L_t^{(n')} - L_t^{(N)}) f(x(t))\, dt\right]\right|$$

$$+ \left|E^P\left[\Phi \int_{t_1}^{t_2} (L_t - L_t^{(N)}) f(x(t))\, dt\right]\right|,$$

since

$$E^{P^{(n')}}\left[\Phi \int_{t_1}^{t_2} L_t^{(N)} f(x(t))\, dt\right] \to E^P\left[\Phi \int_{t_1}^{t_2} L_t^{(N)} f(x(t))\, dt\right]$$

as $n' \to \infty$. But there is a constant M depending only on the bounds on Φ and the second derivatives of f such that

$$\left|E^{P^{(n')}}\left[\Phi \int_{t_1}^{t_2} (L_t^{(n')} - L_t^{(N)})(f(x(t))\, dt\right]\right|$$

$$\le M E^{P^{(n')}}\left[\int_{t_1}^{t_2} \chi_{\text{supp}(f)}(x(t))\|a_{n'}(t, x(t)) - a_N(t, x(t))\|\, dt\right]$$

$$\le 2M A_d(p, \lambda, \Lambda) t_2^{1-(d+2)/2p}$$

$$\times \left(\int_s^{t_2} dt \int_{\text{supp}(f)} \|a_{n'}(t, y) - a_N(t, y)\|^p\, dy\right)^{1/p},$$

where we have used Lemma 7.1.4 in the derivation of the last line. Thus, by (1.10), if $n' \geq N$, then

$$\left| E^{P^{(n')}} \left[\Phi \int_{t_1}^{t_2} (L_t^{(n')} - L_t^{(N)}) f(x(t)) \, dt \right] \right| \leq 2M A_d(p, \lambda, \Lambda) t_2^{1-(d+2)/2p} \varepsilon.$$

Since it is obvious that

$$\left(\int_s^{t_2} dt \int_{\text{supp}(f)} \| a(t, y) - a_N(t, y) \|^p \, dy \right)^{1/p} \leq \varepsilon,$$

we can derive exactly the same estimate for

$$\left| E^P \left[\Phi \int_{t_1}^{t_2} (L_t - L_t^{(N)}) f(x(t)) \, dt \right] \right|$$

once we have proved that (1.3) obtains for our P. But (1.3) holds when the P there is replaced by $P^{(n')}$, and therefore it must also hold for the weak limit of $\{P^{(n')}\}$; that is, for our P. We have therefore derived the following estimate:

$$\limsup_{n' \to \infty} \left| E^{P^{(n')}} \left[\Phi \int_{t_1}^{t_2} L^{(n')} f(x(t)) \, dt \right] - E^P \left[\Phi \int_{t_1}^{t_2} L f(x(t)) \, dt \right] \right|$$
$$\leq 4M A_d(p, \lambda, \Lambda) \, t_2^{1-(d+2)/2p} \varepsilon.$$

Since $\varepsilon > 0$ was arbitrary, our proof of existence is now complete.

In view of Corollary 6.2.6, uniqueness of solutions as well as the measurability of $\{P_{s, x} : (s, x) \in [0, \infty) \times R^d\}$ will follow as soon as we prove that (1.8) holds when $P_{s, x}$ is an arbitrary solution starting from (s, x). Given a solution P starting from (s, x) and $T > s$, we know, from Lemma 7.1.4, that there is a $\varphi \in L^{p'}([s, t] \times R^d)$ such that

$$E^P \left[\int_s^T f(t, x(t)) \, dt \right] = \int_s^T dt \int_{R^d} f(t, y) \varphi(t, y) \, dy$$

for all $f \in C_0^\infty([s, T] \times R^d)$. Given $f \in C_0^\infty([s, T] \times R^d)$, we now have from Lemma 7.1.1:

$$-G_c^T f(s, x) = -E^P \left[\int_s^T f(t, x(t)) \, dt \right] + E^P \left[\int_s^T K^T f(t, x(t)) \, dt \right]$$
$$= -\int_s^T dt \int_{R^d} [(I - K^T) f](t, y) \varphi(t, y) \, dy.$$

Since this equation holds for all $f \in C_0^\infty([s, T] \times R^d)$ and both sides are continuous with respect to $L^p([s, T] \times R^d)$-convergence, we conclude that

$$G_c^T f(s, x) = \int_s^T dt \int_{R^d} [(I - K^T)f](t, y)\varphi(t, y) \, dy$$

for all $f \in L^p([s, T] \times R^d)$. But $(I - K^T)$ is invertible on $L^p([s, T] \times R^d)$, and therefore

$$E^P\left[\int_s^T f(t, x(t)) \, dt\right] = \int_s^T dt \int_{R^d} f(t, y)\varphi(t, y) \, dy$$
$$= G_c^T \circ (I - K^T)^{-1} f(s, x)$$

for all $f \in C_0^\infty([s, T] \times R^d)$. The proof is now complete. $\square$

7.1.7 Corollary. *Let $a: [0, \infty) \times R^d \to S_d$ be a measurable function satisfying the conditions of Theorem 7.1.6 and let $b: [0, \infty) \times R^d \to R^d$ be bounded and measurable. Then the martingale problem for a and b is well-posed and the associated family $\{P_{s, x}: (s, x) \in [0, \infty) \times R^d\}$ is measurable and has the strong Markov property.*

Proof. This result is an immediate consequence of Theorem 7.1.6, Theorem 6.4.3, and Theorem 6.2.2. $\square$

We have not as yet taken full advantage of (1.8). Obviously (1.8) implies regularity properties for the dependence of $P_{s, x}$ on (s, x). To be precise, suppose that $a: [0, \infty) \times R^d \to S_d$ satisfies the conditions of Theorem 7.1.6, let $\{P_{s, x}: (s, x) \in [0, \infty) \times R^d\}$ denote the associated family of solutions to the martingale problem for a and 0, and denote by $P(s, x; t, \cdot)$ the corresponding transition probability function. Given $T > 0$ and a bounded measurable $\varphi: R^d \to \mathbb{C}$, set

(1.11) $\qquad f(s, x) = \int \varphi(y)P(s, x; T, dy), \qquad 0 \leq s < T \quad \text{and} \quad x \in R^d.$

By the Chapman–Kolmogorov equation, one has:

(1.12) $\qquad\qquad\qquad f(s, x) = \int f(t, y)P(s, x; t, dy)$

for all $s < t < T$. Thus, if $0 < h < T - s$, then

(1.13) $\qquad\qquad\qquad f(s, x) = \frac{1}{h}\int_{T-h}^T dt \int f(t, y)P(s, x; t, dy).$

We are now ready to prove the next result.

7.1.8 Lemma. *Let $P(s, x; t, \cdot)$ be the transition probability function described in the preceding paragraph. Given $h > 0$ and $\varepsilon > 0$, there is a $\delta_h(\varepsilon) > 0$ such that for any $T > h$ and $(s, x), (\sigma, \xi) \in [0, T - h] \times R^d$ satisfying $|\sigma - s| \vee |\xi - x| \leq \delta_h(\varepsilon)$:*

$$\left| \int \varphi(y)P(s, x; T, dy) - \int \varphi(y)P(\sigma, \xi; T, dy) \right| \leq \varepsilon \|\varphi\|$$

for all $\varphi \in B(R^d)$. Besides h and ε, the choice of $\delta_h(\varepsilon)$ can be made to depend only on d, p, λ, and Λ (where p, λ, and Λ are as in Theorem 7.1.6).

Proof. Let $h > 0$ and $(s, x), (\sigma, \xi) \in [0, \infty) \times R^d$ with $\sigma \leq s$ be given. For any $R > 0$, set $\varphi_R(y) = \chi_{B(x, R)}(y)\varphi(y)$. Then

$$(1.14) \quad \left| \int \varphi(y)P(s, x; s + h, dy) - \int \varphi(y)P(\sigma, \xi; s + h, dy) \right|$$

$$\leq \left| \int \varphi_R(y)P(s, x; s + h, dy) - \int \varphi_R(y)P(\sigma, \xi; s + h, dy) \right|$$

$$+ \|\varphi\|(P_{s, x}[|x(s + h) - x| \geq R] + P_{\sigma, \xi}[|x(s + h) - x| \geq R]).$$

Given $\varepsilon > 0$, use (2.1) of Chapter 4, to choose $R > 0$ such that

$$P_{\tau, \eta}[|x(s + h) - x| \geq R] \leq \varepsilon/2$$

for all $(\tau, \eta) \in [(s - 1) \vee 0, s] \times B(x, 1)$. Having chosen this R, define

$$f(t, \cdot) = \chi_{[s, s+h]}(t) \int \varphi_R(y)P(t, \cdot, s + h, dy).$$

Note that by estimate (2.1) of Chapter 4:

$$(1.15) \qquad\qquad \|f\|_{L^p([0, s+h] \times R^d)} \leq A\|\varphi\|,$$

where A depends only on d, p, Λ, R and h. Thus by (1.13) and (1.8),

$$\left| \int \varphi_R(y)P(s, x; s + h, dy) - \int \varphi_R(y)P(\sigma, \xi; s + h, dy) \right|$$

$$= \frac{1}{h} |G_c^{s+h}F(s, x) - G_c^{s+h}F(\sigma, \xi)|$$

where $F = (I - K^{s+h})^{-1}f$. But an easy computation shows that

$$\lim_{\sigma \nearrow s} \lim_{\xi \to y} \int_\sigma^{\sigma+h} dt \int dy |g_c(s, x, t, y) - g_c(\sigma, \xi, t, y)|^{p'} = 0$$

at a rate depending only on d, p, λ, Λ and h (we have extended $g_c(s, x, t, y)$ to be zero for $t < s$). Thus we can find $1 \geq \delta > 0$ so that

(1.16) $$\left| G_c^{s+h} F(s, x) - G_c^{s+h} F(\sigma, \xi) \right| \leq \frac{\varepsilon}{4A} \|F\|_{L^p([0, s+h] \times R^d)}$$

so long as $(\sigma, \xi) \in [(s - \delta) \vee 0, s] \times B(x, \delta)$. Combining (1.15) with (1.16) and substituting them into (1.14), we arrive at:

(1.17) $$\left| \int \varphi(y) P(s, x; s + h, dy) - \int \varphi(y) P(\sigma, \xi; s + h, dy) \right| \leq \varepsilon \|\varphi\|$$

whenever $(\sigma, \xi) \in [(s - \delta) \vee 0, s] \times B(x, \delta)$.

Finally suppose that $0 < h < T$ and (s, x), $(\sigma, \xi) \in [0, T - h] \times R^d$ are given. We may and will assume that $\sigma \leq s$. Given $\varphi \in B(R^d)$ define

$$\psi(\cdot) = \int \varphi(y) P(s + h, \cdot; T, dy).$$

Then if $s - \sigma$ and $|\xi - x|$ are less than δ, we have from (1.17) that

$$\left| \int \varphi(y) P(s, x; T, dy) - \int \varphi(y) P(\sigma, \xi; T, dy) \right|$$

$$= \left| \int \psi(y) P(s, x; s + h, dy) - \int \psi(y) P(\sigma, \xi; s + h, dy) \right|$$

$$\leq \varepsilon \|\psi\| \leq \varepsilon \|\varphi\|$$

as was to be proved. $\square$

7.1.9 Theorem. *Let $a: [0, \infty) \times R^d \to S_d$ be a measurable function which satisfies the conditions of Theorem 7.1.6 for some $0 < \lambda \leq \Lambda < \infty$, $(d+2)/2 < p < \infty$, and measurable $c: [0, \infty) \to S_d$. Let $b: [0, \infty) \times R^d \to R^d$ be a bounded measurable function. Denote by $\{Q_{s, x}: (s, x) \in [0, \infty) \times R^d\}$ the family of solutions to the (well-posed) martingale problem for a and b. Given $h > 0$ and $\varepsilon > 0$, there is a $\delta_h(\varepsilon) > 0$, depending only on d, p, λ, Λ, and $\|b\|$ as well as h and ε, such that for any $T > h$ and (σ, ξ), $(s, x) \in [0, T - h] \times R^d$ satisfying $|\sigma - s| \vee |\xi - x| \leq \delta_h(\varepsilon)$:*

$$\left| E^{Q_{s, x}}[\Phi] - E^{Q_{\sigma, \xi}}[\Phi] \right| \leq \varepsilon \|\Phi\|$$

whenever $\Phi: \Omega \to C$ is a bounded $\mathscr{M}^T (\equiv \sigma(x(t): t \geq T))$-measurable function. In particular, the family $\{Q_{s,x}: (s, x) \in [0, \infty) \times R^d\}$ is strongly Feller continuous.

Proof. Let $\{P_{s, x}: (s, x) \in [0, \infty) \times R^d\}$ be the family of solutions for a and 0. For $(s, x) \in [0, \infty) \times R^d$ and $s \leq u < v$, define

$$R^u(v) = \exp\left[\int_u^v \langle a^{-1} b(t, x(t)), dx(t) \rangle - \frac{1}{2} \int_u^v \langle b, a^{-1} b \rangle(t, x(t)) dt \right]$$

relative to $P_{s,x}$. By Theorem 6.4.3, we know that if $t > s$, then $Q_{s,x} \ll P_{s,x}$ on $\mathcal{M}_t$ and that $R^s(t)$ is the corresponding Radon–Nikodym derivative.

Now let $h > 0$ be given, and suppose that (σ, ξ), $(s, x) \in [0, T - h] \times R^d$ with $\sigma \leq s$. Define

$$\varphi_\alpha(y) = E^{Q_{s+\alpha, y}}[\Phi], \qquad 0 \leq \alpha \leq h.$$

By the Markov property:

$$E^{Q_{t,y}}[\Phi] = E^{Q_{t,y}}[\varphi_\alpha(x(s + \alpha))]$$

for all $t \leq s$ and $0 \leq \alpha \leq h$. Thus:

$$
\begin{aligned}
|E^{Q_{s,x}}[\Phi] &- E^{Q_{\sigma,\xi}}[\Phi]| \\
&= |E^{P_{s,x}}[R^s(s + \alpha)\varphi_\alpha(x(s + \alpha))] - E^{P_{\sigma,\xi}}[R^\sigma(s + \alpha)\varphi_\alpha(x(s + \alpha))]| \\
&\leq |E^{P_{s,x}}[\varphi_\alpha(x(s + \alpha))] - E^{P_{\sigma,\xi}}[\varphi_\alpha(x(s + \alpha))]| \\
&\quad + \|\varphi_\alpha\|(E^{P_{s,x}}[|R^s(s + \alpha) - 1|] + E^{P_{\sigma,\xi}}[|R^\sigma(s + \alpha) - 1|]).
\end{aligned}
$$

Clearly $\|\varphi_\alpha\| \leq \|\Phi\|$,

$$
\begin{aligned}
(E^{P_{s,x}}[|R^s(s + \alpha) - 1|])^2 &\leq E^{P_{s,x}}[(R^s(s + \alpha) - 1)^2] \\
&= E^{P_{s,x}}[(R^2(s + \alpha))^2] - 1 \\
&\leq \exp\{\alpha\|b\|^2/\lambda\} - 1,
\end{aligned}
$$

and similarly:

$$
\begin{aligned}
(E^{P_{\sigma,\xi}}[|R^\sigma(s + \alpha) - 1|])^2 &\leq \exp\{(s + \alpha - \sigma)\|b\|^2/\lambda\} - 1 \\
&= \exp\{\alpha\|b\|^2/\lambda\} - 1 + \exp\{\alpha\|b\|^2/\lambda\}(\exp\{(s - \sigma)\|b\|^2/\lambda\} - 1).
\end{aligned}
$$

Given $\varepsilon > 0$, first choose $0 < \alpha_\varepsilon < h$ so that

$$\exp\{\alpha_\varepsilon\|b\|^2/\lambda\} - 1 < (\varepsilon/3)^2.$$

Then, choose $\beta_\varepsilon > 0$ so that

$$\exp\{\alpha_\varepsilon\|b\|^2/\lambda\}(\exp\{\beta_\varepsilon\|b\|^2/\lambda\} - 1) < (\varepsilon/3)^2.$$

Finally, use Lemma 7.1.8 to choose $0 < \delta_h(\varepsilon) \leq \beta_\varepsilon$ so that

$$|E^{P_{s,x}}[\varphi_{\alpha_\varepsilon}(x(s + \alpha_\varepsilon))] - E^{P_{\sigma,\xi}}[\varphi_{\alpha_\varepsilon}(x(s + \alpha_\varepsilon))]| < \frac{\varepsilon}{3}\|\varphi_{\alpha_\varepsilon}\|$$

whenever $|\sigma - s| \vee |\xi - x| \leq \delta_h(\varepsilon)$. With this choice of $\delta_h(\varepsilon)$, we conclude that (1.18) holds for all (σ, ξ), $(s, x) \in [0, T - h] \times R^d$ satisfying $\sigma \leq s$ and $|\sigma - s| \vee |\xi - x| \leq \delta_h(\varepsilon)$. Clearly, this proves the theorem. $\square$

7.2. Uniqueness: Global Case

Most of the work is now done. In order to obtain the result announced in Section 7.0, all that remains to do is feed the results of the preceding section into the machinery developed in Chapter 6.

7.2.1 Theorem. *Let* $a: [0, \infty) \times R^d \to S_d^+$ *and* $b: [0, \infty) \times R^d \to R^d$ *be bounded measurable functions. Assume that for each* $T > 0$ *and* $x \in R^d$:

$$(2.1) \qquad \inf_{0 \leq s \leq T} \inf_{\theta \in R^d \setminus \{0\}} \langle \theta, a(s, x)\theta \rangle / |\theta|^2 > 0,$$

$$(2.2) \qquad \lim_{y \to x} \sup_{0 \leq s \leq T} \|a(s, y) - a(s, x)\| = 0, \qquad x \in R^d.$$

Then the martingale problem for a *and* b *is well posed and the corresponding family* $\{P_{s, x}: (s, x) \in [0, \infty) \times R^d\}$ *is measurable. In particular, for any* $s \geq 0$ *and any stopping time* $\tau \geq s$, $\{\delta_\omega \otimes_{\tau(\omega)} P_{\tau(\omega), x(\tau, \omega)}\}$ *is a r.c.p.d. of* $P_{s, x} | \mathcal{M}_\tau$.

Proof. We will show that the hypotheses of Theorem 6.6.1 are met by a and b.

Let $(s, x) \in [0, \infty) \times R^d$ be given. Then we can find $\delta > 0$ and $0 < \lambda \leq \Lambda < \infty$ such that

$$\lambda |\theta|^2 \leq \langle \theta, a(t, y)\theta \rangle \leq \Lambda |\theta|^2$$

for all $0 \leq t \leq s + \delta$, $|y - x| \leq \delta$, and $\theta \in R^d$. Let $d + 2/2 < p < \infty$ and denote by $C_d(p, \lambda, \Lambda)$ the constant in (0.4) of the appendix. Then we can find an $0 < \varepsilon < \delta$ so that if $\mathcal{G} = [((s - \varepsilon) \vee 0, s + \varepsilon) \times B(x, \varepsilon)]$, then

$$\sup_{(t, y) \in \mathcal{G}} \|a(t, y) - a(t, x)\| \leq 1/d^2 C_d(p, \lambda, \Lambda).$$

Now let $\tilde{c}: [0, \infty) \to S_d$ be a measurable function satisfying $\lambda |\theta|^2 \leq \langle \theta, \tilde{c}(t)\theta \rangle \leq \Lambda |\theta|^2$ for all $t \geq 0$ and $\theta \in R^d$ and equaling $a(t, x)$ for $t \in [0, s + \delta]$. Next, let $\tilde{a}: [0, \infty) \times R^d \to S_d$ be the measurable function given by

$$\tilde{a}(t, y) = \chi_{\mathcal{G}}(t, y)a(t, y) + (1 - \chi_{\mathcal{G}}(t, y))\tilde{c}(t).$$

Clearly $\lambda |\theta|^2 \leq \langle \theta, \tilde{a}(t, y)\theta \rangle \leq \Lambda |\theta|^2$ for all $(t, y) \in [0, \infty) \times R^d$ and $\theta \in R^d$, and

$$\sup_{(t, y)} \|\tilde{a}(t, y) - \tilde{c}(t)\| \leq 1/d^2 C_d(p, \lambda, \Lambda).$$

Finally, define $\tilde{b} = b$. Then Corollary 7.1.7 applies to $\tilde{a}$ and $\tilde{b}$, and so the martingale problem for $\tilde{a}$ and $\tilde{b}$ is well-posed and the associated family of solutions is measurable. We have therefore shown that the hypotheses of Theorem 6.6.1 are satisfied by a and b. $\square$

7.2.2 Theorem. *Let* $a: [0, \infty) \times R^d \to S_d$ *be a bounded measurable function satisfying* *(2.1) and (2.2) and suppose that* $b: [0, \infty) \times R^d \to R^d$ *is a bounded measurable function such that*

$$\sup_{0 \le t \le T} \sup_{x \in R^d} \langle b(t, x), a^{-1}(t, x)b(t, x) \rangle < \infty \quad \text{for all} \quad T > 0.$$

Given $(s, x) \in [0, \infty) \times R^d$, *denote by* $P_{s, x}$ *and* $Q_{s, x}$, *respectively, the solution to the martingale problem for* a *and* 0 *and* a *and* b *starting from* (s, x). *Then, for each* $T > s$, $Q_{s, x}$ *is absolutely continuous with respect to* $P_{s, x}$ *on* $\mathcal{M}_T$ *and*

$$\exp\left[\int_s^T \langle a^{-1}b(t, x(t)), dx(t) \rangle - \frac{1}{2} \int_s^T \langle b, a^{-1}b \rangle (t, x(t)) \, dt \right]$$

is its Radon–Nikodym derivative.

Proof. In view of Theorem 6.4.2, there is nothing to prove. $\square$

7.2.3 Lemma. *Let* $\tilde{a}: [0, \infty) \times R^d \to S_d$ *and* $\tilde{b}: [0, \infty) \times R^d \to R^d$ *be bounded measurable functions for which the martingale problem is well-posed and the associated family of solutions* $\{\tilde{P}_{s, x}: (s, x) \in [0, \infty) \times R^d\}$ *is measurable. Suppose that* $a:$ $[0, \infty) \times R^d \to S_d$ *and* $b: [0, \infty) \times R^d \to R^d$ *are bounded measurable functions such that* $a = \tilde{a}$ *and* $b = \tilde{b}$ *on* $[s, s + \delta) \times B(x, \delta)$ *for some* $(s, x) \in [0, \infty) \times R^d$ *and* $\delta > 0$, *and let* P *solve the martingale problem for* a *and* b *starting from* (s, x). *If* $0 \le t - s \le \delta/1 + \|\tilde{b}\|$ *and* $f \in B(R^d)$, *then*

$$|E^P[f(x(t))] - E^{\tilde{P}_{s,x}}[f(x(t))]| \le 4d\|f\| \exp\left(-\frac{(\delta - \|\tilde{b}\|(t - s))^2}{2d\tilde{\Lambda}(t - s)} \right),$$

where

$$\tilde{\Lambda} = \sup_{(t, y) \in [0, \infty) \times R^d} \sup_{\theta \in R^d \setminus \{0\}} \langle \theta, \tilde{a}(t, y)\theta \rangle / |\theta|^2.$$

Proof. Let $\tau = (\inf\{t \ge s: |x(t) - x| \ge \delta\}) \wedge (s + \delta)$ and let

$$Q = P \otimes_{\tau(\cdot)} \tilde{P}_{\tau(\cdot), x(\tau(\cdot)), \cdot}.$$

Then Q solves for $\tilde{a}$ and $\tilde{b}$ starting from (s, x), and so $Q = \tilde{P}_{s, x}$. On the other hand, $P = Q$ on $\mathcal{M}_\tau$, and therefore P equals $\tilde{P}_{s, x}$ on $\mathcal{M}_\tau$. This means that

$$|E^P[f(x(t))] - E^{\tilde{P}_{s,x}}[f(x(t))]| \le [P(\tau \le t) + \tilde{P}_{s, x}(\tau \le t)]\|f\|$$
$$= 2\|f\| \tilde{P}_{s, x}(\tau \le t)$$

and the desired estimate follows now from (2.1) of Chapter 4. $\square$

7.2.4 Theorem. *Let* $a: [0, \infty) \times R^d \to S_d$ *and* $b: [0, \infty) \times R^d \to R^d$ *be measurable functions which satisfy the hypotheses of Theorem 7.2.1, and denote by* $\{P_{s, x}: (s, x) \in$

$[0, \infty) \times R^d\}$ the associated family of solutions to the martingale problem for a and b. Given $T > 0$ and $x \in R^d$, let

$$\lambda = \inf_{0 \le t \le T} \inf_{\theta \in R^d \setminus \{0\}} \frac{\langle \theta, a(t, x)\theta \rangle}{|\theta|^2},$$

$$\Lambda = \sup_{0 \le t \le T} \sup_{\theta \in R^d \setminus \{0\}} \frac{\langle \theta, a(t, x)\theta \rangle}{|\theta|^2}$$

and

$$B = \sup_{0 \le t \le T} \sup_{|y - x| \le 1} |b(t, y)|.$$

Also, for $\delta > 0$, define

$$\rho(\delta) = \sup_{0 \le t \le T} \sup_{|y - x| \le \delta} \|a(t, y) - a(t, x)\|.$$

Then for $\varepsilon > 0$ and $0 \le s < T$, there is a $\delta > 0$, depending only on d, λ, Λ, B, $\rho(\cdot)$, and $T - s$ as well as ε, such that

$$|E^{P_{s, x}}[\Phi] - E^{P_{\sigma, \xi}}[\Phi]| \le \varepsilon \|\Phi\|$$

whenever $\Phi : \Omega \to C$ is a bounded $\mathcal{M}^T (\equiv \sigma(x(t) : t \ge T))$-measurable function and $(\sigma, \xi) \in [0, T) \times R^d$ satisfies $|\sigma - s| \vee |\xi - x| \le \delta$. In particular, $\{P_{s,x} : (s, x) \in [0, \infty) \times R^d\}$ is strongly Feller continuous.

Proof. Choose and fix $(d + 2)/2 < p < \infty$ and find $0 < \alpha < 1$ so that

$$\rho(\alpha) \le \left(\frac{\lambda}{2}\right) \wedge \left(\frac{1}{d^2 C_d(p, \lambda/2, 2\Lambda)}\right),$$

where $C_d(p, \lambda/2, 2\Lambda)$ is the constant in (0.4) of the appendix. Then

$$\frac{\lambda}{2} |\theta|^2 \le \langle \theta, a(t, y)\theta \rangle \le 2\Lambda |\theta|^2, \qquad \theta \in R^d,$$

and

$$\|a(t, y) - a(t, x)\| \le \frac{1}{d^2 C_d(p, \lambda/2, 2\Lambda)}$$

for all $0 \le t \le T$ and $|y - x| \le \alpha$. Define

$$\tilde{a}(t, y) = \begin{cases} a(t, y) & \text{if } 0 \le t \le T \text{ and } |y - x| \le \alpha \\ a(t \wedge T, x) & \text{otherwise} \end{cases}$$

and

$$\tilde{b}(t, y) = \begin{cases} b(t, y) & \text{if } 0 \le t \le T \text{ and } |y - x| \le \alpha \\ 0 & \text{otherwise,} \end{cases}$$

and let $\{\tilde{P}_{s,x} : (s, x) \in [0, \infty) \times R^d\}$ be the family of solutions to the martingale problem for $\tilde{a}$ and $\tilde{b}$ (cf. Theorem 7.1.6 or Corollary 7.1.7).

Now suppose that $\varepsilon > 0$ is given. Choose $t \in (s, T)$ so that $(t - s) < \frac{1}{3}\alpha/(1 + B)$ and

$$8d \exp\left[-\left(\frac{\alpha}{2} - B(t - s)\right)^2 / 2d\Lambda(t - s)\right] < \varepsilon/3.$$

Next choose $\gamma > 0$, in accordance with Theorem 7.1.9, so that

$$|E^{\tilde{P}_{s,x}}[\Phi] - E^{\tilde{P}_{\sigma,\xi}}[\Phi]| \le \frac{\varepsilon}{2}\|\Phi\|$$

for all bounded $\Phi : \Omega \to \mathbb{C}$ which are $\mathcal{M}^t$-measurable and $(\sigma, \xi) \in (0, t) \times R^d$ satisfying $|\sigma - s| \vee |\xi - x| \le \gamma$. This γ depends only on $t - s, d, p, \lambda/2, 2\Lambda$, and B as well as $\varepsilon/2$. Finally, choose $\delta > 0$ so that $\delta \le (t - s) \wedge \gamma$ and

$$8d \exp\left[-\left(\frac{\alpha}{2} - B(t - s + \delta)\right)^2 / 2d\,\Lambda(t - s + \delta)\right] < \varepsilon/2.$$

Given a bounded $\Phi : \Omega \to \mathbb{C}$ which is $\mathcal{M}^T$-measurable, set

$$f(y) = E^{P_{t,y}}[\Phi], \qquad y \in R^d.$$

Then for $(\sigma, \xi) \in [0, t] \times R^d$ satisfying $|\sigma - s| \vee |\xi - x| \le \delta$:

$$|E^{P_{\sigma,\xi}}[\Phi] - E^{P_{s,x}}[\Phi]| \le |E^{P_{\sigma,\xi}}[f(x(t))] - E^{\tilde{P}_{\sigma,\xi}}[f(x(t))]|$$
$$+ |E^{\tilde{P}_{\sigma,\xi}}[f(x(t))] - E^{\tilde{P}_{s,x}}[f(x(t))]|$$
$$+ |E^{\tilde{P}_{\sigma,\xi}}[f(x(t))] - E^{\tilde{P}_{s,x}}[f(x(t))]| \quad < \varepsilon\|\Phi\|. \quad \square$$

We have used here Lemma 7.2.3 and the inequality $\|f\| \le \|\Phi\|$.

7.3. Exercises

7.3.1. Let $a : [0, \infty) \times R^d \to S_d$ and $b : [0, \infty) \times R^d \to R^d$ be measurable functions satisfying the hypotheses of Theorem 7.2.1. Denote by $\{P_{s, x} : (s, x) \in [0, \infty) \times R^d\}$ and $\{Q_{s, x} : (s, x) \in [0, \infty) \times R^d\}$ the family of solutions to the martingale problem for a and 0 and a and b, respectively. Show that for each $(s, x) \in [0, \infty) \times R^d$ and

$T > s$, $Q_{s,x}$ on $\mathcal{M}_T$ is absolutely continuous with respect to $P_{s,x}$. The idea is the following. Define $b_n: [0, \infty) \times R^d \to R^d$ for $n \geq 1$ by

$$b_n(t, y) = \begin{cases} b(t, y) & \text{if } |y - x| \leq n \\ 0 & \text{otherwise} \end{cases}$$

and let $\{P_{t,y}^{(n)}: (t, y) \in [0, \infty) \times R^d\}$ be the family associated with a and b_n. By Theorem 7.2.2, $P_{s,x}^{(n)} \ll P_{s,x}$ on $\mathcal{M}_T$. Moreover, $Q_{s,x}$ equals $P_{s,x}^{(n)}$ on $\mathcal{M}_{T \wedge \tau_n}$, where

$$\tau_n = \inf\{t \geq s: |x(t) - x| \geq n\}.$$

From these observations, it is easy to conclude that $Q_{s,x}$ itself is absolutely continuous with respect to $P_{s,x}$ on $\mathcal{M}_T$.

With a little more work, one can identify the Radon–Nikodym derivative of $Q_{s,x}$ with respect to $P_{s,x}$ on $\mathcal{M}_T$. To this end, denote this derivative by R and check that

$$E^{P_{s,x}}[R \,|\, \mathcal{M}_{T \wedge \tau_n}] = \exp\left[\int_s^{T \wedge \tau_n} \langle a^{-1}b(u, x(u)), dx(u) \rangle - \frac{1}{2}\int_s^{T \wedge \tau_n} \langle b, a^{-1}b \rangle(u(x(u))\, du\right].$$

Therefore,

$$R = \lim_{n \to \infty} \exp\left[\int_s^{T \wedge \tau_n} \langle a^{-1}b(u, x(u)), dx(u) \rangle - \frac{1}{2}\int_s^{T \wedge \tau_n} \langle b, a^{-1}b \rangle(u, x(u))\, du\right].$$

If one makes the extension of the stochastic integral calculus suggested in 4.6.10, one can rewrite this as

$$R = \exp\left[\int_s^T \langle a^{-1}b(u, x(u)), dx(u) \rangle - \frac{1}{2}\int_s^T \langle b, a^{-1}b \rangle(u, x(u))\, du\right].$$

7.3.2. Using an estimate of Alexandroff [1963] (see also Pucci [1966] and Krylov [1974]), one can show that if $a: R^d \to S^d$ is a bounded continuous function and if $\Lambda|\theta|^2 \geq \langle \theta, a(x)\theta \rangle \geq \lambda|\theta|^2$ for some $0 < \lambda < \Lambda$ and all $\theta \in R^d$ and $x \in R^d$, then for all $p > d$, $T > 0$, $R > 0$ and $f \in C_0(R^d)$ with $\text{supp}(f) \subseteq B(0, R)$:

$$(3.1) \qquad \left| E^{P_x}\left[\int_0^T f(x(t))\, dt\right] \right| \leq C \|f\|_{L^p(R^d)},$$

where C depends only on λ, Λ, p, T and R (not on the modulus of continuity of a). This estimate enables us to prove existence of solutions to the martingale problem for all bounded measurable $a: R^d \to S^d$ and $b: [0, \infty) \times R^d \to R^d$ such that $a(\cdot)$ is uniformly positive definite on compact subsets of R^d. By the reasoning we have already developed, we need only handle the case when $b \equiv 0$ and $\Lambda|\theta|^2 \geq \langle \theta,$

$a(x)\theta\rangle \geq \lambda |\theta|^2$ for all $\theta \in R^d$ and $x \in R^d$ for some $0 < \lambda \leq \Lambda < \infty$. Given such an $a(\cdot)$, choose $\{a_n(\cdot): n \geq 1\}$ to be a sequence of continuous maps of $R^d \to S^d$ so that $\lambda |\theta|^2 \leq \langle \theta, a_n(x)\theta \rangle \leq \Lambda |\theta|^2$ for all $x \in R^d$, $\theta \in R^d$ and $n \geq 1$, and

$$\int_{B(0, R)} \|a_n(x) - a(x)\|^q \, dx \to 0$$

for all $1 \leq q < \infty$ and $R < \infty$. Given $x \in R^d$, let $P_x^{(n)}$, for $n \geq 1$, be the solutions to the martingale problem for

$$L^{(n)} = \frac{1}{2} \sum_{i, j=1}^{d} a_n^{i, j}(x) \frac{\partial^2}{\partial x_i \, \partial x_j}$$

starting from $(0, x)$. Note that $\{P_x^{(n)}; n \geq 1\}$ is weakly conditionally compact and we can assume without loss of generality that it has a weak limit P as $n \to \infty$. Using (3.1), show that if $f \in C_0^\infty(R^d)$ and $\text{supp}(f) \subset B(0, R)$, then

$$(3.2) \qquad \left| E^P\left[\int_0^T f(x(t)) \, dt \right] \right| \leq C \|f\|_{L^p(R^d)},$$

where C is the same constant as in (3.1). Extend (3.1) and (3.2) to bounded measurable f having support in $B(0, R)$. Finally use these estimates to show that for all $0 \leq t_1 < t_2$, bounded continuous $\Phi: \Omega \to T$ and $f \in C_0^\infty(R^d)$:

$$(3.3) \qquad E^P\left[\Phi \int_{t_1}^{t_2} (Lf)(x(t)) \, dt \right] = \lim_{n \to \infty} E^{P_x^{(n)}}\left[\Phi \int_{t_1}^{t_2} (L^{(n)}f)(x(t)) \, dt \right]$$

where $L = \frac{1}{2} \sum a^{ij}(x) \, \partial^2/\partial x_i \, \partial x_j$. Conclude that P is a solution for L starting from x.

7.3.3. When $d = 1$, one can show that the martingale problem is well posed for any bounded measurable coefficients $a: [0, \infty) \times R^1 \to (0, \infty)$ and $b: [0, \infty) \times R^1 \to R^1$ such that a is uniformly positive on compact sets. As usual, we need only look at the case in which $b \equiv 0$ and a satisfies $\lambda \leq a(t, y) \leq \Lambda$ for all $(t, y) \in [0, \infty) \times R^d$, for some $0 < \lambda \leq \Lambda < \infty$. Given such an a, the idea is to show that Theorem 7.1.6 can be applied. Indeed, since $2 > (d + 2)/d$ when $d = 1$, we can take $p = 2$. Next observe that $C_d(2, \Lambda, \Lambda)$ in (0.4) of the Appendix is precisely $2/\Lambda$ when $d = 1$ and that $\sup_{(t, y)} |a(t, y) - c(t)| \leq \Lambda - \lambda$ if $c(\cdot) \equiv \Lambda$. That is

$$\sup_{(t, y)} |a(t, y) - c(t)| \leq 2\left(1 - \frac{\lambda}{\Lambda}\right) \bigg/ d^2 C_d(2, \Lambda, \Lambda).$$

If $2(1 - (\lambda/\Lambda)) \leq 1$, then Theorem 7.1.6 applies directly as it is stated. In general, note that the proof of Theorem 7.1.6 goes through essentially unaltered if one replaces (1.6) by

$$\sup_{(t,\, y)} \|a(t,\, y) - c(t)\| \leq \alpha/d^2 C_d(p,\, \lambda,\, \Lambda)$$

for some $\alpha < 2$.

The above method establishes the uniqueness; and existence can be seen from Exercise 7.3.2. Alternatively, consider a sequence $a_n(t, x)$ of continuous coefficients satisfying uniformly the bounds $0 < \lambda \leq a_n(t, x) \leq \Lambda < \infty$ for all $n \geq 1$, and converging to $a(t, x)$ in the sense:

$$\lim_{n \to \infty} \int_0^T dt \int_{-M}^M |a_n(t,\, x) - a(t,\, x)|^q \, dx = 0$$

for every $1 \leq q < \infty$, $T > 0$ and $M > 0$. Use the procedure described in the earlier part to derive the following bounds:

$$(3.4) \qquad \left| E^{P_{s,x}^n} \left[\int_s^T f(t,\, x(t)) \, dt \right] \right| \leq C \|f\|_{L^2[[s,\, T] \times R^1]}$$

where C depends only on λ, Λ and T [here $P_{s,x}^n$ is the solution corresponding to $a_n(\cdot, \cdot)$ starting from (s, x)]. Now use (3.4) instead of (3.1), and proceed as in Exercise 7.3.2.

7.3.4. It turns out that there is another case in which the type of reasoning used in 7.3.3 yields an interesting result. Namely, when $d = 2$ and $a: R^2 \to S^2$ is bounded and measurable and is uniformly positive definite on compact subsets of R^2, the martingale problem for a and any bounded measurable $b: [0, \infty) \times R^2 \to R^2$ is well posed. The existence of course is covered by Exercise 7.3.2. The proof of uniqueness rests on the following procedure initially carried out by Krylov [1969]. Again we need only look at the case when $b \equiv 0$ and for some $0 < \lambda \leq \Lambda < \infty$, $\lambda |\theta|^2 \leq \langle \theta, a(x)\theta \rangle \leq \Lambda |\theta|^2$ for all $\theta \in R^d$ and $x \in R^d$. In addition, the results of Section 6.4 allow us to assume that Trace $a(x) \equiv 2$ for all $x \in R^2$. Given such an a, set $L = \frac{1}{2} \sum_{i, j=1}^2 a^{ij}(x) \, \partial^2/\partial x_i \, \partial x_j$ and define

$$(G_\mu f)(x) = \int_0^\infty \frac{e^{-\mu t}}{2\pi t} \int_{R^2} e^{-|x - y|^2/2t} f(y) \, dy$$

for $\mu > 0$, $x \in R^2$ and $f \in C_0^\infty(R^d)$. The first important remark is that for each $\mu > 0$, there is an extension of G_μ which is a continuous map from $L^2(R^d)$ into $C_b(R^d)$. Next, define

$$K_\mu f = (L - \tfrac{1}{2}\Delta)G_\mu f$$

for $\mu > 0$ and $f \in C_0^\infty(R^d)$. Note that

$$|(K_\mu f)(x)| = \tfrac{1}{2}\left|(a^{11}(x) - 1)\frac{\partial^2 G_\mu f(x)}{\partial x_1^2}\right.$$

$$+ 2a^{12}(x)\frac{\partial^2 G_\mu f(x)}{\partial x_1\,\partial x_2} + (a^{22}(x) - 1)\frac{\partial^2 G_\mu f(x)}{\partial x_2^2}\Big|$$

$$= \tfrac{1}{2}\left|(a^{11}(x) - 1)\left(\frac{\partial^2 G_\mu f}{\partial x_1^2} - \frac{\partial^2 G_\mu f}{\partial x_2^2}\right)(x) + 2a^{12}(x)\frac{\partial^2 G_\mu f}{\partial x_1\,\partial x_2}(x)\right|$$

$$\leq \tfrac{1}{2}[(a^{11}(x) - 1)^2 + (a^{12}(x))^2]^{\frac{1}{2}}$$

$$\times \left[\left(\frac{\partial^2 G_\mu f}{\partial x_1^2}(x) - \frac{\partial^2 G_\mu f}{\partial x_2^2}(x)\right)^2 + 4\left(\frac{\partial^2 G_\mu f}{\partial x_1\,\partial x_2}(x)\right)^2\right]^{\frac{1}{2}}.$$

By elementary Fourier analysis:

$$\int_{R^2}\left[\left(\frac{\partial^2 G_\mu f}{\partial x_1^2}(x) - \frac{\partial^2 G_\mu f}{\partial x_2^2}(x)\right)^2 + 4\left(\frac{\partial^2 G_\mu f}{\partial x_1\,\partial x_2}\right)^2\right]dx$$

$$= \frac{1}{(2\pi)^2}\int_{R^2}\frac{(\xi_1^2 - \xi_2^2)^2 + 4\xi_1^2\xi_2^2}{(\mu + \tfrac{1}{2}\|\xi\|)^2}\,|\hat{f}(\xi)|^2\,d\xi$$

$$\leq \frac{1}{(2\pi)^2}\int_{R^2}|\hat{f}(\xi)|^2\,d\xi = \|f\|_{L^2(R^2)}^2.$$

Moreover, since $a^{11}(x) + a^{22}(x) = 2$,

$$(a^{11}(x) - 1)^2 + (a^{12}(x))^2 = \det|I - a(x)| \leq (1 - \lambda(x))^2$$

where $\lambda(x)$ is the smaller of the two eigenvalues of $a(x)$. Since $\lambda \leq \lambda(x) \leq 1$, we conclude that

(3.5) $\|K_\mu f\|_{L^2(R^2)} \leq (1 - \lambda)^2\|f\|_{L^2(R^2)}.$

Starting with (3.5) and the fact that G_μ maps $L^2(R^2)$ boundedly onto $C_b(R^2)$, one can repeat the arguments leading up to Theorem 7.1.6, only now one should use the functionals $\int_0^\infty e^{-\mu t}f(x(t))\,dt$ instead of $\int_0^T f(x(t))\,dt$. The reader should arrive at the conclusion that the martingale problem is well posed for such an a, and the solution P_x starting from $(0, x)$ satisfies

$$E^{P_x}\left[\int_0^\infty e^{-\mu t}f(x(t))\,dt\right] = G_\mu \circ (I - K_\mu)^{-1}f(x).$$

7.3.5. Show that the families of solutions constructed in 7.3.3–7.3.4 are strongly Feller continuous in the sense described in Theorem 7.1.9.

Chapter 8

Itô's Uniqueness and Uniqueness to the Martingale Problem

8.0. Introduction

The contents of this chapter are based on the work of Watanabe and Yamada [1971]. What we will be trying to do is give a careful comparison of the notion of uniqueness natural to the martingale problem as opposed to the notion of uniqueness inherent in Itô's method. We have already seen indications that there is a distinction between these two notions (cf. Exercise 5.4.2) and that the distinction is intimately connected with questions of measurability of the path $x(\cdot)$ with respect to $\beta(\cdot)$ in the stochastic integral equation

$$(0.1) \qquad x(t) = x + \int_s^{t \vee s} \sigma(u, x(u)) \, d\beta(u) + \int_s^{t \vee s} b(u, x(u)) \, du.$$

Also, we have reason to suspect that the origin of any difference which exists has to do with the fact that Itô's notion of uniqueness is tied to the choice of σ in (0.1) satisfying $a = \sigma\sigma^*$, whereas σ does not appear in the martingale formulation. All in all, there are several unresolved questions originating from such considerations, and it is our purpose in this chapter to deal with some of them.

8.1. Results of Watanabe and Yamada

The first thing that we have to do is give a precise definition of "Itô uniqueness." Unfortunately, the definition is somewhat awkward. Let $\sigma: [0, \infty) \times R^d \to R^d \times R^d$ and $b: [0, \infty) \times R^d \to R^d$ be bounded measurable functions and $(s, x) \in [0, \infty) \times R^d$. We will say that σ and b satisfy Itô's uniqueness condition starting from (s, x) if and only if for every probability space $(E, \mathscr{F}, \mu)$, every non-decreasing family $\{\mathscr{F}_t: t \geq 0\}$ of sub σ-algebras of $\mathscr{F}$, and every triple of right-continuous, μ-almost surely continuous progressively measurable functions $\beta: [0, \infty) \times E \to R^d$, $\xi: [0, \infty) \times E \to R^d$, and $\eta: [0, \infty) \times E \to R^d$ such that $(\beta(t), \mathscr{F}_t, \mu)$ is a d-dimensional Brownian motion and the equations

$$\xi(t) = x + \int_s^{s \vee t} \sigma(u, \xi(u)) \, d\beta(u) + \int_s^{s \vee t} b(u, \xi(u)) \, du, \qquad t \geq 0,$$

and

$$\eta(t) = x + \int_s^{s \vee t} \sigma(u, \eta(u)) \, d\beta(u) + \int_s^{s \vee t} b(u, \eta(u)) \, du, \qquad t \geq 0,$$

hold μ-almost surely, one has $\xi(\cdot) = \eta(\cdot)$ μ-almost surely. Although the preceding is entirely natural, it has an inherent deficiency. Namely, it is not clear how one can use it to compare solutions to Itô's equations involving different Brownian motions. In particular, how does one show that if σ and b satisfy Itô's uniqueness condition starting from (s, x) then there is at most one solution to the martingale problem for $\sigma\sigma^*$ and b starting from (s, x)? Moreover, if Itô uniqueness obtains, must it be true that the solution is measurable with respect to the underlying Brownian motion? The answers to these questions comprise the content of this section. As usual, it is enough to treat the case in which $s = 0$; and so we will restrict our attention to this case.

Our first step is to give a characterization in terms of distributions of what it means for a process to be the solution of a stochastic differential equation.

8.1.1 Theorem. *Let $\sigma: [0, \infty) \times R^d \to R^d \otimes R^d$ and $b: [0, \infty) \times R^d \to R^d$ be bounded measurable functions and define $\tilde{a}: [0, \infty) \times R^{2d} \to S_{2d}$ and $\tilde{b}: [0, \infty) \times R^{2d} \to R^{2d}$ by*

$$(1.1) \qquad\qquad \tilde{a}(t, (y, z)) = \begin{pmatrix} \sigma\sigma^*(t, y) & \sigma(t, y) \\ \sigma^*(t, y) & I \end{pmatrix}$$

and

$$(1.2) \qquad\qquad \tilde{b}(t, (y, z)) = \begin{pmatrix} b(t, y) \\ 0 \end{pmatrix}.$$

Let $(E, \mathscr{F}, \mu)$ be a probability space and suppose that $\xi: [0, \infty) \times E \to R^d$ and $\beta: [0, \infty) \times E \to R^d$ are $\mathscr{B}_{[0, \infty)} \times \mathscr{F}$-measurable, right-continuous, μ-almost surely continuous functions. Set $\mathscr{F}_t = \sigma((\xi(s), \beta(s)): 0 \leq s \leq t)$. Then $(\beta(t), \mathscr{F}_t, \mu)$ is a d-dimensional Brownian motion and

$$(1.3) \qquad \xi(t) = y + \int_0^t \sigma(u, \xi(u)) \, d\beta(u) + \int_0^t b(u, \xi(u)) \, du, \qquad t \geq 0$$

μ-almost surely if and only if the distribution $\tilde{P}$ on Ω_{2d} of the pair $(\xi(\cdot), \beta(\cdot))$ solves the martingale problem for $\tilde{a}$ and $\tilde{b}$ starting from $(0, (y, 0))$.

Proof. We first prove the "only if" assertion. To this end, define $\tilde{\sigma}: [0, \infty) \times E \to R^{2d} \otimes R^d$ by

$$\tilde{\sigma}(t) = \begin{pmatrix} \sigma(t, \xi(t)) \\ I \end{pmatrix}.$$

Since $(\beta(t), \mathscr{F}_t, \mu)$ is a d-dimensional Brownian motion, we know, from Theorem 4.3.8, that $\int_0^{\cdot} \tilde{\sigma}(u)\,d\beta(u) \sim \mathscr{I}_{2d}^0(\tilde{\sigma}\tilde{\sigma}^*(\cdot), 0)$ with respect to $(E, \mathscr{F}_t, \mu)$. But $\tilde{\sigma}\tilde{\sigma}^*(t) = \tilde{a}(t, (\xi(t), \beta(t)))$ and

$$\begin{pmatrix} \xi(t) \\ \beta(t) \end{pmatrix} = \begin{pmatrix} y \\ 0 \end{pmatrix} + \int_0^t \tilde{\sigma}(u)\,d\beta(u) + \int_0^t \tilde{b}(u, (\xi(u), \beta(u)))\,du, \qquad t \geq 0$$

μ-almost surely. Thus, $(\xi(\cdot), \beta(\cdot)) \sim \mathscr{I}_{2d}^0(\tilde{a}(\cdot, \xi(\cdot), \beta(\cdot)), \tilde{b}(\cdot, \xi(\cdot), \beta(\cdot)))$ with respect to $(E, \mathscr{F}_t, \mu)$, and $\mu[(\xi(0), \beta(0)) = (y, 0)] = 1$. It is therefore obvious that $\tilde{P}$ solves the martingale problem for $\tilde{a}$ and $\tilde{b}$ starting from $(0, (y, 0))$.

Now suppose that $\tilde{P}$ solves the martingale problem for $\tilde{a}$ and $\tilde{b}$ starting from $(0, (y, 0))$. Then $\mu[(\xi(0), \beta(0)) = (y, 0)] = 1$ and

$$(\xi(\cdot), \beta(\cdot)) \sim \mathscr{I}_{2d}^0(\tilde{a}(\cdot, (\xi(\cdot), \beta(\cdot))),$$

$\tilde{b}(\cdot, (\xi(\cdot), \beta(\cdot))))$ with respect to $(E, \mathscr{F}_t, \mu)$. In particular, $(\beta(\cdot), \mathscr{F}_t, \tilde{P})$ is a d-dimensional Brownian motion. Moreover, if $\theta \in R^d$ is given and $\tilde{\theta} : [0, \infty) \times E \to R^{2d}$ is defined by:

$$\tilde{\theta}(t) = \begin{pmatrix} \theta \\ -\sigma^*(t, \xi(t))\theta \end{pmatrix}$$

then, by Theorem 4.3.7,

$$\exp\left[\langle \theta, \xi(t) - y - \int_0^t b(u, \xi(u))\,du - \int_0^t \sigma(u, \xi(u))\,d\beta(u) \right.$$
$$\left. -\frac{1}{2}\int_0^t \langle \tilde{\theta}(u), \tilde{a}(u, (\xi(u), \beta(u)))\tilde{\theta}(u)\rangle\,du\right]$$

is a martingale relative to $(E, \mathscr{F}_t, \mu)$. But a simple computation yields

$$\langle \tilde{\theta}(\cdot), \tilde{a}(\cdot, (\xi(\cdot), \beta(\cdot)))\tilde{\theta}(\cdot)\rangle \equiv 0,$$

and so

$$\xi(t) = y + \int_0^t \sigma(u, \xi(u))\,d\beta(u) + \int_0^t b(u, \xi(u))\,du, \qquad t \geq 0$$

μ-almost surely. This completes the proof. $\square$

8.1.2 Corollary. *Let σ, b, $\tilde{a}$, and $\tilde{b}$ be as in Theorem 8.1.1 and set $a = \sigma\sigma^*$. Given a solution P to the martingale problem for a and b starting from $(0, y)$, there is a solution $\tilde{P}$ to the martingale problem for $\tilde{a}$ and $\tilde{b}$ starting from $(0, (y, 0))$ such that $P = \tilde{P} \circ \Pi^{-1}$, where $\Pi : \Omega_{2d} \to \Omega_d$ is defined as the natural projection of Ω_{2d} onto its first component when Ω_{2d} is represented as $\Omega_d \times \Omega_d$.*

Proof. This result is an immediate consequence of Theorems 4.5.2 and 8.1.1.   ☐

We next want to give a distributional characterization of Itô's uniqueness condition.

8.1.3 Lemma. *Given bounded measurable functions* $\sigma: [0, \infty) \times R^d \to R^d \otimes R^d$ *and* $b: [0, \infty) \times R^d \to R^d$, *define* $\hat{a}: [0, \infty) \times R^{2d} \to S_{2d}$ *and* $\hat{b}\, [0, \infty) \times R^{2d} \to R^{2d}$ *by:*

$$(1.4) \qquad \hat{a}(t, x) = \begin{pmatrix} \sigma(t, y^1)\sigma^*(t, y^1) & \sigma(t, y^1)\sigma^*(t, y^2) \\ \sigma(t, y^2)\sigma^*(t, y^1) & \sigma(t, y^2)\sigma^*(t, y^2) \end{pmatrix}$$

and

$$(1.5) \qquad\qquad\qquad \hat{b}(t, x) = \begin{pmatrix} b(t, y^1) \\ b(t, y^2) \end{pmatrix}$$

for $t \geq 0$ *and* $x = (y^1, y^2) \in R^d \times R^d \ (= R^{2d})$. *Then* σ *and* b *satisfy Itô's uniqueness condition starting from* $(0, y)$ *if and only if any solution* $\hat{P}$ *to the martingale problem for* $\hat{a}$ *and* $\hat{b}$ *starting from* $(0, (y, y))$ *has the property that* $\hat{P}(y^1(t) = y^2(t), t \geq 0) = 1$. (*Here* $y^1(\cdot) = (x_1(\cdot), \ldots, x_d(\cdot))$ *and* $y^2(\cdot) = (x_{d+1}(\cdot), \ldots, x_{2d}(\cdot))$).

Proof. First suppose that σ and b satisfy Itô's uniqueness condition starting from $(0, y)$. Let $\hat{P}$ solve the martingale problem for $\hat{a}$ and $\hat{b}$ starting from $(0, (y, y))$. Define $\hat{\sigma}: [0, \infty) \times R^{2d} \to R^{2d} \otimes R^{2d}$ by

$$\hat{\sigma}(t, (y^1, y^2)) = \begin{pmatrix} \sigma(t, y^1) & 0 \\ \sigma(t, y^2) & 0 \end{pmatrix}$$

Define $\tilde{a}: [0, \infty) \times R^{4d} \to S_{4d}$ and $\tilde{b}: [0, \infty) \times R^{4d} \to R^{4d}$ by

$$\tilde{a}(t, ((y^1, y^2), (z^1, z^2))) = \begin{pmatrix} \hat{a}(t, (y^1, y^2)) & \hat{\sigma}(t, (y^1, y^2)) \\ \hat{\sigma}^*(t, (y^1, y^2)) & I \end{pmatrix}$$

and

$$\tilde{b}(t, ((y^1, y^2), (z^1, z^2))) = \begin{pmatrix} \hat{b}(t, (y^1, y^2)) \\ 0 \end{pmatrix}.$$

By Corollary 8.1.2, there is a solution $\tilde{P}$ on $\Omega_{4d} \ (= \Omega_{2d} \times \Omega_{2d})$ to the martingale problem for $\tilde{a}$ and $\tilde{b}$ starting from $(0, ((y, y), (0, 0)))$ such that $\hat{P} = \tilde{P} \circ \Pi^{-1}$, where Π projects $\Omega_{2d} \times \Omega_{2d}$ onto its first component. Thus, we need only show that $\tilde{P}(y^1(t) = y^2(t), t \geq 0) = 1$. But $(z^1(\cdot), z^2(\cdot))$ is a $2d$-dimensional Brownian motion with respect to $(\Omega_{4d}, \mathcal{M}_t, \tilde{P})$ and, by Theorem 8.1.1,

$$\begin{pmatrix} y^1(t) \\ y^2(t) \end{pmatrix} = \begin{pmatrix} y \\ y \end{pmatrix} + \int_0^t \hat{\sigma}(u, (y^1(u), y^2(u))) \, d\begin{pmatrix} z^1(u) \\ z^2(u) \end{pmatrix}$$

$$+ \int_0^t \hat{b}(u, (y^1(u), y^2(u))) \, du, \qquad t \geq 0$$

$\tilde{P}$-almost surely. That is,

$$y^1(t) = y + \int_0^t \sigma(u, y^1(u))\, dz^1(u) + \int_0^t b(u, y^1(u))\, du, \qquad t \geq 0$$

and

$$y^2(t) = y + \int_0^t \sigma(u, y^2(u))\, dz^1(u) + \int_0^t b(u, y^2(u))\, du, \qquad t \geq 0,$$

$\tilde{P}$-almost surely. Thus, by Itô's uniqueness condition, $\tilde{P}(y^1(t) = y^2(t),\, t \geq 0) = 1$.

Next suppose that for any $\hat{P}$, $\hat{P}(y^1(t) = y^2(t),\, t \geq 0) = 1$. Given a probability space $(E, \mathscr{F}, \mu)$, a non-decreasing family $\{\mathscr{F}_t : t \geq 0\}$, and right-continuous, μ-almost surely continuous progressively measurable functions $\xi^1 : [0, \infty) \times E \to R^d$, $\xi^2 : [0, \infty) \times E \to R^d$ and $\beta : [0, \infty) \times E \to R^d$ such that $(\beta(t), \mathscr{F}_t, \mu)$ is a d-dimensional Brownian motion and

$$\xi^i(t) = y + \int_0^t \sigma(u, \xi^i(u))\, d\beta(u) + \int_0^t b(u, \xi^i(u))\, du, \qquad t \geq 0,$$

μ-almost surely for $i = 1, 2$, note that

$$\begin{pmatrix} \xi^1(t) \\ \xi^2(t) \end{pmatrix} = \begin{pmatrix} y \\ y \end{pmatrix} + \int_0^t \begin{pmatrix} \sigma(u, \xi^1(u)) \\ \sigma(u, \xi^2(u)) \end{pmatrix} d\beta(u)$$

$$+ \int_0^t \begin{pmatrix} b(u, \xi^1(u)) \\ b(u, \xi^2(u)) \end{pmatrix} du, \qquad t \geq 0,$$

μ-almost surely. Thus, $\mu((\xi^1(0), \xi^2(0)) = (y, y)) = 1$ and $(\xi^1(\cdot), \xi^2(\cdot)) \sim \mathscr{I}_{2d}^0(\hat{a}(\cdot, (\xi^1(\cdot), \xi^2(\cdot))), \hat{b}(\cdot, (\xi^1(\cdot), \xi^2(\cdot))))$ with respect to $(E, \mathscr{F}_t, \mu)$. But this means that the distribution of $(\xi^1(\cdot), \xi^2(\cdot))$ on Ω_{2d} solves the martingale problem for $\hat{a}$ and $\hat{b}$ starting from $(0, (y, y))$, and therefore $\mu(\xi^1(t) = \xi^2(t),\, t \geq 0) = 1$. $\square$

We now have all the preliminaries. What we still have to do is learn how to compare solutions coming from different underlying Brownian motions. It is the technique for accomplishing this that is at the heart of Watanabe and Yamada's work. We begin with the following simple observation.

8.1.4 Lemma. *Let* $\sigma : [0, \infty) \times R^d \to R^d \otimes R^d$ *and* $b : [0, \infty) \times R^d \to R^d$ *be bounded measurable functions and define* $\tilde{a}$ *and* $\tilde{b}$ *by* (1.1) *and* (1.2), *respectively. Given a solution* $\tilde{P}$ *to the martingale problem for* $\tilde{a}$ *and* $\tilde{b}$ *starting from* $(0, (y, 0))$, *let* $\{\tilde{P}_\omega\}$ *be a r.c.p.d. of* $\tilde{P} | \sigma(z(u) : u \geq 0)$, *where* $z(\cdot) = (x_{d+1}(\cdot), \ldots, x_{2d}(\cdot))$. *Then for all* $t \geq 0$ *and* $A \in \sigma(y(u) : 0 \leq u \leq t)$, *where* $y(\cdot) = (x_1(\cdot), \ldots, x_d(\cdot))$, $\tilde{P}_\cdot(A) = \tilde{P}(A | \sigma(z(u) : 0 \leq u \leq t))$ $\tilde{P}$-almost surely.*

Proof. Let $A \in \sigma(y(u): 0 \le u \le t)$, $B \in \sigma(z(u): 0 \le u \le t)$, and $C \in \sigma(z(u) - z(t)): u \ge t)$. Since $\tilde{P}(C \mid \mathcal{M}_t) = \tilde{P}(C)$ (a.s., $\tilde{P}$), we have:

$$E^P[\tilde{P}.(A), B \cap C] = \tilde{P}(A \cap B \cap C) = \tilde{P}(A \cap B)\tilde{P}(C)$$
$$= E^P[\tilde{P}(A \mid \sigma(z(u): 0 \le u \le t)), B]\tilde{P}(C)$$
$$= E^P[\tilde{P}(A \mid \sigma(z(u): 0 \le u \le t)), B \cap C].$$

Thus, since $\tilde{P}.(A)$ and $\tilde{P}(A \mid \sigma(z(u): 0 \le u \le t))$ are both $\sigma(z(u): u \ge 0)$-measurable, $\tilde{P}.(A) = \tilde{P}(A \mid \sigma(z(u): 0 \le u \le t))$ (a.s., $\tilde{P}$). $\square$

Before stating the next theorem, we need to introduce some notation. Given $\omega \in \Omega_{3d}$, let

$$y^1(\cdot, \omega) = (x_1(\cdot, \omega), \ldots, x_d(\cdot, \omega)),$$
$$y^2(\cdot, \omega) = (x_{d+1}(\cdot, \omega), \ldots, x_{2d}(\cdot, \omega)),$$
$$z(\cdot, \omega) = (x_{2d+1}(\cdot, \omega), \ldots, x_{3d}(\cdot, \omega)).$$

Define the σ-algebras $\mathcal{M}_t^i$, $\mathcal{M}^i$, $\mathcal{N}_t$, and $\mathcal{N}$ over Ω_{3d} by

$$\mathcal{M}_t^i = \sigma(y^i(u): 0 \le u \le t),$$
$$\mathcal{M}^i = \sigma(y^i(u): u \ge 0),$$
$$\mathcal{N}_t = \sigma(z(u): 0 \le u \le t),$$
$$\mathcal{N} = \sigma(z(u): u \ge 0).$$

Given $\omega \in \Omega_{2d}$, define

$$y(\cdot, \omega) = (x_1(\cdot, \omega), \ldots, x_d(\cdot, \omega))$$
$$z(\cdot, \omega) = (x_{d+1}(\cdot, \omega), \ldots, x_{2d}(\cdot, \omega)).$$

Finally, define $\Pi_i: \Omega_{3d} \to \Omega_{2d}$, $i = 1, 2$, so that

$$y(\cdot, \Pi_i\omega) = y^i(\cdot, \omega)$$
$$z(\cdot, \Pi_i\omega) = z(\cdot, \omega).$$

8.1.5 Theorem. *Let $\sigma: [0, \infty) \times R^d \to R^d \otimes R^d$ and $b: [0, \infty) \times R^d \to R^d$ be bounded measurable functions, and define $\tilde{a}$ and $\tilde{b}$ accordingly as in (1.1) and (1.2). Let $\tilde{P}$ and $\tilde{Q}$ on Ω_{2d} each solve the martingale problem for $\tilde{a}$ and $\tilde{b}$ starting from $(0, (y, 0))$. Then there is a solution R on Ω_{3d} to the martingale problem for*

$$\begin{pmatrix} \sigma\sigma^*(t, y^1) & \sigma(t, y^1)\sigma^*(t, y^2) & \sigma(t, y^1) \\ \sigma(t, y^2)\sigma^*(t, y^1) & \sigma\sigma^*(t, y^2) & \sigma(t, y^2) \\ \sigma^*(t, y^1) & \sigma^*(t, y^2) & 1 \end{pmatrix}$$

and

$$\begin{pmatrix} b(t, y^1) \\ b(t, y^2) \\ 0 \end{pmatrix}$$

starting from $(0, (y, y, 0))$ *such that* $\tilde{P} = R \circ \Pi_1^{-1}$ *and* $\tilde{Q} = R \circ \Pi_2^{-1}$. *Moreover, if* $\{\tilde{P}_\omega\}$ *is a r.c.p.d. of* $\tilde{P} | \sigma(z(u): u \geq 0)$ *and* $A \in \sigma(y(u): u \geq 0)$, *then* $R(\Pi_1^{-1}A | \sigma(\mathcal{M}^2 \cup \mathcal{N})) = \tilde{P}_{\Pi_1(\cdot)}(A)$ (a.s., R).

Proof. Define $\Phi_j \colon \Omega_{2d} \to \Omega_{3d}$, $j = 1, 2$, by

$$y^i(\cdot, \Phi_j \omega) = \begin{cases} y(\cdot, \omega) & \text{if } i = j \\ 0 & \text{if } i \neq j \end{cases}$$

$$z(\cdot, \Phi_j \omega) = z(\cdot, \omega).$$

Set $Q' = \tilde{Q} \circ \Phi_2^{-1}$. Let $\{\tilde{P}_\omega\}$ be a r.c.p.d. of $\tilde{P} | \sigma(z(u): u \geq 0)$, and define $P' = \tilde{P}_{\Pi_1(\cdot)} \circ \Phi_1^{-1}$. Clearly $\omega \to P'_\omega(A)$ is $\mathcal{N}$-measurable for all measurable $A \subseteq \Omega_{3d}$. Hence, there is a unique R on Ω_{3d} such that R equals Q' on $\sigma(\mathcal{M}^2 \cup \mathcal{N})$ and $R(A | \sigma(\mathcal{M}^1 \cup \mathcal{N})) = P'_\cdot(A)$ for all $A \in \mathcal{M}^1$. Simply take $R(A \cap B \cap C) = E^{Q'}[P'_\cdot(A), B \cap C]$ for $A \in \mathcal{M}^1$, $B \in \mathcal{M}^2$, and $C \in \mathcal{N}$.

Since $\Pi_j \circ \Phi_j$ equals the identity on Ω_{2d}, $j = 1, 2$, it is obvious that $\tilde{Q} = R \circ \Pi_2^{-1}$. To show that $\tilde{P} = R \circ \Pi_1^{-1}$, observe that Q' coincides with $P \circ \Phi_1^{-1}$ on $\mathcal{N}$. Thus, if $A \in \mathcal{M}^1$ and $C \in \mathcal{N}$, then

$$R(A \cap C) = E^{Q'}[P'_\cdot(A), C] = E^{P \circ \Phi_1^{-1}}[P'_\cdot(A), C]$$
$$= E^P[\tilde{P}_\cdot(\Phi_1^{-1}A), \Phi_1^{-1}C] = \tilde{P}(\Phi_1^{-1}(A \cap C))$$
$$= \tilde{P} \circ \Phi_1^{-1}(A \cap C);$$

and so R equals $\tilde{P} \circ \Phi_1^{-1}$ on $\sigma(\mathcal{M}^1 \cup \mathcal{N})$. This shows that $\tilde{P} = R \circ \Pi_1^{-1}$.

It remains only to show that R solves the asserted martingale problem. From $\tilde{P} = R \circ \Pi_1^{-1}$, $\tilde{Q} = R \circ \Pi_2^{-1}$, and Theorem 8.1.1, we know that $z(\cdot)$ is a d-dimensional Brownian motion with respect to both $(\Omega_{3d}, \sigma(\mathcal{M}_t^1 \cup \mathcal{N}_t), R)$ and $(\Omega_{3d}, \sigma(\mathcal{M}_t^2 \cup \mathcal{N}_t), R)$ and that

$$y^i(t) = y + \int_0^t \sigma(u, y^i(u)) \, dz(u) + \int_0^t b(u, y^i(u)) \, du, \qquad t \geq 0,$$

R-almost surely, $i = 1, 2$. Thus, if we can show that $z(\cdot)$ is a d-dimensional Brownian motion with respect to $(\Omega_{3d}, \mathcal{M}_t, R)$ ($\mathcal{M}_t = \sigma((y^1(u), y^2(u), z(u)): 0 \leq u \leq t))$, then

$$\begin{pmatrix} y^1(t) \\ y^2(t) \\ z(t) \end{pmatrix} = \begin{pmatrix} y \\ y \\ 0 \end{pmatrix} + \int_0^t \begin{pmatrix} \sigma(u, y^1(u)) \\ \sigma(u, y^2(u)) \\ I \end{pmatrix} dz(u) + \int_0^t \begin{pmatrix} b(u, y^1(u)) \\ b(u, y^2(u)) \\ 0 \end{pmatrix} du, \qquad t \geq 0$$

R-almost surely relative to $(\Omega_{3d}, \mathcal{M}_t, R)$; and the desired result is immediate from this plus Theorem 4.3.8. To prove that $z(\cdot)$ is a d-dimensional Brownian motion with respect to $(\Omega_{3d}, \mathcal{M}_t, R)$, let $\theta \in R^d$, $0 \le t_1 < t_2$, $A \in \mathcal{M}_{t_1}^1$, $B \in \mathcal{M}_{t_1}^2$, and $C \in \mathcal{N}_{t_1}$ be given. Note that by Lemma 8.1.4, $\tilde{P}.(\Phi_1^{-1}(A)) = \tilde{P}(A \mid \sigma(z(u)\colon 0 \le u \le t_1))$ (a.s., $\tilde{P}$), and therefore there is a bounded $\mathcal{N}_{t_1}$-measurable function F on Ω_{3d} such that $P'.(A) = F$ (a.s., R). Hence

$$
E^R\left[\exp\left[i\langle\theta, z(t_2)\rangle + \frac{|\theta|^2}{2}t_2\right], A \cap B \cap C\right]
$$

$$
= E^R\left[\exp\left[i\langle\theta, z(t_2)\rangle + \frac{|\theta|^2}{2}t_2\right]F, B \cap C\right]
$$

$$
= E^R\left[\exp\left[i\langle\theta, z(t_1)\rangle + \frac{|\theta|^2}{2}t_1\right]F, B \cap C\right]
$$

$$
= E^R\left[\exp\left[i\langle\theta, z(t_1)\rangle + \frac{|\theta|^2}{2}\right], A \cap B \cap C\right]
$$

and so the proof is complete. $\square$

8.1.6 Corollary (Watanabe and Yamada). *Given bounded measurable functions $\sigma\colon [0, \infty) \times R^d \to R^d \otimes R^d$ and $b\colon [0, \infty) \times R^d \to R^d$ satisfying Itô's uniqueness condition starting from (s, y), set $a = \sigma\sigma^*$. Then there is at most one solution to the martingale problem for a and b starting from (s, y). In fact, for any $z \in R^d$, there is at most one solution to the martingale problem for $\tilde{a}$ and $\tilde{b}$ starting from $(s, (y, z))$, where $\tilde{a}$ and $\tilde{b}$ are given by (1.1) and (1.2).*

Proof. Without loss of generality, we may and will assume that $s = 0$. Moreover, in view of Corollary 8.1.2, the first assertion follows from the second. Finally, we need only prove the second assertion when $z = 0$.

Suppose that $\tilde{P}$ and $\tilde{Q}$ are two solutions to the martingale problem for $\tilde{a}$ and $\tilde{b}$ starting from $(0, (y, 0))$, and let R be the associated measure on Ω_{3d} given in Theorem 8.1.5. Note that (in the notation introduced just before Theorem 8.1.5) the distribution of the pair $(y^1(\cdot), y^2(\cdot))$ under R solves the martingale problem for $\hat{a}$ and $\hat{b}$ (cf. (1.4) and (1.5)) starting from $(0, (y, y))$. Thus, by Lemma 8.1.3, $R(y^1(t) = y^2(t), t \ge 0) = 1$. But this means that $(y^1(\cdot), z(\cdot))$ and $(y^2(\cdot), z(\cdot))$ have the same distribution under R, and therefore $\tilde{P} = R \circ \Pi_1^{-1} = R \circ \Pi_2^{-1} = \tilde{Q}$. $\square$

8.1.7 Corollary. *Let $\sigma\colon [0, \infty) \times R^d \to R^d \otimes R^d$ and $b\colon [0, \infty) \times R^d \to R^d$ be bounded measurable functions and define $\hat{a}$ and $\hat{b}$ accordingly as in (1.4) and (1.5). Then σ and b satisfy Itô's uniqueness condition starting from (s, y) if and only if there is at most one solution to the martingale problem for $\hat{a}$ and $\hat{b}$ starting from $(s, (y, y))$.*

Proof. Again we take $s = 0$. Assume that Itô's uniqueness condition holds and let $\hat{P}$ and $\hat{Q}$ be solutions to the martingale problem for $\hat{a}$ and $\hat{b}$ starting from $(0, (y, y))$. Then (in the notation of Lemma 8.1.3), $\hat{P}(y^1(t) = y^2(t),\ t \geq 0) = \hat{Q}(y^1(t) = y^2(t),\ t \geq 0) = 1$. Moreover, the distribution of $y^1(\cdot)$ under both $\hat{P}$ and $\hat{Q}$ solves the martingale problem for $\sigma\sigma^*$ and b starting at $(0, y)$. Thus, by Corollary 8.1.6, $y^1(\cdot)$ has the same distribution under $\hat{P}$ as it does under $\hat{Q}$. Combining these remarks, we see that $\hat{P} = \hat{Q}$.

Next suppose that there is at most one $\hat{P}$. Because of Lemma 8.1.3, it suffices for us to show that if $\hat{P}$ exists, then $\hat{P}(y^1(t) = y^2(t),\ t \geq 0) = 1$. Given $\hat{P}$, let $\hat{Q}$ be defined on Ω_{2d} by $\hat{Q}(y^1(t_1) \in \Gamma_1, \ldots, y^1(t_n) \in \Gamma_n;\ y^2(t_1) \in \Delta_1, \ldots, y^2(t_n) \in \Delta_n) = \hat{P}(y^1(t_1) \in \Gamma_1 \cap \Delta_1, \ldots, y(t_n) \in \Gamma_n \cap \Delta_n)$ for $0 \leq t_1 < \cdots < t_n$, $\Gamma_1, \ldots, \Gamma_n \in \mathscr{B}_{R^d}$, and $\Delta_1, \ldots, \Delta_n \in \mathscr{B}_{R^d}$. It is easy to check that $\hat{Q}$ again solves the martingale problem for $\hat{a}$ and $\hat{b}$ starting from $(0, (y, y))$; and therefore $\hat{P} = \hat{Q}$. On the other hand, it is obvious that $\hat{Q}(y^1(t) = y^2(t),\ t \geq 0) = 1$; and so the proof is complete. $\square$

8.1.8 Corollary. *Let* $\sigma: [0, \infty) \times R^d \to R^d \otimes R^d$ *and* $b: [0, \infty) \times R^d \to R^d$ *be bounded measurable functions which satisfy Itô's uniqueness condition starting from* (s, y). *Suppose that* $\tilde{P}$ *is a solution to the martingale problem for* $\tilde{a}$ *and* $\tilde{b}$ *(cf. (1.1) and (1.2)) starting from* $(0, (y, 0))$. *Then, not only is there only one such* $\tilde{P}$, *but also*

$$y(t) = E^P[y(t) \mid \sigma(z(u): u \geq 0)]$$

$\tilde{P}$-almost surely for all $t \geq 0$. In other words, if $(\beta(t),\ \mathscr{F}_t,\ \mu)$ is a d-dimensional Brownian motion in some probability space $(E,\ \mathscr{F},\ \mu)$ and if

$$\xi(t) = y + \int_s^{t \vee s} \sigma(u, \xi(u))\, d\beta(u) + \int_s^{t \vee s} b(u, \xi(u))\, du, \qquad t \geq 0$$

μ-almost surely, then, for each $t \geq 0$, $\xi(t)$ differs from a $\sigma(\beta(u):\ 0 \leq u \leq t)$-measurable function on at most a set of μ-measure zero.

Proof. Assume that $s = 0$. In view of Theorem 8.1.1, the second part follows from the first. To prove the first part, construct R on Ω_{3d} as in Theorem 8.1.5 with $\hat{Q} = \tilde{P}$. Then $R(y^1(t) = y^2(t),\ t \geq 0) = 1$. Moreover, the construction of R in Theorem 8.1.5 is such that $R(\Pi_1^{-1}A \mid \sigma(\mathscr{M}^2 \cup \mathscr{N})) = \tilde{P}_{\Pi_1(\cdot)}(A)$ (a.s., R) for $A \in \sigma(y(u): u \geq 0)$, where $\{P_\omega\}$ is a r.c.p.d. of $\tilde{P} \mid \sigma(z(u): u \geq 0)$. Thus, if $\Gamma \in \mathscr{B}_{R^d}$, then

$$E^P[(\chi_\Gamma(y(t)) - \tilde{P}_\cdot(y(t) \in \Gamma))^2]$$
$$= \tilde{P}(y(t) \in \Gamma) - E^P[\tilde{P}_\cdot(y(t) \in \Gamma),\ y(t) \in \Gamma]$$
$$= 0$$

since

$$E^P[\tilde{P}.(y(t) \in \Gamma), y(t) \in \Gamma]$$
$$= E^Q[\tilde{P}.(y(t) \in \Gamma), y(t) \in \Gamma]$$
$$= E^R[\tilde{P}_{\Pi_1(\cdot)}(y^1(t) \in \Gamma), y^2(t) \in \Gamma]$$
$$= R(y^1(t) \in \Gamma, y^2(t) \in \Gamma) = R(y^1(t) \in \Gamma)$$
$$= \tilde{P}(y(t) \in \Gamma).$$

Thus, by Lemma 8.1.4,

$$\chi_\Gamma(y(t)) = \tilde{P}.(y(t) \in \Gamma) = \tilde{P}(y(t) \in \Gamma \,|\, \sigma(z(u): 0 \le u \le t))$$

$\tilde{P}$-almost surely, and so

$$y(t) = E^P[y(t) \,|\, \sigma(z(u): 0 \le u \le t)] \text{ (a.s., } \tilde{P}\text{).} \quad \Box$$

8.2. More on Itô Uniqueness

We are now going to develop a criterion for Itô uniqueness which, in certain special situations, turns out to provide the easiest proof of martingale uniqueness. The criterion which we have in mind as well as its proof is again due to Watanabe and Yamada [1971].

8.2.1 Theorem. *Let* $\rho: (0, \infty) \to (0, \infty)$ *be an increasing function satisfying:*

$$\lim_{\varepsilon \searrow 0} \int_\varepsilon^1 \frac{1}{\rho^2(u)} \, du = \infty.$$

Let $\sigma: [0, \infty) \times R^1 \to R^1 \otimes R^1$ *and* $b: [0, \infty) \times R^1 \to R^1$ *be bounded measurable functions for which there is an* $M < \infty$ *such that*

$$|\sigma(s, x) - \sigma(s, y)| \le M\rho(|x - y|)$$

and

$$|b(s, x) - b(s, y)| \le M|x - y|$$

for all $s \ge 0$ *and* $x, y \in R^1$. *Then* σ *and* b *satisfy Itô's uniqueness condition starting from each* $(s, x) \in [0, \infty) \times R^1$. *In particular, if* $a = \sigma\sigma^*$, *then the martingale problem for* a *and* b *is well-posed and the associated family* $\{P_{s, x}: (s, x) \in [0, \infty) \times R^d\}$ *is Feller continuous.*

Proof. Suppose that the first part has been proved and let us check that the second part follows easily. Existence of solutions is obvious from the continuity in x of $\sigma(t, x)$ and $b(t, x)$ (cf. Theorem 6.1.7). Moreover, uniqueness is a consequence of Corollary 8.1.6. Finally, Feller continuity follows from uniqueness and the continuity of $\sigma(t, \cdot)$ and $b(t, \cdot)$.

We now turn to the proof of Itô uniqueness. As usual, we assume that $s = 0$. Let $(E, \mathscr{F}, \mu)$ be a probability space, $\{\mathscr{F}_t : t \geq 0\}$ a non-decreasing family of sub σ-algebras, and $\xi^1 : [0, \infty) \times E \to R^1$, $\xi^2 : [0, \infty) \times E \to R^1$, and $\beta : [0, \infty) \times E \to R^1$ right-continuous, μ-almost surely continuous progressively measurable functions such that $(\beta(t), \mathscr{F}_t, \mu)$ is a 1-dimensional Brownian motion and

$$\xi^i(t) = x + \int_0^t \sigma(u, \xi^i(u)) \, d\beta(u) + \int_0^t b(u, \xi^i(u)) \, du, \qquad t \geq 0,$$

μ-almost surely, $i = 1, 2$. Set

$$\eta(\cdot) = \xi^1(\cdot) - \xi^2(\cdot).$$

Then

$$\eta(t) = \int_0^t \left(\sigma(u, \xi^1(u)) - \sigma(u, \xi^2(u)) \right) d\beta(u)$$

$$+ \int_0^t \left(b(u, \xi^1(u)) - b(u, \xi^2(u)) \right) du, \qquad t \geq 0$$

μ-almost surely. In particular, if $f \in C_b^2(R)$, then

$$E[f(\eta(t))] - f(0) = \tfrac{1}{2} E\left[\int_0^t (\sigma(u, \xi^1(u)) - \sigma(u, \xi^2(u)))^2 f''(\eta(u)) \, du \right]$$

$$+ E\left[\int_0^t (b(u, \xi^1(u)) - b(u, \xi^2(u))) f'(\eta(u)) \, du \right]$$

$$\leq \frac{M}{2} E\left[\int_0^t \rho^2(|\eta(u)|) |f''(\eta(u))| \, du \right]$$

$$+ M E\left[\int_0^t |\eta(u)| \, |f'(\eta(u))| \, du \right].$$

We now choose the sequence $\{\alpha_k\}_0^\infty \subseteq (0, 1)$ so that $a_k \downarrow 0$ and

$$\int_{\alpha_k}^{\alpha_{k-1}} \frac{1}{\rho^2(u)} \, du = k$$

for $k \geq 1$. For this choice of α_k's, define $\{\varphi_k''\}_1^\infty \subseteq C_0([0, \infty))$ so that $\varphi_k''(x) = 0$ for $0 \leq x \leq \alpha_k$, $0 \leq \varphi_k''(x) \leq 2/k\rho^2(x)$ for $\alpha_k \leq x \leq \alpha_{k-1}$, $\varphi_k''(x) = 0$ for $x \geq \alpha_{k-1}$, and $\int_0^\infty \varphi_k''(x)\, dx = 1$. Set

$$\varphi_k(x) = \int_0^{|x|} dt \int_0^t \varphi_k''(s)\, ds.$$

Clearly $\varphi_k \in C^2(R)$. On the other hand, φ_k is unbounded. Nonetheless, $|\varphi_k(x)| \leq |x|$, and so an easy limiting procedure can be used to check that

$$E[\varphi_k(\eta(t))] \leq \frac{M}{2} E\left[\int_0^t \rho^2(|\eta(u)|)\,|\varphi_k''(\eta(u))|\, du\right]$$

$$+ ME\left[\int_0^t |\eta(u)|\,|\varphi_k'(\eta(u))|\, du\right].$$

Thus

$$E[\varphi_k(\eta(t))] \leq ME\left[\int_0^t |\eta(u)|\, du\right] + \frac{Mt}{k}$$

for all $k \geq 1$. Letting $k \to \infty$ and noting that $\varphi_k(x) \to |x|$ as $k \to \infty$, we conclude that

$$E[|\eta(t)|] \leq M \int_0^t E[|\eta(u)|]\, du, \qquad t \geq 0.$$

Thus, by Gromwall's inequality, $E[|\eta(t)|] = 0$ for all $t \geq 0$, and so $\xi^1(\cdot) = \xi^2(\cdot)$ (a.s., μ). □

It is interesting to see exactly how good this result is. First, let $\sigma: R^1 \to [0, \infty)$ be a bounded, uniformly positive, smooth function away from 0 such that $\sigma(x) = |x|^{\alpha/2}$ for $|x| < 1$ and some $\alpha > 0$ and $|\sigma'(x)| \leq M$ for $|x| \geq 1$ and some $M < \infty$. If $\alpha \geq 1$, then it is clear from the preceding that σ and 0 satisfy Itô's uniqueness condition starting from any point, and therefore the martingale problem for σ^2 and 0 is well-posed. On the other hand, if $0 < \alpha < 1$, then Theorem 8.2.1 does not apply. In fact, for α's in this range, as we saw in Exercise 6.7.7, the martingale problem for σ^2 and 0 is ill-posed. In Exercise 8.3.2 below, the reader is asked to demonstrate the invalidity of the natural analogue of Theorem 8.2.1 in higher dimensions. The idea is to exploit the same reasoning as that used in Exercise 6.7.7 and thereby show that in two or more dimensions the martingale problem for $\frac{1}{2}\sigma^2(|x|)\Delta$ starting from $(0, 0)$ has at least two solutions as soon as α is less than 2. Combining these remarks, we see that the situations amenable to the techniques in Theorem 8.2.1 are quite special; but when they apply, such techniques can yield essentially optimal results.

8.3. Exercises

8.3.1. Let $\sigma: [0, \infty) \times R^1 \to R^1$ and $b: [0, \infty) \times R^1 \to R^1$ be bounded measurable functions which satisfy Itô's uniqueness condition starting at any point. Let $(E, \mathscr{F}, \mu)$ be a probability space, $\{\mathscr{F}_t: t \geq 0\}$ a non-decreasing family of sub σ-algebras of $\mathscr{F}$, and $\beta: [0, \infty) \times E \to R^1$ a 1-dimensional Brownian motion with respect to $(E, \mathscr{F}_t, \mu)$. Suppose that $x \leq y$ and that

$$\xi(t) = x + \int_0^t \sigma(u, \xi(u)) \, d\beta u + \int_0^t b(u, \xi(u)) \, du, \qquad t \geq 0,$$

and

$$\eta(t) = y + \int_0^t \sigma(u, \eta(u)) \, d\beta u + \int_0^t b(u, \xi(u)) \, du, \qquad t \geq 0,$$

μ-almost surely. Show that $\mu(\xi(t) \leq \eta(t), t \geq 0) = 1$. A version of this result was proved using a very clever argument by A. V. Skorohod [1961]. The method we have in mind consists of the following steps:

(i) let $\hat{P}$ on Ω_2 be the distribution of $(\xi(\cdot), \eta(\cdot))$ and show that $\hat{P}$ solves the martingale problem for $\hat{a}$ and $\hat{b}$ (cf. (1.4) and (1.5)) starting from $(0, (x, y))$;

(ii) let $\tau = \inf\{t \geq 0: y^1(t) = y^2(t)\}$ and, for $T > 0$, let $\{\hat{P}_\omega\}$ be a r.c.p.d. of $\hat{P}|\mathscr{M}_{\tau \wedge T}$. Argue that for $\hat{P}$-almost all ω satisfying $\tau(\omega) \leq T$, $\hat{P}_\omega(y^1(t) \leq y^2(t), \tau(\omega) \leq t \leq T) = 1$.

8.3.2. Let $d \geq 2$ and suppose that $a: R^d \to [0, \infty)$ is a bounded function which is smooth for x away from 0 and equal to $|x|^\alpha$ for $|x| \leq 1$ and some $\alpha > 0$. Show that the martingale problem for a and 0 starting at $(0, 0)$ has exactly one solution if and only if $\alpha \geq 2$.

Chapter 9

Some Estimates on the
Transition Probability Functions

9.0. Introduction

In the derivation of our basic uniqueness theorem, Theorem 7.2.1, we made essential use of an analytic representation for the transition probability function associated with diffusion coefficients which are nearly independent of the spatial variables. We have already seen that this representation not only allows us to prove uniqueness, but also leads to regularity properties of the transition probability function as a function of the "backwards" variables. The purpose of the present chapter is to exploit this same representation to conclude properties about the transition probability function as a function of its "forward" variables. More specifically, we will show that the transition probability function has a density with respect to Lebesgue measure and that this density satisfies certain L^p-estimates.

The procedure which we will use should be becoming familiar. That is, we start with the analytic representation obtained by perturbation theory. This gives us local estimates. We then use a probabilistic procedure to convert the local results into global ones. The estimates at which we eventually arrive in this way are not new to analysts. Fabes [1966] and Fabes and Riviére [1966] proved essentially the same estimates for parabolic equations of any order (not just second order); and their estimates together with our uniqueness results can be combined to yield most of the estimates which we will derive below. Thus our approach not only isn't the only one, it may not even be the most efficient one. Nonetheless, it is our belief that the approach that we give below is the most readily accessible one for probabilists and is of interest in and of itself.

A word of warning is in order. Because we want to keep rather careful track of constants and their dependence on various parameters, we will put the equation number in which a constant first appears as a subscript. Thus $C_{1.5}$ means the constant which appears for the first time in Equation 1.5. If an explanation of the manner in which a constant depends on various parameters is required, such an explanation will be found somewhere in the paragraph containing the equation in which that constant is introduced. We hope that this procedure helps to clarify more than it obscures.

9.1. The Inhomogeneous Case

Let $c: [0, \infty) \to S_d^+$ be a measurable function for which there exists numbers $\lambda > 0$ and $\Lambda < \infty$ such that

(1.1) $\qquad \lambda |\theta|^2 < \langle \theta, c(t)\theta \rangle \le \Lambda |\theta|^2, \qquad t \ge 0 \quad \text{and} \quad \theta \in R^d.$

Define

(1.2) $\qquad\qquad C(s, t) = \int_s^t c(u)\, du, \qquad 0 \le s \le t,$

and

(1.3) $\quad g^{(c)}(s, x; t, y) = \dfrac{1}{(2\pi)^{d/2} |C(s, t)|^{1/2}} \exp \left| \dfrac{\langle y - x, C(s, t)^{-1}(y - x) \rangle}{2} \right|$

for $0 \le s < t$ and $x, y \in R^d$. It is easy to check that:

(1.4) $\qquad\qquad g^{(c)}(s, x; t, y) \le (\Lambda/\lambda)^{d/2} g^{(\Lambda)}(t - s, y - x)$

where

(1.5) $\qquad g^{(\Lambda)}(t - s, y - x) = \dfrac{1}{(2\pi\Lambda(t - s))^{d/2}} \exp \left[\dfrac{-|y - x|^2}{2\Lambda(t - s)} \right].$

Moreover, a simple computation shows that:

(1.6) $\qquad\qquad \int_0^T dt \int_{R^d} dy\, |g^{(\Lambda)}(t, y)|^r = C_{1.6} T^{(d/2)[(d + 2)/d - r]}$

for $1 \le r < (d + 2)/d$ and $T > 0$, where

$$C_{1.6} = \left[(r(2\pi\Lambda)^{r-1})^{d/2} \frac{d}{2}\left(\frac{d + 2}{d} - r \right) \right]^{-1}.$$

Next, define

(1.7) $\quad G_c^T f(s, x) = \int_s^T dt \int_{R^d} dy\, g^{(c)}(s, x; t, y) f(t, y), (s, x) \in [0, T] \times R^d,$

for $f \in L^\infty([0, T] \times R^d)$. We now have the following simple estimate.

9.1.1 Lemma. *If $r_1, r_2 \in [1, \infty]$ and*

(1.8) $\qquad\qquad 0 \le \dfrac{1}{r_1} - \dfrac{1}{r_2} < \dfrac{2}{d + 2},$

then for $T > 0$ and $f \in L^1([0, T] \times R^d) \cap L^\infty([0, T] \times R^d)$:

(1.9) $$\|G_c^T f\|_{L^2([0,T] \times R^d)} \leq C_{1.9} \|f\|_{L^{r_1}([0,T] \times R^d)},$$

where

$$C_{1.9} = (\Lambda/\lambda)^{d/2} C_{1.6}^{1/r_3}$$

with $C_{1.6}$ computed for $r = r_3$ and r_3 is determined from

$$\frac{1}{r_3} = \frac{1}{r_2} - \frac{1}{r_1} + 1.$$

Proof. Given f, note that for $(s, x) \in [0, T] \times R^d$:

$$|G_c^T f(s, x)| \leq (\Lambda/\lambda)^{d/2} \varphi * \psi(s, x)$$

where

$$\varphi(t, y) = \chi_{[0, T]}(-t) g^{(\Lambda)}(-t, y)$$

and

$$\psi(t, y) = \chi_{[0, T]}(t) |f(t, y)|.$$

Thus

$$\|G_c^T f\|_{L^2([0,T] \times R^d)} \leq \left(\frac{\Lambda}{\lambda}\right)^{d/2} \|\varphi * \psi\|_{L^2(R \times R^d)}.$$

Now define r_3 by

$$\frac{1}{r_3} = \frac{1}{r_2} - \frac{1}{r_1} + 1,$$

and note that $1 \leq r_3 < (d + 2)/d$. Hence, by (1.6),

$$\|\varphi\|_{L^{r_3}(R \times R^d)} = C_{1.6}^{1/r_3}$$

and clearly

$$\|\psi\|_{L^{r_1}(R \times R^d)} = \|f\|_{L^{r_1}([0,T] \times R^d)}$$

Finally, by Young's inequality,

$$\|\varphi * \psi\|_{L^2(R^1 \times R^d)} \leq \|\varphi\|_{L^{r_3}(R^1 \times R^d)} \|\psi\|_{L^{r_1}(R \times R^d)},$$

and so the proof is complete. $\square$

Next note that, by estimate (0.4) in the appendix, for each $1 < r < \infty$ there is a $C_d(r, \lambda, \Lambda) < \infty$ such that

(1.10)
$$\left\| \frac{\partial^2 G_c^T f}{\partial x_i \, \partial x_j} \right\|_{L^r([0,\,T] \times R^d)} \leq C_d(r, \lambda, \Lambda) \| f \|_{L^r([0,\,T] \times R^d)}$$

for all $f \in C_0^\infty([0, T] \times R^d)$ and $1 \leq i, j \leq d$. Remembering that the L^r-bound of a linear operator is a log-convex function of $1/r$, we see that for all $1 < r_1 < r_2 < \infty$ there is an

(1.11)
$$0 < \varepsilon = \varepsilon_d(r_1, r_2, \lambda, \Lambda) < \lambda$$

such that

(1.12)
$$\varepsilon d^2 C_d(r, \lambda, \Lambda) \leq 1, \qquad r_1 < r \leq r_2 .$$

Thus, if $a: [0, \infty) \times R^d \to S_d^+$ is a measurable function satisfying

$$\sup_{(s,\,x)} \| a(s, x) - c(s) \| \leq \varepsilon$$

and

(1.13)
$$D_{a,\,c}^T f(s, x) \equiv \frac{1}{2} \sum_{i,\,j=1}^d (a^{ij}(s, x) - c^{ij}(s)) \frac{\partial^2 G_c^T f}{\partial x_i \, \partial x_j}(s, x),$$

then

$$\| D_{a,\,c}^T f \|_{L^r([0,\,T] \times R^d)} \leq \frac{\varepsilon}{2} d^2 C_d(r, \lambda, \Lambda) \| f \|_{L^r([0,\,T] \times R^d)}$$

$$\leq \tfrac{1}{2} \| f \|_{L^r([0,\,T] \times R^d)}$$

for $r_1 \leq r \leq r_2$. It follows that for r's in this range, $(I - D_{a,\,c}^T)^{-1}$ exists as a continuous operator on $L^r([0, T] \times R^d)$ into itself having bound less than or equal to 2. Moreover, as pointed out in the proof of Lemma 7.1.5, the fact that $(I - D_{a,\,c}^T)^{-1}$ is given by a Neumann series shows that it is consistently defined for these r's.

Given $0 < \lambda < \Lambda$ and $1 < r < \infty$, let $\mathscr{A}(r, \lambda, \Lambda)$ stand for the class of measurable $a: [0, \infty) \times R^d \to S_d^+$ with the properties that:

$$\lambda |\theta|^2 \leq \langle \theta, a(s, x)\theta \rangle \leq \Lambda |\theta|^2,$$

for all $(s, x) \in [0, \infty) \times R^d$ and $\theta \in R^d$, and there is an $x^0 \in R^d$ for which

$$\varepsilon = \sup_{(s,\,x)} \| a(s, x) - \bar{a}(s) \|$$

satisfies

$$(1.14) \qquad \varepsilon^2 C_d(\rho, \lambda, \Lambda) \le 1, \qquad r \le \rho \le \left(\frac{d+4}{2} \right) \vee r,$$

where $\bar{a}(\cdot) = a(\cdot, x^0)$. For $a \in \mathscr{A}(r, \lambda, \Lambda)$, $(I - D_{a,\bar{a}})^{-1}$ is consistently defined as a bounded operator on $L^\rho([0, T] \times R^d)$ into itself for $r \le \rho \le ((d+4)/2) \vee r$. Thus, for $r \le \rho \le ((d+4)/2) \vee r$,

$$(1.15) \qquad K_a^T f \equiv G_{\bar{a}}^T \circ (I - D_{a,\bar{a}}^T)^{-1}$$

is consistently defined as a bounded operator on $L^\rho([0, T] \times R^d)$ into $L^\sigma([0, T] \times R^d)$, where

$$0 \le \frac{1}{\rho} - \frac{1}{\sigma} < \frac{2}{d+2}.$$

We are now ready to prove the next lemma.

9.1.2 Lemma. *Let $a \in \mathscr{A}(r, \lambda, \Lambda)$ and set*

$$(1.16) \qquad N = \left[\frac{d+2}{2} \frac{1}{r} \right]$$

Then $(K_a^T)^{N+1}$ maps $L([0, T] \times R^d)$ into $C_b([0, T] \times R^d)$; and, in fact,

$$(1.17) \qquad |(K_a^T)^{N+1} f(s, x)| \le C_{1.17} \|f\|_{L^r([0, T] \times R^d)},$$

where $C_{1.17}$ depends only on d, T, r, λ, and Λ.

Proof. First observe that for $\rho > (d+2)/2$, $G_{\bar{a}}^T$ maps $L^\rho([0, \infty) \times R^d)$ continuously into $C_b([0, T] \times R^d)$ with a bound depending on d, T, ρ, λ, and Λ. Thus if $\rho \ge r$ and $(d+2)/2 < \rho \le ((d+4)/2) \vee r$, so does K_a^T. In particular, there is nothing to prove if $r > (d+2)/2$. Assume that $r \le (d+2)/2$, and choose $(d+2)/2 < r_N < (d+4)/2$ so that

$$\frac{d+2}{2} \left(\frac{1}{r} - \frac{1}{r_N} \right) < \left[\frac{d+2}{2} \frac{1}{r} \right].$$

Then we can find $r = r_0 < \cdots < r_N$ so that $1/r_{k-1} - 1/r_k = 1/N(1/r - 1/r_N)$. By Lemma 9.1.1, K_a^T maps $L^{r_{k-1}}([0, T] \times R^d)$ continuously into $L^{r_k}([0, T] \times R^d)$ with a bound depending only on d, T, $1/N(1/r - 1/r_N)$, λ, and Λ. Therefore, $(K_a^T)^N$ maps $L([0, T] \times R^d)$ continuously into $L^{r_N}([0, T] \times R^d)$ with a bound depending only on d, T, $1/r - 1/r_N$, λ, and Λ. Combining this with the first part of the proof, one arrives at the desired result. $\square$

In order to translate these results into estimates on transition probability functions, we need the following simple observation.

9.1.3 Lemma. *Let $P(s, x; t, \cdot)$ be a transition probability function with the property that $(s, x; t) \to P(s, x; t, \Gamma)$ is measurable on $\{(s, x; t) \in [0, \infty) \times R^d \times [0, \infty): s < t\}$ for all $\Gamma \in \mathscr{B}_{R^d}$. Given $T > 0$, define P^T on $B([0, \infty) \times R^d)$ by*

$$P^T f(s, x) = \int_s^T dt \int_{R^d} P(s, x; t, dy) f(t, y).$$

Then

(1.18) $$(P^T)^{N+1} f(s, x) = \frac{1}{N!} \int_s^T (t - s)^N \, dt \int_{R^d} P(s, x; t, dy) f(t, y).$$

Proof. There is nothing to prove if $N = 0$. Assuming (1.18) for $N - 1$, we have:

$$(P^T)^{N+1} f(s, x) = \frac{1}{(N-1)!} \int_s^T (t - s)^{N-1} \, dt \int_{R^d} P(s, x; t, dy) P^T f(t, y)$$

$$= \frac{1}{(N-1)!} \int_s^T (t - s)^{N-1} \, dt \int_t^T d\tau \int_{R^d} \int_{R^d} P(s, x; t, dy) P(t, y; \tau, d\xi) f(\tau, \xi)$$

$$= \frac{1}{(N-1)!} \int_s^T (t - s)^{N-1} \, dt \int_t^T d\tau \int_{R^d} P(s, x; \tau, d\xi) f(\tau, \xi)$$

$$= \frac{1}{N!} \int_s^T (\tau - s)^N \, d\tau \int_{R^d} P(s, x; \tau, d\xi) f(\tau, \xi). \quad \square$$

We are at last ready to give our preliminary estimates. Let $a \in \mathscr{A}(r, \lambda, \Lambda)$ for some $1 < r < \infty$. Then, by Theorem 7.1.6, the martingale problem for

$$L_t = \frac{1}{2} \sum_{i, j=1}^d a^{ij}(t, \cdot) \frac{\partial^2}{\partial x_i \, \partial x_j}$$

is well-posed. Denote by $\{P_{s, x}: (s, x) \in [0, \infty) \times R^d\}$ the corresponding family of solutions. Then $\{P_{s, x}: (s, x) \in [0, \infty) \times R^d\}$ forms a strong Markov family which is strongly Feller continuous. In fact, if $P(s, x; t, \cdot)$ is the associated transition probability function, then

(1.19) $$P^T f(s, x) \equiv \int_s^T dt \int_{R^d} P(s, x; t, dy) f(t, y) \, dt = K_a^T f(s, x)$$

if $f \in L^\rho([0, T] \times R^d) \cap B([0, T] \times R^d)$ for some $r \leq \rho \leq ((d + 4)/2) \vee r$. Hence

(1.20) $$(P^T)^{N+1} f(s, x) = (K_a^T)^{N+1} f(s, x)$$

for any $N \geq 0$ and any $f \in C_0([0, T] \times R^d)$. In particular, if $N = [(d + 2)/2r]$, then

(1.21) $$|(P^T)^{N+1}f(s, x)| \leq C_{1.17}\|f\|_{L^r([0, T] \times R^d)}.$$

9.1.4 Lemma. *If* $a \in \mathscr{A}(r, \lambda, \Lambda)$ *and* $P(s, x; t, \cdot)$ *is the transition probability function determined by*

$$L_t = \frac{1}{2} \sum_{i, j=1}^{d} a^{ij}(t, \cdot) \frac{\partial^2}{\partial x_i \, \partial x_j},$$

then for $r \leq \rho \leq \infty$

(1.22) $$\left| \int_s^T (t - s)^N \, dt \int P(s, x; t, dy)f(t, y) \right| \leq C_{1.22}\|f\|_{L^\rho([0, \infty) \times R^d)},$$

where

(1.23) $$N = \left[\frac{d + 2}{2} \frac{1}{r} \right]$$

and $C_{1.22} = (N!\,C_{1.17}) \vee (T^{N+1}/(N + 1))$. *Moreover, for each* $\delta > 0$ *and* $r < \rho \leq \infty$

(1.24) $$\left| \int_s^T dt \int_{R^d \setminus B(x, \delta)} P(s, x; t, dy)f(t, y) \, dy \right| \leq C_{1.24}\|f\|_{L^\rho([0, T] \times R^d)},$$

where $C_{1.24}$ *depends only on* $d, T, \delta, \rho, r, \lambda,$ *and* Λ.

Proof. Combining (1.21), (1.18), and the fact that $\int_{R^d} P(s, x; t, dy) = 1$ for $t > s$, we arrive at (1.22) by an easy interpolation argument.

To prove (1.24), we assume that $T \geq s + 1$ and note that

$$\left| \int_s^T dt \int_{R^d \setminus B(x, \delta)} P(s, x; t, dy)f(t, y) \right|$$

$$\leq \sum_{n=1}^{\infty} \int_{s+1/(n+1)}^{s+(1/n)} dt \int_{R^d \setminus B(x, \delta)} P(s, x; t, dy)|f(t, y)|$$

$$+ \int_{s+1}^T (t - s)^N \, dt \int_{R^d} P(s, x; t, dy)|f(t, y)|.$$

The final term is estimated by (1.22). To handle the terms in the sum, define

$$\Lambda_{s, x}^{(n)} \varphi = \int_{s+1/(n+1)}^{s+(1/n)} dt \int_{R^d \setminus B(x, \delta)} P(s, x; t, dy)\varphi(t, y).$$

As a linear functional on $L([0, T] \times R^d)$, $\Lambda^{(n)}_{s, x}$ is bounded by $N!(n + 1)^N C_{1.17}$. On the other hand, if $\varphi \in B([0, T] \times R^d)$, then

$$|\Lambda^{(n)}_{s, x}\varphi| \leq \frac{1}{n}\|\varphi\|P_{s, x}\left(\sup_{s \leq t \leq s+(1/n)} |x(t) - x| \geq \delta\right)$$

$$\leq \frac{2d}{n}\|\varphi\|e^{-n\delta^2/2\Lambda d}.$$

Thus, by interpolation, if $1/\rho = (1 - \theta)/r$, then

$$|\Lambda^{(n)}_{s, x}\varphi| \leq (N!(n + 1)^N C_{1.17})^{1-\theta}\left(\frac{2d}{n}e^{-n\delta^2/2\Lambda d}\right)^\theta \|\varphi\|_{L^\rho([0, T] \times R^d)}$$

and clearly this proves (1.24). □

In order to get away from coefficients which are nearly independent of x, we will use the following localization procedure. The results which we are going to prove in this connection are slightly more refined than are absolutely needed in the present context; however, the generality in which we prove them is useful when dealing with unbounded coefficients of the sort often encountered when dealing with diffusions on a manifold.

9.1.5 Lemma. *Let* $\tilde{a}: [0, \infty) \times R^d \to S_d$ *and* $\tilde{b}: [0, \infty) \times R^d \to R^d$ *be bounded measurable functions with the property that the martingale problem for*

$$\tilde{L}_t = \frac{1}{2}\sum_{i, j=1}^{d} \tilde{a}^{ij}(t, \cdot)\frac{\partial^2}{\partial x_i \partial x_j} + \sum_{i=1}^{d} \tilde{b}^i(t, \cdot)\frac{\partial}{\partial x_i}$$

is well-posed and determines a measurable strong Markov family $\{\tilde{P}_{s, x}: (s, x) \in [0, \infty) \times R^d\}$. *Let* $a: [0, \infty) \times R^d \to S_d$ *and* $b: [0, \infty) \times R^d \to R^d$ *be bounded measurable functions such that* $a = \tilde{a}$ *and* $b = \tilde{b}$ *on* $[s, T] \times G$, *where* $0 \leq s < T$ *and* $G \subseteq R^d$ *is open. If* P *solves the martingale problem for*

$$L_t = \frac{1}{2}\sum_{i, j=1}^{d} a^{ij}(t, \cdot)\frac{\partial}{\partial x_i \partial x_j} + \sum_{i=1}^{d} b^i(t, \cdot)\frac{\partial}{\partial x_i}$$

starting at (s, x) *and* $\tau = \inf\{t \geq s: x(t) \notin G\}$, *then*

$$P(x(t) \in \Gamma, \tau > t) \leq \tilde{P}_{s, x}(x(t) \in \Gamma \cap G)$$

for $s \leq t \leq T$ *and* $\Gamma \in \mathcal{B}_{R^d}$.

Proof. Put $R = P \otimes_{\tau \wedge T}^{0} \tilde{P}_{\tau \wedge T, \, x(\tau \wedge T)}$. Then $R = \tilde{P}_{s, \, x}$ and R equals P on $\mathcal{M}_{\tau \wedge T}$. Thus, if $s \leq t \leq T$ and $\Gamma \in \mathscr{B}_{R^d}$, then

$$P(x(t) \in \Gamma, \, \tau > t) = R(x(t) \in \Gamma, \, \tau > t) \leq R(x(t) \in \Gamma \cap G)$$
$$= \tilde{P}_{s, \, x}(x(t) \in \Gamma \cap G). \quad \square$$

9.1.6 Lemma. *Let* $a: [s^0, \infty) \times \Omega \to S_d$ *and* $b: [s^0, \infty) \times \Omega \to R^d$ *be progressively measurable functions after time* s^0 *and suppose that* P *is a probability measure on* $(\Omega, \mathcal{M})$ *such that* $x(\cdot) \sim \mathscr{I}_d^{s^0}(a, b)$. *Given* $0 < R_1 < R_2$ *and* $x^0 \in R^d$, *define* $\tau_{-1} \equiv s^0$ *and*

$$\tau_{2n} = \inf\{t \geq \tau_{2n-1}: \, |x(t) - x^0| = R_2\}, \qquad n \geq 0$$
$$\tau_{2n+1} = \inf\{t \geq \tau_{2n}: \, |x(t) - x^0| = R_1\}, \qquad n \geq 0.$$

If $T > s^0$ *and*

$$A = \sup\{\|a(t, \omega)\|; \, (t, \omega): s^0 \leq t \leq T \quad and \quad |x(t, \omega) - x^0| \leq R_2\}$$
$$B = \sup\{|b(t, \omega)|; \, (t, \omega): s^0 \leq t \leq T \quad and \quad |x(t, \omega) - x^0| \leq R_2\},$$

then

$$(1.25) \qquad\qquad E^P\left[\sum_{0}^{\infty} \chi_{[s^0, \, T]}(\tau_{2n})\right] \leq C_{1.25},$$

where $C_{1.25}$ *can be chosen to depend only on* d, T, $R_2 - R_1$, A, *and* B.

Proof. Let σ be a bounded stopping time after time s^0 and define

$$\zeta_\sigma = \inf\{t \geq \sigma: \, |x(t) - x(\sigma)| = R_2\}.$$

We will show that if $\delta > 0$ is chosen so that

$$2d \exp\left[\frac{-(R_2 - R_1 - B\delta)^2}{2dA\delta}\right] \leq \frac{1}{2}$$

and $B\delta < R_2 - R_1$, then for any $H \in \mathcal{M}_\sigma$ satisfying $H \subseteq \{|x(\sigma) - x^0| \leq R_1\}$

$$(1.26) \qquad\qquad P(\{\zeta_\sigma - \sigma < \delta\} \cap H) \leq \tfrac{1}{2}P(H).$$

To do this, let $\{P_\omega\}$ be a r.c.p.d. of $P|\mathcal{M}_\sigma$ and define

$$\bar{x}_\omega(t) = x(t \wedge \zeta_{\sigma(\omega)}) - x(\sigma(\omega), \omega) - \int_{\sigma(\omega)}^{t \wedge \zeta_{\sigma(\omega)}} b(u) \, du.$$

Then for P-almost all ω, $\bar{x}_\omega(\cdot) \sim \mathscr{I}_d^{\sigma(\omega)}(\chi_{[\sigma(\omega),\,\zeta_{\sigma(\omega)})}(\cdot)a(\cdot), 0)$ under P_ω. Thus, for P-almost all $\omega \in H$:

$$P_\omega(\zeta_\sigma - \sigma \leq \delta) = P_\omega(\zeta_{\sigma(\omega)} - \sigma(\omega) \leq \delta)$$

$$\leq P_\omega\left(\sup_{\sigma(\omega) \leq t \leq \sigma(\omega) + \delta} |\bar{x}_\omega(t)| \geq R_2 - R_1 - B\delta\right)$$

$$\leq 2 \, d \, \exp[-(R_2 - R_1 - B\delta)^2/2 \, dA\delta] \leq \tfrac{1}{2},$$

and clearly (1.26) follows from this.

We now use (1.26) to prove that

$$(1.27) \qquad\qquad P(\tau_{2n} \leq s^0 + \delta) \leq (\tfrac{1}{2})^n, \; n \geq 0.$$

There is nothing to prove if $n = 0$. Assuming (1.27) for n, we have:

$$P(\tau_{2n+2} \leq s^0 + \delta) = P(\tau_{2n+2} \leq s^0 + \delta, \, \tau_{2n+1} < s^0 + \delta)$$

$$\leq P(\{\zeta_\sigma - \sigma \leq \delta\} \cap \{\sigma < s^0 + \delta\})$$

where $\sigma = \tau_{2n+1} \wedge (s^0 + \delta)$. Since $\mathscr{M}_\sigma \ni \{\sigma < s^0 + \delta\} \subseteq \{|x(\sigma) - x^0| = R_1\}$, (1.26) implies that

$$P(\tau_{2n+2} \leq s^0 + \delta) \leq \tfrac{1}{2}P(\sigma < s^0 + \delta) \leq \tfrac{1}{2}P(\tau_{2n} \leq s^0 + \delta).$$

and so (1.27) follows by induction.

Next set $k_0 = [T - s^0/\delta] + 1$ and choose $n_0 \geq 1$ so that $k_0(\tfrac{1}{2})^{n_0} \leq \tfrac{1}{2}$. We will show that

$$(1.28) \qquad\qquad P(\tau_{2kn_0} \leq s^0 + k\delta) \leq k(\tfrac{1}{2})^{n_0}, \qquad k \geq 1.$$

When $k = 1$, (1.28) is the same as (1.27) with $n = n_0$. Assume (1.28) for k. Then

$$P(\tau_{2(k+1)n_0} \leq s^0 + (k+1)\delta) \leq P(\tau_{2kn_0} \leq s^0 + k\delta)$$

$$+ P(\tau_{2(k+1)n_0} \leq s^0 + (k+1)\delta, \, \tau_{2kn_0} > s^0 + k\delta).$$

The first term on the right is less than or equal to $k(\tfrac{1}{2})^{n_0}$. As for the second, let $\{P_\omega\}$ be a r.c.p.d. of $P \mid \mathscr{M}_\sigma$ with $\sigma = \tau_{2kn_0} \wedge (s^0 + (k+1)\delta)$ and for each ω define τ_n^ω, $n \geq -1$, the same way as we did τ_n, $n \geq -1$, only now with s^0 replaced by $\sigma(\omega)$. Then $\tau_{(2k+1)n_0} = \tau_{n_0}^\omega$ (a.s., P_ω) if $\tau_{2kn_0}(\omega) < s^0 + (k+1)\delta$; and for P-almost all ω, $x(\cdot) \sim \mathscr{I}_d^{\sigma(\omega)}(a, b)$ under P_ω. Thus $P_\omega(\tau_{n_0}^\omega \leq \sigma(\omega) + \delta) \leq (\tfrac{1}{2})^{n_0}$ (a.s., P) and therefore

$$P(\tau_{2(k+1)n_0} \leq s^0 + (k+1)\delta, \, \tau_{2kn_0} > s^0 + k\delta)$$

$$= E^P[P_\cdot(\tau_{n_0}^\cdot \leq \sigma(\omega) + \delta)] \leq (\tfrac{1}{2})^{n_0}.$$

This completes the proof of (1.28). In particular, if $k = k_0$ in (1.28), we arrive at

$$(1.29) \qquad\qquad P(\tau_{2N} \le T) \le \tfrac{1}{2},$$

where $N = k_0 n_0$.

Proceeding in exactly the same way as we did in the derivation of (1.26) from the estimate $P(\tau_2 \le s^0 + \delta) \le \tfrac{1}{2}$, we conclude from (1.29) that

$$(1.30) \qquad\qquad P(\tau_{2nN} \le T) \le (\tfrac{1}{2})^n, \qquad n \ge 0.$$

Since

$$E^P\left[\sum_0^\infty \chi_{[s^0,\,T]}(\tau_{2n})\right] \le N E^P\left[\sum_0^\infty \chi_{[s^0,\,T]}(\tau_{2nN})\right]$$

$$= N \sum_0^\infty P(\tau_{2nN} \le T),$$

(1.25) is an immediate consequence of (1.30). $\quad\square$

9.1.7 Lemma. *Let $a: [0, \infty) \times R^d \to S_d$ and $b: [0, \infty) \times R^d \to R^d$ be bounded measurable functions and let P solve the martingale problem for*

$$L_t = \frac{1}{2} \sum_{i,\,j=1}^d a^{ij}(t, \cdot)\frac{\partial^2}{\partial x_i\,\partial x_j} + \sum_{i=1}^d b^i(t, \cdot)\frac{\partial}{\partial x_i}$$

starting from (s^0, x^1). Suppose that $\tilde{a}: [0, \infty) \times R^d \to S_d$ and $\tilde{b}: [0, \infty) \times R^d \to R^d$ are bounded measurable functions such that the martingale problem for

$$\tilde{L}_t = \frac{1}{2} \sum_{i,\,j=1}^d \tilde{a}^{ij}(t, \cdot)\frac{\partial^2}{\partial x_i\,\partial x_j} + \sum_{i=1}^d \tilde{b}^i(t, \cdot)\frac{\partial}{\partial x_i}$$

is well-posed and determines the measurable strong Markov family $\{\tilde{P}_{s,\,x}: (s, x) \in [0, \infty) \times R^d\}$. Finally, assume that for some $T > s^0$ and $R > 0$, $a(t, y) = \tilde{a}(t, y)$ and $b(t, y) = \tilde{b}(t, y)$ if $s^0 \le t \le T$ and $y \in B(x^0, R)$. Then for each $0 < \delta < R$,

$$(1.31) \qquad \left|E^P\left[\int_{s^0}^T f(t, x(t))\,dt\right]\right| \le E^{P_{s^0,\,x^1}}\left[\int_{s^0}^T |f(t, x(t))|\,dt\right]$$

$$+ C_{1.31} \sup_{\substack{s^0 \le s \le T \\ |x - x^0| = R - \delta}} E^{P_{s,\,x}}\left[\int_s^T |f(t, x(t))|\,dt\right]$$

for all $f \in C_0([s^0, T] \times B(x^0, R - \delta))$. The constant $C_{1.31}$ can be chosen to depend only on d, $T - s^0$, δ,

$$\sup_{\substack{s^0 \le t \le T \\ |y - x^0| \le R}} \|a(t, y)\|, \quad \text{and} \quad \sup_{\substack{s^0 \le t \le T \\ |y - x^0| \le R}} |b(t, y)|.$$

Proof. Let $R_2 = R$, and $R_1 = R - \delta$, and define τ_n for $n \geq -1$ accordingly as in Lemma 9.1.6. Then

$$\left| E^P \left[\int_{s^0}^T f(t, x(t))\, dt \right] \right| \leq E^P \left[\int_{s^0}^{\tau_0 \wedge T} |f(t, x(t))|\, dt \right]$$

$$+ \sum_{n=1}^\infty E^P \left[\int_{\tau_{2n-1} \wedge T}^{\tau_{2n} \wedge T} |f(t, x(t))|\, dt \right].$$

Note that by Lemma 9.1.5

$$E^P \left[\int_{s^0}^{\tau_0 \wedge T} |f(t, x(t))|\, dt \right] = \int_{s^0}^T E^P[|f(t, x(t))|,\ \tau_0 > t]\, dt$$

$$\leq \int_{s^0}^T E^{P_{s^0, x^1}}[|f(t, x(t))|]\, dt$$

$$= E^{P_{s^0, x^1}} \left[\int_{s^0}^T |f(t, x(t))|\, dt \right].$$

Also, if $\{P_\omega^{(n)}\}$ is a r.c.p.d. of $P \mid \mathcal{M}_{\tau_{2n-1} \wedge T}$, then by Lemma 9.1.5

$$E^P \left[\int_{\tau_{2n-1} \wedge T}^{\tau_{2n} \wedge T} |f(t, x(t))|\, dt \right]$$

$$\leq E^P \left[\int_{\tau_{2n-1}}^T E^{P_\omega^{(n)}}[|f(t, x(t))|,\ \tau_{2n} > t]\, dt,\ \tau_{2n-1} \leq T \right]$$

$$\leq E^P \left[E^{P_{\tau_{2n-1}(\cdot),\ x(\tau_{2n-1}(\cdot))}} \left[\int_{\tau_{2n-1}(\cdot)}^T |f(t, x(t))|\, dt \right],\ \tau_{2n-1}(\cdot) \leq T \right].$$

since $\delta_{x(\tau_{2n-1}(\omega)),\ \omega} \otimes_{\tau_{2n-1}(\omega)}^0 P_\omega^{(n)}$ solves the martingale problem for L_t starting from $(\tau_{2n-1}(\omega), x(\tau_{2n-1}(\omega), \omega))$ for P-almost all ω satisfying $\tau_{2n-1}(\omega) \leq T$. Hence

$$\left| E^P \left[\int_{s^0}^T f(t, x(t))\, dt \right] \right| \leq E^{P_{s^0, x^1}} \left[\int_{s^0}^T |f(t, x(t))|\, dt \right]$$

$$+ \sup_{\substack{s^0 \leq s \leq T \\ |x - x^0| = R - \delta}} E^{P_{s, x}} \left[\int_s^T |f(t, x(t))|\, dt \right] \sum_1^\infty P(\tau_{2n-1} \leq T);$$

and so Lemma 9.1.6 can be used to complete the proof. $\square$

9.1.8 Lemma. *Let* $a: [0, \infty) \times R^d \to S_d$ *be a bounded measurable function,* $(s^0, x^0) \in [0, \infty) \times R^d$, $T > 0$, *and* $R > 0$ *such that the function*

$$\tilde{a}(s, x) = \begin{cases} a(s, x) & \text{if } s^0 \leq s \leq T \text{ and } |x - x^0| < R \\ a(s^0, x^0) & \text{otherwise} \end{cases}$$

is an element of $\mathcal{A}(r, \lambda, \Lambda)$ *for some* $1 < r < \infty$ *and* $0 < \lambda < \Lambda$. *Let* P *solve the martingale problem for*

$$L_t = \frac{1}{2} \sum_{i,j=1}^{d} a^{ij}(t, \cdot) \frac{\partial^2}{\partial x_i \, \partial x_j}$$

starting at (s^0, x^1). *Then for each* $0 < \alpha < R$ *and* $r < \rho \le \infty$

(1.32)
$$\left| E^P \left[\int_{s^0}^{T} (t - s)^N f(t, x(t)) \, dt \right] \right| \le C_{1.32} \| f \|_{L^\rho([s^0, T] \times R^d)}$$

for all $f \in C_0([s^0, T] \times B(x^0, \alpha))$, *where* $N = [(d + 2)/2r]$ *and* $C_{1.32}$ *depends only on* $d, T, r, \rho, R - \alpha, \lambda$, *and* Λ. *Also, if* $0 < \alpha < R, r < \rho \le \infty$, *and* $|x^1 - x^0| > \alpha$, *then*

(1.33)
$$\left| E^P \left[\int_{s^0}^{T} f(t, x(t)) \, dt \right] \right| \le C_{1.33} \| f \|_{L^\rho([s^0, T] \times R^d)}$$

for all $f \in C_0([s^0, T] \times B(x^0, \alpha))$, *where* $C_{1.33}$ *depends only on* $d, T, r, \rho, R - \alpha$, $|x^1 - x^0| - \alpha, \lambda$, *and* Λ.

Proof. Let $\{\tilde{P}_{s,x} : (s, x) \in [0, \infty) \times R^d\}$ be the measurable strong Markov process determined by

$$\tilde{L}_t = \frac{1}{2} \sum_{i,j=1}^{d} \tilde{a}^{ij}(t, \cdot) \frac{\partial^2}{\partial x_i \, \partial x_j}.$$

Taking $\delta = (R - \alpha)/2$ in Lemma 9.1.7, we see that:

$$\left| E^P \left[\int_{s^0}^{T} \varphi(t, x(t)) \, dt \right] \right| \le E^{\tilde{P}_{s0,x1}} \left[\int_{s^0}^{T} |\varphi(t, x(t))| \, dt \right]$$

$$+ C_{1.31} \sup_{\substack{s^0 \le s \le T \\ |x - x^0| = R - \delta}} E^{\tilde{P}_{s,x}} \left[\int_{s}^{T} |\varphi(t, x(t))| \, dt \right]$$

for all $\varphi \in C_0([s^0, T] \times B(x^0, \alpha))$. We can therefore apply Lemma 9.1.4 to complete the proof. $\square$

We have at last arrived at the main result of this section.

9.1.9 Theorem. *Let* $a: [0, \infty) \times R^d \to S_d^+$ *and* $b: [0, \infty) \times R^d \to R^d$ *be bounded measurable functions. For each* $T > 0$, *assume that there exist numbers*

$0 < \lambda_T \le \Lambda_T < \infty$ and $B_T < \infty$ and a non-decreasing function $\delta_T \colon (0, \infty) \to (0, \infty)$ such that:

$$\lambda_T |\theta|^2 \le \langle \theta, a(s, x)\theta \rangle \le \Lambda_T |\theta|^2,$$

$$(s, x) \in [0, T] \times R^d \quad \text{and} \quad \theta \in R^d,$$

$$|b(s, x)| \le B_T, (s, x) \in [0, T] \times R^d,$$

and

$$\sup_{\substack{0 \le s \le T \\ |x^1 - x^2| \le \delta_T(\varepsilon)}} \|a(s, x^1) - a(s, x^2)\| < \varepsilon, \varepsilon > 0.$$

Set

$$L_t = \frac{1}{2} \sum_{i, j=1}^{d} a^{ij}(t, \cdot) \frac{\partial^2}{\partial x_i \, \partial x_j} + \sum_{i=1}^{d} b^i(t, \cdot) \frac{\partial}{\partial x_i}$$

and let $P(s, x; t, \cdot)$ be the transition probability function determined by L_t. Then $P(s, x; t, \cdot)$ admits a density in the sense that there exists a non-negative measurable function $p(s, x; t, y)$ on $\{(s, x; t, y) \in ([0, \infty) \times R^d)^2 \colon s < t\}$ such that for each $(s, x) \in [0, \infty) \times R^d$

$$(1.34) \qquad P(s, x; t, \Gamma) = \int_{\Gamma} p(s, x; t, y) \, dy, \qquad \Gamma \in \mathscr{B}_{R^d},$$

for almost every $t > s$. Moreover, if $0 \le s < T$ and $1 \le q < \infty$, then

$$(1.35) \qquad \left(\int_s^T (t - s)^\alpha \, dt \int_{R^d} |p(s, x; t, y)|^q \, dy \right)^{1/q} \le C_{1.35}.$$

where $\alpha = (d + 2)(q - 1)/2$ and $C_{1.35}$ depends only on d, T, q, λ_T, Λ_T, B_T, and $\delta_T(\cdot)$. Finally, if $0 \le s < T$ and $1 \le q < \infty$, then for each $\delta > 0$

$$(1.36) \qquad \left(\int_s^T dt \int_{R^d \setminus B(x, \delta)} |p(s, x; t, y)|^q \, dy \right)^{1/q} \le C_{1.36},$$

where $C_{1.36}$ depend only on d, T, q, λ_T, Λ_T, B_T, $\delta_T(\cdot)$, and δ.

Proof. We will first assume that $b \equiv 0$. Given $1 < q < \infty$, let q' be the conjugate of q and choose $1 < r < q'$ so that

$$N \equiv \left[\frac{d + 2}{2} \frac{1}{r} \right] = \left[\frac{d + 2}{2} \frac{1}{q'} \right].$$

Let ε be defined by

$$\frac{1}{\varepsilon} = C_d(r, \lambda_T, \Lambda_T) \vee C_d\left(\left(\frac{d+2}{2} \vee r\right) + 1, \lambda_T, \Lambda_T\right)$$

Then ε satisfies (1.14). Thus, by Lemma 9.1.8, for any $(s^0, x^0) \in [0, T] \times R^d$ and any $x \in R^d$.

(1.37)
$$\left|\int_s^T (t - s)^N \, dt \int_{B(x^0, \gamma)} P(s, x; t, dy) f(t, y)\right| \le C_{1.37} \|f\|_{L^p([s, T) \times R^d)}$$

where $\gamma = \frac{1}{2}\delta_T(\varepsilon)$, $\rho = (r + q')/2$, and $C_{1.37}$ depends only on d, T, r, ρ, λ_T, and Λ_T. In particular, if

$$Q_{\vec{k}} \equiv \{x + y : |y_j - k_j\gamma/d^{\frac{1}{2}}| \le \gamma/d^{\frac{1}{2}}, \ 1 \le j \le d\}$$

for $\vec{k} = (k_1, \ldots, k_d) \in Z^d$, then

(1.38)
$$\left|\int_s^T (t - s)^N \, dt \int_{Q_{\vec{k}}} P(s, x; t, dy) f(t, y)\right| \le C_{1.37} \|f\|_{L^p([0, T] \times R^d)}$$

On the other hand, by (2.1) in Chapter 4, we know that

$$P(s, x; t, Q_{\vec{k}}) \le 2 d \exp\left[-\frac{|\vec{k}|^2\gamma^2}{2d^2\Lambda_T T}\right]$$

for $s \le t \le T$; and therefore

(1.39)
$$\left|\int_s^T (t - s)^N \, dt \int_{Q_T} P(s, x; t, dy) f(t, y)\right| \le C_{1.39} \exp\left[-\frac{|\vec{k}|^2\gamma^2}{2d^2\Lambda_T T}\right] \|f\|$$

where $C_{1.39}$ depends only on d, T, and N. Combining (1.38) and (1.39), we know by interpolation that

(1.40)
$$\left|\int_s^T (t - s)^N \, dt \int_{Q_{\vec{k}}} P(s, x; t, dy) f(t, y)\right|$$

$$\le (C_{1.38})^{1-\theta}(C_{1.39})^\theta \exp\left[-\frac{\theta|\vec{k}|^2\gamma^2}{2d^2\Lambda_T T}\right] \|f\|_{L^{q'}([0, T] \times R^d)}$$

where $0 < \theta < 1$ is defined by the equation

$$\frac{1}{q'} = \frac{1 - \theta}{\rho}.$$

Summing over $\vec{k} \in Z^d$ in (1.40), we arrive at

$$(1.41) \qquad \left| \int_s^T (t-s)^N \, dt \int_{R^d} P(s, x; t, dy) f(t, y) \right| \leq C_{1.41} \| f \|_{L^{q'}([s, T] \times R^d)},$$

where $C_{1.41}$ is determined by d, θ, $C_{1.38}$, $C_{1.39}$, γ and Λ_T in the obvious way. We now remove the assumption that $b \equiv 0$. Let

$$L_t^0 = \frac{1}{2} \sum_{i, j=1}^d a^{ij}(t, \cdot) \frac{\partial^2}{\partial x_i \, \partial x_j}$$

be the second order part of L_t and define $\{P_{s, x}^0 : (s, x) \in [0, \infty) \times R^d\}$ and $P^0(s, x; t, \cdot)$ accordingly. By (1.41), if $1 < q < \infty$ and $N = [(d+2)/2q']$, then

$$(1.42) \qquad \left| \int_s^T (t-s)^N \, dt \int_{R^d} P^0(s, x; t, dy) f(t, y) \right| \leq C_{1.41} \| f \|_{L^{q'}([s, T] \times R^d)}$$

for all $0 \leq s \leq T$ and $x \in R^d$. Note that

$$P(s, x; t, \Gamma) = E^{P_{s, x^0}}[X^s(T), x(t) \in \Gamma]$$

for $0 \leq s \leq t < T$ and $\Gamma \in \mathcal{B}_{R^d}$, where

$$X^s(T) = \exp\left[\int_s^T \langle b(t, x(t)) \, dx(t) \rangle - \frac{1}{2} \int_s^T \langle b(t, x(t)), a^{-1}(t, x(t)) b(t, x(t)) \rangle \, dt \right]$$

As we have seen before:

$$(1.43) \qquad\qquad E^{P_{s, x^0}}[(X^s(T))^2] \leq \exp\left[\frac{B_T^2}{\lambda_T} T\right] \equiv C_{1.43}.$$

Thus, if $1 < q < \infty$ and $N = [(d+2)/2q')]$, then

$$\left| \int_s^T (t-s)^{N/2} \, dt \int_{R^d} P(s, x; t, dy) f(t, y) \right|$$

$$= E^{P_{s, x^0}} \left[\left| \int_s^T (t-s)^{N/2} f(t, x(t)) \, dt X^s(T) \right| \right]$$

$$\leq (T-s)^{\frac{1}{2}} C_{1.43}^{\frac{1}{2}} \left(E^{P_{s, x^0}} \left[\int_s^T (t-s)^N | f(t, x(t))|^2 \, dt \right] \right)^{\frac{1}{2}}$$

$$\leq (T-s)^{\frac{1}{2}} C_{1.43}^{\frac{1}{2}} C_{1.41}^{\frac{1}{2}} \| |f|^2 \|_{L^{q'/2}([s, T] \times R^d)},$$

where the $C_{1.41}$ here is the one in (1.42) when q' is replaced by $q'/2$. We have therefore shown that if $1 < q < \infty$ and $N = [(d + 2)/2q')]$, then

$$(1.44) \qquad \left| \int_s^T (t - s)^{N/2} \, dt \int_{R^d} P(s, x; t, dy) f(t, y) \right| \leq C_{1.44} \| f \|_{L^{q'}([s, T] \times R^d)},$$

where $C_{1.44}$ depends only on d, T, q, λ_T, Λ_T, B_T, and $\delta_T(\cdot)$.

We next show that $p(s, x; t, y)$ exists. For $n \geq 1$, define $\mu^n_{s, x}$ on $([0, \infty) \times R^d$, $\mathcal{B}_{[0, \infty) \times R^d})$ by

$$\mu^n_{s, x}([t_1, t_2] \times \Gamma) = \int_{s + (1/n)}^{s + n} \chi_{[t_1, t_2]}(t) P(s, x; t, \Gamma) \, dt.$$

Clearly $(s, x) \to \mu^n_{s, x}$ is measurable; and, by (1.44), $\mu^n_{s, x} \ll dt \times dy$. Thus, by standard results, there is a measurable non-negative function φ_n on $\{(s, x; t, y) \in ([0, \infty) \times R^d)^2 : s < t\}$ such that

$$\frac{d\mu^n_{s, x}}{dt \times dy}(t, y) = \varphi_n(s, x; t, y).$$

We define

$$p(s, x; t, y) = \varphi_{n+1}(s, x; t, y)$$

for $(s, x; t, y)$ such that $1/(n + 1) \leq t - s < 1/n$ or $n \leq t - s < n + 1$. It is easy to check that $p(s, x; t, y)$ is the desired density.

From (1.44) it is immediate that

$$\left(\int_s^T (t - s)^{Nq/2} \, dt \int_{R^d} |p(s, x; t, y)|^q \, dy \right)^{1/q} \leq C_{1.44}.$$

Since

$$\frac{Nq}{2} = \frac{1}{2} \left[\frac{d + 2}{2} \frac{2}{q'} \right] q \leq \frac{d + 2}{2} \left(1 - \frac{1}{q} \right) q = \frac{d + 2}{2} (q - 1),$$

(1.35) follows.

The proof of (1.36) can be accomplished in either one of two ways. One can proceed in exactly the same manner as we just have in the derivation of (1.35) from (1.32), only this time using (1.33); or one can derive (1.36) directly from (1.35) in the same way as we got (1.24) from (1.22) in the proof of Lemma 9.1.4. The details are left to the reader. $\square$

9.1.10 Corollary. *Let* $a: [0, \infty) \times R^d \to S_d^+$ *and* $b: [0, \infty) \times R^d \to R^d$ *be bounded measurable functions. For each* $T > 0$ *and* $R > 0$, *assume that there exist numbers* $0 < \lambda_{T, R} \leq \Lambda_{T, R} < \infty$ *and* $B_{T, R} < \infty$ *and a non-decreasing function* $\delta_{T, R}$:

$(0, \infty) \to (0, \infty)$ *such that*

$$\lambda_{T,R} |\theta|^2 \le \langle \theta, a(s, x)\theta \rangle \le \Lambda_{T,R} |\theta|^2,$$

$$(s, x) \in [0, T] \times B(0, R) \quad \text{and} \quad \theta \in R^d,$$

$$|b(s, x)| \le B_{T,R}, \qquad (s, x) \in [0, T] \times B(0, R),$$

and

$$\sup_{\substack{0 \le s \le T}} \sup_{\substack{x^1, x^2 \in B(0, R) \\ |x^1 - x^2| < \delta_{T,R}(\varepsilon)}} \|a(s, x^1) - a(s, x^2)\| < \varepsilon, \; \varepsilon > 0.$$

Set

$$L_t = \frac{1}{2} \sum_{i,j=1}^{d} a^{ij}(t, \cdot) \frac{\partial^2}{\partial x_i \, \partial x_j} + \sum_{i=1}^{d} b^i(t, \cdot) \frac{\partial}{\partial x_i}$$

and let $P(s, x; t, \cdot)$ *be the transition probability function determined by* L_t. *Then* $P(s, x; t, \cdot)$ *admits a density* $p(s, x; t, y)$ *in the sense described in Theorem 9.1.9. Moreover, if* $0 < r < R, 0 \le s < T$ *and* $1 \le q < \infty$, *then*

$$(1.45) \qquad \left(\int_s^T (t - s)^\alpha \, dt \int_{B(0, r)} |p(s, x; t, y)|^q \, dy \right)^{1/q} \le C_{1.45},$$

where $\alpha = ((d + 2)/2)(q - 1)$ *and* $C_{1.45}$ *depends only on* $d, T, q, R - r, \lambda_{T,R}, \Lambda_{T,R},$ $B_{T,R},$ *and* $\delta_{T,R}(\cdot)$. *Finally, if* $0 < r < R, 0 \le s < T,$ *and* $1 \le q < \infty$, *then for each* $R > 0$ *and* $\delta > 0$:

$$(1.46) \qquad \left(\int_s^T dt \int_{B(0, r) \setminus B(x, \delta)} |p(s, x; t, y)|^q \, dy \right)^{1/q} \le C_{1.46},$$

where $C_{1.46}$ *depends only on* $d, T, q, R - r, \lambda_{T,R}, \Lambda_{T,R}, B_{T,R},$ *and* $\delta_{T,R}(\cdot)$ *as well as* δ.

Proof. Given $T > 0$ and $0 < r < R$, choose measurable $\tilde{a}: [0, \infty) \times R^d \to S_d^+$ and $\tilde{b}: [0, \infty) \times R^d \to R^d$ so that $\tilde{a} = a$ and $\tilde{b} = b$ on $[0, T] \times \overline{B(0, r)}$ and $\tilde{a}$ and $\tilde{b}$ satisfy the conditions of Theorem 9.1.9 with $\lambda_T = \lambda_{T,R}, \Lambda_T = \Lambda_{T,R}, B_T = B_{T,R}$ and $\delta_T(\cdot) = \delta_{T,R}(\cdot)$. Set

$$\tilde{L}_t = \frac{1}{2} \sum_{i,j=1}^{d} \tilde{a}^{ij}(t, \cdot) \frac{\partial^2}{\partial x_i \, \partial x_j} + \sum_{i=1}^{d} \tilde{b}^i(t, \cdot) \frac{\partial}{\partial x_i}$$

and define $\tilde{P}(s, x; t, \cdot)$ *accordingly. By Lemma 9.1.7,*

$$\left| \int_s^T dt \int_{B(0, r)} P(s, x; t, dy) f(t, y) \right| \le \int_s^T dt \int \tilde{P}(s, x; t, dy) |f(t, y)|$$

$$+ C_{1.31} \sup_{\substack{s \le u \le T \\ |z| = r}} \int_u^T dt \int \tilde{P}(u, z; t, dy) |f(t, y)|$$

where $C_{1.31}$ depends only on d, T, $R - r$, $\Lambda_{T, R}$ and $B_{T, R}$. Thus (1.45) and (1.46) follow easily from Theorem 9.1.9. $\square$

One of the applications of Corollary 9.1.10 is to the connection between the diffusion process associated with L_t and the known analytic facts about solutions to the backwards equation. To be precise, we give the next theorem.

9.1.11 Theorem. Let $a: [0, \infty) \times R^d \to S_d^+$ and $b: [0, \infty) \times R^d \to R^d$ satisfy the conditions of Corollary 9.1.10 and let $\{P_{s, x}: (s, x) \in [0, \infty) \times R^d\}$ and $P(s, x; t, \cdot)$ be defined accordingly. Suppose that $\{f_n\}_1^\infty \subseteq C_b^{1, 2}([s, T] \times R^d)$ has the properties that $f_n \to \varphi \in C_b([s, T] \times R^d)$ boundedly and pointwise in $[s, T] \times R^d$ while

$$\frac{\partial f_n}{\partial t} + L_t f_n \to \psi \in B([s, T] \times R^d)$$

in $L_{\text{loc}}^\rho((s, T) \times R^d)$ for some $1 < \rho \le \infty$. Then, for each $x \in R^d$, $\varphi(t \wedge T, x(t \wedge T)) - \int_s^{t \wedge T} \psi(u, x(u)) \, du$ is a $P_{s, x}$-martingale after time s. In particular,

$$(1.47) \quad \varphi(s, x) = \int P(s, x; T, dy)\varphi(T, y) - \int_s^T dt \int_{R^d} P(s, x; t, dy)\psi(t, y).$$

Proof. Clearly (1.47) will follow if we prove the first assertion. To do this, let $s < t_1 < t_2 < T$ be given and define

$$\tau_R = \inf\{t > t_1: |x(t)| \ge R\}, \qquad R > 0.$$

Then for any $A \in \mathscr{M}_{t_1}$:

$$E^{P_{s, x}}[f_n(t_2 \wedge \tau_R, x(t_2 \wedge \tau_R)) - f_n(t_1 \wedge \tau_R, x(t_1 \wedge \tau_R)), A]$$
$$= E^{P_{s, x}}\left[\int_{t_1 \wedge \tau_R}^{t_2 \wedge \tau_R} \left(\frac{\partial}{\partial t} + L_t\right) f_n(t, x(t)) \, dt, A\right].$$

As $n \to \infty$,

$$E^{P_{s, x}}[f_n(t \wedge \tau_R, x(t \wedge \tau_R)), A] \to E^{P_{s, x}}[\varphi(t \wedge \tau_R)), A].$$

and by (1.45)

$$\left| E^{P_{s, x}}\left[\int_{t_1 \wedge \tau_R}^{t_2 \wedge \tau_R} \left(\left(\frac{\partial}{\partial t} + L_t\right) f_n(t, x(t)) - \psi(t, y(t))\right) dt, A\right]\right|$$
$$\le \int_{t_1}^{t_2} dt \int_{B(0, R)} \left|\left(\frac{\partial}{\partial t} + L_t\right) f_n(t, y)\right) - \psi(t, y)\right| p(s, x; t, y) \, dy \to 0$$

where $p(s, x; t, y)$ is the density of $P(s, x; t, \cdot)$. Thus

$$E^{P_{s,x}}[\varphi(t_2 \wedge \tau_R, x(t_2 \wedge \tau_R)) - \varphi(t_1 \wedge \tau_R, x(t_1 \wedge \tau_R)), A]$$

$$= E^{P_{s,x}}\left[\int_{t_1 \wedge \tau_R}^{t_2 \wedge \tau_R} \psi(t, x(t)) \, dt, A\right].$$

Letting $R \nearrow \infty$, we see that $\varphi(t, x(t)) - \int_s^t \psi(u, x(u)) \, du$ is indeed a $P_{s,x}$-martingale. $\square$

Before one can make use of Theorem 9.1.11 it is necessary to have existence theorems from analysis. Fortunately, such existence theorems are often known. For instance, if a is uniformly positive definite and uniformly continuous, then Fabes and Riviére [1966] have shown that each $\psi \in C_0([0, T] \times R^d)$ is the limit in every $L^q([0, T] \times R^d)$, for $1 < q < \infty$, of a sequence ψ_n given by $\psi_n = (\partial f_n/\partial t) + L_t f_n$, where $f_n(\cdot, T) \equiv 0, f_n \in C_b^{1,2}([0, T] \times R^d)$ and the functions f_n converge uniformly to a limit f. With a little more effort one can extend their results to mesh with the more general set-up considered in Theorem 9.1.11.

There is one more type of estimate about $p(s, x; t, y)$ which we want. The origin of our interest in this estimate will not become apparent until Chapter 11.

Given $0 < \lambda < \Lambda$, recall that $\mathscr{A}(d + 3, \lambda, \Lambda)$ stands for the set of measurable functions $a: [0, \infty) \times R^d \to S_d$ such that

$$\lambda |\theta|^2 \leq \langle \theta, a(s, x)\theta \rangle \leq \Lambda |\theta|^2, \qquad (s, x) \in [0, \infty) \times R^d \quad \text{and} \quad \theta \in R^d,$$

and, for some $x^0 \in R^d$,

$$\|a(s, x) - \bar{a}(s)\| \leq \frac{1}{d^2} C_d(d + 3, \lambda, \Lambda), \quad (s, x) \in [0, \infty) \times R^d,$$

where $\bar{a}(\cdot) = a(\cdot, x^0)$. Given $a \in \mathscr{A}(d + 3, \lambda, \Lambda)$ and $T > 0$, define $D_{a, \bar{a}}^T$ as in (1.13), and recall that $(I - D_{a, \bar{a}}^T)^{-1}$ exists as an operator with norm at most 2 on $L^{d+3}([0, T] \times R^d)$ into itself.

9.1.12 Lemma. *Given $a \in \mathscr{A}(d + 3, \lambda, \Lambda)$, define K_a^T on $L^{d+3}([0, T] \times R^d)$ by (1.15). Then for $f \in L^{d+3}([0, T] \times R^d)$, $K_a^T f \in C_b([0, T] \times R^d)$, and, in fact,*

$$(1.48) \qquad \|K_a^T f\| \leq C_{1.48} \|f\|_{L^{d+3}([0, T] \times R^d)}$$

and

$$(1.49) \qquad |K_a^T f(s, x^1) - K_a^T f(s, x^2)| \leq C_{1.49} |x^1 - x^2| \|f\|_{L^{d+3}([0, T] \times R^d)},$$

where the constants $C_{1.48}$ and $C_{1.49}$ depend only on d, T, λ, and Λ. Moreover, if $h \in R^d$ and $a'(s, x) = a(s, x + h)$, $(s, x) \in [0, \infty) \times R^d$, then $a' \in \mathscr{A}(d + 3, \lambda, \Lambda)$ and

$$(1.50) \quad \| K_a^T f - K_{a'}^T f \| \leq C_{1.50} \sup_{(s, x) \in [0, \infty) \times R^d} \| a(s, x + h) - a(s, x) \| \| f \|_{L^{d+3}(R^d)}$$

where $C_{1.50}$ depends only on d, T, λ, and Λ.

Proof. Clearly (1.48) is an immediate consequence of (1.9) (with $r_1 = d + 3$ and $r_2 = \infty$) and the boundedness of $(I - D_{a,\bar{a}}^T)^{-1}$. Moreover, by the same sort of reasoning as led to Lemma 9.1.1, it is easy to see that $(\partial G_{\bar{a}}^T f)/\partial x_i \in C_b([0, T] \times R^d)$, $1 \leq i \leq d$, for $f \in L^{d+3}([0, T] \times R^d)$ and that

$$(1.51) \qquad \left\| \frac{\partial G_{\bar{a}}^T f}{\partial x_i} \right\| \leq C_{1.51} \| f \|_{L^{d+3}([0, T] \times R^d)},$$

where $C_{1.51}$ depends only on d, T, λ, and Λ. Again, because $(I - D_{a,\bar{a}}^T)^{-1}$ is bounded on $L^{d+3}([0, T] \times R^d)$, this proves (1.49). Finally, to prove (1.50), note that we can use the same $\bar{a}$ in defining K_a^T, as we use for $K_{a'}^T$. Since

$$\| D_{a,\bar{a}}^T f - D_{a',\bar{a}}^T f \|_{L^{d+3}([0, T] \times R^d)} \leq \frac{\eta}{2} d^2 C_d(d + 3, \lambda, \Lambda) \| f \|_{L^{d+3}([0, T] \times R^d)}$$

where $\eta = \sup_{(s, x) \in [0, \infty) \times R^d} \| a(s, x + h) - a(s, x) \|$ and because

$$\| D_{a,\bar{a}}^T f \|_{L^{d+3}([0, T] \times R^d)} \vee \| D_{a',\bar{a}}^T f \|_{L^{d+3}([0, T] \times R^d)} \leq \tfrac{1}{2} \| f \|_{L^{d+3}([0, T] \times R^d)},$$

we see that

$$\| (D_{a,\bar{a}}^T)^n f - (D_{a',\bar{a}}^T)^n f \|_{L^{d+3}([0, T] \times R^d)} \leq \frac{n}{2^n} \eta d^2 C_d(d + 3, \lambda, \Lambda) \| f \|_{L^{d+3}([0, T] \times R^d)}.$$

Thus

$$(1.52) \quad \| (I - D_{a,\bar{a}}^T)^{-1} f - (I - D_{a',\bar{a}}^T)^{-1} f \|_{L^{d+3}([0, T] \times R^d)} \leq C_{1.52} \eta \| f \|_{L^{d+3}([0, T] \times R^d)},$$

where $C_{1.52}$ depends only on d, λ, and Λ. Using (1.9) once more, we now get (1.5) from (1.52). $\square$

9.1.13 Lemma. *Let $a \in \mathscr{A}(d + 3, \lambda, \Lambda)$ and denote by $P(s, x; t, \cdot)$ the transition probability function determined by the martingale problem for*

$$L_t = \frac{1}{2} \sum_{i, j = 1}^d a^{ij}(t, \cdot) \frac{\partial^2}{\partial x_i \, \partial x_j}.$$

Given $T > 0$ and $f \in C_0^\infty([0, T) \times R^d)$, define $f_h(s, x) = f(s, x - h)$. Then

$$(1.52) \quad \left| \int_s^T dt \int_{R^d} (f(t, y) - f_h(t, y)) P(s, x; t, dy) \right|$$

$$\leq C_{1.53} \left(|h| + \sup_{(t, y) \in [0, \infty) \times R^d} \|a(t, y) - a(t, y + h)\| \right) \|f\|_{L^{d+3}([0, T) \times R^d)}.$$

Proof. Set $a'(s, x) = a(s, x + h)$, $(s, x) \in [0, \infty) \times R^d$, and let $P'(s, x; t, \cdot)$ be the associated transition probability function. It is easy to check that

$$\int_{R^d} \varphi_h(y) P(s, x + h; t, dy) = \int_{R^d} \varphi(y) P'(s, x; t, dy)$$

for $\varphi \in C_b(R^d)$. Thus:

$$\left| \int_s^T dt \int_{R^d} (f_h(t, y) - f(t, y)) P(s, x; t, dy) \right|$$

$$\leq \left| \int_s^T dt \int_{R^d} f_h(t, y) (P(s, x; t, dy) - P(s, x + h; t, dy)) \right|$$

$$+ \left| \int_s^T dt \int_{R^d} f(t, y) (P'(s, x; t, dy) - P(s, x; t, dy)) \right|$$

$$= |K_a^T f_h(s, x + h) - K_a^T f_h(s, x)| + |K_{a'}^T f(s, x) - K_a^T f(s, x)|$$

$$\leq (C_{1.49} |h| + C_{1.50} \eta) \|f\|_{L^{d+3}([0, T) \times R^d)},$$

where $\eta = \sup_{(t, y) \in [0, \infty) \times R^d} \|a(t, y + h) - a(t, y)\|$. $\quad \square$

9.1.14 Lemma. *Let $a: [0, \infty) \times R^d \to S_d$ be a bounded measurable function, and suppose that for some $(s^0, x^0) \in [0, \infty) \times R^d$, $T > s^0$, and $R > 0$, there exists an $\tilde{a} \in \mathscr{A}(d + 3, \lambda, \Lambda)$ such that $a = \tilde{a}$ on $[s^0, T) \times B(x^0, R)$. Let P be a solution to the martingale problem for*

$$L_t = \frac{1}{2} \sum_{i, j = 1}^d a^{ij}(t, \cdot) \frac{\partial^2}{\partial x_i \, \partial x_j}$$

starting from (s^0, x^0). Then for any $0 < r < R$ and all $0 < |h| < r$ and $f \in C_0^\infty([s^0, T) \times B(x^0, r - |h|))$:

$$(1.54) \quad \left| E^P \left[\int_{s^0}^T (f_h(t, x(t)) - f(t, x(t))) \, dt \right] \right|$$

$$\leq C_{1.54} \left(|h| + \sup_{(t, y) \in [0, \infty) \times R^d} \|a(t, y + h) - a(t, y)\| \right) \|f\|,$$

where $C_{1.54}$ depends only on d, T, λ, Λ, R, and r.

Proof. Define τ_n, $n \geq -1$, as in Lemma 9.1.6 with $R_1 = r$ and $R_2 = R$. By the familiar reasoning, if $g \in C_0([s^0, T] \times B(x^0, r))$, then:

$$\left| E^P \left[\int_{s^0}^{T} g(t, x(t)) \, dt \right] \right| \leq \sum_{n=0}^{\infty} \left| E^P \left[\int_{\tau_{2n-1}}^{\tau_{2n} \wedge T} g(t, x(t)) \, dt, \tau_{2n-1} \leq T \right] \right|$$

$$\leq \sum_{n=0}^{\infty} \int_{\tau_{2n-1}(\omega) \leq T} \left| E^{\tilde{P}_\omega^n} \left[\int_{\tau_{2n-1}(\omega)}^{\tau_{2n}^\omega} g(t, x(t)) \, dt \right] \right| P(d\omega)$$

where $\tilde{P}_\omega^n \equiv \tilde{P}_{\tau_{2n-1}(\omega), x(\tau_{2n-1}(\omega), \omega)}$ and $\tilde{P}_{s,x}$ solves the martingale problem for

$$\tilde{L}_t = \frac{1}{2} \sum_{i,j=1}^{d} \tilde{a}(t, \cdot) \frac{\partial^2}{\partial x_i \, \partial x_j}$$

starting from (s, x) and

$$\tau_{2n}^\omega = \inf\{t \geq \tau_{2n-1}(\omega): |x(t) - x^0| \geq R\}.$$

But if $\tau = \inf\{t \geq s: |x(t) - x^0| \geq R\}$, then

$$\tilde{P}(s, x; t, \Gamma) = \tilde{P}_{s,x}(x(t) \in \Gamma, \tau > t)$$
$$+ E^{\tilde{P}_{s,x}}[\tilde{P}(\tau, x(\tau); t, \Gamma), \tau \leq t];$$

and so

$$\left| E^{\tilde{P}_{s,x}} \left[\int_s^{\tau \wedge T} g(t, x(t)) \, dt \right] \right| \leq \int_s^T dt \int_{R^d} |g(t, y)| \tilde{P}(s, x; t, dy)$$
$$+ E^{\tilde{P}_{s,x}} \left[\int_s^T dt \int_{R^d} |g(t, y)| \tilde{P}(s, x(\tau); t, dy), \tau \leq T \right].$$

Taking $g = f_h - f$ and applying Lemma 9.1.13, we arrive at

$$E^P \left[\int_{s^0}^T (f_h(t, x(t)) - f(t, x(t)) \, dt \right]$$

$$\leq \left(\sum_{n=0}^{\infty} P(\tau_{2n-1} \leq T) \right)$$

$$\times 2C_{1.53} \left(|h| + \sup_{(t,\, y)} \|a(t, y + h) - a(t, y)\| \right) \|f\|_{L^{d+3}([0, T] \times R^d)}.$$

We can now use Lemma 9.1.6 and the obvious estimate of $\|f\|_{L^{d+3}([0, T] \times R^d)}$ in terms of r and $\|f\|$ to get (1.54). $\square$

9.1.15 Theorem. *Let* $a: [0, \infty) \times R^d \to S_d^+$ *and* $b: [0, \infty) \times R^d \to R^d$ *be bounded measurable functions. For each* $T > 0$, *assume that there exist numbers* $0 < \lambda_T \le \Lambda_T < \infty$ *and* $B_T < \infty$ *and a nondecreasing function* $\delta_T: (0, \infty) \to (0, \infty)$ *such that* $\lim_{\varepsilon \searrow 0} \delta_T(\varepsilon) = 0$ *and*

$$\lambda_T |\theta|^2 \le \langle \theta, a(s, x)\theta \rangle \le \Lambda_T |\theta|^2, \ (s, x) \in [0, T] \times R^d \quad \text{and} \quad \theta \in R^d,$$
$$|b(s, x)| \le B_T, \ (s, x) \in [0, T] \times R^d,$$

and

$$\sup_{\substack{0 \le s \le T \\ |x^1 - x^2| \le \delta_T(\varepsilon)}} \|a(s, x^1) - a(s, x^2)\| \le \varepsilon, \ \varepsilon > 0.$$

Let

$$L_t = \frac{1}{2} \sum_{i, j=1}^{d} a^{ij}(t, \cdot) \frac{\partial^2}{\partial x_i \, \partial x_j} + \sum_{i=1}^{d} b^i(t, \cdot) \frac{\partial}{\partial x_i}$$

and define $P(s, x; t, \cdot)$ *and* $p(s, x; t, y)$ *accordingly as in Theorem 9.1.9. Then for each* $T > 0$ *there is a non-decreasing function* $\Phi: (0, \infty) \to (0, \infty)$ *depending only on* d, T, λ_T, Λ_T *and* $\delta_T(\cdot)$ *such that* $\lim_{\varepsilon \searrow 0} \Phi(\varepsilon) = 0$ *and*

$$(1.55) \qquad \int_s^T dt \int |p(s, x; t, y + h) - p(s, x; t, y)| \, dy \le \Phi(|h|).$$

Proof. Assume that we have proved (1.55) in the case when $b \equiv 0$. Set

$$L_t^0 = \frac{1}{2} \sum_{i, j=1}^{d} a^{ij}(t, \cdot) \frac{\partial^2}{\partial x_i \, \partial x_j}$$

and let $\{P_{s, x}^0: (s, \dot{x}) \in [0, \infty) \times R^d\}$, be the associated family of solutions to the martingale problem. Then our assumption implies that

$$\left| E^{P_{s, x}^0} \left[\int_s^T (f_h(t, x(t)) - f(t, x(t)) \, dt \right] \right| \le \Phi(|h|) \|f\|$$

for $f \in C_0^\infty([0, T) \times R^d)$. By Theorem 7.2.2,

$$\left| E^{P_{s, x}} \left[\int_s^T (f_h(t, x(t)) - f(t, x(t))) \, dt \right] \right|$$
$$= \left| E^{P_{s, x}^0} \left[\int_s^T (f_h(t, x(t)) - f(t, x(t))) \, dt X^s(T) \right] \right|$$
$$\le E^{P_{s, x}^0} \left[\left(\int_s^T (f_h(t, x(t)) - f(t, x(t))) \, dt \right)^2 \right]^{\frac{1}{2}} E^{P_{s, x}^0} [(X^s(T))^2]^{\frac{1}{2}},$$

where $P_{s,x}$ solves for L_t starting from (s,x) and

$$X^s(T) = \exp\left[\int_s^T \langle a^{-1}b(x(u)), dx(u)\rangle - \frac{1}{2}\int_s^T \langle b, a^{-1}b\rangle(u, x(u))\,du\right].$$

Since, as we have already seen, $E^{P^0_{s,x}}[(X^s(T))^2]$ is bounded by $\exp[B_T^2(T-s)/2\lambda_T]$, it remains only to estimate

$$E^{P^0_{s,x}}\left[\left(\int_s^T (f_h(t, x(t)) - f(t, x(t)))\,dt\right)^2\right]^{\frac{1}{4}}.$$

But

$$E^{P^0_{s,x}}\left[\left(\int_s^T (f_h(t, x(t)) - f(t, x(t)))\,dt\right)^2\right]$$

$$= 2E^{P^0_{s,x}}\left[\int_s^T (f_h(u, x(u)) - f(u, x(u)))\,du \int_u^T (f_h(v, x(v)) - f(v, x(v)))\,dv\right]$$

$$= 2E^{P^0_{s,x}}\left[\int_s^T (f_h(u, x(u)) - f(u, x(u)))\,du\, E^{P^0_{u, x(u)}}\left[\int_u^T (f_h(v, x(v)) - f(v, x(v)))\,dv\right]\right]$$

$$\le 2\|f\|\Phi(|h|)E^{P^0_{s,x}}\left[\int_s^T |f_h(u, x(u)) - f(u, x(u))|\,du\right]$$

$$\le 4\|f\|^2(T-s)\Phi(|h|).$$

Thus,

$$\left|E^{P_{s,x}}\left[\int_s^T (f_h(t, x(t)) - f(t, x(t)))\,dt\right]\right|$$

$$\le 2\exp\left[\frac{B_T^2}{\lambda_T}(T-s)\right]\|f\|(T-s)^{\frac{1}{2}}(\Phi(|h|))^{\frac{1}{2}}.$$

It is easy to see from the preceding that (1.54) now holds with $2T^{\frac{1}{2}}\exp[B_T^2 T/2\lambda_T]\Phi(|h|)^{\frac{1}{2}}$ replacing $\Phi(h)$ on the right-hand side. These considerations show that we can assume that $b \equiv 0$.

Assuming that $b \equiv 0$, we see from Lemma 9.1.14 that there is a $\rho > 0$, depending only on $d, \delta_T(\cdot), \lambda_T$, and Λ_T, such that for all $(s, x) \in [0, T) \times R^d, |h| < \rho$, and bounded measurable f having support in $[0, T] \times Q(x^0, \rho)$ where $Q(x^0, \rho) = \{x : x_i^0 - \rho \le x_i < x_i^0 + \rho, 1 \le i \le d\}$, for some $x^0 \in R^d$:

$$\left|E^{P_{s,x}}\left[\int_s^T (f_h(t, x(t)) - f(t, x(t)))\,dt\right]\right| \le C_{1.54}(|h| + \delta_T(h))\|f\|,$$

where $C_{1.54}$ depends only on d, T, λ_T, Λ_T, and ρ. By the estimate in (2.1) of Chapter 4, we can now find A depending on d, ρ, and $C_{1.54}$ and $\alpha > 0$ depending on d, Λ_T, and ρ such that for any $f \in C_b([0, T] \times R^d)$ and $|h| < \rho$:

$$\left| E^{P_{s,x}} \left[\int_s^T (f_h(t, x(t)) - f(t, x(t))) \, dt \right] \right| \leq A(n^d(|h| + \delta_T(h)) + e^{-\alpha n}) \|f\|$$

for all $n \geq 1$. Since there is a C depending on d, A, and α such that

$$\inf_n A(n^d(|h| + \delta_T(h)) + e^{-\alpha n}) \leq C(|h| + \delta_T(h))^{\frac{1}{2}},$$

and (1.55) follows. $\square$

9.2. The Homogeneous Case

In this section we will show how to refine the results of the preceding section in the time homogeneous case. To be precise, let $a: R^d \to S_d^+$ be a bounded continuous function, $b: R^d \to R^d$ a bounded measurable function, and

$$L = \frac{1}{2} \sum_{i, j = 1}^d a^{ij}(\cdot) \frac{\partial^2}{\partial x_i \, \partial x_j} + \sum_{i=1}^d b^i(\cdot) \frac{\partial}{\partial x_i}.$$

The martingale problem for L is well-posed. Let $\{P_{s,x}: (s, x) \in [0, \infty) \times R^d\}$ be the associated family of solutions. At the heart of the aforementioned refinements is the observation made in the next lemma.

9.2.1 Lemma. *For $s \geq 0$ define $\Phi_s: \Omega \to \Omega$ so that $x(t, \Phi_s(\omega)) = x((t - s) \vee 0, \omega)$, $t \geq 0$. Then $P_{s,x} = P_{0,x} \circ \Phi_s^{-1}$. In particular, if $P(s, x; t, \cdot)$ is the transition probability function associated with L, then $P(s, x; t, \cdot) = P(0, x; t - s, \cdot)$ for $0 \leq s < t$.*

Proof. The first assertion is immediate from the fact that $P_{0,x} \circ \Phi_s^{-1}$ solves the martingale problem for L starting from (s, x). To prove the second part, note that from $P_{s,x} = P_{0,x} \circ \Phi_s^{-1}$ we have

$$P(s, x; t, \Gamma) = P_{s,x}(x(t) \in \Gamma) = P_{0,x}(x(t - s) \in \Gamma) = P(0, x; t - s, \Gamma).   \square$$

In view of the preceding lemma, it is natural to adopt the following conventions when dealing with time-independent L's. In the first place, we don't really need $P_{s,x}$ for any s other than $s = 0$; and so we use P_x to denote $P_{0,x}$ and abandon the special notation for solutions starting after time 0. Secondly, we define

$$P(t, x, \cdot) = P(0, x; t, \cdot)$$

and call $P(t, x, \cdot)$ the *(time homogeneous) transition probability function* associated with L. Observe that the Chapman-Kolmogorov equation now reads:

$$(2.1) \qquad P(t + s, x, \Gamma) = \int P(t, y, \Gamma) P(s, x, dy).$$

The fact that the transition probability function possesses a special form in the time homogeneous case enables us to refine the results of the preceding section. The next lemma is typical of the sort of improvement that we can expect to make.

9.2.2 Lemma. *For each $\Gamma \in \mathscr{B}_{R^d}$, $(s, x) \rightarrow P(s, x, \Gamma)$ is continuous on $(0, \infty) \times R^d$. In particular, for all $(s, x) \in (0, \infty) \times R^d$, $P(s, x, \cdot)$ possesses a density $p(s, x, y)$ with respect to Lebesgue measure.*

Proof. Let $s > 0$ be given and let $(s_n, x_n) \rightarrow (s, x)$ with $0 \leq s_n < 2s$ for all n. Then

$$P(s_n, x_n, \Gamma) = P(2s - (2s - s_n), x_n, \Gamma)$$

$$= P(2s - s_n, x_n; 2s, \Gamma)$$

$$\rightarrow P(s, x; 2s, \Gamma) = P(s, x, \Gamma),$$

where we have used the strong Feller property. Finally, we know from Corollary 9.1.10 that $P(s, x, \cdot) = P(0, x; s, \cdot)$ has a density for almost every $s > 0$. In particular, if Γ has Lebesgue measure zero, then $P(s, x, \Gamma) = 0$ for almost every $s > 0$. Thus, by the preceding, $P(s, x, \Gamma) = 0$ for all $s > 0$, and so the proof is complete. $\square$

We now want to obtain L^q-estimates on $p(s, x, \cdot)$ for $1 < q < \infty$. Unfortunately, this will require our proceeding, as in Section 9.1, from the case when $b \equiv 0$ and a is a small perturbation. However, much of the procedure is essentially identical to the one in Section 9.1, and therefore at times we will not bother again with every detail but, instead, will simply refer to the appropriate part of Section 9.1.

Since we are now dealing with time independent coefficients, nothing will be lost by restricting our attention to perturbations off of constant $c \in S_d^+$. Thus, we introduce the notation:

$$(2.2) \qquad g^{(c)}(s, x) = \frac{1}{((2\pi s)^d |c|)^{\frac{1}{2}}} \exp[-\tfrac{1}{2} \langle x, c^{-1} x \rangle / 2s]$$

for $c \in S_d^+$, $s > 0$, and $x \in R^d$. Obviously, if $c(\cdot) \equiv c$, then

$$g^{(c)}(t - s, y - x) = g^{(c)}(s, x; t, y)$$

where the right-hand side is given by (1.3). In particular, if $0 < \lambda < \Lambda < \infty$ are the lower and upper bounds on the eigenvalues of c, and if

$$G_c f(s, x) = \int_s^1 dt \int_{R^d} g^{(c)}(t - s, y - x) f(t, y) \, dy, \qquad 0 \le s \le 1,$$

then for $1 \le r_1 \le r_2 \le \infty$ satisfying $1/r_1 - 1/r_2 < 2/(d + 2)$:

(2.3) $\|G_c f\|_{L^{r_2}([0,1] \times R^d)} \le C_{2.3} \|f\| \, \|_{L^{r_1}([0,1] \times R^d)},$

where $C_{2.3}$ depends only on d, r_1, r_2, λ, and Λ. Next set

$$g^{(c)}_{,i}(t, y) = \frac{\partial g^{(c)}}{\partial y_i}(t, y), \qquad 1 \le i \le d.$$

Then an easy computation yields:

$$\left(\int_0^1 dt \int_{R^d} |g^{(c)}_{,i}(t, y)|^q \, dy \right)^{1/q}$$

is finite for $1 \le q < (d + 2)/(d + 1)$ and is bounded by a constant depending only on d, q, λ, and Λ. Hence, by Young's inequality,

(2.4) $\left| \dfrac{\partial G_c f}{\partial x_i}(s, x) \right| \le C_{2.4} \|f\|_{L^r([0, 1] \times R^d)}$

for $d + 2 < r < \infty$, where $C_{2.4}$ depends only on d, r, λ, and Λ.

Given $0 < \lambda < \Lambda$ and $1 < r < \infty$, let $\mathscr{A}_0(r, \lambda, \Lambda)$ be the set of continuous functions $a : R^d \to S_d^+$ such that

$$\lambda |\theta|^2 \le \langle \theta, a(x)\theta \rangle \le \Lambda |\theta|^2, \qquad x \in R^d \quad \text{and} \quad \theta \in R^d$$

and, for some $x^0 \in R^d$,

$$\sup_{x \in R^d} \|a(x) - \bar{a}\| d^2 C_d(\rho, \lambda, \Lambda) \le 1, \qquad r \le \rho \le (d + 3) \vee r$$

where $\bar{a} = a(x^0)$ and $C_d(\rho, \lambda, \Lambda)$ is the constant given in (0.4) of the appendix. For $a \in \mathscr{A}_0(r, \lambda, \Lambda)$ we define

$$D_{a, \bar{a}} f(s, x) = \frac{1}{2} \sum_{i, j = 1}^d (a^{ij}(x) - \bar{a}^{ij}) \frac{\partial^2 G_{\bar{a}} f}{\partial x_i \, \partial x_j}(s, x)$$

and

$$K_a = G_{\bar{a}} \cdot (I - D_{a, \bar{a}})^{-1}$$

on $L^\rho([0, 1] \times R^d)$ for $r \le \rho \le (d + 3) \vee r$. The comments about K_a^T as defined in (1.15) apply equally well to K_a defined above.

With these preliminaries, we are ready to begin.

9.2.3 Lemma. *Let* $0 < \lambda < \Lambda$ *and* $1 < q < \infty$ *be given. If* $a \in \mathscr{A}_0(r, \lambda, \Lambda)$, *where* $r = q'$, *then the associated density function* $p(s, x, y)$ *satisfies:*

$$(2.6) \qquad \left(\int_0^1 s^\alpha \, ds \int_{R^d} |p(s, x, y)|^q \, dy \right)^{1/q} \le C_{2.6}$$

and

$$(2.7) \qquad \left(\int_0^1 s^\alpha \, ds \int_{R^d} |p(s, x^1, y) - p(s, x^2, y)|^q \, dy \right)^{1/q} \le C_{2.7} |x^1 - x^2|$$

for some $\alpha \ge 0$ *depending only on* d *and* q *and some* $C_{2.6}$ *and* $C_{2.7}$ *depending only on* d, q, λ, *and* Λ.

Proof. Set $N = [(d + 2)/2r]$. By Lemma 9.1.1, $K_a^{N+1} f \in C_b([0, 1] \times R^d)$ with

$$(2.8) \qquad |K_a^{N+1} f(s, x)| \le C_{2.8} \|f\|_{L^r([0, 1] \times R^d)},$$

where $C_{2.8}$ depends on d, r, λ and Λ. Also, since K_a is continuous on $L^r([0, 1] \times R^d)$ into itself with a bound of 2,

$$\|K_a^{N+1} f\|_{L^r([0, 1] \times R^d)} \le 2^{N+1} \|f\|_{L^r([0, 1] \times R^d)}$$

Thus if $\rho = (d + 3) \vee r$, then

$$(2.9) \qquad \|K_a^{N+1} f\|_{L^\rho([0, 1] \times R^d)} \le C_{2.9} \|f\|_{L^r([0, 1] \times R^d)},$$

where $C_{2.9}$ depends only on d, r, λ, and Λ. Since K_a has a bound of 2 on $L^\rho([0, 1] \times R^d)$ into itself, we conclude that

$$\left| \frac{\partial K_a^{N+2} f}{\partial x_i}(0, x) \right| \le 2C_{2.4} \|K_a^{N+1} f\|_{L^\rho([0, 1] \times R^d)}$$

$$\le 2C_{2.4} C_{2.9} \|f\|_{L^r([0, 1] \times R^d)},$$

and so

$$(2.10) \qquad |K_a^{N+2} f(0, x^1) - K_a^{N+2} f(0, x^2)| \le C_{2.10} |x^1 - x^2| \, \|f\|_{L^r([0, 1] \times R^d)}$$

Using Lemma 9.1.2, we have from (2.8):

$$\left| \int_0^1 s^{N+1} \, ds \int_{R^d} p(s, x, y) f(s, y) \, dy \right|$$

$$\leq \int_0^1 s^{N+1} \, ds \int_{R^d} p(s, x, y) |f(s, y)| \, dy$$

$$= N! (K_a^{N+1} |f|)(0, x) \leq N! C_{2.8} \|f\|_{L^r([0, 1] \times R^d)};$$

and from (2.10):

$$\left| \int_0^1 s^{N+1} \, ds \int_{R^d} (p(s, x^1, y) - p(s, x^2, y)) f(s, y) \, dy \right|$$

$$\leq (N+1)! |x^1 - s^2| |K_a^{N+2} f(0, x^1) - K_a^{N+2} f(0, x^2)|$$

$$\leq (N+1)! |x^1 - x^2| C_{2.10} \|f\|_{L^r([0, 1] \times R^d)}.$$

From these it is easy to get (2.6) and (2.7) with $\alpha = (N+1)q$. □

We are now going to see how estimates like those in Lemma 9.2.3 can be used to yield L^q-estimates on $p(s, x, \cdot)$. The idea is familiar to experts in the theory of partial differential equations and comes under the general heading of "Sobolev inequalities"; although what we need is, more precisely, a "Morrey inequality." In any case, the general principle is that uniform estimates can be derived from mean estimates plus sufficiently good estimates on the modulus of continuity. Our next theorem makes these comments precise.

9.2.4 Theorem. *Let $(E, \mathcal{F}, P)$ be a probability space and $F: R^d \times R^n \times E \to R^1$ a measurable function. Assume that*

(i) for each $q \in E$ and $\varphi \in C_0(R^n)$, $x \to L_\varphi(x, q)$ is continuous, where

$$L_\varphi(x, q) = \int_{R^n} F(x, y, q) \varphi(y) \, dy;$$

(ii) there is a $q \geq 1$, a $B < \infty$, and an $\alpha > 0$ such that

$$E\left[\int |F(x^1, y) - F(x^2, y)|^q \, dy \right] \leq B |x^1 - x^2|^{d + \alpha}.$$

Then for each $\beta > (d + \alpha + 1)/q$ there is an $M = M(d, q, \alpha, \beta)$ such that

$$P\left(\sup_{x \in R^d} \frac{(\int |F(x, y) - F(x^0, y)|^q \, dy)^{1/q}}{(1 + |x - x^0|)^\beta} \geq L \right) \leq MB/L^q, \qquad L > 0,$$

for all $x^0 \in R^d$.

Proof. It suffices to handle the case when $x^0 = 0$. Let $\eta = 2d + \alpha/2$. Then

$$H \equiv \int_{B(0,\,1)} \int_{B(0,\,1)} |x^1 - x^2|^{d+\alpha-\eta} \, dx_1 \, dx_2 < \infty.$$

Hence, for each $R > 0$:

$$E\left[\int_{B(0,\,R)} \int_{B(0,\,R)} \left(\frac{\|F(x^1) - F(x^2)\|_q}{|x^1 - x^2|^{\eta/q}} \right)^q \, dx^1 \, dx^2 \right] \le BHR^{d+\alpha/2}.$$

where

$$\|F(x^1) - F(x^2)\|_q = \left(\int_{R^N} |F(x^1, y) - F(x^2, y)|^q \, dy \right)^{1/q}$$

From this we know that

$$P\left(\exists \varphi \in C_0(R^n)\colon \int_{B(0,\,R)} \int_{B(0,\,R)} \right.$$
$$\times \frac{|L_\varphi(x^1) - L_\varphi(x^2)|^q}{|x^1 - x^2|^{\eta/q}} \, dx^1 \, dx^2 \ge \lambda \|\varphi\|_{q'}^q \right)$$
$$\le \frac{BHR^{d+\alpha/2}}{\lambda}.$$

Applying Exercise 2.4.1, we can now assert that

$$P\left(\exists \varphi \in C_0(R^n)\colon \sup_{x^1,\,x^2 \in B(0,\,R)} \right.$$
$$\times \frac{|L_\varphi(x^1) - L_\varphi(x^2)|}{|x^1 - x^2|^\delta} > A\lambda^{1/q} \|\varphi\|_q \right)$$
$$\le \frac{BHR^{d+\alpha/2}}{\lambda},$$

where $\delta = \alpha/2q$ and

$$A = 8\left(\frac{4^{d+1}}{\gamma^2} \right)^{1/q} \frac{2\eta}{\alpha}$$

and $\gamma = \gamma_d$ is explained in Exercise 2.4.1. In particular,

(2.11) $$P\left(\sup_{x \in B(0,\,R)} \frac{\|F(x) - F(0)\|_q}{|x|^\delta} \ge A\lambda^{1/q} \right) \le \frac{BHR^{d+\alpha/2}}{\lambda}$$

for all $R > 0$ and $\lambda > 0$.

Taking $R = N \geq 1$ and choosing λ so that

$$A\lambda^{1/q} = LN^{\beta - \delta}.$$

we see that (2.11) becomes:

$$P\left(\sup_{x \in B(0, N)} \frac{\|F(x) - F(0)\|_q}{|x|^\delta} \geq LN^{\beta - \delta} \right) \leq \left(\frac{A}{L} \right)^q BHN^{d + \alpha - q\beta},$$

and so

$$P\left(\sup_{x \in B(0, N) \setminus B(0, N-1)} \frac{\|F(x) - F(0)\|_q}{(1 + |x|)^\beta} \leq L \right) \leq \frac{A^q BH}{L^q} N^{d + \alpha - q\beta}.$$

Since $\sum_1^\infty N^{d + \alpha - q\beta} < \infty$, the theorem is immediate from this. $\quad\square$

9.2.5 Lemma. *Let* $d < q < \infty$ *and set* $r = q'$. *Given* $a \in \mathscr{A}_0(r, \lambda, \Lambda)$, *let* $p(s, x, y)$ *be the associated density. Then for all* $1 \leq \rho \leq q$, $p(s, x, \cdot) \in L(R^d)$ *for all* $(s, x) \in [0, \infty) \times R^d$. *In fact,*

$$(2.12) \qquad \|p(s, x, \cdot)\|_{L^\rho(R^d)} \leq C_{2.12}(s \wedge 1)^{-\nu},$$

where $C_{2.12}$ *depends only on* $d, q, \rho, \lambda,$ *and* Λ *and* $\nu \geq 0$ *depends only on* $d, \rho,$ *and* q.

Proof. In Theorem 9.2.4, take $E = (0, 1]$, $P(ds) = s^\alpha/(1 + \alpha)\, ds$ with the α in Lemma 9.2.3, and $F(x, y, s) = p(s, x, y)$. By (2.7), we have:

$$\frac{1}{\alpha + 1} \int_0^1 s^\alpha\, ds \int_{R^d} |p(s, x^1, y) - p(s, x^2, y)|^q\, dy \leq \frac{1}{1 + \alpha} C_{2.7}^q |x^1 - x^2|^q$$

and therefore, by Theorem 9.2.4, for any $\beta > 1 + 1/q$:

$$(2.13) \qquad P\left(s: \sup_{x \in R^d} \frac{\|p(s, x, \cdot) - p(s, x^0, \cdot)\|_{L^q}}{(1 + |x - x^0|)^\beta} \geq L \right) \leq \frac{C_{2.13}}{L^q}, \quad L > 0.$$

where $C_{2.13}$ depends only on $d, C_{2.7}^q/(1 + \alpha), q,$ and β. On the other hand, from (2.6) we also know that:

$$(2.14) \qquad P(s: \|p(s, x^0, \cdot)\|_q \geq L) \leq C_{2.14}/L^q,$$

where $C_{2.14} = C_{2.6}^q/(1 + \alpha)$. Thus, if $0 < \sigma < 1$ is given and we take $L = \gamma/\sigma^\nu$, where $\gamma = 4(\alpha + 1)C_{2.13}C_{2.14}$ and $\nu = \alpha/q$, then

$$P\left(s \in (0, \sigma): \|p(s, x^0, \cdot)\|_q \geq \gamma\sigma^{-\nu} \quad \text{or} \right.$$

$$\left. \sup_{x \in R^d} \frac{\|p(s, x, \cdot) - p(s, x^0, \cdot)\|_q}{(1 + |x - x^0|)^\beta} > \gamma\sigma^{-\nu} \right) \leq \tfrac{1}{2}P((0, \sigma)).$$

Thus there is an $s \in (0, \sigma)$ such that

$$\|p(s, x, \cdot)\|_q \leq \gamma \sigma^{-\nu} (2 + |x - x^0|)^\beta, \qquad x \in R^d.$$

Using the continuous version of Minkowski's inequality, we can write:

$$\|p(\sigma, x^0, \cdot)\|_q = \left(\int_{R^d} dy \left| \int_{R^d} p(\sigma - s, x^0, x) p(s, x, y) \, dx \right|^q \right)^{1/q}$$

$$\leq \int_{R^d} p(\sigma - s, x^0, x) \|p(s, x, \cdot)\|_q \, dx$$

$$\leq \gamma \sigma^{-\nu} \int_{R^d} p(\sigma - s, x^0, x)(2 + |x - x^0|^\beta) \, dx.$$

Finally, applying (2.1) of Chapter 4, we see that

$$\int_{R^d} p(\sigma - s, x^0, x)(2 + |x - x^0|)^\beta \, dx$$

is bounded by a constant depending only on d, β and Λ.

We have therefore shown that

(2.15) $\|p(s, x, \cdot)\|_{L^q(R^d)} \leq C_{2.15} s^{-\nu}, \ (s, x) \in (0, 1] \times R^d,$

where $C_{2.15}$ depends only on d, q, λ, and Λ. If $1 \leq \rho < q$, define θ by $1/\rho = (1 - \theta) + (\theta/q)$. Then, since $\|p(s, x, \cdot)\|_{L^1(R^d)} = 1$,

(2.16) $\|p(s, x, \cdot)\|_{L^\rho(R^d)} \leq (C_{2.15})^\theta s^{-\nu\theta}, \ (s, x) \in (0, 1] \times R^d.$

Finally, if $s > 1$, then

$$\|p(s, x^0, \cdot)\|_{L^\rho(R^d)} \leq \int_{R^d} p(s - 1, x^0, x) \|p(1, x, \cdot)\|_\rho \, dx \leq (C_{2.27})^\theta,$$

and so the proof is complete.   □

9.2.6 Theorem. *Let* $a: R^d \to S_d^+$ *be a bounded, uniformly continuous function with* δ: $(0, \infty) \to (0, \infty)$ *as a modulus of continuity and assume that*

$$\lambda |\theta|^2 \leq \langle \theta, a(x)\theta \rangle \leq \Lambda |\theta|^2, \qquad x \in R^d \quad and \quad \theta \in R^d,$$

for some $0 < \lambda < \Lambda$. *Let* $b: R^d \to R^d$ *be a bounded measurable function such that* $|b(x)| \leq B$ *for all* $x \in R^d$. *Set*

$$L = \frac{1}{2} \sum_{i,j=1}^d a^{ij}(\cdot) \frac{\partial^2}{\partial x_i \, \partial x_j} + \sum_{i=1}^d b^i(\cdot) \frac{\partial}{\partial x_i}$$

and let $p(s, x, y)$ denote the density of the transition probability function determined by L. Then for each $1 \leq q < \infty$

$$(2.17) \qquad \left(\int_{R^d} |p(s, x, y)|^q \, dy \right)^{1/q} \leq C_{2.17}(s \wedge 1)^{-\nu}$$

where ν depends on d and q alone while $C_{2.17}$ depends on d, q, $\delta(\cdot)$, λ, Λ, and B. Moreover, if $\alpha > 0$, then

$$(2.18) \qquad \left(\int_{R^d \setminus B(x, \alpha)} |p(s, x, y)|^q \, dy \right)^{1/q} \leq C_{2.18},$$

where $C_{2.18}$ depends only on d, q, α, $\delta(\cdot)$, λ, Λ, and B.

Proof. Because

$$\sup_{x \in R^d} \left(\int_\Gamma |p(s, x, y)|^q \right)^{1/q}, \quad \Gamma \in \mathcal{B}_{R^d},$$

is a non-increasing function of $s \geq 0$, we need only consider $0 < s \leq 1$. Moreover, the Cameron-Martin formula can be used in the same way as we did in the proof of Theorem 9.1.9 to get the general result from the case in which $b \equiv 0$. Finally, the partitioning technique used in the proof of Theorem 9.1.9 can be employed again here to reduce the problem to that of showing that there is a $\gamma > 0$ depending on d, $\delta(\cdot)$, q, λ, and Λ such that

$$(2.19) \qquad \left(\int_{B(x^0, \gamma)} |p(s, x, y)|^q \, dy \right)^{1/q} \leq C_{2.19} s^{-\nu}$$

for all $(s, x) \in (0, 1] \times R^d$ and $x^0 \in R^d$, where ν depends only on d and q while $C_{2.19}$ depends on d, q, λ, and Λ. But (2.19) can be proved from Lemma 9.2.5 by the same localization technique as was used to prove Lemma 9.1.8 from Lemmas 9.1.4, 9.1.5, and 9.1.6. All that is needed is the observation that if $\tilde{a}$ satisfies the hypothesis of Lemma 9.2.5 and $\tilde{p}(s, x, y)$ is the associated density, then for any $\alpha > 0$ and $1 \leq \rho < q$ with $1/\rho = \theta + (1 - \theta)/q$:

$$\left(\int_{R^d \setminus B(x, \alpha)} |\tilde{p}(s, x, y)|^\rho \, dy \right)^{1/\rho}$$

$$\leq \left[\int_{R^d \setminus B(x, \alpha)} \tilde{p}(s, x, y) \, dy \right]^\theta \left[\int_{R^d} |\tilde{p}(s, x, y)|^q \, dy \right]^{(1-\theta)/q}$$

$$\leq (2d \exp[-\alpha^2 / 2ds\Lambda])^\theta (C_{2.12} s^{-\nu})^{1-\theta},$$

and so

$$(2.20) \qquad \left(\int_{R^d \setminus B(x, \alpha)} |\tilde{p}(s, x, y)|^\rho \, dy \right)^{1/\rho} \leq C_{2.20},$$

where $C_{2.20}$ depends on α as well as d, q, ρ, λ, and Λ. With these remarks, the reader can complete the proof of (2.17). As for (2.18), it can be easily derived from (2.17) by the same argument as we just used to get (2.20) from (2.12). $\quad\square$

9.2.7 Corollary. *Let $a\colon R^d \to S_d^+$ be a bounded continuous function and $b\colon R^d \to R^d$ a bounded measurable function. For each $R > 0$, let the numbers $0 < \lambda_R < \Lambda_R$ and B_R satisfy*

$$\lambda_R|\theta| \le \langle \theta, a(x)\theta\rangle \le \Lambda_R|\theta|^2, \qquad x \in B(0, R) \quad and \quad \theta \in R^d,$$

and

$$|b(x)| \le B_R, \, x \in B(0, R);$$

and let $\delta_R(\cdot)$ be a modulus of continuity for a restricted to $B(0, R)$. Then for each $0 < r < R$ and $1 \le q < \infty$:

(2.21)
$$\left(\int_{B(0,\,r)} |p(s, x, y)|^q \, dy\right)^{1/q} \le C_{2.21}(s \wedge 1)^{-v},$$

and for $\alpha > 0$

(2.22)
$$\left(\int_{B(0,\,r)\backslash B(x,\,\alpha)} |p(s, x, y)|^q \, dy\right)^{1/q} \le C_{2.22}.$$

In (2.21) and (2.22), v depends only on d and q, $C_{2.21}$ depends on d, q, $R - r$, $\delta_R(\cdot)$, λ_R, Λ_R, and B_R, and $C_{2.22}$ depends on these as well as α.

Proof. The derivation of this result from Theorem 9.2.6 is the same as that of Corollary 9.1.10 from Theorem 9.1.9. We leave the details to the reader. $\quad\square$

We will close this section with the analogue for the present context of Theorem 9.1.15.

9.2.8   Lemma.   *Let $\, a \in \mathscr{A}_0(d + 3, \lambda, \Lambda)$.   Then   for $f \in L^{d+3}([0, 1) \times R^d)$, $K_a f \in C_b([0, 1] \times R^d)$ and, in fact:*

(2.23)
$$\|K_a f\| \le C_{2.23}\|f\|_{L^{d+3}([0,\,1]\times R^d)}$$

and

(2.24) $|K_a f(s, x^1) - K_a f(s, x^2)| \le C_{2.24}|x^1 - x^2|\|f\|_{L^{d+3}([0,\,1)\times R^d)}$

where $C_{2.23}$ and $C_{2.24}$ depend only on d, λ, and Λ. Moreover, if $h \in R^d$ and $a'(x) = a(x + h)$, $x \in R^d$, then $a' \in \mathscr{A}_0(d + 3, \lambda, \Lambda)$ and

$$(2.25) \qquad \|K_a f - K_{a'} f\|$$

$$\leq C_{2.25} \sup_{x \in R^d} \|a(x + h) - a(x)\| \quad \|f\|_{L^{d+3}([0,\, 1) \times R^d)}$$

and

$$(2.26) \qquad |(K_a f(s, x^1) - K_{a'} f(s, x^1)) - (K_a f(s, x^2) - K_{a'} f(s, x^2))|$$

$$\leq C_{2.26} \sup_{x \in R^d} \|a(x + h) - a(x)\| \, |x^1 - x^2| \, \|f\|_{L^{d+3}([0,\, 1] \times R^d)}$$

where $C_{2.25}$ and $C_{2.26}$ depend only on d, λ, and Λ.

Proof. Clearly (2.23) is a special case of (2.8) and (2.24) follows easily from (2.4) plus the boundedness of $(I - D_{a,\, \bar{a}})^{-1}$ on $L^{d+3}([0,\, 1) \times R^d)$ into itself. Also, (2.25) can be viewed as a special case of (1.50). Thus, we can restrict our attention to the derivation of (2.26).

Just as in the proof of Lemma 9.1.12, we see that if $\eta = \sup_x \|a(x + h) - a(x)\|$ then

$$(2.27) \qquad \|(I - D_{a,\, \bar{a}})^{-1}f - (I - D_{a',\, \bar{a}})^{-1}f\|_{L^{d+3}([0,\, 1) \times R^d)}$$

$$\leq C_{2.27}\eta\|f\|_{L^{d+3}([0,\, 1) \times R^d)},$$

where $C_{2.27}$ depends only on d. Thus, by (2.4):

$$\begin{aligned}
&|(K_a f(s, x^1) - K_{a'} f(s, x^1)) - (K_a(s, x^2) - K_{a'}(s, x^2))| \\
&= |G_{\bar{a}} \circ [(I - D_{a,\, \bar{a}})^{-1} - (I - D_{a'\, \bar{a}})^{-1}]f(s, x^1) \\
&\quad - G_{\bar{a}} \circ [(I - D_{a,\, \bar{a}})^{-1} - (I - D_{a',\, \bar{a}})^{-1}]f(s, x^2)| \\
&\leq C_{2.4}|x^1 - x^2| \|[(I - D_{a,\, \bar{a}})^{-1} - (I - D_{a',\, \bar{a}})^{-1}]f\|_{L^{d+3}([0,\, 1) \times R^d)} \\
&\leq C_{2.4}C_{2.27}|x^1 - x^2| \eta\|f\|_{L^{d+3}([0,\, 1) \times R^d)},
\end{aligned}$$

and so (2.26) is proved. $\square$

9.2.9 Lemma. *Let $a \in \mathscr{A}_0(r, \lambda, \Lambda)$ for some $1 < r < d + 3$ and denote by $p(s, x, y)$ the density function associated with the transition probability function determined by*

$$L = \frac{1}{2} \sum_{i,\, j=1}^{d} a^{ij}(\cdot) \frac{\partial^2}{\partial x_i \, \partial x_j}.$$

Given $r < \rho \le d + 3$, define θ by $1/\rho = \theta/(d + 3) + (1 - \theta)/r$. Then

(2.28) $\quad \left(\int_0^1 s^\alpha \int_{R^d} |p(s, x, y + h) - p(s, x, y)|^{\rho'} \, dy \right)^{1/\rho'}$

$$\le C_{2.28} \left(|h| + \sup_y \|a(y + h) - a(y)\| \right)^\theta$$

and

$$\left(\int_0^1 s^\alpha \int_{R^d} |(p(s, x^1, y + h) - p(s, x^2, y)) \right.$$

(2.29) $$\left. - (p(s, x^2, y + h) - p(s, x^2, y))|^{\rho'} \, dy \right)^{1/\rho'}$$

$$\le C_{2.29} \left(|h| + \sup_y \|a(y + h) - a(y)\| \right)^\theta |x^1 - x^2|,$$

where $\alpha \ge 0$ is the same as in Lemma 9.2.3 and the constants $C_{2.28}$ and $C_{2.29}$ depend only on d, r, ρ, λ, and Λ.

Proof. Suppose we can prove that:

(2.30) $\quad \left(\int_0^1 ds \int_{R^d} |p(s, x, y + h) - p(s, x, y)|^q \, dy \right)^{1/q}$

$$\le C_{2.30} \left(|h| + \sup_y \|a(y + h) - a(y)\| \right)$$

and

$$\left(\int_0^1 ds \int_{R^d} |p(s, x^1, y + h) - p(s, x^1, y)) \right.$$

(2.31) $$\left. - (p(s, x^2, y + h) - p(s, x^2, y))|^q \, dy \right)^{1/q}$$

$$\le C_{2.31} \left(|h| + \sup_y \|a(y + h) - a(y)\| \right) |x^1 - x^2|,$$

where $1/q = 1 - 1/(d + 3)$ and $C_{2.30}$ and $C_{2.31}$ depend only on d, λ, and Λ. We can then use (2.7) together with Hölder's inequality to obtain (2.28) and (2.29), respectively. But (2.30) and (2.31) follow easily from (2.25) and (2.26) in the same way as we got (1.53) from (1.50). $\square$

9.2.10 Lemma. *Let r, a and $p(s, x, y)$ be as in the preceding and assume that $r < d/(d - 1)$. Then there is a $v > 0$, depending on d and r and a non-decreasing* function $\Phi: (0, \infty) \to (0, \infty)$, depending on d, r, λ, and Λ, such that $\lim_{\varepsilon \searrow 0} \Phi(\varepsilon) = 0$ and

$$(2.32) \qquad \int_{R^d} |p(s, x, y + h) - p(s, x, y)| \, dy$$

$$\leq (s \wedge 1)^{-\nu} \Phi\left(|h| + \sup_y \|a(y + h) - a(y)\| \right).$$

Moreover, for each $\delta > 0$ there is a non-decreasing function $\Psi: (0, \infty) \to (0, \infty)$ depending on δ as well as d, r, λ, and Λ such that $\lim_{\varepsilon \searrow 0} \Psi(\varepsilon) = 0$ and

$$(2.33) \qquad \int_{|y - x| \geq \delta} |p(s, x, y + h) - p(s, x, y)| \, dy$$

$$\leq \Psi\left(|h| + \sup_y \|a(y + h) - a(y)\| \right).$$

Proof. Choose $r < \rho < (d - 1)/d$. Then $\rho' > d$. We can therefore apply Theorem 9.2.4 (in the same way as we did in the derivation of Lemma 9.2.5) to (2.28) and (2.29) and thereby obtain

$$(2.34) \qquad \left(\int_{R^d} |p(s, x, y + h) - p(s, x, y)|^{\rho'} \right)^{1/\rho'}$$

$$\leq C_{2.34}(s \wedge 1)^{-\nu} \left(|h| + \sup_y \|a(y + h) - a(y)\| \right)^{\gamma},$$

where $\nu > 0$ and $\gamma > 0$ depend on d and ρ while $C_{2.34}$ depends on d, r, ρ, λ, and Λ. Since

$$\int_{R^d \setminus B(x, \delta)} |p(s, x, y + h) - p(s, x, y)| \, dy \leq 2d e^{-\delta^2/8d\Lambda s}$$

so long as $|h| \leq \delta/2$, we can use Hölder's inequality to show that if $|h| \leq \delta/2$ then

$$(2.35) \qquad \left(\int_{R^d \setminus B(x, \delta)} |p(s, x, y + h) - p(s, x, y)|^{\sigma} \right)^{1/\sigma}$$

$$\leq C_{2.35} \left(|h| + \sup_y \|a(y + h) - a(y)\| \right)^{\gamma/2},$$

where $\sigma = 2\rho'/(\rho' + 1)$. Because

$$\int_{|y - x| \geq R} p(s, x, y) \, dy \to 0$$

as $R \nearrow \infty$ at a rate depending only on d and Λ, it is easy to obtain (2.32) and (2.33) from (2.34) and (2.35), respectively. $\quad\square$

9.2.11 Lemma. *Let $a: R^d \to S_d$ be a bounded measurable function and suppose that there is an $R > 0$, an $x^0 \in R^d$, and an $\tilde{a} \in \mathscr{A}_0(\rho, \lambda, \Lambda)$ with $\rho < (d-1)/d$ such that $a = \tilde{a}$ on $B(x^0, R)$. Let P be a solution to the martingale problem for*

$$L = \frac{1}{2} \sum_{i,j=1}^d a^{ij}(\cdot) \frac{\partial^2}{\partial x_i \, \partial x_j}$$

starting from $(0, x^1)$. Then for $0 < r < R$ there is a $\nu > 0$ depending on $d, \rho, \lambda,$ and Λ a non-decreasing function $\Phi: (0, \infty) \to (0, \infty)$ depending on $d, \rho, \lambda, \Lambda, r,$ and R such that $\lim_{\varepsilon \searrow 0} \Phi(\varepsilon) = 0$ and

$$(2.36) \qquad |E^P[f_h(x(s)) - f(x(s))]| \le s^{-\nu} \Phi\left(|h| + \sup_y \|\tilde{a}(y+h) - \tilde{a}(y)\|\right)\|f\|$$

for $f \in C_0^\infty(B(x^0, r))$. (Recall that $f_h(x) = f(x - h)$.)

Proof. Define τ_n, $n \ge -1$, as in Lemma 9.1.6 with $s^0 = 0$, $R_1 = (R+r)/2$, and $R_2 = R$. If $g \in C_0(B(0, (R+r)/2))$, then

$$|E^P[g(x(s))]|$$

$$\le |E^P[g(x(s)), \tau_0 > s]|$$

$$+ \sum_1^\infty |E^P[g(x(s)), \tau_{2n-1} \le s < \tau_{2n}]|$$

$$\le |E^{\tilde{P}_{x.1}}[g(x(s)), \tau_0 > s]|$$

$$+ \sum_1^\infty \int_{\tau_{2n-1} \le s} |E^{\tilde{P}_{x(\tau_{2n-1}(\omega),\omega)}}[g(x(s - \tau_{2n-1}(\omega))), \tau_0 > s - \tau_{2n-1}(\omega)]|P(d\omega)$$

where $\{\tilde{P}_x: x \in R^d\}$ is the family associated with

$$\tilde{L} = \frac{1}{2} \sum_{i,j=1}^d \tilde{a}^{ij}(\cdot) \frac{\partial^2}{\partial x_i \, \partial x_j}$$

Noting that

$$\tilde{P}_x(x(t) \in \Gamma, \tau_0 > t) = \tilde{P}(t, x, \Gamma) - E^{\tilde{P}_x}[\tilde{P}(t - \tau_0, x(\tau_0), \Gamma), \tau_0 \le t],$$

where $\tilde{P}(t, x, \cdot)$ is the transition probability function determined by $\tilde{L}$, we now see that the desired estimate follows by taking $g = f_h - f$, with $|h| < (R-r)/2$, and then applying (2.32), (2.33), and Lemma 9.1.6. $\quad\square$

Starting from Lemma 9.2.11, we can derive the next result in exactly the same way as Theorem 9.1.15 was obtained from Lemma 9.1.14.

9.2.12 Theorem. *Let* $a: R^d \to S_d^+$ *be a uniformly continuous function with* $\delta(\cdot)$ *as a modulus of continuity and assume that*

$$\lambda |\theta|^2 \leq \langle \theta, a(x)\theta \rangle < \Lambda |\theta|^2, \qquad x \in R^d \quad and \quad \theta \in R^d.$$

Let $p(s, x, y)$ *be the density associated with*

$$L = \frac{1}{2} \sum_{i,\,j=1}^{d} a^{ij}(\cdot) \frac{\partial^2}{\partial x_i \, \partial x_j}$$

Then there is a non-decreasing function $\Phi: (0, \infty) \to (0, \infty)$, *depending only on* d, λ, Λ, *and* $\delta(\cdot)$, *such that* $\lim_{\varepsilon \searrow 0} \Phi(\varepsilon) = 0$ *and*

$$(2.37) \qquad \int_{R^d} |p(s, x, y + h) - p(s, x, y)| \, dy \leq s^{-\nu} \Phi(|h|)$$

where $\nu > 0$ *depends only on* d, λ, *and* Λ.

Chapter 10

Explosion

10.0. Introduction

Up to this point we have discussed the martingale problem only in connection with bounded coefficients. However it should be evident that many of the more difficult aspects of the martingale problem are really local and do not depend on the global properties of the coefficients. Thus, it should be, and indeed it is, a simple exercise to extend most of the results of Chapters 6, 7 and 9 to martingale problems associated with locally bounded coefficients. The one place at which difficulties can arise is the place where our methods were not local, i.e., in the proof of the existence of solutions to the martingale problem. Although the local existence of solutions is determined completely by the local properties of the coefficients, it is impossible to predict on the basis of local considerations whether a diffusion "runs out to infinity" in a finite time. Perhaps the best way to see this is to look at the simple ordinary differential equation

$$(0.1) \qquad \frac{dx}{dt} = x^2(t), \qquad x(0) = 1.$$

Clearly any solution to the martingale problem for $a(x) \equiv 0$ and $b(x) = x^2$ starting from $x = 1$ at time 0, is concentrated on paths which satisfy (0.1). But this means that for $0 \le t < 1$

$$P\left[x(t) = \frac{1}{1-t}\right] = 1$$

and so

$$P\left[\lim_{t \uparrow 1} x(t) = \infty\right] = 1.$$

In other words the martingale problem has no solution for times after 1.

In Section 10.1 we will reformulate the results of Chapters 6, 7 and 9 when the boundedness assumption on the coefficients is replaced by the assumption that the

associated diffusion does not explode. In Section 10.2 we will find some conditions on the coefficients that ensure that the process does not explode, i.e., does
not run away to infinity in a finite time.

10.1. Locally Bounded Coefficients

Let $a: [0, \infty) \times R^d \to S_d$ and $b: [0, \infty) \times R^d \to R^d$ be locally bounded measurable
functions. Define, as usual,

$$(1.1) \qquad L_t = \frac{1}{2} \sum_{i,j=1}^{d} a^{ij}(t, \cdot) \frac{\partial^2}{\partial x_i \, \partial x_j} + \sum_{i=1}^{d} b^i(t, \cdot) \frac{\partial}{\partial x_i}.$$

We say that the probability measure P on $(\Omega, \mathcal{M})$ solves the martingale problem
for L_t, starting from (s, x) if $P[x(t) = x, \, 0 \le t \le s] = 1$ and

$$f(x(t)) - \int_s (L_u f)(x(u)) \, du$$

is $(\Omega, \mathcal{M}_t, P)$ martingale for times $t \ge s$, for all functions f in $C_0^\infty(R^d)$. The essential
fact on which our analysis of such martingale problems rests is contained in the
following theorem.

10.1.1 Theorem. *Let $a: [0, \infty) \times R^d \to S_d$ and $b: [0, \infty) \times R^d \to R^d$ be locally
bounded measurable coefficients and let L_t be defined by (1.1). Given a solution P to
the martingale problem for L_t starting from (s, x), a stopping time $\tau: \Omega \to (s, \infty)$, and
a r.c.p.d. $\{P_\omega\}$ of P given $\mathcal{M}_\tau$, there is a P-null set $N \in \mathcal{M}_\tau$, such that for all $\omega \notin N$,
$\delta_{x(\tau(\omega), \omega)} \otimes_{\tau(\omega)} P_\omega$ solves the martingale problem for L_t starting from $(\tau(\omega),
x(\tau(\omega), \omega))$. Moreover, if the martingale problem for L_t is well-posed and $\{P_{s,x}: (s, x) \in
[0, \infty) \times R^d\}$ is the associated family of solutions then the map $(s, x) \to P_{s,x}$ is
measurable and forms a strong Markov family. Finally, if the martingale problem for
L_t is well-posed and $\bar{a}: [0, \infty) \times R^d \to S_d$ and $\bar{b}: [0, \infty) \times R^d \to R^d$ are a second set
of locally bounded measurable coefficients such that $\bar{a} = a$ and $\bar{b} = b$ on some
bounded open set $G \subset [0, \infty) \times R^d$, then for $(s, x) \in G$ and any solution $\bar{P}$ to the
martingale problem for $\bar{a}$ and $\bar{b}$ starting from (s, x): $\bar{P}$ equals $P_{s,x}$ on $\mathcal{M}_\tau$, where
$\tau = \inf\{t \ge s: (t, x(t)) \notin G\}$.*

Proof. The first assertion is derived from Theorem 1.2.10 in exactly the same way
as we proved Theorem 6.1.3. To prove the second part, once the first part has been
established, it is obviously enough to check the measurability of $\{P_{s,x}: (s, x) \in
[0, \infty) \times R^d\}$. But the argument given in Exercise 6.7.4 depends in no way on the
coefficients being bounded and therefore applies in the present context as well.
Finally to prove the last assertion, take $P = \bar{P}$ and $Q_\omega = P_{\tau(\omega), x(\tau(\omega), \omega)}$ in Theorem
6.1.2. Then by that theorem $\bar{P} \otimes_{\tau(\cdot)} Q.$ solves the martingale problem for L_t starting from (s, x) and is therefore equal to $P_{s,x}$. At the same time, by construction,
$\bar{P} \otimes_{\tau(\cdot)} Q.$ equals $\bar{P}$ on $\mathcal{M}_\tau$. Thus $\bar{P}$ equals $P_{s,x}$ on $\mathcal{M}_\tau$. $\square$

10.1.2 Corollary. *Let* $a: [0, \infty) \times R^d \to S_d$ *and* $b: [0, \infty) \times R^d \to R^d$ *be locally bounded measurable functions. Assume that there exists an increasing sequence of bounded open sets* $G_n \subset [0, \infty) \times R^d$ *with* $\bigcup_n G_n = [0, \infty) \times R^d$ *and bounded measurable coefficients* $a_n: [0, \infty) \times R^d \to S_d$ *and* $b_n: [0, \infty) \times R^d \to R^d$ *such that* $[0, \infty) \times R^d = \bigcup_n G_n, a_n \equiv a$ *and* $b_n \equiv b$ *on* G_n, *and for each n the martingale problem for* a_n, b_n *is well posed. Then for each* (s, x) *there is at most one solution to the martingale problem for a, b starting from* (s, x). *Moreover if* $P_{s,x}^{(n)}$ *is the solution for* a_n, b_n *starting from* (s, x) *and if* $\tau_n = \inf\{t \geq s: (t, x(t)) \notin G_n\}$, *then a solution for a and b starting from* (s, x) *exists if and only if*

$$(1.2) \qquad\qquad \lim_{n \to \infty} P_{s, x}^{(n)}[\tau_n \leq T] = 0 \quad \text{for each} \quad T \geq s.$$

Finally if (1.2) obtains, then the unique solution P for a and b starting from (s, x) *equals* $P_{s, x}^{(n)}$ *on* $\mathcal{M}_{\tau_n}$ *for every n.*

Proof. In view of Theorem 10.1.1 there is really very little to be done. Indeed, if a solution P for a and b starting from (s, x) exists, then for every n, P must agree with $P_{s, x}^{(n)}$ on $\mathcal{M}_{\tau_n}$. Since $P[\tau_n \leq T] \to 0$ as $n \to \infty$ for every $T \geq s$, this proves the uniqueness of P as well as the necessity of condition (1.2) for the existence of P. Finally, to prove the sufficiency of (1.2), observe that $P_{s, x}^{(n+1)}$ equals $P_{s, x}^{(n)}$ on $\mathcal{M}_{\tau_n}$. Thus, by Theorem 1.3.5, if (1.2) holds there is a P which equals $P_{s, x}^{(n)}$ on $\mathcal{M}_{\tau_n}$ for every n. Clearly such a P is a solution to the martingale problem for a, b starting from (s, x). $\square$

We now specialize these results to the case in which we can use the results of Chapter 7. Namely, let $a: [0, \infty) \times R^d \to S_d$ and $b: [0, \infty) \times R^d \to R^d$ be locally bounded measurable functions and assume that for every $T > 0$ and $x \in R^d$:

$$(1.3) \qquad\qquad \inf_{0 \leq s \leq T} \inf_{|\theta| = 1} \langle \theta, a(s, x)\theta \rangle > 0$$

and

$$(1.4) \qquad\qquad \lim_{y \to x} \sup_{0 \leq s \leq T} \|a(s, y) - a(s, x)\| = 0.$$

Given $n \geq 1$, let $G_n = [0, n) \times B(0, n)$ and choose $\psi_n \in C_0^\infty([0, \infty) \times R^d)$ so that $0 \leq \psi_n \leq 1$, $\psi_n \equiv 1$ on G_n and $\psi_n \equiv 0$ outside G_{n+1}. Set

$$(1.5) \qquad\qquad a_n = \psi_n a + (1 - \psi_n)I$$

and

$$(1.6) \qquad\qquad b_n = \psi_n b.$$

Then it is clear that a_n and b_n satisfy the hypothesis of Theorem 7.2.1, and therefore the associated martingale problem is well-posed. Denote by $\{P_{s,\,x}^{(n)}: (s, x) \in [0, \infty) \times R^d\}$ the corresponding family of solutions. (We note that the measurability of $\{P_{s,\,x}^{(n)}\}$ is established independently of Exercise 6.7.4.) Using Corollary 7.1.2, we have now proved the next theorem.

10.1.3 Theorem. *Let* $a: [0, \infty) \times R^d \to S_d$ *and* $b: [0, \infty) \times R^d \to R^d$ *be locally bounded measurable functions and assume that* a *satisfies* (1.3) *and* (1.4) *for each* $T > 0$ *and* $x \in R^d$. *Then for each* $(s, x) \in [0, \infty) \times R^d$ *there is at most one solution to the martingale problem for* a *and* b *starting from* (s, x). *Moreover, if* a_n *and* b_n *are defined by* (1.5) *and* (1.6), *respectively, and if* $\{P_{s,\,x}^{(n)}: (s, x) \in [0, \infty) \times R^d\}$ *denotes the family of solutions to the corresponding martingale problem, then a solution for* a *and* b *starting from* (s, x) *exists if and only if*

$$(1.7) \qquad \lim_{n \to \infty} P_{s,\,x}^{(n)}\left[\sup_{s \leq t \leq T} |x(t)| \geq n\right] = 0.$$

Finally, if (1.7) *holds, then the unique solution* $P_{s,\,x}$ *satisfies*

$$(1.8) \qquad P_{s,\,x} = P_{s,\,x}^{(n)} \quad \text{on} \quad \mathcal{M}_{\tau_n}$$

for every $n \geq s$, *where* $\tau_n = (\inf\{t \geq s: |x(t)| \geq n\}) \wedge n$.

10.1.4 Corollary. *Let* a *and* b *be as in Theorem 10.1.3 and assume that* (1.7) *holds for each* $(s, x) \in [0, \infty) \times R^d$. *Then the martingale problem for* a *and* b *is well-posed and the associated set* $\{P_{s,\,x}: (s, x) \in [0, \infty) \times R^d\}$ *of solutions form a measurable strong Markov family. Moreover, given* $T > 0$ *and* $x \in R^d$ *let*

$$\lambda = \inf_{0 \leq t \leq T} \inf_{|\theta| = 1} \langle \theta, a(t, x)\theta \rangle$$

$$\Lambda = \sup_{0 \leq t \leq T} \sup_{|\theta| = 1} \langle \theta, a(t, x)\theta \rangle$$

$$B = \sup_{0 \leq t \leq T} \sup_{|y - x| \leq 1} |b(t, y)|$$

and for $0 < \delta \leq 1$, *define*

$$\rho(\delta) = \sup_{0 \leq t \leq T} \sup_{y:\,|y - x| \leq \delta} \|a(t, y) - a(t, x)\|.$$

Then for each $s \in [0, T]$ *and* $\varepsilon > 0$, *there is a* $\delta_\varepsilon > 0$, *depending only on* d, λ, Λ, B, $\rho(\cdot)$, $T - s$ *and* ε, *such that*

$$(1.9) \qquad |E^{P_{s,\,x}}[\Phi] - P^{P_{\sigma,\,y}}[\Phi]| \leq \varepsilon\|\Phi\|$$

where $\Phi: \Omega \to R$ is a bounded $\mathcal{M}^T(\equiv \sigma(x(t)): t \geq T)$ measurable map and $(\sigma, y) \in [0, T] \times R^d$ satisfies $|\sigma - s| < \delta_\varepsilon$, and $|y - x| < \delta_\varepsilon$. In particular, $\{P_{s, x}: (s, x) \in [0, \infty) \times R^d\}$ is strongly Feller continuous. Finally, if $P(s, x, t, \cdot)$ denotes the transition probability function associated with $\{P_{s, x}: (s, x) \in [0, \infty) \times R^d\}$, then $P(s, x, t, \cdot)$ admits a density $p(s, x, t, y)$ in the sense described in Theorem 9.1.9. In fact, if $0 < r < R$, $0 < s < T$ and $1 \leq q < \infty$, then

(1.10)
$$\left(\int_s^T (t - s)^\alpha \, dt \int_{B(0, r)} |p(s, x, t, y)|^q \, dy \right)^{1/q} \leq C_{1.10},$$

where $\alpha = (d + 2)(q - 1)/2$ and $C_{1.10}$ depends only on $d, T, q, R - r$,

$$\lambda_{T, R} = \inf_{\substack{0 \leq s \leq T \\ x \in B(0, R)}} \inf_{|\theta| = 1} \langle \theta, a(s, x)\theta \rangle$$

$$\Lambda_{T, R} = \sup_{\substack{0 \leq s \leq T \\ x \in B(0, R)}} \sup_{|\theta| = 1} \langle \theta, a(s, x)\theta \rangle$$

$$B_{T, R} = \sup_{\substack{0 \leq s \leq T \\ x \in B(0, R)}} |b(s, x)|$$

and the function $\rho_{T, R}: (0, 1] \to (0, \infty)$ given by

$$\rho_{T, R}(\delta) = \sup_{\substack{0 \leq s \leq T \\ x^1, x^2 \in B(0, R) \\ |x^1 - x^2| \leq \delta}} \|a(s, x_1) - a(s, x_2)\|,$$

and if $0 < \gamma < r$, then

(1.11)
$$\left(\int_s^T dt \int_{B(0, r) \backslash B(x, \gamma)} |p(s, x, t, y|^q \, dy \right)^{1/q} \leq C_{1.11}$$

where $C_{1.11}$ depends on $d, T, q, \lambda_{T, R}, \Lambda_{T, R}, \rho_{T, R}(\cdot)$ and γ.

Proof. The first assertion follows immediately from Theorem 10.1.3 and Theorem 10.1.1. (Avoiding Exercise 6.7.4 if desired.) The proof of (1.9) follows immediately from (1.8) and Theorem 7.2.4 applied to a_n and b_n. Similarly, (1.10) and (1.11) follow at once from Corollary 9.1.10. $\square$

10.1.5 Corollary. Let $a: R^d \to S_d^+$ be continuous and $b: R^d \to R^d$ be locally bounded and measurable. Let $\{\psi_n: n \geq 1\}$ be a sequence of non-negative functions in $C_0^\infty(R^d)$ such that $\chi_{B(0, n)} \leq \psi_n \leq \chi_{B(0, n+1)}$ and set $a_n = \psi_n a + (1 - \psi_n)I$ and $b_n = \psi_n b$. For each $n \geq 1$, let $\{P_x^n: x \in R^d\}$ be the family of diffusions associated with a_n and b_n (as in Lemma 9.2.1). Then the martingale problem for a and b is well posed if and only if $\lim_{n \to \infty} P_x^{(n)}(\sup_{0 \leq t \leq T} |x(t)| \geq n) = 0$ for each $T > 0$ and $x \in R^d$. Moreover, if $P_{s, x}$ denotes the solution for a and b starting from (s, x), then $P_{s, x} = P_x \circ \Phi_s^{-1}$, where

$P_x = P_{0,x}$ and $\Phi_s\colon \Omega \to \Omega$ is as described in Lemma 9.2.1. Thus $\{P_x\colon x \in R^d\}$ forms a measurable, time homogeneous, strong Markov family with the property that for each $x \in R^d$, $T > 0$ and $\varepsilon > 0$, there exists a $\delta_\varepsilon > 0$, depending only on d, T,

$$\lambda = \inf_{|\theta| = 1} \langle \theta, a(x)\theta \rangle$$

$$\Lambda = \sup_{|\theta| = 1} \langle \theta, a(x)\theta \rangle$$

$$B = \sup_{y\colon |y - x| \le 1} |b(x)|$$

and the function $\rho\colon (0, 1] \to (0, \infty)$ given by

$$\rho(\delta) = \sup_{y\colon |y - x| \le \delta} \|a(y) - a(x)\|,$$

such that

$$\left| E^{P_y}[\Phi] - E^{P_x}[\Phi] \right| \le \varepsilon \|\Phi\|$$

whenever $|y - x| < \delta_\varepsilon$ and $\Phi\colon \Omega \to R$ is a bounded $\mathcal{M}^T$ measurable function. Finally, if $P(s, x, \cdot)$ is the time homogeneous transition probability function associated with $\{P_x\colon x \in R^d\}$, then $P(s, x, \cdot)$ has a density $p(s, x, y)$ and for all $1 \le q < \infty$, $s > 0$ and $0 < r < R$

$$(1.12) \qquad \left(\int_{B(0, r)} |p(s, x, y)|^q \, dy \right)^{1/q} \le C_{1.12}(s \wedge 1)^{-v}$$

where v depends only on d and q and $C_{1.12}$ depends only on d, q, $R - r$,

$$\lambda_R = \inf_{|x| \le R} \inf_{|\theta| = 1} \langle \theta, a(x)\theta \rangle$$

$$\Lambda_R = \sup_{|x| \le R} \sup_{|\theta| = 1} \langle \theta, a(x)\theta \rangle$$

$$B_R = \sup_{|x| \le R} |b(x)|$$

and the function $\rho\colon (0, \infty) \to (0, \infty)$ given by

$$\rho(\delta) = \sup_{\substack{|x^1| \le R,\, |x^2| \le R \\ |x^1 - x^2| < \delta}} \|a(x^1) - a(x^2)\|.$$

Proof. Only the last part is not an obvious consequence of Corollary 10.1.4. On the other hand, the last part follows easily from Corollary 9.2.7. □

10.2. Conditions for Explosion and Non-Explosion

In this section we discuss a few simple criteria which guarantee explosion or non-explosion. We begin with the following condition which relates the problem to a purely analytical one.

10.2.1 Theorem. *Let* a, b, a_n *and* b_n *be as in Corollary 10.1.2 with* $G_n = [0, n) \times B(0, n)$. *A sufficient condition that the martingale problem for* a, b *to be well-posed is that for each* $T > 0$, *there exists a number* $\lambda = \lambda_T > 0$ *and a non-negative function* $\varphi = \varphi_T \in C^{1,\,2}([0, T] \times R^d)$ *such that*

$$\lim_{|x| \to \infty} \inf_{0 \le t \le T} \varphi(t, x) = \infty$$

and

$$\left(\frac{\partial}{\partial t} + L_t \right) \varphi \le \lambda \varphi$$

on $[0, T] \times R^d$. *Conversely if* $(s^0, x^0) \in [0, \infty) \times R^d$ *and there exists a number* $\lambda > 0$ *and a bounded function* $\varphi \in C^{1,\,2}([s^0, T] \times R^d)$ *for some* $T > s^0$ *such that,*

$$\varphi(s^0, x^0) > e^{-\lambda(T - s_0)} \sup_{x \in R^d} \varphi(T, x)$$

and

$$\left(\frac{\partial}{\partial t} + L_t \right) \varphi \ge \lambda \varphi$$

on $[s^0, T] \times R^d$ *then there is no solution to the martingale problem starting from* (s^0, x^0) *corresponding to* a *and* b.

Proof. To prove the first part, let $0 \le s \le T$ and $x \in R^d$ be given. Choose $n_0 > T$ so that $(s, x) \in G_n$ for $n \ge n_0$. Let $\varphi = \varphi_T$ be the function in the hypothesis. Since $(\partial \varphi / \partial t) + L_t \varphi - \lambda \varphi \le 0$ on $[0, T] \times R^d$, we observe that

$$e^{-\lambda s} \varphi(s, x) \ge E^{P_{s,x}^{(n)}}[e^{-\lambda(T \wedge \tau_n)} \varphi(T \wedge \tau_n, x(T \wedge \tau_n))]$$

$$\ge e^{-\lambda T} E^{P_{s,x}^{(n)}}[\varphi(\tau_n, x(\tau_n)); \tau_n < T]$$

where $\tau_n = \inf\{t \ge s: (t, x(t)) \notin G_n\}$ and $n \ge n_0$. Since

$$\lim_{n \to \infty} \inf_{0 \le t \le T} \varphi(t, x) = \infty$$

and $|x(\tau_n)| = n$ if $\tau_n < T$, it follows that $\lim_{n \to \infty} P^{(n)}_{s, x}(\tau_n < T) = 0$. Thus the first part has been proved. We now turn to the second part. Choose n_0 so that $(s^0, x^0) \in G_{n_0}$ and define τ_n for $n \geq n_0$ as before. Since we are now assuming that $(\partial\varphi/\partial t) + L_t \varphi - \lambda\varphi \geq 0$ on $[s^0, \infty) \times R^d$, we have for our $T > s^0$ and $n \geq n_0$

$$e^{-\lambda s^0}\varphi(s^0, x^0) \leq e^{-\lambda T}\left(\sup_{x \in R^d} \varphi(T, x)\right)P^{(n)}_{s^0, x^0}[\tau_n > T]$$

$$+ \left(\sup_{s^0 \leq t \leq T} \sup_{x \in R^d} \varphi(t, x)\right)P^{(n)}_{s^0, x^0}[\tau_n \leq T].$$

If $\lim_{n \to \infty} P^{(n)}_{s^0, x^0}[\tau_n \leq T]$ were zero we would then have

$$e^{-\lambda s^0}\varphi(s^0, x^0) \leq e^{-\lambda T} \sup_{x \in R^d} \varphi(T, x)$$

contradicting our assumption. The proof of the second part is therefore complete. $\square$

A rather crude but nevertheless useful application of Theorem 10.2.1 is the following result.

10.2.2 Theorem. *Let a, b, a_n, b_n and G_n be as in Theorem 10.2.1. If for each $T > 0$, there is a $C_T < \infty$ such that*

$$(2.1) \qquad \sup_{0 \leq t \leq T} \|a(t, x)\| \leq C_T(1 + |x|^2), \qquad \text{for } x \in R^d$$

and

$$(2.2) \qquad \sup_{0 \leq t \leq T} \langle x, b(t, x) \rangle \leq C_T(1 + |x|^2), \qquad \text{for } x \in R^d$$

then the martingale problem for a, b is well posed.

Proof. Let $\varphi(x) = 1 + |x|^2$ for $x \in R^d$. Then

$$L_t \varphi(x) = \text{Trace}(a(t, x)) + 2\langle x, b(t, x) \rangle$$

$$\leq (d + 2)C_T(1 + |x|^2) = \lambda_T \varphi(x)$$

where $\lambda_T = (d + 2)C_T$. Thus we can apply the first part of Theorem 10.2.1 with $\varphi_T = 1 + |x|^2$ and $\lambda_T = (d + 2)C_T$. $\square$

The reason that Theorem 10.2.2 is crude is that it does not take into consideration the dimension d of the space or the effect of any balancing between a and b. A more refined result is the test of Hasminskii which we will present below. The treatment is adapted from H. P. McKean [1969].

10.2.3 Theorem. *Let a, b, a_n, b_n and G_n be as in Theorem 10.2.1. Suppose for each $T > 0$ there exists an $r > 0$ and continuous functions A_T: $[r, \infty) \to (0, \infty)$ and B_T: $[r, \infty) \to (0, \infty)$ such that for $\rho \geq (2r)^{\frac{1}{2}}$, $0 \leq t \leq T$ and $|x| = \rho$,*

$$A_T\left(\frac{\rho^2}{2}\right) \geq \langle x, a(t, x)x \rangle$$

$$\langle x, a(t, x)x \rangle B_T\left(\frac{\rho^2}{2}\right) \geq (\text{Trace } a(t, x) + 2\langle x, b(t, x)\rangle)$$

and

$$\int_r^\infty [C_T(\rho)]^{-1} \, d\rho \int_r^\rho [C_T(\sigma)][A_T(\sigma)]^{-1} \, d\sigma = \infty$$

where

$$C_T(\rho) = \exp\left[\int_r^\rho B_T(\sigma) \, d\sigma\right].$$

Then the martingale problem for a, b is well-posed.

Proof. Let $T > 0$ be given. For $\rho \geq r$ define, by induction, $u_0 \equiv 1$ and

$$u_n(\rho) = 2 \int_r^\rho [C(\sigma)]^{-1} \, d\sigma \int_r^\sigma \frac{B(\tau)}{A(\tau)} u_{n-1}(\tau) \, d\tau,$$

where $A = A_T$, $B = B_T$ and $C = C_T$. Using induction, one sees that for all $n \geq 0$, $u_n \geq 0$, $u_n' \geq 0$ and

(2.3) $$u_n(\rho) \leq \frac{2^n}{n!}\left(\int_r^\rho [C(\sigma)]^{-1} \, d\sigma \int_r^\sigma \frac{B(\tau)}{A(\tau)} \, d\tau\right)^n.$$

Finally

(2.4) $$u_n'' = \frac{2}{A} u_{n-1} - B u_n' \quad \text{for} \quad \rho \geq r.$$

From these observations, it is easy to see that the series $\sum_{n=0}^\infty u_n$ converges uniformly on compact subsets of $[r, \infty)$ and can be differentiated twice term by term. The sum u is twice continuously differentiable on $[r, \infty)$ and is such that $u \geq 1$, $u' > 0$ and

(2.5) $$u = \tfrac{1}{2} A(u'' + B u'), \quad \text{for } \rho \geq r.$$

Finally, since $u \geq u_1$, our hypothesis implies that

(2.6) $$\lim_{\rho \uparrow \infty} u(\rho) = \infty.$$

Now choose $\varphi \in C^2(R^d)$ so that $\varphi \geq 1$ and $\varphi(x) = u(|x|^2/2)$ for $|x| \geq (2r)^{\frac{1}{2}}$. Certainly we can find a $\lambda \geq 1$ so that $L_t\varphi(x) \leq \lambda$ for all $0 \leq t \leq T$ and $|x| \leq (2r)^{\frac{1}{2}}$. Also if $0 \leq t \leq T$ and $|x| > (2r)^{\frac{1}{2}}$, then

$$L_t\varphi(x) = \tfrac{1}{2}\langle x, a(t, x)x \rangle u''\left(\frac{|x|^2}{2}\right)$$

$$+ \left(\tfrac{1}{2} \text{ Trace } a(t, x) + \langle x, b(t, x)\rangle\right) u'\left(\frac{|x|^2}{2}\right)$$

$$\leq \tfrac{1}{2}\langle x, a(t, x)x \rangle \left(u''\left(\frac{|x|^2}{2}\right) + B\left(\frac{|x|^2}{2}\right) u'\left(\frac{|x|^2}{2}\right)\right)$$

$$\leq \tfrac{1}{2}A\left(\frac{|x|^2}{2}\right)\left(u''\left(\frac{|x|^2}{2}\right) + B\left(\frac{|x|^2}{2}\right) u'\left(\frac{|x|^2}{2}\right)\right)$$

$$= u\left(\frac{|x|^2}{2}\right) = \varphi(x).$$

We have used (2.5) to obtain the last line and also to show that $u'' + Bu' \geq 0$, thereby justifying the second to the last line. It follows that $L_t\varphi \leq \lambda\varphi$ on $[0, T] \times R^d$, and the first part of Theorem 10.2.1 can now be applied to finish the proof. $\square$

We end this section with a partial converse to Theorem 10.2.3. The idea again goes back to Hasminskii.

10.2.4 Theorem. *Let a, b, a_n, b_n and G_n be as in Theorem 10.2.1. We now assume that for each $R > 0$*

$$\inf_{|\theta| = 1} \inf_{t \geq 0} \inf_{|x| \leq R} \langle \theta, a(t, x)\theta \rangle > 0 \quad and \quad \sup_{t \geq 0} \sup_{|x| \leq R} |b(t, x)| < \infty.$$

In addition, assume that there are continuous functions $A: [\tfrac{1}{2}, \infty) \to (0, \infty)$ and $B: [\tfrac{1}{2}, \infty) \to (0, \infty)$ such that for $\rho \geq 1$, $t \geq 0$ and $|x| = \rho$,

$$A\left(\frac{\rho^2}{2}\right) \leq \langle x, a(t, x)x \rangle,$$

$$\langle x, a(t, x)x \rangle B\left(\frac{\rho^2}{2}\right) \leq (\text{Trace } a(t, x) + 2\langle x, b(t, x)\rangle),$$

and

$$\int_{\frac{1}{2}}^{\infty} [C(\rho)]^{-1} \, d\rho \int_{\frac{1}{2}}^{\rho} \frac{C(\sigma)}{A(\sigma)} \, d\sigma < \infty$$

where

$$C(\rho) = \exp\left[\int_{\frac{1}{2}}^{\rho} B(\sigma) \, d\sigma\right].$$

Then there is no $(s, x) \in [0, \infty) \times R^d$ starting from which the martingale problem for a and b has a solution. In fact, for each $(s, x) \in [0, \infty) \times R^d$,

$$\lim_{T \to \infty} \lim_{n \to \infty} P_{s, x}^{(n)}[\tau_n \leq T] = 1,$$

where

$$\tau_n = \inf[t \geq s: (t, x(t)) \notin G_n].$$

Proof. The first step in the proof is to show that for each $\varepsilon > 0$ there is an $R_\varepsilon > 1$ and a $T_\varepsilon > 0$ such that

(2.7) $$P_{s, x}^{(n)}(\tau_n \leq s + T_\varepsilon) \geq 1 - \varepsilon$$

for all $n \geq 2$, $s \geq 0$ and $|x| \geq R_\varepsilon$. To prove (2.7), define $\{u_n, n \geq 0\}$ and u as in the proof of Theorem 10.2.3 using the functions A and B given in the statement of this theorem. Once again $u \in C^2([\frac{1}{2}, \infty))$, $u \geq 1$, $u' > 0$ and $u = \frac{1}{2}A(u'' + Bu')$. Moreover, by (2.3)

$$u(\rho) \leq \exp\left[2 \int_{\frac{1}{2}}^{\rho} [C(\sigma)]^{-1} \, d\sigma \int_{\frac{1}{2}}^{\sigma} \frac{C(\tau)}{A(\tau)} \, d\tau\right],$$

and so $1 = u(\frac{1}{2}) < u(\rho)$ and $u(\rho)$ increases to a finite limit $u(\infty)$ as $\rho \to \infty$. Define $\varphi(x) = u(|x|^2/2)$ for $x \in R^d \backslash B(0, 1)$ and observe that $L_t \varphi \geq \varphi$ on $[0, \infty) \times R^d \backslash B(0, 1)$. Thus if $n \geq 2$, and $(s, x) \in G_n$ with $|x| > 1$, we have

$$E^{P_{s, x}^{(n)}}[e^{-\sigma \wedge \tau_n} \varphi(x(\sigma \wedge \tau_n))] \geq e^{-s} \varphi(x)$$

where

$$\sigma = \inf\{t \geq s: |x(t)| \leq 1\}.$$

In other words,

$$u\left(\frac{n^2}{2}\right) E^{P_{s, x}^{(n)}}[e^{-\tau_n}, \tau_n < \sigma \wedge n] + u(\tfrac{1}{2}) E^{P_{s, x}^{(n)}}[e^{-\sigma}, \sigma < \tau_n]$$

$$+ e^{-n} E^{P_{s, x}^{(n)}}[\varphi(x(n)), n \leq \sigma \wedge \tau_n] \geq e^{-s} u\left(\frac{|x|^2}{2}\right).$$

Because

$$E^{P_{s, x}^{(n)}}[e^{-\tau_n}, \tau_n < \sigma \wedge n] + E^{P_{s, x}^{(n)}}[e^{-\sigma}, \sigma < \tau_n] \leq 1$$

we obtain from the above

$$(2.8) \qquad E^{P_{s,x}^{(n)}}[e^{-(\tau_n - s)}] \geq \frac{u\left(\dfrac{|x|^2}{2}\right) - u(\frac{1}{2}) - e^{-(n-s)}u(\infty)}{u(\infty) - u(\frac{1}{2})}$$

for all $n \geq 2$ and $(s, x) \in G_n$ with $|x| \geq 1$. Now choose $R_\varepsilon > 1$ so that

$$\frac{u\left(\dfrac{|x|^2}{2}\right) - u(\frac{1}{2})}{u(\infty) - u(\frac{1}{2})} \geq 1 - \frac{\varepsilon}{4} \qquad \text{for all} \quad |x| \geq R_\varepsilon$$

and take $T_\varepsilon > 0$ so that

$$2(1 - e^{-T_\varepsilon}) \geq 1 \quad \text{and} \quad \frac{e^{-T_\varepsilon}u(\infty)}{u(\infty) - u(\frac{1}{2})} < \frac{\varepsilon}{2}.$$

We claim that (2.7) holds with the choice of R_ε and T_ε. Indeed, let $n \geq 2$, $s \geq 0$ and $|x| \geq R_\varepsilon$ be given. If either $|x| \geq n$ or $n \leq s + T_\varepsilon$, (2.7) is trivial. But if $n > |x| \vee (s + T_\varepsilon)$, then by (2.8) and the choice of R_ε and T_ε,

$$E^{P_{s,x}^{(n)}}[e^{-(\tau_n - s)}] \geq 1 - \frac{\varepsilon}{2},$$

and so

$$P_{s,x}^{(n)}(\tau_n > s + T_\varepsilon) \leq \frac{1}{1 - e^{-T_\varepsilon}} E^{P_{s,x}^{(n)}}[1 - e^{-(\tau_n - s)}]$$

$$\leq \frac{\varepsilon}{2(1 - e^{-T_\varepsilon})} \leq \varepsilon.$$

Having established (2.7), we will be done once we show that for each $R > 0$ and $(s, x) \in [0, \infty) \times R^d$

$$(2.9) \qquad \lim_{T \to \infty} \sup_{n \geq R \vee (s+T)} P_{s,x}^{(n)}\left(\sup_{s \leq t \leq s + T} |x(t)| < R\right) = 0.$$

However, (2.9) is an easy consequence of Exercise 3.1 below. □

10.3. Exercises

10.3.1. Let $a: [0, \infty) \times R^d \to S_d$ and $b: [0, \infty) \times R^d \to R^d$ be locally bounded measurable functions. Let $R > 0$ be given and assume that $\lambda = \inf_{t \geq 0} \inf_{|\theta| = 1} \inf_{|x| \leq R} \times \langle \theta, a(t, x)\theta \rangle > 0$ and that $B = \sup_{r \geq 0} \sup_{|x| \leq R} |b(t, x)| < \infty$. Suppose that P solves the martingale problem for a and b starting from $(s, x) \in [0, \infty) \times B(0, R)$ and let $\tau = \inf\{t \geq s: |x(t)| \geq R\}$. Show that there exists an $\varepsilon > 0$ and a $C < \infty$ depending only on d, R, λ and B such that $E^P[\exp[\varepsilon(\tau - s)]] \leq C$. There are many ways in which this can be done. One of the most straightforward is the following. Define $\varphi_\eta(x) = 1 - e^{-(x-\eta)^2/2}$ for x, $\eta \in R^d$. Show that there is an $\eta \in R^d$ and an $\varepsilon > 0$ depending only on d, R, λ and B such that $\varepsilon \leq \inf_{t \geq 0} \inf_{|y| \leq R} L_t \varphi_\eta(y)$. Conclude that $E^P[\exp[\varepsilon(\tau - s)]\varphi_\eta(x(\tau))] \leq \varphi_\eta(x)$.

10.3.2. Let $a: [0, \infty) \times R^d \to S_d$ and $b: [0, \infty) \times R^d \to R^d$ be locally bounded measurable coefficients and let $c: [0, \infty) \times R^d \to R^d$ be a measurable function such that ac is locally bounded. Assume that for some $(s^0, x^0) \in [0, \infty) \times R^d$ the martingale problems for both a, b and a, $b + ac$ starting from (s^0, x^0) have exactly one solution each, say P_{s^0, x^0} and Q_{s^0, x^0} respectively. Show that for each $t > s^0$, P_{s^0, x^0} and Q_{s^0, x^0} are equivalent on $\mathcal{M}_t$ and develop a representation for the Radon-Nikodym derivative $dQ_{s^0, x^0}/dP_{s^0, x^0}$ on $\mathcal{M}_t$ (at least as a limit).

10.3.3. To illustrate how the dimension d enters in the criterion for explosion, let $a: R^d \to (0, \infty)$ be a locally bounded measurable function such that a is uniformly positive on compact sets of R^d. Let $L = \frac{1}{2}a(x)\Delta$. If d equals 1 or 2 show that the martingale problem for L is well-posed. If $d \geq 3$ and if there is a continuous α: $(1, \infty) \to (0, \infty)$ such that $\alpha(\rho) \leq a(x)$, for $\rho \geq 1$ and $|x| = \rho$, and $\int_1^\infty 1/\alpha(\sqrt{\rho}) \, d\rho < \infty$, then the martingale problem for L has no solution starting from any point. The reason for the difference between $d \leq 2$ and $d \geq 3$ can best be understood in terms of the recurrence properties of Brownian motion. Indeed in one and two dimensions Brownian motion is recurrent and so the existence can be seen by the techniques introduced in Section 7.5. On the other hand, if $d \geq 3$, then it is easy to see that a solution P starting from $(0, x) \in R^d$ would have the property that

$$E^P\left[\int_0^\infty \chi_{B(0, R)}(x(t)) \, dt\right] = C_d \int_{B(0, R)} \frac{1}{|x - y|^{d-2}} \cdot \frac{1}{a(y)} \, dy$$

which means that

$$\lim_{R \to \infty} E^P\left[\int_0^\infty \chi_{B(0, R)}(x(t)) \, dt\right] < \infty$$

if there is an $\alpha(\cdot)$ with the stated properties. Of course one can also use Hasminskii's test, i.e., Theorem 10.2.4.

10.3.4. To illustrate how the relations between a and b can influence explosion, let $d \geq 3$ and take $a(x) = (|x| \vee 1)^{2+\varepsilon}$ and $b(x) = -(\beta/2)|x|^\varepsilon x$ for some $\varepsilon > 0$ and $\beta \in R$. Show that if $\beta \geq d - 2$ then the martingale problem is well-posed. If $\beta < d - 2$, show that there is no solution starting from any point.

Chapter 11

Limit Theorems

11.0. Introduction

We have established conditions on coefficients which guarantee existence and uniqueness of solutions to the associated martingale problem. What we want to do in this chapter is to show that these same conditions also guarantee stability properties for these solutions. For example, suppose that $a: [0, \infty) \times R^d \to S_d$ and $b: [0, \infty) \times R^d \to R^d$ are continuous coefficients for which the martingale problem is well-posed. Let $a_n: [0, \infty) \times R^d \to S_d$ and $b_n: [0, \infty) \times R^d \to R^d$, $n \geq 1$, be coefficients which tend to a and b, respectively, uniformly on compact subsets. Assume that P^n is a solution to the martingale problem for a_n and b_n starting from (s_n, x^n) and that $(s_n, x^n) \to (s, x)$. If the a_n's and b_n's are bounded independent of $n \geq 1$, then $\{P^n : n \geq 1\}$ is relatively compact, and it is clear that any limit point must be a solution for a and b starting from (s, x). Hence $P^n \to P_{s, x}$, the unique solution for a and b starting from (s, x). This example is the prototype of one kind of stability result in which we will be interested. A second sort of stability with which we will be concerned has as its prototype the celebrated invariance principle of M. Donsker (cf. Exercise 2.4.2). From the analytic viewpoint, this type of stability entails the approximation of solutions to various differential equations involving the coefficients a and b by solutions to finite difference equations.

Our presentation will be somewhat complicated by our desire to show that these stability results follow from the assumption that the martingale problem for a and b is well-posed and do not depend on any particular condition which might be made to insure that this assumption is justified. In particular, we do not want to assume that a and b are bounded but only that the associated martingale problem admits exactly one solution. The reader should realize that, unlike results about uniqueness, extending stability results from bounded to unbounded coefficients involves proving compactness without the benefit of global bounds on the coefficients. One of the interesting features is that the mere existence of a solution for the limiting coefficients already implies the compactness of any sequence of diffusions corresponding to approximating coefficients.

11.1. Convergence of Diffusion Processes

In this section we are going to give our basic convergence result for diffusion processes. Refinements of this result will be given in Sections 11.3 and 11.4 under more restricted conditions.

In order to have a compactness condition when our coefficients are not necessarily bounded, we will need the following lemma.

11.1.1 Lemma. *Let* $\{P^n: n \geq 1\}$ *be a sequence of probability measures on* Ω *and suppose that* $\{\tau_k: k \geq 1\}$ *is a non-decreasing sequence of lower semicontinuous stopping times increasing to* ∞ *for each* ω. *For each* $n \geq 1$, *let* $\{P^{n,k}: n \geq 1\}$ *be a sequence of probability measures such that for each* $k \geq 1$, $P^{n,k}$ *equals* P^n *on* $\mathcal{M}_{\tau_k}$ *and* $\{P^{n,k}: n \geq 1\}$ *is relatively compact. Finally, assume that* P *is a probability measure on* Ω *such that for each* $k \geq 1$ *and any limit point* Q^k *of* $\{P^{n,k}: n \geq 1\}$, Q^k *equals* P *on* $\mathcal{M}_{\tau_k}$. *Then* $\{P^n: n \geq 1\}$ *converges to* P.

Proof. First observe that for each $k \geq 1$ and $t \geq 0$, $\{\tau_k \leq t\}$ is closed in Ω. Thus, since $\{\tau_k \leq t\} \in \mathcal{M}_{\tau_k}$,

$$(1.1) \qquad\qquad \limsup_{n \to \infty} P^n(\tau_k \leq t) \leq P(\tau_k \leq t)$$

because $P^n(\tau_k \leq t) = P^{n,k}(\tau_k \leq t)$, $\{P^{n,k}: n \geq 1\}$ is relatively compact, and any limit point of $\{P^{n,k}: n \geq 1\}$ equals P on $\mathcal{M}_{\tau_k}$.

Now suppose $\Phi: \Omega \to R^1$ is a bounded continuous $\mathcal{M}_t$-measurable function. Then, since $\Phi\chi_{(t,\infty)}(\tau_k)$ is $\mathcal{M}_{\tau_k}$-measurable,

$$(1.2) \qquad |E^{P^n}[\Phi] - E^P[\Phi]| \leq |E^{P^{n,k}}[\Phi] - E^P[\Phi]| + 2\|\Phi\|P^n(\tau_k \leq t).$$

For fixed $k \geq 1$, choose a convergent subsequence $\{P^{n',k}\}$ of $\{P^{n,k}: n \geq 1\}$ so that

$$\lim_{n' \to \infty} |E^{P^{n',k}}[\Phi] - E^P[\Phi]| = \limsup_{n \to \infty} |E^{P^{n,k}}[\Phi] - E^P[\Phi]|,$$

and let $Q^k = \lim_{n' \to \infty} P^{n',k}$. Since Q^k equals P on $\mathcal{M}_{\tau_k}$, we see that

$$\limsup_{n \to \infty} |E^{P^{n,k}}[\Phi] - E^P[\Phi]| = |E^{Q^k}[\Phi] - E^P[\Phi]| \leq 2\|\Phi\|P(\tau_k \leq t).$$

Combining this with (1.1) and (1.2), we arrive at:

$$\limsup_{n \to \infty} |E^{P^n}[\Phi] - E^P[\Phi]| \leq 4\|\Phi\|P(\tau_k \leq t)$$

for all $k \geq 1$. Since $P(\tau_k \leq t)\searrow 0$ as $k \to \infty$, this completes our proof. $\square$

11.1.2 Lemma. *Let* P *be a probability measure on* Ω *and for each* $R > 0$ *let*

$$\tau_R = \inf\{t \geq 0: |x(t)| \geq R\}.$$

Then τ_R is lower semicontinuous for all $R > 0$. Moreover, with the exception of at most countably many R's, for P-almost all $\omega \in \{\tau_R < \infty\}$, τ_R is continuous at ω.

Proof. It is clear that τ_R is lower semicontinuous. It is equally clear that $\tau_R^+ = \inf\{t \geq 0: |x(t)| > R\}$ is upper semi-continuous and that $\tau_R^+ = \lim_{S \searrow R} \tau_S$. Since $\tau_R^+ \geq \tau_R$, we will be done once we have shown that $E^P[e^{-\tau_R^+}] = E^P[e^{-\tau_R}]$ for all but a countable number of R's, because when this equality holds $\tau_R^+ = \tau_R$, P-almost surely on $\{\tau_R < \infty\}$. But $R \to E^P[e^{-\tau_R}]$ is a non-negative, non-increasing function, and so

$$E^P[e^{-\tau_R^+}] = \lim_{S \searrow R} E^P[e^{-\tau_S}] = E^P[e^{-\tau_R}]$$

for all but a countable number of R's. $\square$

11.1.3 Lemma. *For each $n \geq 1$ let P^n be a probability measure on Ω and let $X_n: [0, \infty) \times \Omega \to R^1$ be a continuous progressively measurable function. Assume that there is an $R_0 > 0$ such that if $\zeta = \inf\{t \geq 0: |x(t)| \geq R_0\}$ then*

$$\sup_{n \geq 1} \sup_{(t, \omega)} |X_n(t \wedge \zeta(\omega), \omega)| < \infty$$

and for all $n \geq 1$, $\langle X_n(t \wedge \zeta), \mathcal{M}_t, P^n \rangle$ is a martingale. Finally, suppose that P is a probability measure on Ω and that $X: [0, \infty) \times \Omega \to R^1$ is a progressively measurable function which is jointly continuous in (t, ω) such that $P_n \to P$ and $X_n \to X$ uniformly on compact subsets of $[0, \infty) \times \Omega$. Then $\langle X(t \wedge \zeta), \mathcal{M}_t, P \rangle$ is a martingale.

Proof. First assume that at P-almost all $\omega \in \{\zeta < \infty\}$, ζ is continuous. Then $\zeta \wedge t$ is P-almost surely continuous for all $t \geq 0$. Let $0 \leq t_1 < t_2$ and a bounded continuous $\mathcal{M}_{t_1}$-measurable $\Phi: \Omega \to R^1$ be given. Then $X(t_i \wedge \zeta)\Phi$, $i = 1, 2$, is bounded and P-almost surely continuous. Thus

$$(1.3) \qquad E^P[X(t_i \wedge \zeta)\Phi] = \lim_{n \to \infty} E^{P^n}[X(t_i \wedge \zeta)\Phi], \qquad i = 1, 2.$$

On the other hand,

$$|E^{P^n}[X(t_i \wedge \zeta)\Phi] - E^{P^n}[X_n(t_i \wedge \zeta)\Phi]| \leq \|\Phi\| E^{P^n}[|X(t_i \wedge \zeta) - X_n(t_i \wedge \zeta)|],$$

and it is easy to see from our assumption that $E^{P^n}[|X(t_i \wedge \zeta) - X_n(t_i \wedge \zeta)|] \to 0$ as $n \to \infty$. Thus we may replace $X(t_i \wedge \zeta)$ by $X_n(t_i \wedge \zeta)$ on the right-hand side of (1.3). In particular,

$$\begin{aligned} E^P[X(t_2 \wedge \zeta)\Phi] &= \lim_{n \to \infty} E^{P^n}[X_n(t_2 \wedge \zeta)\Phi] \\ &= \lim_{n \to \infty} E^{P^n}[X_n(t_1 \wedge \zeta)\Phi] \\ &= E^P[X(t_1 \wedge \zeta)\Phi]. \end{aligned}$$

To eliminate the assumption about the continuity of ζ, we use Lemma 11.1.2 to find $R_n \nearrow R_0$ such that $\zeta_n \equiv \inf\{t \geq 0: |x(t)| \geq R_n\}$ is P-almost surely continuous on $\{\zeta_n < \infty\}$. By the preceding, $\langle X(t \wedge \zeta_n), \mathcal{M}_t, P \rangle$ is a martingale for all $n \geq 1$; and therefore, since $X(t)$ is bounded and continuous and $\zeta_n \nearrow \zeta$, $\langle X(t \wedge \zeta), \mathcal{M}_t, P \rangle$ must be a martingale. □

We are now ready to prove the main result of this section. The reader should be aware that the preceding preparatory discussion is necessary only in the case when our coefficients are unbounded. For uniformly bounded coefficients, the next theorem is nearly obvious.

11.1.4 Theorem. *Let* $a: [0, \infty) \times R^d \to S_d$ *and* $b: [0, \infty) \times R^d \to R^d$ *be locally bounded measurable functions which are continuous in x for each $t \geq 0$, and assume that for each $(s, x) \in [0, \infty) \times R^d$ the martingale problem for a and b starting from (s, x) has exactly one solution $P_{s, x}$. Suppose that for each $n \geq 1$, $a_n: [0, \infty) \times R^d \to S_d$ and $b_n: [0, \infty) \times R^d \to R^d$ are measurable functions, and assume that for all $T > 0$ and $R > 0$:*

$$\sup_{n \geq 1} \sup_{0 \leq s \leq T} \sup_{|x| \leq R} \|a_n(s, x)\| + |b_n(s, x)| < \infty$$

and

$$\lim_{n \to \infty} \int_0^T \sup_{|x| \leq R} (\|a(s, x) - a_n(s, x)\| + |b_n(s, x) - b(s, x)|) \, ds = 0.$$

Let P^n, $n \geq 1$, *be a solution to the martingale problem for a_n and b_n starting from (s_n, x^n). If $(s_n, x^n) \to (s, x)$, then $P^n \to P_{s, x}$.*

Proof. For $k > \sup(|s_n| + |x^n|)$, define

$$\zeta_k = \inf\{t \geq 0: |x(t)| \geq k\}$$

and $\tau_k = \zeta_k \wedge k$. Let $y^k(t) = x(t \wedge \tau_k)$, $t \geq 0$, and denote by $P^{n, k}$ the distribution of $y^k(\cdot)$ under P^n. Clearly $P^{n, k} = P^n$ on $\mathcal{M}_{\tau_k}$. We are going to show that for each k, $\{P^{n, k}: n \geq 1\}$ is relatively compact and that any limit point coincides with $P_{s, x}$ on $\mathcal{M}_{\tau_k}$. By Lemma 11.1.1, this will complete the proof that $P^n \to P_{s, x}$.

We first check that $\{P^{n, k}: n \geq 1\}$ is relatively compact. To this end, let $f \in C_0^\infty(R^d)$ be given. Clearly

$$f(x(t)) - \int_{s_n}^{(t \vee s_n) \wedge \tau_k} L_u^n f(x(u)) \, du$$

is a $P^{n,k}$-martingale where L_t^n is defined in terms of a_n and b_n. Thus for each f there is a $C_k(f)$, not depending on n but only on the bounds on f and its first two derivatives, such that

$$(f(x(t)) + C_k(f)t, \mathcal{M}_t, P^{n,k})$$

is a submartingale. Therefore, the relative compactness of $\{P^{n,k}: n \geq 1\}$ now follows from Theorem 1.4.6.

We next turn to the identification on $\mathcal{M}_{\tau_k}$ of any limit point of $\{P^{n,k}: n \geq 1\}$. Let $Q^k = \lim_{n' \to \infty} P^{n',k}$, where $\{P^{n',k}\}$ is a subsequence of $\{P^{n,k}: n \geq 1\}$. Clearly $Q^k(x(t) = x, 0 \leq t \leq s) = 1$. We will show that if L_t is defined in terms of a and b, then

$$\left(f(x(t \wedge \tau_k)) - \int_s^{t \wedge \tau_k} L_u f(x(u))\, du, \mathcal{M}_t, Q^k\right)$$

is a martingale after time s for each $f \in C_0^\infty(R^d)$. By Exercise 11.5.1 this will prove that $Q^k = P_{s,x}$ on $\mathcal{M}_{\tau_k}$. But if

$$X_{n'}(t) = f(x(t \wedge k)) - \int_{s_{n'}}^{(t \wedge k) \vee s_{n'}} L_u^n f(x(u))\, du$$

and

$$X(t) = f(x(t \wedge k)) - \int_s^{(t \wedge k) \vee s} L_u f(x(u))\, du,$$

then

$$\sup_{n'} \sup_{(t, \omega)} |X_{n'}(t \wedge \zeta_k(\omega), \omega)| < \infty$$

and $X_{n'} \to X$ uniformly on compact subsets of $[0, \infty) \times \Omega$. Moreover, $(t, \omega) \to X(t, \omega)$ is continuous and

$$(X_{n'}(t \wedge \zeta_k), \mathcal{M}_t, P^{n',k})$$

is a martingale. Thus by Lemma 11.1.3, $(X(t \wedge \zeta_k), \mathcal{M}_t, Q^k)$ is a martingale. That is,

$$f(x(t \wedge \tau_k)) - \int_s^{t \wedge \tau_k} L_u f(x(u))\, du$$

is a Q^k-martingale after time s. The theorem is therefore proved. □

By taking $a_n = a$ and $b_n = b$ for all $n \geq 1$, we obtain the following corollary of Theorem 11.1.4.

11.1.5 Corollary. *Let* $a: [0, \infty) \times R^d \to S_d$ *and* $b: [0, \infty) \times R^d \to R^d$ *be locally bounded measurable functions which are continuous in x for each* $t \geq 0$. *If for each* $(s, x) \in [0, \infty) \times R^d$ *there is exactly one solution* $P_{s, x}$ *to the martingale problem for a and b starting from* (s, x), *then the family* $\{P_{s, x}: (s, x) \in [0, \infty) \times R^d\}$ *is Feller continuous.*

11.2. Convergence of Markov Chains to Diffusions

The purpose of this section is to study the approximation of a diffusion process by Markov chains. For simplicity, we will be assuming that all of our processes are homogeneous in time. However, the context in which our theorems are stated is sufficiently general to allow one to obtain analogous results for time-inhomogeneous processes by simply considering the time-space process.

In order to explain what we are about to do, we will first recast Exercise 2.4.2 into the general setting in which we are going to be working (cf. Exercise 6.7.2). Suppose that $X_1, \ldots, X_n, \ldots$ is a sequence of mutually independent R^d-valued normal random variables having mean 0 and covariance I on some probability space $(E, \mathscr{F}, \mu)$. For each $h > 0$ and $x \in R^d$, define $\Phi_{h, x}: E \to \Omega$ by

$$\Phi_{h, x}(q) = x + h^{\frac{1}{2}} \left[\sum_{k=1}^{[t/h]} X_k(q) + \left(t - h \left[\frac{t}{h} \right] \right) X_{[t/h]+1}(q) \right].$$

(where $\sum_{k=1}^0 X_k \equiv 0$). Denote by P_x^h the probability measure on Ω induced from μ by $\Phi_{h, x}$. An easy computation shows that if Q_x^h is the distribution on $(R^d)^{\{0, \ldots, n, \ldots\}}$ of $\{x(kh): k \geq 0\}$ under P_x^h, then $\{Q_x^h: x \in R^d\}$ is the time-homogeneous Markov chain with transition probability $\Pi_h(x, dy)$ given by:

$$\Pi_h(x, \Gamma) = \frac{1}{(2\pi h)^{d/2}} \int_\Gamma e^{-|y-x|^2/2h} \, dy.$$

In particular, if $A_h f(x) = \int (f(y) - f(x)) \Pi_h(x, dy)$, then

$$\left(f(x(kh)) - \sum_0^{k-1} A_h f(x(jh)), \mathscr{M}_{kh}, P_x^h \right)$$

is a discrete parameter martingale for all $f \in C_0^\infty(R^d)$ (cf. Exercise 6.7.1). Indeed, this fact together with $P_x^h(x(0) = 0) = 1$ and $P_x^h[x(t) = [((k+1)h - t)/h]x(kh) + ((t - kh)/h)x((k+1)h), kh \leq t < (k+1)h] = 1$ for $k \geq 0$ uniquely characterizes P_x^h. Thus, we ought to be able to obtain the conclusion of Exercise 2.4.2 directly from this characterization of P_x^h. Once one realizes that this is a possibility, it is not hard to guess how one should proceed. The crux of the matter is contained in the observation that for each $f \in C_0^\infty(R^d)$:

(2.1) $$\frac{1}{h} A_h f(x) \to \frac{1}{2} \Delta f(x) \quad \text{as} \quad h \searrow 0$$

boundedly and uniformly. With this remark, one can check that the hypotheses of Theorem 1.4.11 are satisfied by $\{P_x^h: h > 0\}$. Hence $\{P_x^h: h > 0\}$ is pre-compact. Moreover, for each $f \in C_0^\infty(R^d)$ and $t \geq 0$:

$$\sum_0^{[t/h]-1} A_h f(x(jh)) \to \int_0^t \tfrac{1}{2} \Delta f(x(u))\, du$$

uniformly on compact subsets on Ω. Thus, any limit point Q of $\{P_x^h: h > 0\}$ has the property that $(f(x(t)) - \int_0^t \tfrac{1}{2} \Delta f(x(u))\, du, \mathcal{M}_t, Q)$ is a martingale for all $f \in C_0^\infty(R^d)$, and so it is obvious that $Q = \mathcal{W}_x^{(d)}$. We have therefore shown that $P_x^h \to \mathcal{W}_x^{(d)}$ directly from the martingale characterization of P_x^h together with (2.1). In particular, the specific form of $\Pi_h(x, \cdot)$ does not matter once one has (2.1). Obviously, one can carry out this line of reasoning in much greater generality, and that is exactly what we intend to do.

Throughout the remainder of this section we will be working with the following set-up. For each $h > 0$, let $\Pi_h(x, \cdot)$ be a transition function on R^d. Given $x \in R^d$, let P_x^h be the probability measure on Ω characterized by the properties that

$$(2.2) \quad \begin{cases} (i)\ P_x^h(x(0) = x) = 1, \\[2mm] (ii)\ P_x^h\left[x(t) = \dfrac{(k+1)h - t}{h}\, x(kh) + \dfrac{t - kh}{h}\, x((k+1)h),\right. \\[2mm] \qquad \left. kh \leq t < (k+1)h\right] = 1 \text{ for all } k \geq 0, \\[2mm] (iii)\ P_x^h(x((k+1)h) \in \Gamma \mid \mathcal{M}_{kh}) = \Pi_h(x(kh), \Gamma)\ \text{(a.s., } P_x^h) \text{ for all } k \geq 0 \text{ and } \\ \qquad \Gamma \in \mathcal{B}_{R^d}. \end{cases}$$

It is easy to see that (i) and (iii) are equivalent to saying that the distribution on $(R^d)^{\{0, \ldots, n, \ldots\}}$ of $x(kh)$, $k \geq 0$, is the time-homogeneous Markov chain starting from x with transition probability $\Pi_h(x, \cdot)$. In particular, from Exercise 6.7.1, (iii) is equivalent to saying that

$$\left(f(x(kh)) - \sum_0^{k-1} A_h f(x(jh)),\ \mathcal{M}_{kh},\ P_x^h\right)$$

is a martingale for all $f \in C_0^\infty(R^d)$, where

$$(2.3) \qquad A_h f(x) = \int (f(y) - f(x)) \Pi_h(x, dy).$$

We are now going to make the certain assumptions about $\Pi_h(x, \cdot)$ as a function of $h > 0$. Namely, define

$$a_h^{ij}(x) = \frac{1}{h} \int_{|y - x| \leq 1} (y_i - x_i)(y_j - x_j) \Pi_h(x, dy),$$

$$b_h^i(x) = \frac{1}{h} \int_{|y - x| \leq 1} (y_i - x_i) \Pi_h(x, dy),$$

and

$$\Delta_h^\varepsilon(x) = \frac{1}{h} \Pi_h(x, R^d \backslash B(x, \varepsilon)).$$

What we are going to assume is that for all $R > 0$:

(2.4) $\lim_{h \searrow 0} \sup_{|x| \le R} \|a_h(x) - a(x)\| = 0,$

(2.5) $\lim_{h \searrow 0} \sup_{|x| \le R} |b_h(x) - b(x)| = 0,$

(2.6) $\lim_{h \searrow 0} \sup_{|x| \le R} \Delta_h^\varepsilon(x) = 0, \qquad \varepsilon > 0,$

where $a: R^d \to S_d$ and $b: R^d \to R^d$ are continuous functions. The origin of these conditions is made clear by the following Lemma.

11.2.1 Lemma. *The conditions* (2.4), (2.5), *and* (2.6) *are equivalent to the condition that for each* $f \in C_0^\infty(R^d)$

(2.7) $\frac{1}{h} A_h f \to Lf$

uniformly on compact subsets of R^d, *where*

$$L = \frac{1}{2} \sum_{i, j=1}^d a^{ij}(x) \frac{\partial^2}{\partial x_i \, \partial x_j} + \sum_{i=1}^d b^i(x) \frac{\partial}{\partial x_i}.$$

Proof. First assume that (2.4)–(2.6) hold. Given $f \in C_0^\infty(R^d)$, set

$$H(x, y) = \sum_{i=1}^d (y_i - x_i) \frac{\partial f}{\partial x_i}(x)$$

$$+ \frac{1}{2} \sum_{i, j=1}^d (y_i - x_i)(y_j - x_j) \frac{\partial^2 f}{\partial x_i \, \partial x_j}(x).$$

By Taylor's theorem there is a $C_f < \infty$ such that

$$|f(y) - f(x) - H(x, y)| \le C_f |y - x|^3$$

for all $x, y \in R^d$. Thus if L_h is the operator with coefficients a_h and b_h, then

$$\left| \frac{1}{h} A_h f(x) - L_h f(x) \right| \le C_f \int_{|y-x| \le 1} |y - x|^3 \Pi_h(x, dy)$$

$$+ \int_{|y-x| > 1} |f(y) - f(x)| \Pi_h(x, dy).$$

Since (2.4) and (2.5) guarantee that $L_h f \to Lf$ uniformly on compacts, (2.7) will follow once we show that

$$\int_{|y-x|\leq 1} |y - x|^3 \Pi_h(x, dy) \to 0$$

and

$$\int_{|y-x|>1} |f(y) - f(x)| \Pi_h(x, dy) \to 0$$

uniformly on compacts. But for $0 < \varepsilon < 1$:

$$\int_{|y-x|\leq 1} |y - x|^3 \Pi_h(x, dy) \leq \varepsilon^3 + \Delta_h^\varepsilon(x),$$

and clearly

$$\int_{|y-x|>1} |f(y) - f(x)| \Pi_h(x, dy) \leq 2\|f\| \Delta_h^1(x).$$

Thus (2.6) guarantees the desired convergence.

Now suppose that (2.7) obtains. We will prove (2.6) first. Clearly, it is enough to show that if $x_h \to x_0$ as $h \to 0$, then for each $\varepsilon > 0$:

$$\lim_{h \searrow 0} \Delta_h^\varepsilon(x_h) = 0.$$

To this end, let $\varphi \in C_0^\infty(R^d)$ be chosen so that $0 \leq \varphi \leq 1$, $\varphi \equiv 1$ on $B(x_0, \varepsilon/4)$ and $\varphi \equiv 0$ off $B(x_0, \varepsilon/2)$. Then $\varphi \leq \chi_{B(x, \varepsilon)}$ for all $|x - x_0| \leq \varepsilon/2$; and so if $|x - x_0| < \varepsilon/4$, then

$$-\frac{1}{h} A_h \varphi(x) = \frac{1}{h} \int (1 - \varphi(y)) \Pi_h(x, dy)$$

$$\geq \frac{1}{h} \Pi_h(x, R^d \backslash B(x, \varepsilon)) = \Delta_h^\varepsilon(x).$$

Since $L\varphi(x) = 0$ for $|x - x_0| < \varepsilon/4$, $(A_h \varphi)/h \to 0$ uniformly on $B(x_0, \varepsilon/4)$. Thus $\lim_{h \searrow 0} \Delta_h^\varepsilon(x_h) = 0$, because $x_h \in B(x_0, \varepsilon/4)$ for small $h > 0$. The proofs of (2.4) and (2.5) are now quite easy. In fact, it is clear that all we need to show is that for any $f \in C^\infty(R^d)$ and any $R > 0$:

$$(2.8) \qquad \lim_{h \searrow 0} \sup_{|x| \leq R} \left| \int_{|y-x| \leq 1} (f(y) - f(x)) \Pi_h(x, dy) - Lf(x) \right| = 0.$$

Indeed, we can then simply take f successively equal to x_i and $x_i x_j$, $1 \leq i, j \leq d$, and thereby obtain (2.4) and (2.5) after a little easy algebra. But to prove (2.8) for a given $f \in C^\infty(R^d)$ and $R > 0$, we can replace the given f by one which vanishes outside $B(0, R + 1)$; and therefore, in view of (2.7), it suffices to show that for $f \in C_0^\infty(R^d)$:

$$(2.9) \qquad \lim_{h \searrow 0} \sup_{|x| \leq R} \frac{1}{h} \left| \int_{|y-x| \geq 1} (f(y) - f(x))\Pi_h(x, dy) \right| = 0.$$

But (2.9) is obvious, since f is bounded and we have just seen that (2.7) implies (2.4). Thus the proof is complete. $\square$

11.2.2 Lemma. *Assume that in addition to (2.4) and (2.5) we have*

$$\sup_{h > 0} \sup_{x \in R^d} (\|a_h(x)\| + |b_h(x)|) < \infty,$$

and that instead of (2.6) we have

$$\lim_{h \searrow 0} \sup_{x \in R^d} \Delta_h^\varepsilon(x) = 0, \qquad \varepsilon > 0.$$

Let $x_h \to x_0$ as $h \searrow 0$. Then the set $\{P_{x_h}^h : h > 0\}$ is pre-compact and any limit point as $h \searrow 0$ solves the martingale problem for a and b starting from x_0.

Proof. Let $f \in C_0^\infty(R^d)$. Then

$$\left| \frac{1}{h} A_h f(x) \right| \leq \frac{|\nabla f(x)|}{h} \left| \int_{|y-x| \leq 1} (y - x)\Pi_h(x, dy) \right|$$

$$+ \frac{\sum_{i,j=1}^{d} \|\partial^2 f / \partial x_i \, \partial x_j\|}{2h} \int_{|y-x| \leq 1} |y - x|^2 \Pi_h(x, dy)$$

$$+ 2\|f\| \frac{1}{h} \Pi_h(x, R^d \setminus B(x, 1)).$$

From this and our assumptions, we see that

$$\sup_{h > 0} \frac{1}{h} |A_h f(x)| \leq C_f < \infty,$$

where C_f depends only on the bounds on f and its first two derivatives. Since $(f(kh) - \sum_0^{k-1} A_h f(x(jh)), \mathcal{M}_{kh}, P_{x_h}^h)$ is a discrete parameter martingale, we now see that

$$(f(kh) + C_f kh, \mathcal{M}_{kh}, P_{x_h}^h)$$

is a submartingale. Next note that

$$P^h_{xh}(|x((k+1)h) - x(kh)| \geq \varepsilon) = E^{P^h_{x,h}}[\Pi_h(x(kh), R^d \backslash B(x(kh), \varepsilon))]$$

$$\leq \sup_{x \in R^d} h\Delta^\varepsilon_h(x).$$

Thus, if $T > 0$, then

$$\sum_{0 \leq jh \leq T} P^h_{xh}(|x(k+1)h) - x(kh)| \geq \varepsilon) \leq (T+1) \sup_{x \in R^d} \Delta^\varepsilon_h(x) \to 0$$

as $h \searrow 0$. We can therefore use Theorem 1.4.11 to conclude that $\{P^h_{xh} : h > 0\}$ is pre-compact.

To complete the proof, suppose that $h_n \searrow 0$ is a sequence such that P_n given by $P^h_{x_h}$ with $h = h_n$ converges to P. Clearly $P(x(0) = x_0) = 1$. Given $f \in C^\infty_0(R^d)$, $0 \leq t_1 \leq t_2$, and $\Phi: \Omega \to R^1$ which is bounded continuous and $\mathcal{M}_{t_1}$-measurable, we have

$$E^{P_n}[(f(x(l_n h_n)) - f(x(k_n h_n)) - \sum_{k_n}^{l_n - 1} A_{h_n} f(x(jh_n)))\Phi] = 0$$

where

$$k_n = \left[\frac{t_1}{h_n}\right] + 1 \quad \text{and} \quad l_n = \left[\frac{t_2}{h_n}\right] + 1.$$

Note that by (2.7):

$$f(x(l_n h_n)) - f(x(k_n h_n)) - \sum_{k_n}^{l_n - 1} A_{h_n} f(x(jh_n))$$

$$\to f(x(t_2)) - f(x(t_1)) - \int_{t_1}^{t_2} Lf(x(u))\, du$$

uniformly on compact subsets of Ω and boundedly. Thus, since

$$(f(x(t_2)) - f(x(t_1)) - \int_{t_1}^{t_2} Lf(x(u))\, du)\Phi$$

is continuous, we conclude that

$$E^P[(f(x(t_2)) - f(x(t_1)) - \int_{t_1}^{t_2} Lf(x(u))\, du)\Phi] = 0.$$

That is, $(f(x(t)) - \int_0^t Lf(x(u))\, du, \mathcal{M}_t, P)$ is a martingale. $\square$

11.2.3 Theorem. *Assume that in addition to being continuous the coefficients a and b have the property that for each $x \in R^d$ the martingale problem for a and b has exactly one solution P_x starting from x. Then $P_x^h \to P_x$ as $h \searrow 0$ uniformly on compact subsets of R^d.*

Proof. Let $h_n \searrow 0$ and $x_n \to x_0$ as $n \to \infty$, and set $P^n = P_{x_n}^{h_n}$. We must show that $P^n \to P_{x_0}$ as $n \to \infty$.

For each $k \geq 1$, let $\varphi_k \in C_0^\infty(R^d)$ be chosen so that $0 \leq \varphi_k \leq 1$, $\varphi_k \equiv 1$ on $B(0, k)$, and $\varphi_k \equiv 0$ off $B(0, k + 1)$. Define

$$\Pi_{n,k}(x, \Gamma) = \varphi_k(x)\Pi_{h_n}(x, \Gamma) + (1 - \varphi_k(x))\chi_\Gamma(x),$$

and let $\{P_x^{n,k} : x \in R^d\}$ be defined in terms of $\Pi_{n,k}(x, \cdot)$ by the same procedure as $P_x^{h_n}$ is obtained from $\Pi_{h_n}(x, \cdot)$. Take $P^{n,k} = P_{x_n}^{n,k}$. It is easy to see that if $a_{n,k}(x)$ and $b_{n,k}(x)$ are defined in terms of $\Pi_{n,k}(x, \cdot)$ as $a_{h_n}(x)$ and $b_{h_n}(x)$ are from $\Pi_{h_n}(x, \cdot)$, then for each $k \geq 1$:

$$\sup_{n \geq 1} \sup_{x \in R^d} (\|a_{n,k}(x)\| + |b_{n,k}(x)|) < \infty$$

and as $n \to \infty$,

$$a_{n,k}(x) \to \varphi_k(x)a(x)$$

and

$$b_{n,k}(x) \to \varphi_k(x)b(x)$$

uniformly on compacts. Thus, by Lemma 11.2.2, $\{P^{n,k} : n \geq 1\}$ is pre-compact and any limit point solves the martingale problem for $\varphi_k(\cdot)a(\cdot)$ and $\varphi_k(\cdot)b(\cdot)$ starting from x_0. By Theorem 10.1.1, this means that any limit point of $\{P^{n,k} : n \geq 1\}$ equals P_{x_0} on $\mathcal{M}_{\tau_k}$, where

$$\tau_k = \inf\{t \geq 0 : |x(t)| \geq k\},$$

since $\varphi_k \equiv 1$ on $B(0, k)$. At the same time, it is easy to see that $P^{n,k}$ equals P^n on $\mathcal{M}_{\tau_k}$ for all n. Thus, by Lemma 11.1.1, $P_n \to P_{x_0}$. $\square$

11.3. Convergence of Diffusion Processes: Elliptic Case

In this section we return to questions about convergence of the sort discussed in Section 1. Where our present treatment differs from the one that we gave there is that we now are going to be assuming that the matrix a is strictly positive definite valued. This assumption will allow us to weaken considerably the sense in which the drift coefficients (i.e. the first order coefficients) have to converge in order that the associated diffusions converge. To help in understanding what is happening here, consider the following example. Let $L_t^n = \frac{1}{2}\Delta + \sum_{i=1}^{d} b_n^i(t, \cdot)\, \partial/\partial x_i$, where $b_n\colon [0, \infty) \times R^d \to R^d$ is a bounded measurable function for each $n \geq 1$. If $P_{s, x}^n$ is the solution to the martingale problem for L_t^n starting from (s, x), then we know from Theorem 6.4.3 that for the bounded $\mathcal{M}_t$-measurable $\Phi\colon \Omega \to C$:

$$E^{P_{s, x}^n}[\Phi] = E^{W_{s, x}^{(d)}}[R_n^s(t)\Phi],$$

where

$$R_n^s(t) = \exp\left[\int_s^{t \vee s} \langle b_n(u, x(u)),\, dx(u)\rangle - \frac{1}{2}\int_s^{t \vee s} |b_n(u, x(u))|^2\, du\right].$$

In particular, if $b_n \to b$ boundedly and pointwise, then, since

$$E^{W_{s, x}^{(d)}}[(R_n^s(t))^2] \leq e^{\frac{1}{2}(t \vee s - s)\|b_n\|^2},$$

$$R_n^s(t) \to R^s(t) = \exp\left[\int_s^{t \vee s} \langle b(u, x(u)),\, dx(u)\rangle - \frac{1}{2}\int_s^{t \vee s} |b(u, x(u))|^2\, du\right]$$

in $L^1(W_{s, x}^{(d)})$; and so it is clear that $P_{s, x}^n \to P_{s, x}$ (in variation) on $\mathcal{M}_t$, where $P_{s, x}$ solves for $L_t = \frac{1}{2}\Delta + \sum_{i=1}^{d} b^i(t, \cdot)\, \partial/\partial x_i$ starting from (s, x). What is not so obvious is that we can weaken even further the sense in which $b_n \to b$. The point is that as long as the b_n's all share a common bound, it is quite simple to check (using the Cameron-Martin formula and the explicit expression for the transition probability function going with $\{W_{s, x}\colon (s, x) \in [0, \infty) \times R^d\}$) that transition probability function $P_n(s, x; t, \cdot)$ determined by $L_t^n \equiv \frac{1}{2}\Delta + \sum_1^{d} b_n^i(t, \cdot)\, \partial/\partial x_i$ admits a density $P_n(s, x; t, \cdot)$ and that $\{P_n(s, x; t, \cdot)\}_1^\infty$ is pre-compact in $L^1(R^d)$ for each $(s, x) \in [0, \infty) \times R^d$ and that $t > s$. Thus, if $\{b_n\}_1^\infty$ is uniformly bounded and $b_n \to b$ in the weak topology on $L^\infty(R^d)$, then for all $\varphi \in C_0^\infty(R^d)$:

$$\lim_{n \to \infty} \int_s^T dt \int_{R^d} L_t^n \varphi(y) p_n(s, x; t, y)\, dy = \int_s^T dt \int_{R^d} L_t \varphi(y) p(s, x; t, y)\, dy$$

where $L_t = \frac{1}{2}\Delta + \sum_{i=1}^{d} b^i(t, \cdot)\, \partial/\partial x_i$ and $p(s, x; t, \cdot)$ is defined in terms of L_t. From this observation, one can easily conclude that the associated diffusions converge. The first one to point out that weak convergence of the b_n's to b guarantees convergence of the associated diffusions was M. Freidlin [1967]. It is our aim in this section to carry out the proof of Freidlin's idea in considerable generality.

Given $0 < \lambda \leq \Lambda < \infty$ and a non-decreasing function $\delta\colon (0, \infty) \to (0, \infty)$ such that $\lim_{\varepsilon \searrow 0} \delta(\varepsilon) = 0$, let $\mathcal{A}_d(\lambda, \Lambda, \delta(\cdot))$ be the set of all pairs $\{a, b\}$ where $a\colon [0, \infty) \times R^d \to S_d$ and $b\colon [0, \infty) \times R^d \to R^d$ are measurable functions satisfying:

$$(3.1) \qquad \lambda|\theta|^2 \leq \langle\theta, a(s, x)\theta\rangle \leq \Lambda|\theta|^2, \qquad (s, x) \in [0, \infty) \times R^d \quad \text{and} \quad \theta \in R^d,$$

$$(3.2) \qquad \|a(s, x) - a(s, y)\| \leq \delta(|x - y|), \qquad s \geq 0 \quad \text{and} \quad x, y \in R^d,$$

$$(3.3) \qquad |b(s, x)| \leq \Lambda, \qquad (s, x) \in [0, \infty) \times R^d.$$

For each $\{a, b\} \in \mathscr{A}(\lambda, \Lambda, \delta(\cdot))$, the corresponding martingale problem is well-posed and the associated family of solutions $\{P^{a,\,b}_{s,\,x} : (s, x) \in [0, \infty) \times R^d\}$ is strongly Markov and strongly Feller continuous (cf. Theorems 7.2.1 and 7.2.4). Let $\{a_n, b_n\} \in \mathscr{A}_d(\lambda, \Lambda, \delta(\cdot))$, $n \geq 1$, be a sequence such that for some $\{a, b\} \in \mathscr{A}_d(\lambda, \Lambda, \delta(\cdot))$:

(3.4)
$$\lim_{n \to \infty} \int_0^\infty \int_{R^d} a_n^{ij}(s, x)\varphi(s, x) \, ds \, dx$$

$$= \int_0^\infty \int_{R^d} a^{ij}(s, x)\varphi(s, x) \, ds \, dx, \qquad 1 \leq i, j \leq d,$$

and

(3.5)
$$\lim_{n \to \infty} \int_0^\infty \int_{R^d} b_n^i(s, x)\varphi(s, x) \, ds \, dx$$

$$= \int_0^\infty \int_{R^d} b^i(s, x)\varphi(s, x) \, ds \, dx, \qquad 1 \leq i \leq d,$$

for all $\varphi \in C_0^\infty([0, \infty) \times R^d)$. What we wish to show is that if $(s_n, x_n) \to (s, x)$, then $P^{a_n,\,b_n}_{s_n,\,x_n} \to P^{a,\,b}_{s,\,x}$. Notice that stating the sense in which $a_n \to a$ as we have in (3.4) is something of a hoax, since the fact that $\{\{a_n, b_n\}\}_1^\infty \subseteq \mathscr{A}_d(\lambda, \Lambda, \delta(\cdot))$ already implies that for all $0 \leq s < T$ the family $\{\int_0^\infty a_n^{ij}(s, x)\varphi(s, x) \, ds\}_1^\infty$ is pre-compact with respect to uniform convergence in x. Thus (3.4) should be viewed as simply a method of identifying a as the limit of the a_n's rather than really describing the manner in which the limit is taken.

11.3.1 Lemma. *Let $\varphi \in C_0^\infty(R^d)$ be a non-negative function such that $\int_{R^d} \varphi(x) \, dx = 1$. For $g \in B(R^d)$, define $g_\varepsilon(s, x) = \int \varphi_\varepsilon(x - y)g(t, y) \, dy$ where $\varphi_\varepsilon(x) = \varepsilon^{-d}\varphi(x/\varepsilon)$. Then for each $T > 0$,*

$$\frac{1}{1 + \|g\|} E^{P^{a,\,b}_{s,\,x}}\left[\left| \int_s^T (g - g_\varepsilon)(t, x(t)) \, dt \right|\right]$$

tends to zero as $\varepsilon \searrow 0$ at a rate which is independent of $\{a, b\} \in \mathscr{A}_d(\lambda, \Lambda, \delta(\cdot))$, $(s, x) \in [0, T] \times R^d$, and $g \in B([0, \infty) \times R^d)$.

Proof. For $\{a, b\} \in \mathscr{A}_d(\lambda, \Lambda, \delta(\cdot))$, let $p_{a,\,b}(s, x; t, y)$ be the associated density function described in Theorem 9.1.9. From Theorem 9.1.15, we know that:

$$\int_s^T dt \int_{R^d} |p_{a,\,b}(s, x; t, y + h) - p_{a,\,b}(s, x; t, y)| \, dy$$

tends to 0 as $|h| \to 0$ at a rate which is independent of $(s, x) \in [0, T] \times R^d$ and $\{a, b\} \in \mathscr{A}_d(\lambda, \Lambda, \delta(\cdot))$. Since

$$\left| \int_s^T dt \int_{R^d} \dot{p}_{a,b}(s, x; t, y)(g - g_\varepsilon)(t, y)\, dy \right|$$

$$\le \int_{R^d} \varphi_\varepsilon(z)\, dz \left| \int_s^T dt \int_{R^d} (p_{a,b}(s, x; t, y - z) \right.$$

$$\left. - p_{a,b}(s, x; t, y))g(t, y)\, dy \right|$$

$$\le \|g\| \sup_{z \in \mathrm{supp}(\varphi_\varepsilon)} \int_s^T dt \int_{R^d} |p_{a,b}(s, x; t, y - z)$$

$$- p_{a,b}(s, x; t, y)|\, dy,$$

it follows that

$$\frac{1}{1 + \|g\|} \left| \int_s^T dt \int_{R^d} \dot{p}_{a,b}(s, x; t, y)(g - g_\varepsilon)(t, y)\, dy \right|$$

tends to 0 as $\varepsilon \searrow 0$ at a rate which does not depend on $\{a, b\} \in \mathscr{A}_d(\lambda, \Lambda, \delta(\cdot))$, $(s, x) \in [0, T] \times R^d$, or $g \in B([0, \infty] \times R^d)$. But

$$E^{P_{s;x}^{a,b}} \left[\left(\int_s^T (g - g_\varepsilon)(t, x(t))\, dt \right)^2 \right]$$

$$= 2 E^{P_{s;x}^{a,b}} \left[\int_s^T (g - g_\varepsilon)(t, x(t))\, dt \int_t^T (g - g_\varepsilon)(u, x(u))\, du \right]$$

$$\le 4\|g\| \int_s^T \sup_{z \in R^d} \left| \int_t^T du \int_{R^d} p_{a,b}(t, z; u, y)(g - g_\varepsilon)(u, y)\, dy \right| dt$$

$$\le 4(T - s)\|g\| \sup_{(t, z) \in [s, T] \times R^d}$$

$$\times \left| \int_t^T du \int_{R^d} p_{a,b}(t, z; u, y)(g - g_\varepsilon)(u, y)\, dy \right|,$$

and so

$$\frac{1}{1 + \|g\|} E^{P_{s;x}^{a,b}} \left[\left| \int_s^T (g - g_\varepsilon)(t, x(t))\, dt \right| \right]$$

$$\le \frac{(T - s)^{1/2}}{1 + \|g\|} E^{P_{s;x}^{a,b}} \left[\left(\int_s^T (g - g_\varepsilon)(t, x(t))\, dt \right)^2 \right]^{\frac{1}{2}}$$

$$\le 2(T - s) \left(\frac{1}{1 + \|g\|} \sup_{(t, z) \in [s, T] \times R^d} \right.$$

$$\left. \times \left| \int_t^T du \int_{R^d} p_{a,b}(t, z; u, y)(g - g_\varepsilon)(u, y)\, du \right| \right)^{\frac{1}{2}}$$

tends to 0 as $\varepsilon \searrow 0$ at a rate which is independent of $\{a, b\} \in \mathscr{A}_d(\lambda, \Lambda, \delta(\cdot))$, $(s, x) \in [0, T] \times R^d$, and $g \in B([0, T] \times R^d)$. $\quad\square$

11.3.2 Lemma. Let $\{a, b\} \in \mathscr{A}_d(\lambda, \Lambda, \delta(\cdot))$ and $(s_n, x_n) \in [0, \infty) \times R^d$, $n \geq 1$. Set $P^n = P_{s_n, x_n}^{a_n, b_n}$ and suppose that $s_n \to s_0$ and $P^n \to P$ as $n \to \infty$. Let $\{g_n\}_1^\infty \subseteq B([0, \infty] \times R^d)$ be a uniformly bounded sequence such that for some $g \in B([0, \infty] \times R^d)$ and all $\varphi \in C_0^\infty([0, \infty) \times R^d)$:

$$\lim_{n \to \infty} \int_0^\infty \int_{R^d} g_n(s, x)\varphi(s, x) \, ds \, dx = \int_0^\infty \int_{R^d} g(s, x)\varphi(s, x) \, ds \, dx.$$

Then for all $s_0 \leq t_1 < t_2$ and bounded continuous $\mathscr{M}_{t_1}$-measurable $\Phi: \Omega \to R^1$:

$$E^P\left[\Phi \int_{t_1}^{t_2} g(t, x(t)) \, dt\right] = \lim_{n \to \infty} E^{P^n}\left[\Phi \int_{t_1}^{t_2} g_n(t, x(t)) \, dt\right].$$

Proof. If the g_n's were continuous in x and converged uniformly in x to g for each t, there would be no problem. Thus we introduce

$$g_{n, \varepsilon}(s, x) = \int \varphi_\varepsilon(x - y)g_n(s, y) \, dy$$

and

$$g_\varepsilon(s, x) = \int \cdot \varphi_\varepsilon(x - y)g(s, y) \, dy.$$

It is easy to check that for each $\varepsilon > 0$,

$$\int_{t_1}^{t_2} g_{n, \varepsilon}(t, x) \, dt \to \int_{t_1}^{t_2} g_\varepsilon(t, x) \, dt$$

boundedly and uniformly with respect to x. Thus, by Corollary 1.1.5,

$$E^P\left[\Phi \int_{t_1}^{t_2} g_\varepsilon(t, x(t)) \, dt\right] = \lim_{n \to \infty} E^{P^n}\left[\Phi \int_{t_1}^{t_2} g_{n, \varepsilon}(t, x(t)) \, dt\right]$$

It is therefore enough for us to show that

$$E^P\left[\Phi \int_{t_1}^{t_2} g(t, x(t)) \, dt\right] = \lim_{\varepsilon \searrow 0} E^P\left[\Phi \int_{t_1}^{t_2} g_\varepsilon(t, x(t)) \, dt\right]$$

and that

$$\limsup_{\varepsilon \searrow 0} \limsup_{n \to \infty} \left| E^{P^n}\left[\Phi \int_{t_1}^{t_2} g_n(t, x(t)) \, dt\right] - E^{P^n}\left[\Phi \int_{t_1}^{t_2} g_{n, \varepsilon}(t, x(t)) \, dt\right] \right| = 0.$$

To prove the first of these, choose $\{h_k\}_1^\infty \subseteq C_b([0, \infty) \times R^d)$ such that $\|h_k\| \le \|g\|$, $h_k \to g$ in $L^1([t_1, t_2] \times R^d)$, and

$$E^P\left[\Phi \int_{t_1}^{t_2} g(t, x(t))\, dt\right] = \lim_{k \to \infty} E^P\left[\Phi \int_{t_1}^{t_2} h_k(t, x(t))\, dt\right].$$

Then, by Lemma 11.3.1:

$$\left|E^P\left[\Phi \int_{t_1}^{t_2} (g - g_\varepsilon)(t, x(t))\, dt\right]\right|$$

$$= \lim_{k \to \infty} \left|E^P\left[\Phi \int_{t_1}^{t_2} (h_k - h_{k, \varepsilon})(t, x(t))\, dt\right]\right|$$

$$= \lim_{k \to \infty} \lim_{n \to \infty} \left|E^{P^n}\left[\Phi \int_{s_n \vee t_1}^{s_n \vee t_2} (h_k - h_{k, \varepsilon})(t, x(t))\, dt\right]\right|$$

$$\le \|\Phi\| \sup_{k \ge 1} \sup_{n \ge 1} E^{P^n}\left[E_{s_n \vee t_1, x(s_n \vee t_1)}^{P^{a_n, b_n}}\left[\left|\int_{s_n \vee t_1}^{s_n \vee t_2} (h_k - h_{k,\varepsilon})(t, x(t))\, dt\right|\right]\right]$$

$$\to 0$$

as $\varepsilon \searrow 0$. To prove the second, observe that:

$$\limsup_{\varepsilon \searrow 0} \limsup_{n \to \infty} \left|E^{P^n}\left[\Phi \int_{t_1}^{t_2} (g_n - g_{n, \varepsilon})(t, x(t))\, dt\right]\right|$$

$$= \limsup_{\varepsilon \searrow 0} \overline{\lim}_{n \to \infty} \left|E^{P^n}\left[\Phi \int_{s_n \vee t_1}^{s_n \vee t_2} (g_n - g_{n, \varepsilon})(t, x(t))\, dt\right]\right|$$

$$\le \limsup_{\varepsilon \searrow 0} \sup_{n \ge 1} \|\Phi\| E^{P^n}\left[E_{s_n \vee t_1, x(s_n \vee t_1)}^{P^{a_n, b_n}} \times \left[\left|\int_{s_n \vee t_1}^{s_n \vee t_2} (g_n - g_{n, \varepsilon})(t, x(t))\, dt\right|\right]\right]$$

$$= 0,$$

again by Lemma 11.3.1. $\square$

11.3.3 Theorem. *For each* $n \ge 1$, *let* $\{a_n, b_n\} \in \mathscr{A}_d(\lambda, \Lambda, \delta(\cdot))$ *and* $(s_n, x_n) \in [0, \infty) \times R^d$. *Assume that* $(s_n, x_n) \to (s_0, x_0)$ *and that there is an* $\{a, b\} \in \mathscr{A}_d(\lambda, \Lambda, \delta(\cdot))$ *such that* (3.4) *and* (3.5) *hold. Then* $P_{s_n, x_n}^{a_n, b_n} \to P_{s_0, x_0}^{a, b}$ *as* $n \to \infty$.

Proof. Let $P^n = P_{s_n, x_n}^{a_n, b_n}$. By Theorem 1.4.6, $\{P^n : n \ge 1\}$ is pre-compact. Thus, all that we need to show is that if $\{P^{n'}\}$ is a subsequence of $\{P^n\}$ which converges to P, then P solves the martingale problem for a and b starting from (s_0, x_0). Clearly $P(x(t) = x_0, \ 0 \le t \le s_0) = 1$. To complete the proof, let $s_0 \le t_1 < t_2$ and $\Phi : \Omega \to R^1$ be bounded, continuous, and $\mathscr{M}_{t_1}$-measurable. Given $f \in C_0^\infty(R^d)$, we must show that

$$E^P\left[\Phi\left(f(x(t_2)) - f(x(t_1)) - \int_{t_1}^{t_2} L_t f(x(t))\, dt\right)\right] = 0,$$

where

$$L_t = \frac{1}{2} \sum_{i,j=1}^{d} a^{ij}(t, \cdot) \frac{\partial^2}{\partial x_i \partial x_j} + \sum_{i=1}^{d} b^i(t, \cdot) \frac{\partial}{\partial x_i}.$$

Since

$$E^{P^{n'}} \left[\Phi(f(x(t_2 \vee s_{n'})) - f(x(t_1 \vee s_{n'})) - \int_{t_1 \vee s_{n'}}^{t_2 \vee s_{n'}} L_t^{n'} f(x(t)) \, dt \right] = 0$$

for all n', it is clear that we need only show that

$$E^P \left[\Phi \int_{t_1}^{t_2} L_t^{n'} f(x(t)) \, dt \right] = \lim_{n' \to \infty} E^{P^{n'}} \left[\Phi \int_{t_1 \vee s_{n'}}^{t_2 \vee s_{n'}} L_t^{n'} f(x(t)) \, dt \right].$$

But this last fact is an immediate consequence of (3.4), (3.5), and Lemma 11.3.2. Thus the theorem is proved. □

It should be clear that the conclusion of Theorem 11.3.3 ought to remain true even after we replace our assumptions by their local analogues. To be precise, suppose that $R \to \lambda_R$ and $R \to \Lambda_R$ are functions on $(0, \infty)$ such that $0 < \lambda_R \leq \Lambda_R$ for all R, and let $R \to \delta_R(\cdot)$ be a mapping of $(0, \infty)$ into non-increasing functions from $(0, \infty) \to (0, \infty)$ such that $\lim_{\varepsilon \to 0} \delta_R(\varepsilon) = 0$ for all $R > 0$. We will say that the pair $\{a, b\}$ is an element of $\mathscr{A}_d^{loc}(\lambda_R, \Lambda_R, \delta_R(\cdot))$ if $a: [0, \infty) \times R^d \to S_d$ and $b: [0, \infty) \times R^d \to R^d$ are measurable functions satisfying: for $R > 0$, (s, x), $(s, y) \in [0, R] \times \overline{B(0, R)}$, and $\theta \in R^d$,

(3.6) $\lambda_R |\theta|^2 \leq \langle \theta, a(s, x)\theta \rangle \leq \Lambda_R |\theta|^2,$

(3.7) $\|a(s, x) - a(s, y)\| \leq \delta_R(|x - y|),$

(3.8) $|b(s, x)| \leq \Lambda_R,$

and the martingale problem for a and b is well-posed. Note that, given (3.6), (3.7) and (3.8), the martingale problem for a and b starting from (s, x) has at most one solution (cf. Theorem 7.3.1). Thus the martingale problem for a and b is well-posed if solutions exist at all; and according to Corollary 10.1.2, existence is equivalent to non-explosion. Finally, if $\{a, b\} \in \mathscr{A}_d^{loc}(\lambda_R, \Lambda_R, \delta_R(\cdot))$, then the associated family of solutions is strongly Markovian and strongly Feller continuous (cf. Corollary 10.1.4).

11.3.4 Theorem. *For each $n \geq 1$, let $\{a_n, b_n\} \in \mathscr{A}_d^{loc}(\lambda_R, \Lambda_R, \delta_R(\cdot))$ and $(s_n, x_n) \in [0, \infty) \times R^d$. Assume that $(s_n, x_n) \to (s_0, x_0)$ and suppose that there exists $\{a, b\} \in \mathscr{A}_d^{loc}(\lambda_R, \Lambda_R, \delta_R(\cdot))$ for which (3.4) and (3.5) hold. If P^n is the solution to the martingale problem for a_n and b_n starting from (s_n, x_n) and P is the solution for a and b starting from (s_0, x_0), then $P^n \to P$.*

Proof. We will reduce the present case to the situation in Theorem 11.3.3. For each $k \geq \sup_n (|x_n| + s_n)$, let $\eta_k \in C_0^\infty([0, \infty) \times R^d)$ be chosen so that $0 \leq \eta_k \leq 1$ and $\eta_k \equiv 1$ on $[0, k) \times \overline{B(0, k)}$. Define $a_{n, k} = \eta_k a_n + (1 - \eta_k)I$ and $b_{n, k} = \eta_k b_n$ for $n \geq 1$, and set $a_k = \eta_k a + (1 - \eta_k)I$ and $b_k = \eta_k b$. For each $k \geq 1$, $\{\{a_{n, k}, b_{n, k}\}\}_{n=1}^\infty$ and $\{a_k, b_k\}$ satisfy the conditions of Theorem 11.3.3. Thus, if $P^{n, k} = P_{s_n, x_n}^{a_{n, k}, b_{n, k}}$ and $P^k = P_{s_0, x_0}^{a_k, b_k}$, then $P^{n, k} \to P^k$ as $n \to \infty$. Finally, if $\tau_k = \inf \{t \geq 0 : t \geq k$ or $|x(t)| \geq k\}$, then $P^{n, k}$ equals $P^n \otimes_{\tau_k(\cdot)} P_{\tau_k(\cdot), x(\tau_k(\cdot)), \cdot}^{a_{n, k}, b_{n, k}}$ and P^k equals $P \otimes_{\tau_k(\cdot)} P_{\tau_k(\cdot), x(\tau_k(\cdot)), \cdot}^{a_k, b_k}$. Thus P^n equals $P^{n, k}$ on $\mathcal{M}_{\tau_k}$ and P equals P^k on $\mathcal{M}_{\tau_k}$. By Lemma 11.1.1, we therefore know that $P^n \to P$. $\square$

11.4. Convergence of Transition Probability Densities

In the preceding sections we have seen that solutions to the martingale problem vary continuously with the coefficients and starting place. The type of convergence which we have imposed on the coefficients is different in Section 3 from that in Section 1, but the sense in which the associated solutions converge is the one dictated by the weak topology on $M(\Omega)$. It is not hard to see that the weak topology is the strongest one which we can afford in general, since, even if the starting point is held fixed, solutions corresponding to different second order coefficients will be singular. Nonetheless, if we do not look at the solution itself but only at the marginal distributions derived from it, we can sometimes do better. In this section we will illustrate this point by looking at the transition probability function for time-homogeneous coefficients which satisfy the conditions of the preceding section.

   Given $0 < \lambda \leq \Lambda$ and a non-decreasing function $\delta \colon (0, \infty) \to (0, \infty)$ such that $\lim_{\varepsilon \searrow 0} \delta(\varepsilon) = 0$, let $\mathcal{H}_d(\lambda, \Lambda, \delta(\cdot))$ stand for the class of $\{a, b\} \in \mathcal{A}_d(\lambda, \Lambda, \delta(\cdot))$ such that a and b are time-independent. For $\{a, b\} \in \mathcal{H}_d(\lambda, \Lambda, \delta(\cdot))$, we know that the martingale problem is well-posed and that the associated family of solutions $\{P_{s, x}^{a, b} \colon (s, x) \in [0, \infty) \times R^d\}$ is time-homogeneous in the sense that $P_{s, x}^{a, b} = P_x^{a, b} \circ \Phi_s^{-1}$, where $P_x^{a, b} = P_{0, x}^{a, b}$ and $\Phi_s \colon \Omega \to \Omega$ is defined by $x(t, \Phi_s(\omega)) = x((t - s) \vee 0, \omega)$, $t \geq 0$. Furthermore, we saw in Section 9.2 that the time-homogeneous transition probability function $P_{a, b}(s, x, \cdot)$ admits a density $p_{a, b}(s, x, y)$ with respect to Lebesgue measure and that $p_{a, b}(s, x, \cdot) \in L^p(R^d)$ for all $1 \leq p < \infty$. What we are going to prove now is that $p_{a_n, b_n}(s, x, \cdot) \to p_{a, b}(s, x, \cdot)$ in $L^p(R^d)$, $1 \leq p < \infty$, if $\{a_n, b_n\} \to \{a, b\}$ in the sense of the preceding section.

11.4.1 Lemma. *Let $\{f_n\}_1^\infty$ be a sequence of non-negative $\mathcal{B}_{R^d}$-measurable functions such that $\int f_n(x)\, dx = 1$ for all $n \geq 1$ and $\lim_{|h| \to 0} \sup_{n \geq 1} \int |f_n(x + h) - f_n(x)|\, dx = 0$. Assume that there is an $f \in L^1(R^d)$ such that*

$$\int f(x)\psi(x)\, dx = \lim_{n \to \infty} \int f_n(x)\psi(x)\, dx$$

for all $\psi \in C_b(R^d)$. Then $f_n \to f$ in $L^1(R^d)$.

Proof. First observe that for any $\varepsilon > 0$ we can find an R_ε such that

$$\sup_{n \geq 1} \int_{|x| \geq R_\varepsilon} |f_n(x) - f(x)| \, dx \leq \sup_{n \geq 1} \int_{|x| \geq R_\varepsilon} |f_n(x)| \, dx + \int_{|x| \geq R_\varepsilon} |f(x)| \, dx$$

$$\leq \frac{\varepsilon}{2} + \frac{\varepsilon}{2} = \varepsilon.$$

Thus, we need only show that $f_n \to f$ in $L^1(B(0, R))$ for each $R > 0$. To this end, let $\varphi \in C_0^\infty(R^d)$ be chosen so that $\varphi \geq 0$, $\varphi(-x) = \varphi(x)$, and $\int \varphi(x) \, dx = 1$. Define $\varphi_\varepsilon(x) = \varepsilon^{-d}\varphi(x/\varepsilon)$ for $\varepsilon > 0$ and set $f_{n,\varepsilon} = \varphi_\varepsilon * f_n$ and $f_\varepsilon = \varphi_\varepsilon * f$. Given $\psi \in C_b(R^d)$, we have:

$$\int f_{n,\varepsilon}(x)\psi(x) \, dx = \int f_n(x)\psi_\varepsilon(x) \, dx \to \int f(x)\psi_\varepsilon(x) \, dx$$

$$= \int f_\varepsilon(x)\psi(x) \, dx,$$

where $\psi_\varepsilon = \varphi_\varepsilon * \psi$. Moreover, it is easy to check that $\sup_{n \geq 1} (\|f_{n,\varepsilon}\| + \|\nabla f_{n,\varepsilon}\|) < \infty$ for each $\varepsilon > 0$. Thus by the Arzela-Ascoli theorem, $f_{n,\varepsilon} \to f_\varepsilon$ uniformly on compacts for each $\varepsilon > 0$. Given $R > 0$, we now have:

$$\limsup_{n \to \infty} \int_{B(0, R)} |f_n(x) - f(x)| \, dx$$

$$\leq \limsup_{n \to \infty} \int_{B(0, R)} |f_n(x) - f_{n,\varepsilon}(x)| \, dx + \int_{B(0, R)} |f(x) - f_\varepsilon(x)| \, dx$$

for all $\varepsilon > 0$. But

$$\int |f(x) - f_\varepsilon(x)| \, dx \to 0$$

as $\varepsilon \searrow 0$, and

$$\sup_{n \geq 1} \int |f_n(x) - f_{n,\varepsilon}(x)| \, dx = \sup_{n \geq 1} \int dx \, |(f_n(x) - f_n(x - y))\varphi_\varepsilon(y) \, dy|$$

$$\leq \sup_{n \geq 1} \int \varphi_\varepsilon(y) \, dy \int |f_n(x + y) - f_n(x)| \, dx$$

$$\leq \sup_{n \geq 1} \sup_{y \in \text{supp}(\varphi_\varepsilon)} \int |f_n(x + y) - f_n(x)| \, dx \to 0$$

as $\varepsilon \searrow 0$. This completes the proof. $\square$

11.4.2 Theorem. *For each* $n \geq 1$, *let* $\{a_n, b_n\} \in \mathcal{H}_d(\lambda, \Lambda, \delta(\cdot))$ *and suppose that* $\{a, b\} \in \mathcal{H}_d(\lambda, \Lambda, \delta(\cdot))$ *satisfies:*

$$\lim_{n \to \infty} \int a_n^{ij}(x)\varphi(x)\, dx = \int a^{ij}(x)\varphi(x)\, dx, \qquad 1 \leq i, j \leq d,$$

$$\lim_{n \to \infty} \int b_n^i(x)\varphi(x)\, dx = \int b^i(x)\varphi(x)\, dx, \qquad 1 \leq i \leq d,$$

for all $\varphi \in C_0^\infty(R^d)$. *Finally, assume that* $\{(s_n, x_n)\}_1^\infty \subseteq (0, \infty) \times R^d$ *tends to* (s_0, x_0) $\in (0, \infty) \times R^d$. *Then, for each* $1 \leq p < \infty$, $p_{a_n, b_n}(s_n, x_n, \cdot) \to p_{a, b}(s_0, x_0, \cdot)$ *in* $L^p(R^d)$.

Proof. By Theorem 11.3.3, we know that

$$\int \psi(y) p_{a, b}(s_0, x_0, y)\, dy = \lim_{n \to \infty} \int \psi(y) p_{a_n, b_n}(s_n, x_n, y)\, dy$$

for all $\psi \in C_b(R^d)$. Moreover, by Theorem 9.2.12,

$$\lim_{|h| \searrow 0} \sup_{n \geq 1} \int \big| p_{a_n, b_n}(s_n, x_n, y + h) - p_{a_n, b_n}(s_n, x_n, y) \big|\, dy = 0.$$

Thus, by Lemma 11.4.1, $p_{a_n, b_n}(s_n, x_n, \cdot) \to p_{a, b}(s_0, x_0, \cdot)$ in $L^1(R^d)$. Since, by Theorem 9.2.6, we also know that $\sup_{n \geq 1} \int |p_{a_n, b_n}(s_n, x_n, y)|^p\, dy$ and $\int |p_{a, b}(s_0, x_0, y)|^p\, dy$ are finite for all $1 \leq p < \infty$, an easy application of Hölder's inequality completes the proof. $\square$

Finally, we give the localized version of the preceding. Recall the definition of $\mathcal{A}_d^{loc}(\lambda_R, \Lambda_R, \delta_R(\cdot))$ given in Section 11.3, and denote by $\mathcal{H}_d^{loc}(\lambda_R, \Lambda_R, \delta_R(\cdot))$ the subset of $\mathcal{A}_d^{loc}(\lambda_R, \Lambda_R, \delta_R(\cdot))$ consisting of time independent elements. It is easy to see from Corollary 10.1.5 that is $\{a, b\} \in \mathcal{H}_d^{loc}(\lambda_R, \Lambda_R, \delta_R(\cdot))$, then the associated family $\{P_{s, x}^{a, b} : (s, x) \in [0, \infty) \times R^d\}$ of solutions to the martingale problems forms a strongly Feller continuous, time-homogeneous Markov family. Moreover, by Corollary 10.1.5, we know that the time-homogeneous transition probability function $P_{a, b}(s, x, \cdot)$ admits a density $p_{a, b}(s, x, y)$ for $s > 0$ such that $p_{a, b}(s, x, \cdot) \in L^{p, loc}(R^d)$ for all $1 \leq p < \infty$. Finally, for any $1 \leq p < \infty$, $0 \leq r < R$, and $\delta > 0$,

$$(4.1) \qquad \sup_{\substack{s \geq \delta \\ |x| \leq r}} \quad \sup_{\{a, b\} \in \mathcal{H}_d^{loc}(\lambda_R, \Lambda_R, \delta_R(\cdot))} \int_{B(0, R)} |p_{a, b}(s, x, y)|^p\, dy < \infty.$$

11.4.3 Theorem. *For each $n \geq 1$, let $\{a_n, b_n\} \in \mathcal{H}_d^{loc}(\lambda_R, \Lambda_R, \delta_R(\cdot))$ and suppose that there is an $\{a, b\} \in \mathcal{H}_d^{loc}(\lambda_R, \Lambda_R, \delta_R(\cdot))$ such that*

$$\int a^{ij}(x)\varphi(x)\, dx = \lim_{n \to \infty} \int a_n^{ij}(x)\varphi(x)\, dx, \qquad 1 \leq i, j \leq d,$$

and

$$\int b^i(x)\varphi(x)\, dx = \lim_{n \to \infty} \int b_n^i(x)\varphi(x)\, dx, \qquad 1 \leq i \leq d,$$

for all $\varphi \in C_0^\infty(R^d)$. If $\{(s_n, x_n)\}_1^\infty \subseteq (0, \infty) \times R^d$ and $(s_n, x_n) \to (s_0, x_0) \in (0, \infty) \times R^d$, then $p_{a_n, b_n}(s_n, x_n, \cdot) \to p_{a, b}(s_0, x_0, \cdot)$ in $L^1(R^d)$ and in $L^{p, loc}(R^d)$ for each $1 < p < \infty$.

Proof. In view of (4.1), it suffices to prove that $p_{a_n, b_n}(s_n, x_n, \cdot) \to p_{a, b}(s_0, x_0, \cdot)$ in $L^1(R^d)$. We will do this by reducing this case to the one just handled in Theorem 11.4.2. For $k \geq 1$, choose $\eta_k \in C_0^\infty(R^d)$ so that $0 \leq \eta_k \leq 1$ and $\eta_k \equiv 1$ on $B(0, k)$. Set

$$a_{n, k} = \eta_k a_{n, k} + (1 - \eta_k)I$$

and

$$b_{n, k} = \eta_k b_{n, k}.$$

Clearly, for each $k \geq 1$, Theorem 11.4.2 applies to $\{\{a_{n, k}, b_{n, k}\}: n \geq 1\}$ and

$$a_k = \eta_k a + (1 - \eta_k)I$$

and

$$b_k = \eta_k b.$$

Thus, the proof will be complete once we show that

$$\lim_{k \nearrow \infty} \sup_{n \geq 1} \int \left| p_{a_{n, k}, b_{n, k}}(s_n, x_n, y) - p_{a_n, b_n}(s_n, x_n, y) \right| dy = 0.$$

But

$$\int \left| p_{a_{n, k}, b_{n, k}}(s_n, x_n, y) - p_{a_n, b_n}(s_n, x_n, y) \right| dy \leq 2 P_{x_n}^{a_n, b_n}(\tau_k \leq s_n),$$

where $\tau_k = \inf\{t \geq 0: |x(t)| \geq k\}$. Since $P_{x_n}^{a_n, b_n} \to P_x^{a, b}$ and

$$\lim_{k \nearrow \infty} P_x^{a, b}(\tau_k \leq \sup_{n \geq 1} s_n) = 0,$$

it is easy to conclude that $\lim_{k \nearrow \infty} \sup_{n \leq 1} P_{s_n, x_n}^{a_n, b_n}(\tau_k \leq s_n) = 0$. $\quad\square$

11.5. Exercises

11.5.1. Let $a: [0, \infty) \times R^d \to S_d$ and $b: [0, \infty) \times R^d \to R^d$ be locally bounded measurable coefficients. Assume that the martingale problem for a, b is well-posed in that for each starting point (s, x) there is exactly one solution to the martingale problem corresponding a, b starting from (s, x). Let $(s^0, x^0) \in [0, \infty) \times R^d$ be given and let G be any open set in $[0, \infty) \times R^d$ containing (s^0, x^0). Define $\tau = \{\inf t \geq s^0 : (t, x(t)) \notin G\}$. If P is any measure on $(\Omega, \mathcal{M})$ such that $P[x(t) = x^0$ for $0 \leq t \leq s^0] = 1$ and

$$f(x(\tau \wedge t)) - \int_{s^0}^{\tau \wedge t} (L_u f)(x(u))\, du$$

is a $(\Omega, \mathcal{M}_t, P)$ martingale for $t \geq s^0$, then show that $P = P_{s^0, x^0}$ on $\mathcal{M}_\tau$ where P_{s^0, x^0} is the unique solution to the martingale problem for a, b starting from (s^0, x^0).

11.5.2. Let $a: R^d \to S_d$ and $b: R^d \to R^d$ be locally bounded measurable coefficients such that the martingale problem for the time-homogeneous coefficients a, b is well posed and admits a family $\{P_x\}$ of solutions starting from x which forms a continuous family, i.e. the mapping $x \to P_x$ is weakly continuous. If $F \in \mathcal{M}_{0+} = \bigcap_{t>0} \mathcal{M}_t$ then show that for each $x \in R^d$, $P_x(F) = 0$ or 1. This property is known as Blumenthal's zero-one law, and can be established as follows. Let ψ be a bounded continuous map from $(R^d)^N$ into R. Let $t_N > t_{N-1} \cdots > t_1 > 0$. Consider $\Phi(\omega) = \psi(x(t_1), \ldots, x(t_N))$ mapping $\Omega \to R$. For small ε, since $F \in \mathcal{M}_\varepsilon$, the Markov property implies

$$\int_F \Phi(\omega) dP_x = \int_F E^{P_{x(\varepsilon)}}[\psi(x(t_1 - \varepsilon), \ldots, x(t_N - \varepsilon))] dP_x \,.$$

Fixing $t_1, \ldots, t_N$ and letting $\varepsilon \to 0$ conclude that

$$\int_F \Phi(\omega)\, dP_x = P_x(F) E^{P_x}[\Phi(\omega)].$$

Now conclude that F is independent of itself and therefore $P_x(F)$ is 0 or 1.

11.5.3. Let $a: R^d \to S_d$ and $b: R^d \to R^d$ be bounded and continuous. Assume that $a(x)$ is positive definite for each $x \in R^d$. We then know that there is a Feller continuous family $\{P_x : x \in R^d\}$ of solutions. Let $G \subset R^d$ be a smooth region in R^d and $x^0 \in \partial G$, a point on the boundary ∂G of G. Let $\tau'(\omega) = \{\inf t \geq 0 : x(t, \omega) \notin \bar{G}\}$. Show that $P_{x^0}[\tau(\omega) = 0] = 1$. Hint: Let $\varphi(x)$ be the defining function, i.e. $\partial G = \{x : \varphi(x) = 0\}$, $\varphi > 0$ on $\bar{G}^c$ and $(\nabla \varphi)(x^0) \neq 0$. Consider the map $\Omega_d \to \Omega_{d+1}$ defined by $y_j(t) = (x_j(\varepsilon t) - x_j^0)/\sqrt{\varepsilon}$ for $1 \leq j \leq d$ and $y_0(t) = \varphi(x(\varepsilon t))/\sqrt{\varepsilon}$. Show that the process induced from P_{x^0}, is a diffusion and compute its coefficients $\tilde{a}_\varepsilon^{ij}(y)$ and $\tilde{b}_\varepsilon^j(y)$. Let $\varepsilon \to 0$ and conclude that the induced process converges to a suitable constant coefficient diffusion. It will now follow that $\lim_{t\to 0} P_{x_0}[\varphi(x(t)) > 0] = \frac{1}{2}$. Use Exercise 11.5.2 to conclude that $P_{x_0}[\tau'(\omega) = 0] = 1$.

11.5.4. Using the same notation as in Exercise 11.5.3, let $\tau(\omega) = \inf[t \geq 0 : x(t) \notin G]$ and conclude that for each $x \in G$, $P_x[\tau(\omega) < \infty, \tau'(\omega) > \tau(\omega)] = 0$. Hint: Use $P_x[\tau(\omega) < \infty, \tau'(\omega) > \tau(\omega)] = E^{P_x}[P_{x(\tau)}[\tau'(\omega) > 0], \tau(\omega) < \infty]$.

11.5.5. Using the same notation, assume in addition that G is bounded. By Exercise 10.3.1, $\tau(\omega)$ is finite almost surely with respect to P_x for each $x \in G$. Use Exercise 11.5.4 to conclude that $\tau(\omega)$ is almost surely continuous with respect to P_x. Hint: Show that $\tau(\omega)$ is lower semi-continuous but $\tau'(\omega)$ is upper semi-continuous.

11.5.6. If f is a bounded continuous function on ∂G then $u(x) = E^{P_x}[f(x(\tau))]$ is the solution to Dirichlet's problem, i.e., $Lu = 0$ in G (in some generalized sense if L does not have smooth coefficients) and $u = f$ on ∂G. Under the same assumptions as in Exercise 11.5.5, prove that $u(x)$ is continuous in x on $\bar{G}$. Also, if $P_x^{(n)}$ are diffusions approaching P_x in the sense that $P_x^{(n)} \to P_x$ weakly as $n \to \infty$, then $\lim_{n\to\infty} E^{P_x^{(n)}}[f(x(\tau))] = E^{P_x}[f(x(\tau))]$.

Chapter 12

The Non-unique Case

12.0 Introduction

The last few chapters have been devoted to finding out when the martingale problem is well-posed and what can be said when it is. In this chapter, we are going to examine whether anything useful can still be proved when we drop the assumption of uniqueness. In particular, we want to show that even in this case it may be possible to "select" a family of solutions in such a way that the strong Markov property prevails. We also want to describe the structure of the set of solutions associated with given coefficients and a given initial condition.

The idea behind our procedure for selecting strongly Markovian versions is the following. Given coefficients a and b, let $\mathscr{C}(s, x)$ denote the set of solutions to the martingale problem for a and b starting from (s, x). Assuming, as we will be, that a and b are bounded and continuous, it is easy to check that $\mathscr{C}(s, x)$ is a (weakly) compact, convex set in $M(\Omega)$, the set of probability measures on Ω. Thus, when we try to make a nice selection, it is natural to attempt isolating a unique point from each $\mathscr{C}(s, x)$ by insisting that it be that element of $\mathscr{C}(s, x)$ which is characterized by some extremal property. Moreover, since we would like the resulting selection to fit together in a Markovian way, the extremal property that we choose should be preserved under conditioning. Finally, we would like to make the selection as well-behaved as possible with respect to changes in (s, x), and so the extremal property should be one which behaves well under weak convergence. That such a program exists and yields the desired results was realized by N. Krylov [1973]. Our treatment does not differ substantially from his.

As for the structure of $\mathscr{C}(s, x)$, what we want to show is that every element of $\mathscr{C}(s, x)$ can be obtained by "piecing together" elements of $\mathscr{C}(s, x)$ which are members of strongly Markovian selections. What we have in mind is made precise in Section 12.3.

One final comment. We are going to assume that the coefficients a and b are independent of time. However, this is not really a restriction, since by considering the "time-space" process, one can easily handle coefficients which are time-dependent as if they were time-independent. See Exercise 12.4.1 below for more details on this point.

12.1. Existence of Measurable Choices

Before we can prove the main theorems of this chapter we need some facts about the space of compact subsets of a separable metric space. This section is a digression in that direction. Among other things, we will prove some results about the possibility of choosing measurable selections from set-valued maps.

Let X be a separable metric space with a metric d. We denote by $\text{comp}(X)$ the space of all compact subsets of X and define a metric $\rho(K_1, K_2)$ between two points K_1, K_2 of $\text{comp}(X)$ by

$$(1.1) \qquad \rho(K_1, K_2) = \inf[\varepsilon > 0 \colon K_1 \subseteq K_2^\varepsilon \quad \text{and} \quad K_2 \subseteq K_1^\varepsilon].$$

Here, for any set $A \in X$, A^ε denotes $A^\varepsilon = \{y \colon d(x, y) < \varepsilon \text{ for some } x \in A\}$; in other words, A^ε is the sphere around A of radius ε. It is easily verified that ρ is a metric on $\text{comp}(X)$. Moreover if two points x, y of X are viewed as the single point sets $\{x\}$ and $\{y\}$ in $\text{comp}(X)$, then

$$\rho(\{x\}, \{y\}) = d(x, y).$$

That is to say, X is embedded isometrically in $\text{comp}(X)$. We will state a series of lemmas about the metric space $(\text{comp}(X), \rho)$.

12.1.1 Lemma. $(\text{comp}(X), \rho)$ *is a separable metric space.*

Proof. Let $X_0 \subset X$ be a countable dense subset of X. One can check easily that the class of all finite subsets of X_0 is a countable dense subset of $\text{comp}(X)$. $\quad\square$

12.1.2 Lemma. *If $C \subseteq X$ is closed in X then $\{K \colon K \subseteq C\}$ is closed in $\text{comp}(X)$. If $G \subseteq X$ is open in X then $\{K \colon K \subseteq G\}$ is open in $\text{comp}(X)$.*

Proof. This is obvious and is left as an exercise. $\quad\square$

For any set $A \subseteq X$, we denote by $J(A)$ the set $\{K \colon K \subseteq A\}$ in $\text{comp}(X)$.

12.1.3 Lemma. *If $F \subseteq X$ is any finite set, then the ball $S(F, \varepsilon)$ of radius ε around F in $\text{comp}(X)$ is given by*

$$S(F, \varepsilon) = J(F^\varepsilon) \cap \left\{ \bigcap_{x \in F} [J(B(x, \varepsilon)^c)]^c \right\}.$$

Here $B(x, \varepsilon)$ is the ball in X of radius ε around x.

Proof. Clearly

$$K \in S(F, \varepsilon) \Leftrightarrow \rho(K, F) < \varepsilon$$
$$\Leftrightarrow K \subseteq F^{\varepsilon'} \quad \text{and} \quad F \subseteq K^{\varepsilon'} \quad \text{for some} \quad \varepsilon' < \varepsilon$$
$$\Leftrightarrow K \subseteq F^{\varepsilon} \quad \text{and} \quad F \subseteq K^{\varepsilon}.$$

Now $K \subset F^{\varepsilon}$ if and only if $K \in J(F^{\varepsilon})$. On the other hand,

$$F \subseteq K^{\varepsilon} \Leftrightarrow x \in K^{\varepsilon} \quad \text{for each} \quad x \in F$$
$$\Leftrightarrow K \cap B(x, \varepsilon) \neq \phi \quad \text{for each} \quad x \in F$$
$$\Leftrightarrow K \notin J(B(x, \varepsilon)^c) \quad \text{for each} \quad x \in F$$
$$\Leftrightarrow K \in [J(B(x, \varepsilon)^c)]^c \quad \text{for each} \quad x \in F$$
$$\Leftrightarrow K \in \bigcap_{x \in F} [J(B(x, \varepsilon)^c)]^c. \quad \square$$

12.1.4 Lemma. *The class of sets of the form $J(A)$ where A runs over all the sets A in X which are either closed or open in X, generates the Borel σ-field of comp(X).*

Proof. It is clear from Lemma 12.1.3 that the σ-field generated by $J(A)$ as A runs over the class of sets that are either open or closed in X contains sets of the form $S(F, \varepsilon)$ for any $\varepsilon > 0$ and any finite set F. The class of sets of the form $S(F, \varepsilon)$ as F runs over finite subsets of X and ε runs over all positive numbers is clearly a basis for the topology of comp(X). Any σ-field that contains a basis in a separable metric space must contain the Borel σ-field. $\square$

12.1.5 Lemma. *If $G \subseteq X$ is open, then $J(G)$ is a countable union of the form $\bigcup_n J(C_n)$ with C_n closed in X. If $C \subseteq X$ is closed then $J(C)$ is a countable intersection of the form $\bigcap_n J(G_n)$ with G_n open in X. Therefore, either one of the classes $\{J(A): A$ open in $X\}$ or $\{J(A): A$ closed in $X\}$ generates the Borel σ-field of comp(X).*

Proof. If $G \subseteq X$ is open we define

$$C_n = \left\{ x: B\left(x, \frac{1}{n}\right) \subseteq G \right\}.$$

Then each C_n is closed and $\bigcup_n C_n = G$. Moreover, since K is compact, it is easy to check that, if $K \subset G$, then $K \subset C_n$ for some n. Therefore

$$J(G) = \bigcap J(C_n).$$

On the other hand, if we write

$$C = \bigcap_n G_n \quad \text{with} \quad G_n \quad \text{open in} \quad X,$$

then

$$J(C) = \bigcap_n J(G_n). \quad \square$$

12.1.6 Lemma. *Let $A \subset X$ be a closed subset of X. Let $\mathscr{E} = \{K: K \cap A \neq \varnothing\}$. Then $\mathscr{E}$ is a closed subset of $comp(X)$ and the map $K \to K \cap A$ is a Borel map of $\mathscr{E}$ into $comp(X)$.*

Proof. Since $\mathscr{E} = [J(A^c)]^c$, it is closed. In order to prove the measurability of the map $K \to K \cap A$, it is enough to show that for any $G \subseteq X$ which is open in X, $\{K: K \cap A \subseteq G\}$ is a Borel subset of $comp(X)$. Clearly $K \cap A \subseteq G$ if and only if $K \subseteq G \cup A^c$. Therefore $\{K: K \cap A \subseteq G\}$ is the same as $\{K: K \subseteq G \cup A^c\} = J(G \cup A^c)$, and $J(G \cup A^c)$ is open in $comp(X)$. $\quad \square$

12.1.7 Lemma. *Let $f(x)$ be a real valued upper semi-continuous function on X. For $K \in comp(X)$, set*

$$f_K = \sup_{x \in K} f(x),$$

and define $f : comp(X) \to comp(X)$ by

$$f(K) = \{y: y \in K \text{ and } f(y) = f_K\}.$$

Then the maps $K \to f_K$ and $K \to f(K)$ are Borel maps of $comp(X)$ into R and $comp(X)$, respectively.

Proof.
$$\{K: f_K < l\} = \left\{K: \sup_{x \in K} f(x) < l\right\}$$

$$= \{K: K \subseteq \{x: f(x) < l\}\}$$

$$= J(G_l)$$

where $G_l = \{x: f(x) < l\}$ is open in X. $J(G_l)$ is therefore open in $comp(X)$ and therefore $K \to f_K$ is a Borel map. On the other hand, if $G \subseteq X$ is open, then

$$\{K: f(K) \subseteq G\} = \{K: K \subseteq G\} \cup \left\{K: K \cap G^c \neq \varnothing, \sup_{x \in K \cap G^c} f(x) < \sup_{x \in K} f(x)\right\}$$

$$= J(G) \cup \left[\{K: K \cap G^c \neq \varnothing\} \cap \left\{K: f_{K \cap G^c} < f_K\right\}\right].$$

By Lemma 12.1.6, $K \to K \cap G^c$ is a Borel map of comp(X) into itself on $\{K: K \cap G^c \neq \varnothing\}$. From the first part of the lemma, $K \to f_K$ is a Borel map of comp(X) into R. Therefore $\{K: f(K) \subseteq G\}$ is a Borel set in comp(X) and we are done. $\quad\square$

12.1.8 Lemma. *Let Y be a metric space and $\mathscr{B}$ its Borel σ-field. Let $y \to K_y$ be a map of Y into comp(X) for some separable metric space X. Suppose for any sequence $y_n \to y$ and $x_n \in K_{y_n}$, it is true that x_n has a limit point x in K_y. Then the map $y \to K_y$ is a Borel map of Y into comp(X).*

Proof. Let $G \subseteq X$ be open. We will show that $\{y: K_y \subseteq G\}$ is open in Y; or equivalently that $\{y: K_y \cap G^c \neq \varnothing\}$ is closed in Y. Let $K_{y_n} \cap G^c \neq \varnothing$ for any n and $y_n \to y$. Choose $x_n \in K_{y_n} \cap G^c$. By our assumption, x_n will have a limit point x which will be in K_y. But, since G^c is closed, $x \in G^c$ as well. Therefore $K_y \cap G^c$ is non-empty and we are done. $\quad\square$

12.1.9 Lemma. *Let $(E, \mathscr{F})$ be a measurable space. Let $q \to h(q)$ and $q \to K_q$ be measurable maps of E into some separable metric space X and comp(X), respectively. Then the set $\{q: h(q) \in K_q\}$ is a measurable subset of E.*

Proof. It is enough to verify that the set of pairs (x, K) in $X \times$ comp(X) with $x \in K$ is a Borel subset of the product space. But it is in fact closed, because if $x_n \to x$ and $K_n \to K$ with $x_n \in K_n$, it follows from the definition of $\rho(K, K_n)$ that $d(x_n, K) \to 0$. Therefore $x \in K$. $\quad\square$

We will complete this section with a theorem which will be important for us in the rest of this chapter.

12.1.10 Theorem. *Let $(E, \mathscr{F})$ be any measurable space and $q \to K_q$ a measurable map of E into comp(X), where X is some separable metric space. Then there is a measurable map $q \to h(q)$ of E into X such that $h(q) \in K_q$ for every $q \in E$.*

Proof. Let $f_1, f_2, \dots, f_n, \dots$ be a countable sequence of continuous functions on X which separate points in X. Take $K_q^{(0)} = K_q$ and define inductively

$$K_q^{(n)} = f_n(K_q^{(n-1)}).$$

(The notation here is that of Lemma 12.1.7.) By Lemma 12.1.7, for each $n, q \to K_q^{(n)}$ is a measurable map of E into comp(X). Moreover for each $q, K_q^{(n)}$ is a non-increasing sequence of non-empty compact subsets of X. Therefore $K_q^\infty \equiv \bigcap_n K_q^{(n)}$ is non-empty and one can verify that K_q^∞ is the limit of $K_q^{(n)}$ as $n \to \infty$, in the space comp(X). Therefore the map $q \to K_q^\infty$ is also a measurable map of E into comp(X). We will complete the prooof by showing that for each q, K_q^∞ consists of exactly one point of X. If we denote that point by $h(q)$, then $q \to h(q)$ will be a measurable map of E into X (remember that X is embedded isometrically in comp(X) and $h(q) \in K_q$ for each q).

To see that K_q^∞ consists of exactly one point, we reason as follows. By our construction, $f_n(x)$ is constant on $K_q^{(n)}$. Since K_q^∞ is contained in $K_q^{(n)}$ for every n, all the functions $f_n(x)$ are constant on K_q^∞. Since $f_1, f_2, \ldots$ separate points in X, it follows that K_q^∞ can have at most one point for each q. But K_q^∞ is non-empty for each q and must therefore consist of exactly one point. This completes the proof. □

12.2. Markov Selections

We are now going to adapt the preceding results to the situation arising in the study of the martingale problem.

For each $(s, x) \in [0, \infty) \times R^d$, suppose that $\mathscr{C}(s, x)$ is a non-empty compact subset of $M(\Omega)$, the space of probability measures on Ω, and assume that the family $\{\mathscr{C}(s, x) : (s, x) \in [0, \infty) \times R^d\}$ satisfies the following conditions:

   (a) $(s, x) \to \mathscr{C}(s, x)$ is measurable on $[0, \infty) \times R^d$ into comp$(M(\Omega))$,
   (b) $P \in \mathscr{C}(0, x)$ if and only if $P \circ \Phi_s^{-1} \in \mathscr{C}(s, x)$ for all $s \geq 0$, where $\Phi_s : \Omega \to \Omega$ is defined by

(2.1) $$x(t, \Phi_s \omega) = x((t - s) \vee 0, \omega); \quad t \geq 0,$$

   (c) if $P \in \mathscr{C}(0, x)$, $\tau : \Omega \to [0, \infty)$ is a finite stopping time, and $\{P_\omega\}$ is a r.c.p.d. of $P | \mathscr{M}_\tau$, then there is a P-null set $N \in \mathscr{M}_\tau$ such that $\delta_{x(\tau(\omega), \omega)} \otimes_{\tau(\omega)} P_\omega \in \mathscr{C}(\tau(\omega), x(\tau(\omega), \omega))$ for each $\omega \notin N$,
   (d) if $P \in \mathscr{C}(0, x)$, τ is a finite stopping time, and $\omega \to Q_\omega$ is an $\mathscr{M}_\tau$-measurable map such that $\delta_{x(\tau(\omega), \omega)} \otimes_{\tau(\omega)} Q_\omega \in \mathscr{C}(\tau(\omega), x(\tau(\omega), \omega))$ for all ω, then $P \otimes_\tau Q. \in \mathscr{C}(0, x)$.

12.2.1 Lemma. *Suppose that $a: R^d \to S_d$ and $b: R^d \to R^d$ are bounded continuous functions. For each $(s, x) \in [0, \infty) \times R^d$, let $\mathscr{C}(s, x)$ be the set of all solutions to the martingale problem for a and b starting from (s, x). Then each $\mathscr{C}(s, x)$ is a non-empty compact subset of $M(\Omega)$ and the family $\{\mathscr{C}(s, x) \in [0, \infty) \times R^d\}$ satisfies conditions (a) through (d) of (2.1).*

Proof. The fact that $\mathscr{C}(s, x)$ is a non-empty compact subset of $M(\Omega)$ is an immediate consequence of Theorem 6.1.7, the compactness criterion given in Theorem 1.4.6, plus the observation that the continuity of a and b guarantee that $\mathscr{C}(s, x)$ is closed. Condition (a) follows from Lemma 12.1.8 and the fact that if $(s_n, x_n) \to (s, x)$, then $\bigcup_1^\infty \mathscr{C}(s_n, x_n)$ is relatively compact (cf. Theorem 1.4.6) and any convergent subsequence of a sequence $\{P_n\}_1^\infty$ satisfying $P_n \in \mathscr{C}(s_n, x_n)$, $n \geq 1$, converges to an element of $\mathscr{C}(s, x)$. The condition (b) is just a restatement of the observation on which the proof of Lemma 6.5.1 turns. Finally, conditions (c) and (d) are proved by applying Theorems 6.2.1 and 6.1.2, respectively. □

12.2.2 Lemma. *Let* $\{\mathscr{C}(s, x): (s, x) \in [0, \infty) \times R^d\}$ *satisfy* (a) *through* (d) *of* (2.1). *Given* $\lambda > 0$ *and a bounded upper semi-continuous function* $f: R^d \to R$, *define*

$$u(s, x) = \sup_{P \in \mathscr{C}(s, x)} E^P \left[\int_0^\infty e^{-\lambda t} f(x(t + s)) \, dt \right].$$

Then $u(s, x) = u(0, x)$ *for all* $s \geq 0$ *and* $x \in R^d$. *For each* (s, x), *set*

$$\mathscr{C}'(s, x) = \left| P \in \mathscr{C}(s, x): E^P \left[\int_0^\infty e^{-\lambda t} f(x(t + s)) \, dt \right] = u(s, x) \right|.$$

Then $\mathscr{C}'(s, x) \in comp(M(\Omega))$ *for all* (s, x) *and the family* $\{\mathscr{C}'(s, x): (s, x) \in [0, \infty) \times R^d\}$ *again satisfies* (a)–(d) *of* (2.1).

Proof. That $u(s, x) = u(0, x)$ is obvious from (b) for $\{\mathscr{C}(s, x): (s, x) \in [0, \infty) \times R^d\}$. Each $\mathscr{C}'(s, x)$ is in $comp(M(\Omega))$ and $(s, x) \to \mathscr{C}'(s, x)$ is measurable because of Lemma 12.1.7 and the easily verified fact that

$$(s, P) \to E^P \left[\int_0^\infty e^{-\lambda t} f(x(t + s)) \, dt \right]$$

is upper-semicontinuous. Condition (b) for $\{\mathscr{C}'(s, x): (s, x) \in [0, \infty) \times R^d\}$ is immediate from (b) for $\{\mathscr{C}(s, x): (s, x) \in [0, \infty) \times R^d\}$ plus the identity $u(\cdot, x) \equiv u(0, x)$.

To prove that (c) is satisfied, let $P \in \mathscr{C}'(0, x^0)$ and a stopping time τ be given. Suppose $\{P_\omega\}$ is an r.c.p.d. of $P|\mathscr{M}_\tau$ and define

$$N = \{\omega: P_\omega \notin \mathscr{C}(\tau(\omega), x(\tau(\omega), \omega))\}$$

and

$$A = \{\omega \in N^c: P_\omega \notin \mathscr{C}'(\tau(\omega), x(\tau(\omega), \omega))\}.$$

By (c) for $\{\mathscr{C}(s, x): (s, x) \in [0, \infty) \times R^d\}$, $N \in \mathscr{M}_\tau$ and $P(N) = 0$. Moreover, $N \cup A = \{\omega: P_\omega \notin \mathscr{C}'(\tau(\omega), x(\tau(\omega), \omega))\}$, and therefore, by Lemma 12.1.9, $N \cup A \in \mathscr{M}_\tau$. Thus $A \in \mathscr{M}_\tau$. We want to show that $P(A) = 0$. To this end, use Theorem 12.1.10 to choose a measurable map $(s, x) \to R_{s, x}$ so that $R_{s, x} \in \mathscr{C}'(s, x)$ for all (s, x). Then $\omega \to R_\omega$ given by

$$R_\omega = \delta_\omega \otimes_{\tau(\omega)} R_{\tau(\omega), \, x(\tau(\omega), \, \omega)}$$

is $\mathscr{M}_\tau$-measurable. Define

$$Q_\omega = \begin{vmatrix} R_\omega & \text{if} & \omega \in N \cup A \\ P_\omega & \text{if} & \omega \notin N \cup A. \end{vmatrix}$$

By (d) for $\{\mathscr{C}(s, x): (s, x) \in [0, \infty) \times R^d\}$, $Q \equiv P \otimes_\tau Q. \in \mathscr{C}(0, x^0)$. Thus

$$
\begin{aligned}
u(0, x^0) &\geq E^Q \left[\int_0^\infty e^{-\lambda t} f(x(t))\, dt \right] \\
&= E^P \left[\int_0^\tau e^{-\lambda t} f(x(t))\, dt \right] \\
&\quad + E^P \left[e^{-\lambda \tau} E^Q \cdot \left[\int_0^\infty e^{-\lambda t} f(x(t + \tau))\, dt \right] \right] \\
&= E^P \left[\int_0^\tau e^{-\lambda t} f(x(t))\, dt \right] \\
&\quad + E^P \left[e^{-\lambda \tau} E^P \cdot \left[\int_0^\infty e^{-\lambda t} f(x(t + \tau))\, dt \right] \right] \\
&\quad + E^P \Big\{ e^{-\lambda \tau} \Big(E^R \cdot \left[\int_0^\infty e^{-\lambda t} f(x(t + \tau))\, dt \right] \\
&\quad - E^P \cdot \left[\int_0^\infty e^{-\lambda t} f(x(t + \tau))\, dt \right] \Big), NUA \Big\} \\
&= u(0, x^0) \\
&\quad + E^P \left[e^{-\lambda \tau} \Big(u(\tau, x(\tau)) - E^P \cdot \left[\int_0^\infty e^{-\lambda t} f(x(t + \tau))\, dt \right] \Big), A \right],
\end{aligned}
$$

since

$$
\begin{aligned}
u(0, x^0) &= E^P \left[\int_0^\infty e^{-\lambda t} f(x(t))\, dt \right] \\
&= E^P \left[\int_0^\tau e^{-\lambda t} f(x(t))\, dt \right] \\
&\quad + E^P \left[e^{-\lambda \tau} E^P \cdot \left[\int_0^\infty e^{-\lambda t} f(x(t + \tau))\, dt \right] \right]
\end{aligned}
$$

and $P(N) = 0$. Thus

$$
E^P \left[e^{-\lambda \tau} \Big(u(\tau, x(\tau)) - E^P \cdot \left[\int_0^\infty e^{-\lambda t} f(x(t + \tau))\, dt \right] \Big), A \right] \leq 0.
$$

On the other hand,

$$
E^{P_\omega} \left[\int_0^\infty e^{-\lambda t} f(x(t + \tau))\, dt \right] < u(\tau(\omega), x(\tau(\omega), \omega))
$$

for $\omega \in A$, and so we conclude that $P(A) = 0$.

Finally, we must show that $\{\mathscr{C}'(s, x): (s, x) \in [0, \infty) \times R^d\}$ satisfies (d) of (2.1). For this purpose, let $P \in \mathscr{C}'(0, x^0)$ and suppose $\omega \to Q_\omega$ is an $\mathcal{M}_\tau$-measurable map such that $\delta_{x(\tau(\omega),\, \omega)} \otimes_{\tau(\omega)} Q_\omega \in \mathscr{C}'(\tau(\omega), x(\tau(\omega), \omega))$ for all ω. Set $Q = P \otimes_\tau Q_\cdot$. By (d) for $\{\mathscr{C}(s, x): (s, x) \in [0, \infty) \times R^d\}$, $Q \in \mathscr{C}(0, x^0)$. Moreover,

$$
\begin{aligned}
u(0, x^0) &\geq E^Q \left[\int_0^\infty e^{-\lambda t} f(x(t)) \, dt \right] \\
&= E^P \left[\int_0^\tau e^{-\lambda t} f(x(t)) \, dt \right] \\
&\quad + E^P \left[e^{-\lambda \tau} E^Q \cdot \left[\int_0^\infty e^{-\lambda t} f(x(t + \tau)) \, dt \right] \right] \\
&= E^P \left[\int_0^\tau e^{-\lambda t} f(x(t)) \, dt \right] + E^P [e^{-\lambda \tau} u(\tau, x(\tau))] \\
&\geq E^P \left[\int_0^\tau e^{-\lambda t} f(x(t)) \, dt \right] \\
&\quad + E^P \left[e^{-\lambda \tau} E^P \cdot \left[\int_0^\infty e^{-\lambda t} f(x(t + \tau)) \, dt \right] \right] \\
&= E^P \left[\int_0^\infty e^{-\lambda t} f(x(t)) \, dt \right] \\
&= u(0, x^0),
\end{aligned}
$$

since, by (c) for $\{\mathscr{C}(s, x): (s, x) \in [0, \infty) \times R^d\}$,

$$
u(\tau, x(\tau)) \geq E^P \cdot \left[\int_0^\infty e^{-\lambda t} f(x(t + \tau)) \, dt \right] \qquad \text{(a.s., } P).
$$

We have therefore shown that $Q \in \mathscr{C}'(0, x^0)$. $\square$

We are now ready to prove the main theorem of this section.

12.2.3 Theorem. *Let* $a: R^d \to S_d$ *and* $b: R^d \to R^d$ *be bounded continuous functions. Given* $\lambda > 0$ *and a bounded upper semi-continuous function* $f: R^d \to R^1$, *there is a measurable map* $(s, x) \to P_{s, x}$ *of* $[0, \infty) \times R^d$ *into* $M(\Omega)$ *such that*

(i) $P_{s, x}$ *solves the martingale problem for* a *and* b *starting from* (s, x),
(ii) $P_{s, x} = P_{0, x} \cdot \Phi_s^{-1}$, $s \geq 0$, *where* $\Phi_s: \Omega \to \Omega$ *is the map described in b) of (2.1)*,
(iii) *if* $\tau: \Omega \to [s, \infty)$ *is a stopping time and* $\{P_\omega\}$ *is a r.c.p.d. of* $P_{s, x} | \mathcal{M}_\tau$, *then there is a P-null set* $N \in \mathcal{M}_\tau$ *such that* $P_\omega = \delta_\omega \otimes_{\tau(\omega)} P_{\tau(\omega),\, x(\tau(\omega),\, \omega)}$ *for all* $\omega \notin N$,
(iv) *for each* (s, x), $E^{P_{s, x}} [\int_0^\infty e^{-\lambda t} f(x(t + s)) \, dt]$ *maximizes* $E^P [\int_0^\infty e^{-\lambda t} f(x(t + s)) \, dt]$ *as* P *varies over solutions to the martingale problem for* a *and* b *starting from* (s, x).

Proof. Let $\{\sigma_n\}_1^\infty$ be a dense subset of $(0, \infty)$ and $\{\varphi_n\}_1^\infty$ a dense subset of $C_0(R^d)$, and let $\{(\lambda_N, f_N): N \geq 1\}$ be an enumeration of $\{(\sigma_m, \varphi_n): m \geq 1 \text{ and } n \geq 1\}$. Set $\lambda_0 = \lambda$ and $f_0 = f$. Take $\mathscr{C}_0(s, x) = \{P: P \text{ solves the martingale problem for } a \text{ and } b \text{ starting from } (s, x)\}$ and

$$u_0(s, x) = \sup\left\{E^P\left[\int_0^\infty e^{-\lambda_0 t} f_0(x(t + s))\, dt\right]: P \in \mathscr{C}_0(s, x)\right\};$$

and define inductively:

$$\mathscr{C}_{N+1}(s, x) = \left\{P \in \mathscr{C}_N(s, x): E^P\left[\int_0^\infty e^{-\lambda_N t} f_N(x(t + s))\, dt\right] = u_N(s, x)\right\}$$

and

$$u_{N+1}(s, x) = \sup\left\{E^P\left[\int_0^\infty e^{-\lambda_{N+1} t} f_{N+1}(x(t + s))\, dt\right]: P \in \mathscr{C}_{N+1}(s, x)\right\}.$$

By Lemmas 12.2.1 and 12.2.2, for each $N \geq 0$, $\{\mathscr{C}_N(s, x): (s, x) \in [0, \infty) \times R^d\}$ is a subset of $\text{comp}(M(\Omega))$ which satisfies (a)–(d) of (2.1). Since $C_{N+1}(s, x) \subseteq \mathscr{C}_N(s, x)$, it is clear that $\mathscr{C}_\infty(s, x) = \bigcap_0^\infty \mathscr{C}_N(s, x) \in \text{comp}(M(\Omega))$ for all (s, x) and that $\{\mathscr{C}_\infty(s, x): (s, x) \in [0, \infty) \times R^d\}$ satisfies (a)–(d) of (2.1). Thus, if we can show that $\mathscr{C}_\infty(s, x)$ has at most one member for each (s, x), the theorem will be proved.

Suppose that $P, Q \in \mathscr{C}_\infty(s, x)$. Then, for all $m \geq 1$ and $n \geq 1$,

$$E^P\left[\int_0^\infty e^{-\lambda_m t} f_n(x(t + s))\, dt\right] = E^Q\left[\int_0^\infty e^{-\lambda_m t} f_n(x(t + s))\, dt\right].$$

Since $\{\lambda_m\}_1^\infty$ is dense in $(0, \infty)$, it follows from the uniqueness of the Laplace transform plus the continuity of $E^P[f_n(x(\cdot))]$ and $E^Q[f_n(x(\cdot))]$ that

$$E^P[f_n(x(t))] = E^Q[f_n(x(t))], \qquad t \geq 0,$$

for all $n \geq 1$. But $\{f_n\}_1^\infty$ is dense in $C_0(R^d)$ and so

$$E^P[f(x(t))] = E^Q[f(x(t))], \qquad t \geq 0,$$

for all $f \in B(R^d)$. We can now proceed in exactly the same way as we did in the proof of Theorem 6.2.3 to prove that each $\mathscr{C}_\infty(s, x)$ contains exactly one element. In fact, the proof here is easier since we know *a priori* that for each $t \geq 0$ and $f \in B(R^d)$ the function $u(s, x)$ such that $E^P[f(x(t))] = u(s, x)$ for all $P \in \mathscr{C}_\infty(s, x)$ is measurable. The details are left to the reader. $\square$

An interesting consequence of the preceding is the following.

12.2.4 Theorem. *Let* $a: R^d \to S_d$ *and* $b: R^d \to R^d$ *be bounded continuous functions. Then the martingale problem for* a *and* b *is well-posed if and only if there is exactly one strong Markov, time homogeneous measurable Markov family* $\{P_x : x \in R^d\}$ *such that for each* $x \in R^d$, P_x *is a solution to the martingale problem for* a *and* b *starting from* x.

Proof. We already know that the "only if" assertion is true. To prove the "if" assertion, note that if the martingale problem is not well-posed, then by Corollary 6.2.4 and Lemma 6.5.1 we can find an $x^0 \in R^d$, a $\lambda > 0$, an $f \in C_0(R^d)$, and solutions P and Q to the martingale problem for a and b starting from $(0, x^0)$ such that

$$E^P\left[\int_0^\infty e^{-\lambda t} f(x(t))\, dt\right] > E^Q\left[\int_0^\infty e^{-\lambda t} f(x(t))\, dt\right].$$

Applying Theorem 12.2.3 with this λ and f, we know that there is a strong Markov, time-homogeneous measurable family of solutions $\{P_x^{(1)} : x \in R^d\}$ such that

$$E^{P_{x^0}^{(1)}}\left[\int_0^\infty e^{-\lambda t} f(x(t))\, dt\right] \geq E^P\left[\int_0^\infty e^{-\lambda t} f(x(t))\, dt\right].$$

Applying Theorem 12.2.3 with this λ and $-f$, we get a second such family $\{P_x^{(2)} : x \in R^d\}$ such that

$$E^{P_{x^0}^{(2)}}\left[\int_0^\infty e^{-\lambda t} f(x(t))\, dt\right] \leq E^Q\left[\int_0^\infty e^{-\lambda t} f(x(t))\, dt\right].$$

In particular, $\{P_x^{(1)} : x \in R^d\}$ cannot equal $\{P_x^{(2)} : x \in R^d\}$, and so the proof is complete. $\square$

Less striking, but nonetheless useful for the purposes of the next section, is the following refinement of Theorem 12.2.3.

12.2.5 Theorem. *Let* $a: R^d \to S_d$ *and* $b: R^d \to R^d$ *be bounded continuous functions. Given* $\lambda > 0$ *and a bounded upper semi-continuous function* $f : R^N \times R^d \to R^1$, *there is a measurable map* $(s, z, x) \to P_{s,x}^z$ *of* $[0, \infty) \times R^N \times R^d$ *into* $M(\Omega)$ *such that for each* z:

(i) $P_{s,x}^z$ *solves the martingale problem for* a *and* b *starting from* (s, x),

(ii) $P_{s,x}^z = P_{0,x}^z \circ \Phi_s^{-1}$, $s \geq 0$, *where* $\Phi_s: \Omega \to \Omega$ *is the map described in* (b) *of* (2.1)

(iii) *if* $\tau: \Omega \to [s, \infty)$ *is a stopping time and* $\{P_\omega\}$ *is a r.c.p.d. of* $P_{s,x}^z \,|\, \mathcal{M}_\tau$, *then there is a* $P_{s,x}^z$-*null* $N \in \mathcal{M}_\tau$ *such that* $P_\omega = \delta_\omega \otimes_{\tau(\omega)} P_{\tau(\omega), x(\tau(\omega), \omega)}^z$ *for* $\omega \notin N$,

(iv) *for each* (s, x), $E^{P_{s,x}^z}[\int_0^\infty e^{-\lambda t} f(z, x(t + s)) dt)]$ *maximizes* $E^P[\int_0^\infty e^{-\lambda t} \times f(z, x(t + s)) dt]$ *as* P *varies over solutions to the martingale problem for* a *and* b *starting from* (s, x).

Proof. One can obtain this result as a consequence of Theorem 12.2.3 by the following trick. Define coefficients $\bar{a}: R^{N+d} \to S_{N+d}$ and $\bar{b}: R^{N+d} \to R^d$ by

$$\bar{a}^{ij}(\bar{x}) = \begin{cases} a^{ij}(x) & \text{if} \quad N+1 \le i, \; j \le N+d \\ 0 & \text{otherwise} \end{cases}$$

and

$$\bar{b}^i(\bar{x}) = \begin{cases} b^i(x) & \text{if} \quad N+1 \le i \le N+d \\ 0 & \text{otherwise,} \end{cases}$$

where $\bar{x} = (z, x) \in R^N \times R^d$. Clearly $\bar{a}$ and $\bar{b}$ satisfy the conditions of Theorem 12.4.3. Thus we can find a strong Markov, time-homogeneous, measurable family $\{\bar{P}_{s,\bar{x}} : (s, \bar{x}) \in [0, \infty) \times R^{N+d}\}$ of solutions for $\bar{a}$ and $\bar{b}$ such that $E^{s,\bar{x}} [\int_0^\infty e^{-\lambda t} f(\bar{x}(t + s)) dt]$ maximizes $E^P [\int_0^\infty e^{-\lambda t} f(\bar{x}(t + s)) dt]$ over all P which solve the martingale problem for $\bar{a}$ and $\bar{b}$ starting from $(s, \bar{x})$. But, for each $z \in R^N$, there is an obvious one-to-one correspondence between solutions for a and b starting from a given point (s, x) and those for $\bar{a}$ and $\bar{b}$ starting from $(s, (z, x))$. Using this correspondence, the proof is completed by simply setting $P^z_{s,x} = \bar{P}_{s,(z,x)}$. $\square$

12.3. Reconstruction of All Solutions

Let $a: R^d \to S_d$ and $b: R^d \to R^d$ be given bounded continuous functions. For each $x \in R^d$, let $\mathscr{C}(x)$ denote the set of all solutions to the martingale problem for a and b starting from x. We have just seen that there is at least one measurable selection $x \to P_x \in \mathscr{C}(x)$ such that the resulting family $\{P_x : x \in R^d\}$ forms a time-homogenious strong Markov process. We have also seen that there is exactly one such selection if and only if each $\mathscr{C}(x)$ contains only one element. The purpose of the present section is to refine these statements by showing that in a certain sense every element of $\mathscr{C}(x)$ can be obtained by "piecing together" homogeneous strong Markov selections. We now give a precise formulation.

Let $\mathscr{A}$ be an index set for the time-homogeneous strong Markov selections. That is, for each $\alpha \in \mathscr{A}$ there is exactly one time-homogeneous strong Markov selection $\{P^\alpha_x : x \in R^d\}$. A mapping φ from a measurable space $(E, \mathscr{F})$ into $\mathscr{A}$ will be said to be measurable if $(q, x) \to P^{\alpha(q)}_x$ is measurable on $(E \times R^d, \mathscr{F} \times \mathscr{B}_{R^d})$ into $M(\Omega)$. Given $x \in R^d$, $n \ge 1$, $0 < t_1 < \cdots < t_n$, $\alpha_0 \in \mathscr{A}$, and measurable maps $\alpha_j: (\Omega, \mathscr{M}_{t_j}) \to \mathscr{A}$, $1 \le j \le n$, consider the measure

$$(3.1) \qquad P^{\alpha_0}_x \otimes_{t_1} P^{\alpha_1(\cdot)}_{t_1, x(t_1, \cdot)} \otimes_{t_2} P^{\alpha_2(\cdot)}_{t_2, x(t_2, \cdot)} \otimes \cdots \otimes P^{\alpha_n(\cdot)}_{t_n, x(t_n, \cdot)},$$

where $P^\alpha_{t, y} = P^\alpha_y \circ \Phi_t^{-1}$ for any $\alpha \in \mathscr{A}$, $t \ge 0$, and $y \in R^d$ (cf. *b*) of (2.1) for the definition of Φ_t). By Theorem 6.1.2, the measure in (3.1) belongs to $\mathscr{C}(x)$. We will denote by $\mathscr{D}_x$ the set of all measures of the form given in (3.1) generated by $n \ge 1$, $0 < t_1 < \cdots < t_n$, $\alpha_0 \in \mathscr{A}$, and the maps $\alpha_j: (\Omega, \mathscr{M}_{t_j}) \to \mathscr{A}$. It is our purpose to prove the following theorem.

12.3.1 Theorem. *For each* $x \in R^d$, $\mathscr{C}(x)$ *coincides with the closed convex hull* $\hat{\mathscr{D}}_x$ *of* $\mathscr{D}_x$.

We first observe that proving Theorem 12.3.1 reduces to showing that for any $N \geq 1, 0 < t_1 < \cdots < t_N$, and bounded continuous $f: (R^d)^N \to R^1$:

$$(3.2) \qquad \sup_{P \in \mathscr{C}(x)} E^P[f(x(t_1), \ldots, x(t_N))] = \sup_{P \in \mathscr{D}_x} E^P[f(x(t_1), \ldots, x(t_N))].$$

To see that (3.2) is enough, we invoke the Hahn–Banach theorem for locally convex topological spaces to show that $\mathscr{C}(x) = \hat{\mathscr{D}}_x$ is equivalent to

$$\sup_{P \in \mathscr{C}(x)} E^P[F] = \sup_{P \in \mathscr{D}_x} E^P[F]$$

for all bounded continuous $F: \Omega \to R^1$. If one now notices that there is for each $\varepsilon > 0$ a compact set K in Ω such that $P(\Omega \backslash K) \leq \varepsilon$ for all $P \in \mathscr{C}(x)$ (and hence for $P \in \hat{\mathscr{D}}_x$) and that by the Stone–Weierstrass theorem, every bounded continuous function on Ω can be approximated uniformly well on K by a function of the form in (3.2) having the same bound, it is clear that (3.2) suffices. We will prove (3.2) in two steps.

12.3.2 Lemma. *Let* $f: R^N \times R^d \to R^1$ *be a bounded continuous function. Given* $\lambda > 0$ *and* $n \geq 2$, *there is a measurable map*

$$(s_1, \ldots, s_{n-1}, z, x) \to P_x^{s_1, \ldots, s_{n-1}, z} \in \mathscr{D}_x$$

of $[0, \infty)^{n-1} \times R^N \times R^d$ *into* $M(\Omega)$ *such that the function*

$$(3.3) \qquad \Phi_n(z, x) \equiv \lambda^n \int_0^\infty \cdots \int_0^\infty E_x^{P_x^{s_1, \ldots, s_{n-1}, z}}[f(z, x(s_1 + \cdots + s_n))]$$
$$\times e^{-\lambda(s_1 + \cdots + s_n)} ds_1 \cdots ds_n$$

is upper semicontinuous and satisfies:

$$(3.4) \qquad \Phi_n(z, x) \geq \sup_{P \in \mathscr{C}(x)} E^P\left[\frac{\lambda^n}{(n-1)!} \int_0^\infty s^{n-1} e^{-\lambda s} f(z, x(s)) ds\right].$$

Proof. Define $S_\lambda \varphi$ for bounded upper semicontinuous $\varphi: R^N \times R^d \to R^1$ by:

$$S_\lambda \varphi(z, x) = \sup_{P \in \mathscr{C}(x)} \lambda E^P\left[\int_0^\infty e^{-\lambda s} \varphi(z, x(s)) ds\right].$$

Clearly $S_\lambda \varphi$ is upper semicontinuous and is bounded by the same bound as φ. Thus, by Theorem 12.2.5, we can find for each $1 \le m \le n$ a measurable map $\alpha_m : R^N \to \mathscr{A}$ such that

$$S_\lambda^{n-m+1} f(z, x) = \lambda E^{P_x^{\alpha_m(z)}} \left[\int_0^\infty e^{-\lambda s} S_\lambda^{n-m} f(z, x(s)) \, ds \right].$$

Define

$$P_x^{s_1, \ldots, s_{n-1}, z} = P_x^{\alpha_1(z)} \otimes_{t_1} P_{t_1, x(t_1, \cdot)}^{\alpha_2(z)}$$

$$\otimes_{t_2} \cdots \otimes_{t_{n-1}} P_{t_{n-1}, x(t_{n-1}, \cdot)}^{\alpha_n(z)},$$

where $t_k = \sum_1^k s_j$ and $P_{t_k, y}^{\alpha_k(z)} = P_y^{\alpha_k(z)} \circ \Phi_{t_k}^{-1}$. It is easy to see that $(s_1, \ldots, s_{n-1}, z, x)$ $\to P_x^{s_1, \ldots, s_{n-1}, z}$ is measurable. We will next show that

$$(3.5) \qquad \lambda^n \int_0^\infty \cdots \int_0^\infty E^{P_x^{s_1, \ldots, s_{n-1}, z}} [f(z, x(s_1 + \cdots + s_n)))]$$

$$\times e^{-\lambda(s_1 + \cdots + s_n)} \, ds_1 \cdots ds_n = S_\lambda^n f(z, x).$$

This will certainly prove that the Φ_n in (3.3) is upper semi-continuous. It will also enable us to prove (3.4).

To prove (3.5), we will show that for $1 \le m \le n - 1$:

$$\lambda^m \int_0^\infty \cdots \int_0^\infty E^{P_{m, x}^{s_1, \ldots, s_{m-1}, z}} [S_\lambda^{n-m} f(z, x(s_1 + \cdots + s_m))]$$

$$\times e^{-\lambda(s_1 + \cdots + s_m)} \, ds_1 \cdots ds_m$$

$$(3.6) \qquad = \lambda^{m+1} \int_0^\infty \cdots \int_0^\infty E^{P_{m+1, x}^{s_1, \ldots, s_m, z}}$$

$$\times [S_\lambda^{n-m-1} f(z, x(s_1 + \cdots + s_{m+1}))]$$

$$\times e^{-\lambda(s_1 + \cdots + s_{m+1})} \, ds_1 \cdots ds_{m+1}$$

where

$$P_{m, x}^{s_1, \ldots, s_{m-1}, z} = P_x^{\alpha_1(z)} \otimes_{t_1} P_{t_1, x(t_1, \cdot)}^{\alpha_2(z)} \otimes_{t_2} \cdots \otimes_{t_{m-1}} P_{t_{m-1}, x(t_{m-1}, \cdot)}^{\alpha_m(z)}.$$

Since $P_x^{s_1, \ldots, s_{n-1}, z} = P_{n, x}^{s_1, \ldots, s_{n-1}, z}$ and

$$S_\lambda^n f(z, x) = \lambda E^{P_x^{\alpha_1(z)}} \left[\int_0^\infty e^{-\lambda t} S_\lambda^{n-1} f(z, x(s)) \, ds \right],$$

(3.5) certainly follows from (3.6). To see (3.6), note that $P_{m+1,\,x}^{s\,1,\,\ldots,\,s\,m,\,z}$ equals $P_{m,\,x}^{s\,1,\,\ldots,\,s\,m-1,\,z}$ on $\mathcal{M}_{t_m}$. Thus:

$$\lambda^{m+1}\int_0^\infty \cdots \int_0^\infty E^{P_{m+1,\,x}^{s\,1,\,\ldots,\,s\,m,\,z}}[S_\lambda^{n-m-1}f(z,\,x(s_1 + \cdots + s_{m+1}))]$$

$$\times\, e^{-\lambda(s_1 + \cdots + s_{m+1})}\, ds_1 \cdots ds_{m+1}$$

$$= \lambda^m \int_0^\infty \cdots \int_0^\infty E^{P_{m,\,x}^{s\,1,\,\ldots,\,s\,m-1}}$$

$$\times \left[\lambda E^{P_{t_m,\,x(t_m)}^{2m+1(z)}} \left[\int_0^\infty e^{-\lambda s} S^{n-m-1} f(z,\,x(t_m + s))\, ds \right] \right]$$

$$\times\, e^{-\lambda(s_1 + \cdots + s_m)}\, ds_1 \cdots ds_m$$

$$= \lambda^m \int_0^\infty \cdots \int_0^\infty E^{P_{m,\,x}^{s\,1,\,\ldots,\,s\,m-1,\,z}}[S_\lambda^{n-m}f(z,\,x(s_1 + \cdots + s_m))]$$

$$\times\, e^{-\lambda(s_1 + \cdots + s_m)}\, ds_1 \cdots ds_m.$$

In view of (3.5), (3.4) will be proved once we have shown that if $m \geq 1$, then

$$S_\lambda^m \varphi(z,\,x) \geq E^P\left[\frac{\lambda^m}{(m-1)!} \int_0^\infty s^{m-1} e^{-\lambda s} \varphi(z,\,x(s))\, ds \right]$$

for all $P \in \mathscr{C}(x)$ and bounded upper semi-continuous $\varphi: R^N \times R^d \to R^1$. This is obvious for $m = 1$. Assuming it for m, we have for $P \in \mathscr{C}(x)$:

$$E^P\left[\frac{\lambda^{m+1}}{m!} \int_0^\infty s^m e^{-\lambda s} \varphi(z,\,x(s))\, ds \right]$$

$$= \frac{\lambda^m}{(m-1)!} \int_0^\infty t^{m-1} e^{-\lambda t} E^P\left[\lambda E^{P_t^!}\left[\int_0^\infty e^{-\lambda s} \varphi(z,\,x(t+s))\, ds \right] \right] dt$$

$$\leq \frac{\lambda^m}{(m-1)!} \int_0^\infty t^{m-1} e^{-\lambda t} E^P[S_\lambda \varphi(z,\,x(t))]\, dt \leq [S_\lambda^m(S_\lambda \varphi)](z,\,x),$$

where $\{P_\omega^t\}$ is a r.c.p.d. of $P|\mathcal{M}_t$. We have therefore completed the proof of (3.4). $\square$

12.3.3 Lemma. *For every $N \geq 1$, $0 < t_1 < \cdots < t_N$, and bounded continuous $f: (R^d)^N \to R^1$, (3.2) holds.*

Proof. Clearly it suffices to show that the left hand side of (3.2) dominates the right. We will do so by induction on N.

If $N = 1$ and if $t > 0$ and a bounded continuous $f: R^d \to R^1$ are given, choose $(s_1, \ldots, s_{n-1}, x) \to P_x^{s_1, \ldots, s_{n-1}}$ as in Lemma 12.3.2 relative to f and $\lambda_n = n/t$; and define

$$Q_x^n = \lambda_n^{n-1} \int_0^\infty \cdots \int_0^\infty e^{-\lambda_n(s_1 + \cdots + s_{n-1})} P_x^{s_1, \ldots, s_{n-1}} \, ds_1 \cdots ds_{n-1}.$$

Note that $Q_x^n \in \hat{\mathscr{D}}_x$ and that

$$\Big| E^{Q_x^n}[f(x(t))] - \lambda_n^n \int_0^\infty \cdots \int_0^\infty e^{-\lambda_n(s_1 + \cdots + s_n)}$$

$$\times E_x^{P_{s_1}^{s_1, \ldots, s_{n-1}}}[f(x(s_1 + \cdots + s_n))] \, ds_1 \cdots ds_{n-1} \Big|$$

$$\leq \lambda_n^n \int_0^\infty \cdots \int_0^\infty E_x^{P_{s_1}^{s_1, \ldots, s_{n-1}}}[|f(x(t)) - f(x(s_1 + \cdots + s_n))|]$$

$$\times e^{-\lambda_n(s_1 + \cdots + s_n)} \, ds_1 \cdots ds_n.$$

By the weak law of large numbers:

$$\lambda_n^n \int_{\{|t - \sum_1^n s_j| \geq \varepsilon\}} \cdots \int e^{-\lambda_n(s_1 + \cdots + s_n)} \, ds_1 \cdots ds_n \to 0$$

as $n \to \infty$ for all $\varepsilon > 0$. Thus, since $\mathscr{C}(x)$ is compact,

$$(3.7) \qquad \lambda_n^n \int_0^\infty \cdots \int_0^\infty \sup_{P \in \mathscr{C}(x)} E^P[|f(x(t)) - f(x(s_1 + \cdots + s_n))|]$$

$$\times e^{-\lambda_n(s_1 + \cdots + s_n)} \, ds_1 \cdots ds_n \to 0$$

as $n \to \infty$. In particular,

$$\sup_{P \in \hat{\mathscr{D}}_x} E^P[f(x(t))] \geq \limsup_{n \to \infty} E^{Q_x^n}[f(x(t))]$$

$$= \limsup_{n \to \infty} \lambda_n^n \int_0^\infty \cdots \int_0^\infty E_x^{P_{s_1}^{s_1, \ldots, s_{n-1}}}[f(x(s_1 + \cdots + s_n))]$$

$$\times e^{-\lambda_n(s_1 + \cdots + s_n)} \, ds_1 \cdots ds_n$$

$$\geq \limsup_{n \to \infty} E^P\Big[\frac{\lambda_n^n}{(n-1)!} \int_0^\infty s^{n-1} f(x(s)) e^{-\lambda_n s} \, ds\Big]$$

for all $P \in \mathscr{C}(x)$. Since

$$\frac{\lambda_n^n}{(n-1)!} \int_{\{|s - t| \geq \varepsilon\}} s^{n-1} e^{-\lambda_n s} \, ds \to 0$$

as $n \to \infty$, we have now proved (3.2) when $N = 1$.

Next assume that (3.2) holds for N and let $0 < t_1 < \cdots < t_{N+1}$ and a bounded continuous $f \colon (R^d)^{N+1} \to R^1$ be given. By the induction hypothesis we know that if $\mathscr{C}_x(t_1, \ldots, t_N)$ and $\hat{\mathscr{D}}_x(t_1, \ldots, t_N)$ denote, respectively, the images of $\mathscr{C}(x)$ and $\hat{\mathscr{D}}_x$ under the mapping induced by $\omega \to (x(t_1, \omega), \ldots, x(t_N, \omega))$ then $\mathscr{C}_x(t_1, \ldots, t_N) = \hat{\mathscr{D}}_x(t_1, \ldots, t_N)$. Now choose $(s_1, \ldots, s_{n-1}, z, x) \to P_s^{s_1, \ldots, s_{n-1}, z}$ for f and $\lambda_n = n/(t_{N+1} - t_N)$ with $z = (x^1, \ldots, x^N)$ and $x = x^{N+1}$, and define

$$\Phi_n(z, x) = \lambda_n^n \int_0^\infty \cdots \int_0^\infty E^{P_x} E^{P_x^{s_1, \ldots, s_{n-1}, z}} [f(z, x(s_1 + \cdots s_n))]$$

$$\times e^{-\lambda_n(s_1 + \cdots + s_n)} \, ds_1 \cdots ds_n.$$

Since Φ_n is bounded and upper semicontinuous, we can find $P_x^n \in \hat{\mathscr{D}}_x$ such that

$$E^{P_x^n}[\Phi_n(x(t_1), \ldots, x(t_N), x)] = \sup_{P \in \mathscr{C}(x)} E^P[\Phi_n(x(t_1), \ldots, x(t_N), x)].$$

Next, define

$$Q_\omega^{s_1, \ldots, s_{n-1}} = \delta_\omega \otimes_{t_N} \left(P_{x(t_N), \omega}^{s_1, \ldots, s_{n-1}, (x(t_1, \omega), \ldots, x(t_N, \omega))} \circ \Phi_{t_N}^{-1} \right).$$

It is obvious that

$$Q_x^n \equiv \lambda_n^{n-1} \int_0^\infty \cdots \int_0^\infty e^{-\lambda_n(s_1 + \cdots + s_{n-1})}$$

$$\times P_x^n \otimes_{t_N} Q^{s_1, \ldots, s_{n-1}} \, ds_1 \cdots ds_{n-1}$$

is an element of $\hat{\mathscr{D}}_x$. Moreover, by the same argument as we used when $N = 1$ to prove (3.7), we have

$$\lim_{n \to \infty} |E^{Q_x^n}[f(x(t_1), \ldots, x(t_{N+1}))] - E^{P_x^n}[\Phi_n(x(t_1), \ldots, x(t_N), x)]| = 0.$$

Thus, it only remains to check that

$$\limsup_{n \to \infty} E^{P_x^n}[\Phi_n(x(t_1), \ldots, x(t_N), x)] \geq \sup_{P \in \mathscr{C}(x)} E^P[f(x(t_1), \ldots, x(t_{N+1}))].$$

But if $P \in \mathscr{C}(x)$, then

$$E^{P_x^n}[\Phi_n(x(t_1), \ldots, x(t_N), x)] \geq E^P[\Phi_n(x(t_1), \ldots, x(t_N), x)]$$

$$\geq E^P \left[E^{P^{t_N}} \left[\frac{\lambda_n^n}{(n-1)!} \int_0^\infty s^{n-1} e^{-\lambda_n s} \right. \right.$$

$$\left. \left. \times f(x(t_1, \cdot), \ldots, x(t_N, \cdot), x(t_N + s)) \, ds \right] \right]$$

where $\{P_\omega^{t_N}\}$ is a r.c.p.d. of $P|\mathcal{M}_{t_N}$. That is, for $P \in \mathscr{C}(x)$

$$E^{P_x^n}[\Phi_n(x(t_1), \dots, x(t_N), x)]$$

$$\geq E^P\left[\frac{\lambda_n^n}{(n-1)!}\int_0^\infty s^{n-1}e^{-\lambda_n s}f(x(t_1), \dots, x(t_N), x(t_N + s))\, ds\right],$$

and so the desired result follows from the easily seen fact that

$$\lim_{n\to\infty} E^P\left[\frac{\lambda_n^n}{(n-1)!}\int_0^\infty s^{n-1}e^{-\lambda_n s}f(x(t_1), \dots, x(t_N), x(t_N + s))\, ds\right]$$

$$= E^P[f(x(t_1), \dots, x(t_N), x(t_{N+1}))].$$

We have therefore completed the proof of (3.2). $\square$

12.4. Exercises

12.4.1. Use Exercise 6.7.2 and the results of Section 12.2 to show that for any bounded continuous $a: [0, \infty) \times R^d \to S_d$ and $b: (0, \infty) \times R^d \to R^d$ there is a measurable family $\{P_{s,x}: (s, x) \in [0, \infty) \times R^d\}$ of solutions to the martingale problem for a and b such that for any (s, x) and any finite stopping time $\tau \geq s$: $\{\delta_\omega \otimes_{\tau(\omega)} P_{\tau(\omega), x(\tau(\omega), \omega)}\}$ is a r.c.p.d. of $P_{s,x}|\mathcal{M}_\tau$. Also, note that all solutions can be recovered from members of such strong Markov selections in the sense described in Section 12.3. Finally, observe that one does not really need a and b to be continuous with respect to t so long as they are continuous with respect to x. However continuity in t is needed if one wants to reduce the inhomogeneous case to the homogeneous case.

12.4.2. It is reasonable to ask if one cannot always make a Feller continuous selection of solutions when the coefficients are continuous. After all, uniqueness guarantees Feller continuity automatically. However the following counterexample shows that this may not always be possible. Let $d = 1$, $a \equiv 0$ and $b(x)$ any bounded continuous function which equals $x/|x|^{\frac{1}{2}} = (\mathrm{sgn}\, x)|x|^{\frac{1}{2}}$ for $|x| \leq 1$ and which is continuously differentiable for $|x| \geq 1$. Show that if $x \neq 0$, the solution to the martingale problem for a, b starting from x has exactly one solution P_x. Next show that P_x has a limit as $x \to 0$ through positive or negative values, but that these limits are different. Conclude from this that no Feller continuous choice exists.

12.4.3. Let $a: R^d \to S_d$ and $b: R^d \to R^d$ be bounded measurable functions such that $a(\cdot)$ is uniformly positive definite. Show that there is a measurable strong Markov family $\{P_x: x \in R^d\}$ of solutions to the martingale problem for a, b. To do this, it suffices to treat the case of $b \equiv 0$. Using 7.3.2, choose $C(T, R) < \infty$ for $T, R > 0$

such that for all $x \in R^d$ there is a solution to the martingale problem from $(0, x)$ satisfying

(4.1)
$$E^P \left[\int_0^T f(x(t)) \, dt \right] \le C(T, R) \| f \|_{L^d(R^d)}$$

for all $f \in C_0(B(0, R))$. For each $x \in R^d$ let $\mathscr{C}(x)$ denote the set of solutions satisfying (4.1). Let $\mathscr{C}(s, x) = \{ P \circ \Phi_s^{-1} : P \in \mathscr{C}(x) \}$ for $(s, x) \in [0, \infty) \times R^d$ (cf. property b of (2.1)) and check that $\{ \mathscr{C}(s, x) : (s, x) \in [0, \infty) \times R^d \}$ satisfies the properties listed in (2.1).

Appendix

A.0. Introduction

This appendix is devoted to the derivation of the inequalities on which the results of Chapters 7 and 9 depend. To be precise, let $c: [0, \infty) \to S_d$ be a measurable function for which there exist $0 < \lambda < \Lambda < \infty$ with the property that

$$(0.1) \qquad \lambda |\theta|^2 \leq \langle \theta, c(t)\theta \rangle \leq \Lambda |\theta|^2, \qquad t \geq 0 \quad \text{and} \quad \theta \in R^d.$$

Define $C(s, t) = \int_s^t c(u)\, du$ for $0 \leq s \leq t$ and

$$(0.2) \qquad g^{(c)}(s, x; t, y) = (2\pi)^{-d/2}(\det C(s, t))^{-\frac{1}{2}}$$

$$\times \exp\left[- \frac{\langle y - x, C(s, t)^{-1}(y - x) \rangle}{2} \right]$$

for $0 \leq s < t$ and $x, y \in R^d$. For $f \in C_0^\infty([0, \infty) \times R^d)$ and $T > 0$, set

$$(0.3) \qquad G_{(c)}^T f(s, x) = \int_s^T dt \int_{R^d} f(t, y)g^{(c)}(s, x; t, y)\, dy$$

on $[0, T] \times R^d$. What we have to prove is that

$$(0.4) \qquad \left\| \frac{\partial^2 G_{(c)}^T f}{\partial x_i \, \partial x_j} \right\|_{L^p([0, T] \times R^d)} \leq C_d(p, \lambda, \Lambda) \|f\|_{L^p([0, T) \times R^d)}$$

for all $1 \leq i, j \leq d$ and $1 < p < \infty$. The constant $C_d(p, \lambda, \Lambda)$ in (0.4) depends only on d, p, λ, and Λ and not on either $T > 0$ or the particular $c: [0, \infty) \to S_d^+$ satisfying (0.1) (cf. Theorem A.2.4).

In order to appreciate what the proof of an inequality like (0.4) entails, let us look at the special case when $d = 1$ and $c(\cdot) \equiv 1$. For this case, after we cavalierly interchange the order of integration and differentiation, we see that (0.4) comes down to proving

$$(0.5) \qquad \|k * f\|_{L^p(R \times R)} \leq C_p \|f\|_{L^p(R \times R)}, \qquad 1 < p < \infty,$$

where

$$(0.6) \qquad k(s, x) = \chi_{(-\infty, 0)}(s)\left[\left(\frac{x^2}{s^2} + \frac{1}{s}\right)\frac{1}{(2\pi s)^{\frac{1}{2}}} e^{x^2/2s}\right].$$

An indication that (0.5) cannot be entirely elementary is contained in the observation that "$k * f$" cannot be given an immediate meaning. Indeed, k is not an integrable function and so the integral defining $k * f$ is not a Lebesgue integral and must be defined via a principle value procedure. To be precise, set $k^{(\varepsilon)}(s, x) = \chi_{(-1/\varepsilon, -\varepsilon)}(s)k(s, x)$ for $\varepsilon > 0$. Then $k^{(\varepsilon)} \in L^1(R \times R)$ and so $k^{(\varepsilon)} * f$ not only makes sense, but also, by Young's inequality, satisfies

$$(0.7) \qquad \|k^{(\varepsilon)} * f\|_{L^p(R \times R)} \leq \|k^{(\varepsilon)}\|_{L^1(R \times R)}\|f\|_{L^p(R \times R)}, \qquad 1 \leq p \leq \infty.$$

What we want to do is define

$$(0.8) \qquad k * f = \lim_{\varepsilon \searrow 0} k^{(\varepsilon)} * f.$$

However, it is not clear that the limit in (0.8) exists, since the right hand side of (0.7) explodes as $\varepsilon \searrow 0$. It should by now be clear that in order to get anywhere we are going to be forced to take advantage of cancellations; putting in absolute values is just too crude. To see how cancellations enter, we compute $\widehat{k^{(\varepsilon)} * f}$. (Throughout we will use the convention:

$$\hat{f}(\xi) \equiv \int_{R^d} e^{i\langle \xi, x \rangle}f(x)\, dx$$

and

$$f^{\vee}(\xi) = \frac{1}{(2\pi)^d}\int_{R^d} e^{-i\langle \xi, x \rangle}f(x)\, dx$$

for $f \in L^1(R^d)$.) A simple computation yields:

$$\widehat{k^{(\varepsilon)}}(\tau, \xi) = \frac{-2|\xi|^2}{2i\tau - |\xi|^2}\left(e^{\varepsilon(2i\tau - |\xi|^2)/2} - e^{(2i\tau - |\xi|^2)/2\varepsilon}\right).$$

In particular,

$$\lim_{\varepsilon \searrow 0} \widehat{k^{(\varepsilon)} * f}(\tau, \xi) = \frac{-2|\xi|^2}{2i\tau - |\xi|^2}\hat{f}(\tau, \xi)$$

exists in $L^2(R \times R)$ for $f \in L^2(R \times R)$, and we have

$$(0.9) \qquad \sup_{\varepsilon > 0} \|k^{(\varepsilon)} * f\|_{L^2(R \times R)} \leq 2\|f\|_{L^2(R \times R)}.$$

Of course, the derivation of (0.9) involves Parseval's equality and the cancellations missed by (0.7) are hidden in the Fourier analysis. Nonetheless, this simple computation demonstrates that there are cancellations to be exploited if one is clever enough to do so.

The archetypical example of this sort is the Hilbert transform; and it was through a thorough understanding of the Hilbert transform that Calderon and Zygmund [1952] were able to make their original discoveries in the area which is now known as the *theory of singular integrals*. The type of singular integral which we want to study does not fit into the scheme of Calderon and Zygmund but is very closely related. A way to handle them was first given by B. Frank Jones [1964] and was fully developed by Fabes and Riviére [1966a]. The approach that we are going to adopt is based on the ideas of Fefferman and Stein [1972] and is an adaptation of the presentation in Stroock [1973]. We are grateful to E. Fabes who communicated the proof of Lemma A.2.2.

A.1. L_p Estimates for Some Singular Integral Operators

In the preceding section we saw that, with the aid of Fourier analysis, it is possible to prove estimates of the form (0.4) when $p = 2$. In order to get other p's, we want to apply interpolation theory. However, (0.4) is false for $p = 1$ and $p = \infty$; and therefore the Riesz-Thorin theorem is not directly applicable. One is therefore forced to adopt another form of interpolation theorem. The one usually used is that of Marcinkiewicz for which one must establish a weak-type $(1, 1)$ estimate. For an elegant treatment of this method, the paper of Riviére [1971] is recommended. In order to introduce what we consider to be an interesting technique to a new audience, we have decided not to take the Marcinkiewicz route but, instead, the newer Fefferman-Stein approach. As indicated by Exercise A.3.2 below and the papers of Burkholder and Gundy [1970] and Garsia [1973], these ideas ought to prove useful to probabilists in the future. We have therefore made our presentation look as much as possible like "probability theory."

A.1.1 Lemma. *Let $(E, \mathscr{F}, \mu)$ be a probability space and $\{\mathscr{F}_n\}_0^\infty$ a non-decreasing sequence of sub σ-algebras. For each $n \geq 0$, let X_n be an $\mathscr{F}_n$-measurable random variable. Assume that there exists a μ-integrable non-negative random variable Y such that*

$$(1.1) \qquad \sup_{n \geq m} E[\,|X_n - X_{m-1}|\,|\,\mathscr{F}_m] \leq E[Y\,|\,\mathscr{F}_m] \qquad (a.s., \mu)$$

for all $m \geq 1$. Then, for every $\alpha, \beta > 0$:

$$(1.2) \qquad \mu\left(\sup_{n \geq 0}\,|X_n - X_0| \geq \alpha + \beta\right) \leq \frac{2}{\alpha} E\left[Y, \sup_{n \geq 0}\,|X_n - X_0| \geq \beta\right].$$

Proof. Given $N \geq 1$, let

$$\sigma = \min\{n: n \geq N \quad \text{or} \quad |X_n - X_0| \geq \beta\}$$

and

$$\tau = \min\{n: n \geq N \quad \text{or} \quad |X_n - X_0| \geq \alpha + \beta\}.$$

If $B \in \mathscr{F}_\sigma$, then

$$E[|X_\tau - X_{\sigma-1}|, B] \leq E[|X_N - X_{\sigma-1}|, B] + E[|X_N - X_\tau|, B]$$

$$= \sum_{m=1}^N E[|X_N - X_{m-1}|, B \cap \{\sigma = m\}] + \sum_{n=1}^N E[|X_N - X_n|, B \cap \{\tau = n\}]$$

$$\leq \sum_{m=1}^N E[Y, B \cap \{\sigma = m\}] + \sum_{n=1}^N E[Y, B \cap \{\tau = n\}]$$

$$= 2E[Y, B],$$

since $B \cap \{\sigma = m\}$ and $B \cap \{\tau = n\} = (\bigcup_1^n B \cap \{\sigma = m\}) \cap \{\tau = n\}$ are elements of $\mathscr{F}_m$ and $\mathscr{F}_n$, respectively. We have therefore proved that

$$(1.3) \qquad E[|X_\tau - X_{\sigma-1}| \,|\, \mathscr{F}_\sigma] \leq 2E[Y|\mathscr{F}_\sigma] \qquad (\text{a.s.}, \mu).$$

In particular:

$$\mu\left(\sup_{0 \leq n \leq N} |X_n - X_0| \geq \alpha + \beta\right) = \mu(|X_\sigma - X_0| \geq \beta \text{ and } |X_\tau - X_0| \geq \alpha + \beta)$$

$$\leq \mu(|X_\tau - X_{\sigma-1}| \geq \alpha \text{ and } |X_\sigma - X_0| \geq \beta)$$

$$\leq \frac{1}{\alpha} E[|X_\tau - X_{\sigma-1}|, |X_\sigma - X_0| \geq \beta]$$

$$\leq \frac{2}{\alpha} E[Y, |X_\sigma - X_0| \geq \beta]$$

$$= \frac{2}{\alpha} E\left[Y, \sup_{0 \leq n \leq N} |X_n - X_0| \geq \beta\right].$$

Letting $N \nearrow \infty$, we arrive at (1.2). □

A.1.2 Lemma. *Let $(E, \mathscr{F}, \mu)$ and $\{\mathscr{F}_n\}_0^\infty$ be as in Lemma A.1.1. Assume that $\mathscr{F}_0 = \{\phi, E\}$ and $\mathscr{F} = \sigma(\bigcup_0^\infty \mathscr{F}_n)$. Given a μ-integrable random variable X, define*

$X_n = E[X | \mathscr{F}_n]$, $n \geq 0$, and set $X^\dagger = \sup_{n \geq 1} E[|X - X_{n-1}| \,|\, \mathscr{F}_n]$. Then for $1 < p < \infty$:

$$(1.4) \quad \frac{(p-1)^p}{4p^{p+1}} E[|X - E[X]|^p]^{1/p} \leq E[|X^\dagger|^p]^{1/p} \leq \frac{2p}{p-1} E[|X - E[X]|^p]^{1/p}$$

in the sense that if any of these is infinite then they all are.

Proof. First observe that we need only prove (1.4) under the assumption that X is bounded, since we can always reduce to this case by truncation. In particular, we will not worry about any of the quantities in (1.4) being infinite.

Next, note that for any $1 \leq m \leq n$:

$$E[|X_n - X_{m-1}| \,|\, \mathscr{F}_m] \leq E[|X - X_n| \,|\, \mathscr{F}_m] + E[|X - X_{m-1}| \,|\, \mathscr{F}_m]$$

$$= E[E[|X - X_n| \,|\, \mathscr{F}_{n+1}] \,|\, \mathscr{F}_m] + E[|X - X_{m-1}| \,|\, \mathscr{F}_m]$$

$$\leq 2E[X^\dagger | \mathscr{F}_m] \qquad (a.s., \mu).$$

We can therefore take $Y = 2X^\dagger$ in Lemma A.1.1 and thereby obtain:

$$\mu\left(\sup_{n \geq 0} |X_n - X_0| \geq \alpha + \beta\right) \leq \frac{4}{\alpha} E\left[X^\dagger, \sup_{n \geq 0} |X_n - X_0| \geq \beta\right]$$

for any pair $\alpha, \beta \geq 0$. Set $\tilde{X} = \sup_{n \geq 0} |X_n - X_0|$. Then

$$(1.5) \qquad\qquad \mu(\tilde{X} \geq (1 + \alpha)R) \leq \frac{4}{\alpha R} E[X^\dagger, \tilde{X} \geq R]$$

for any $\alpha > 0$ and $R > 0$. Proceeding as in Exercise 1.5.4, we have from (1.5) that

$$(1.6) \qquad\qquad \frac{(p-1)\alpha}{4p(1 + \alpha)^p} E[|\tilde{X}|^p]^{1/p} \leq E[|X^\dagger|^p]^{1/p}$$

for any $1 < p < \infty$ and $\alpha > 0$. Noting that $X - E[X] = \lim_{n \to \infty} X_n - X_0$ (cf. Exercise 1.5.10 and note that $X_0 = E[X]$ (a.s., μ)) and taking $\alpha = 1/(p - 1)$ in (1.6), we obtain the left hand side of (1.4).

To get the right hand side of (1.4), observe that $X^\dagger \leq 2 \sup_n E[|X - X_0| \,|\, \mathscr{F}_n]$ and, therefore, by Doob's inequality:

$$E[|X^\dagger|^p]^{1/p} \leq 2\left\{E\left[\sup_n E[|X - X_0| \,|\, \mathscr{F}_n]^p\right]\right\}^{1/p} \leq \frac{2p}{p-1} E[|X - X_0|^p]^{1/p}. \qquad \square$$

A.1.3 Theorem. Let $(E, \mathscr{F}, \mu)$ be a probability space and $\{\mathscr{F}_n\}$ a non-decreasing sequence of sub σ-algebras such that $\mathscr{F}_0 = \{\phi, E\}$ and $\mathscr{F} = \sigma(\bigcup_0^\infty \mathscr{F}_n)$. Given a

μ-integrable random variable X, set $X^\dagger = \sup_{n \geq 1} E[\,|X - E[X\,|\,\mathscr{F}_{n-1}]|\,|\,\mathscr{F}_n]$. Suppose that T is a linear map defined on $L^2(\mu) \cap L^\infty(\mu)$ into $L^1(\mu)$ and assume that

$$\|(Tf)^\dagger\|_{L^\infty(\mu)} \leq M_\infty \|f\|_{L^\infty(\mu)}, \qquad f \in L^2(\mu) \cap L^\infty(\mu),$$
$$\|Tf\|_{L^2(\mu)} \leq M_2 \|f\|_{L^2(\mu)}, \qquad f \in L^2(\mu) \cap L^\infty(\mu),$$

for some M_∞ and M_2. Then for each $2 < p < \infty$:

(1.7) $$\|Tf - E[Tf]\|_{L^p(\mu)} \leq \frac{4p^{p+1}}{(p-1)^p} M_\infty^{1-(2/p)}(4M_2)^{2/p}\|f\|_{L^p(\mu)}.$$

Proof. Given a $\mathscr{B}_{z^+} \times \mathscr{F}$-measurable $g \colon Z^+ \times E \to \mathbb{C}$ satisfying $|g| \equiv 1$, define

$$U_g(n)f = E[(Tf - E[Tf\,|\,\mathscr{F}_{n-1}])g(n)\,|\,\mathscr{F}_n]$$

for $f \in L^2(\mu) \cap L^\infty(\mu)$. Clearly, $|U_g(n)f| \leq (Tf)^\dagger$. Next, given g and an $\mathscr{F}$-measurable $\tau \colon E \to Z^+$, define $[U_g(\tau)f](q) = [U_g(\tau(q))f](q)$, $q \in E$. Again:

$$|U_g(\tau)f| \leq (Tf)^\dagger.$$

Thus, by assumption,

$$\|U_g(\tau)f\|_{L^\infty(\mu)} \leq M_\infty \|f\|_{L^\infty(\mu)}.$$

Also, by assumption and the right hand side of (1.4),

$$\|U_g(\tau)f\|_{L^2(\mu)} \leq \|(Tf)^\dagger\|_{L^2(\mu)}$$
$$\leq 4\|Tf - E[Tf]\|_{L^2(\mu)}$$
$$\leq 4\|Tf\|_{L^2(\mu)} \leq 4M_2\|f\|_{L^2(\mu)}.$$

Thus, by the Riesz-Thorin interpolation theorem, applied to $U_g(\tau)$,

(1.8) $$\|U_g(\tau)\|_{L^p(\mu)} \leq M_\infty^{1-(2/p)}(4M_2)^{2/p}\|f\|_{L^p(\mu)}$$

for $2 < p < \infty$. Maximizing the left hand of (1.8) over all choices of g and τ, we arrive at

(1.9) $$\|(Tf)^\dagger\|_{L^p(\mu)} \leq M^{1-(2/p)}(4M_2)^{2/p}\|f\|_{L^p(\mu)}$$

for $2 < p < \infty$. If we now apply the left hand side of (1.4), (1.7) results. $\square$

We now want to develop the context in which we will be applying Theorem A.1.3. Given a bounded measurable set $S \subseteq R \times R^d$ and a locally integrable f on $R \times R^d$, define f_S to be the mean value of f on S. (That is, $f_S \equiv 1/|S| \int_S f$, where

$|S|$ is the Lebesgue measure of S.) Let $\mathscr{Q}$ stand for the collection of all "parabolic" cubes Q in $R \times R^d$ of the form

$$\{(t, y) \in R \times R^d \colon s - \delta^2 \le t < s + \delta^2$$

$$\text{and} \quad x_j - \delta \le y_j < x_j + \delta, \ 1 \le j \le d\}$$

for some $(s, x) \in R \times R^d$ and $\delta > 0$. For $f \in L^1_{loc}(R \times R^d)$, define

(1.10)
$$\|f\|_{\mathscr{P}} = \sup_{Q \in \mathscr{Q}} (|f - f_Q|)_Q.$$

Note that $\|\cdot\|_{\mathscr{P}}$ is a semi-norm, and that $\|f\|_{\mathscr{P}} = 0$ if and only if f is equal almost everywhere to a constant. Hence, if $\mathscr{P}(R \times R^d) = \{f \in L^1_{loc}(R \times R^d)/\mathbb{C} \colon \|f\|_{\mathscr{P}} < \infty\}$, then $\mathscr{P}(R \times R^d)$ together with $\|\cdot\|_{\mathscr{P}}$ forms a normed linear space. When talking about elements of $\mathscr{P}(R \times R^d)$, we will indulge in the usual abuse of identifying the equivalence class of f with f itself. To get some feeling for the position of $\mathscr{P}(R \times R^d)$ among more familiar function spaces, note that $L^\infty(R \times R^d)/\mathbb{C} \subseteq \mathscr{P}(R \times R^d)$ (in fact $\|f\|_{\mathscr{P}} \le 2\|f\|_{L^\infty}$); but the opposite inclusion fails since $f(s, x) = \log|x_1|$ is an element of $\mathscr{P}(R \times R^d)$ which certainly is not in $L^\infty(R \times R^d)$. (Exercise A.3.1 shows that $\log|x_1|$ is just about as singular as an element of $\mathscr{P}(R \times R^d)$ can be.) The importance of $\mathscr{P}(R \times R^d)$ to us is contained in the next lemma.

A.1.4 Lemma. *Given* $N \ge 0$, *let* $E^{(N)} = \{(t, y) \in R \times R^d; \ -4^N \le t < 4^N$ *and* $-2^N \le y_j < 2^N, \ 1 \le j < d\}$ *and let* $\mathscr{F}^{(N)}$ *be the Borel field on* $E^{(N)}$. *For* $n \ge 0$, *let* $\mathscr{P}^{(N)}_n$ *be the partition of* $E^{(N)}$ *consisting of sets* Q *having the form*

$$\{(t, y) \colon l_0 4^{N-n} \le t < (l_0 + 1)4^{N-n}$$

$$\text{and} \quad l_j 2^{N-n} \le y_j < (l_j + 1)2^{N-n}, \qquad 1 \le j \le d\}$$

with $\vec{l} = (l_0, \ldots, l_d) \in Z^{d+1}$ *satisfying* $-4^n \le l_0 < 4^n$ *and* $-2^n \le l_j < 2^n, \ 1 \le j \le d$. *Set* $\mathscr{F}^{(N)}_n = \sigma(\mathscr{P}^{(N)}_n)$. *Finally, let* $\mu^{(N)}$ *denote normalized Lebesgue measure on* $(E^{(N)}, \mathscr{F}^{(N)})$ *(i.e.,* $\mu^{(N)}(\Gamma) = |\Gamma|/|E^{(N)}|$ *for* $\Gamma \in \mathscr{F}^{(N)})$. *Then* $F^{(N)}_0 = \{\phi, E^{(N)}\}$, $\mathscr{F}^{(N)}_n \subseteq \mathscr{F}^{(N)}_{n+1}$, *and* $\mathscr{F}^{(N)} = \sigma(\bigcup_0^\infty \mathscr{F}^{(N)}_n)$. *Moreover, if* $f \in L^1(\mu^{(N)})$ *and*

$$f^\dagger_{(N)} \equiv \sup_{n \ge 1} E^{\mu^{(N)}}[|f - E^{\mu^{(N)}}[f | \mathscr{F}^{(N)}_{n-1}]| \, | \, \mathscr{F}^{(N)}_n],$$

then

$$f^\dagger_{(N)} \le 2^{d+2} \sup_{n \ge 0} \sup_{Q \in \mathscr{P}^{(N)}_n} (|f - f_Q|)_Q.$$

Proof. We will drop the superscript N during the proof. The assertions about the $\mathscr{F}_n$'s are obvious. To prove the stated estimate, let $Q_n(t, y)$, $n \ge 0$ and $(t, y) \in E$, denote the unique element of $\mathscr{P}_n$ containing (t, y). Then:

$$E^\mu[\,|\,f - E^\mu[f\,|\,\mathscr{F}_{n-1}]\,|\,|\,\mathscr{F}_n](t, y) = (\,|\,f - f_{Q_{n-1}(t, y)}\,|\,)_{Q_n(t, y)}$$

$$\leq \frac{2^{d+2}}{|Q_{n-1}(t, y)|} \int_{Q_{n-1}(t, y)} |\,f - f_{Q_{n-1}(t, y)}\,|$$

$$= 2^{d+2}(\,|\,f - f_{Q_{n-1}(t, y)}\,|\,)_{Q_{n-1}(t, y)}$$

since $Q_n(t, y) \subseteq Q_{n-1}(t, y)$ and

$$\frac{|Q_{n-1}(t, y)|}{|Q_n(t, y)|} = 2^{d+2}. \quad \square$$

We are now ready to prove the interpolation theorem which we will be needing in the next section.

A.1.5 Theorem. *Let T be a linear mapping of $L^2(R \times R^d) \cap L^\infty(R \times R^d)$ into $L^2(R \times R^d)$ for which there exist M_∞ and M_2 such that*

$$\|Tf\|_{\mathscr{P}(R \times R^d)} \leq M_\infty \|f\|_{L^\infty(R \times R^d)}, \qquad f \in L^2(R \times R^d) \cap L^\infty(R \times R^d),$$

and

$$\|TF\|_{L^2(R \times R^d)} \leq M_2 \|f\|_{L^2(R \times R^d)}, \qquad f \in L^2(R \times R^d) \cap L^\infty(R \times R^d).$$

Then if $2 < p < \infty$:

$$(1.11) \qquad \|Tf\|_{L^p(R \times R^d)} \leq \frac{4p^{p+1}}{(p-1)^p} (2^{d+2} M_\infty)^{1-(2/p)} (4M_2)^{2/p} \|f\|_{L^p(R \times R^d)}.$$

Proof. Given $N \geq 0$, define $T^{(N)}f$ to be the restriction of $T(f\chi_{E^{(N)}})$ to $E^{(N)}$. Then

$$\|T^{(N)}f\|_{L^2(\mu^{(N)})} \leq M_2 \|f\|_{L^2(\mu^{(N)})}$$

and, by Lemma A.1.4,

$$\|(T^{(N)}f)^\dagger_{(N)}\|_{L^\infty(\mu^{(N)})} \leq 2^{d+2} \sup_{n \geq 0} \sup_{Q \in \mathscr{P}_n^{(N)}} (\,|\,T^{(N)}f - (T^{(N)}f)_Q\,|\,)_Q$$

$$\leq 2^{d+2} \|T(f\chi_{E^{(N)}})\|_{\mathscr{P}}$$

$$\leq 2^{d+2} M_\infty \|f\|_{L^\infty(\mu^{(N)})}.$$

Thus, by Theorem A.1.3,

$$\|T^{(N)}f - E^{\mu^{(N)}}[T^{(N)}f]\|_{L^p(\mu^{(N)})} \leq \frac{4p^{p+1}}{(p-1)^p} (2^{d+2} M_\infty)^{1-(2/p)}$$

$$\times (4M_2)^{2/p} \|f\|_{L^p(\mu^{(N)})}$$

for $2 < p < \infty$. In other words

$$\|(T(f \cdot \chi_{E^{(N)}}) - (T(f\chi_{E^{(N)}}))_{E^{(N)}})\chi_{E^{(N)}}\|_{L^p(R \times R^d)}$$

$$\leq \frac{4p^{p+1}}{(p-1)^p} (2^{d+2} M_\infty)^{1-(2/p)} (4M_2)^{2/p} \|f\|_{L^p(R \times R^d)}.$$

Since $(T(f \cdot \chi_{E^{(N)}}))\chi_{E^{(N)}} \to Tf$ in $L^2(R \times R^d)$, (1.1) will follow if we show that

$$\|(T(f\chi_{E^{(N)}}))_{E^{(N)}} \chi_{E^{(N)}}\|_{L^p(R \times R^d)} \to 0$$

as $N \to \infty$. But

$$\|(Tf\chi_{E^{(N)}}))_{E^{(N)}} \chi_{E^{(N)}}\|_{L^p(R \times R^d)} \leq \frac{\int_{E^{(N)}} |T(f\chi_{E^{(N)}})|}{|E^{(N)}|^{1-1/p}}$$

$$\leq \frac{\|T(f\chi_{E^{(N)}})\|_{L^2(R \times R^d)}}{|E^{(N)}|^{\frac{1}{2}-1/p}}$$

$$\leq \frac{M_2 \|f\|_{L^2(R \times R^d)}}{|E^{(N)}|^{\frac{1}{2}-1/p}} \to 0$$

since $p > 2$. Thus the proof is complete. $\square$

We conclude this section with the proof of a criterion which ensures that an operator satisfies the hypotheses of the preceding theorem. Before doing so, we need to introduce the "parabolic metric" on $R \times R^d$. For $(s, x) \in R \times R^d$, define

$$(1.12) \qquad \rho(s, x) = \left(\frac{|x|^2 + (|x|^4 + 4s^2)^{\frac{1}{2}}}{2} \right)^{\frac{1}{2}}.$$

Note that $\rho(s, x)$ is determined by the property that:

$$(1.13) \qquad \frac{s^2}{\rho^4(s, x)} + \frac{|x|^2}{\rho^2(s, x)} = 1;$$

from which it is easy to see that $\rho(s + t, x + y) \leq \rho(s, x) + \rho(t, y)$ and that for each $(s, x) \in R \times R^d \setminus \{(0, 0)\}$ there is a unique $\eta = (\eta_0, \bar{\eta}) \in S^d$ such that $(s, x) = (\rho^2(s, x)\eta_0, \rho(s, x)\bar{\eta})$. Finally, given $(s, x) \in R \times R^d$, the Jacobian of the transformation

$$(t, y) \to (s + \rho^2(t - s, y - x)\eta_0, x + \rho(t - s, y - x)\bar{\eta})$$

is

$$(1.14) \qquad \rho^{d+1} J_d(\eta)$$

where $J_d(\eta)$ is a smooth function on S^d satisfying:

$$(1.15) \qquad\qquad 1 \leq J_d(\eta) \leq M_d, \qquad \eta \in S^d.$$

This metric ρ and its associated "parabolic polar coordinate system" is a useful tool when dealing with singular integrals having parabolic homogeneity. We will be making use of it in the next section; the reason for our introducing these notions here is mostly based on aesthetic considerations. What they do is enable us to make a clean translation of the "almost L^1" condition of Hörmander [1960] into the parabolic context. These tricks are due to Fabes and Riviére [1966a].

A.1.6 Theorem. *Suppose that* $k: (R \times R^d)^2 \to C$ *is a measurable function such that*

$$\int \int |k(s, x; t, y)| \, dt \, dy < \infty$$

for all $(s, x) \in R \times R^d$. *Define the operator* K *on* $L^2(R \times R^d) \cap L^\infty(R \times R^d)$ *by*

$$Kf(s, x) = \int \int k(s, x; t, y) f(t, y) \, dt \, dy,$$

and assume that

$$(1.15) \qquad\qquad \|Kf\|_{L^2(R \times R^d)} \leq A \|f\|_{L^2(R \times R^d)}.$$

If there is a B for which:

$$(1.16) \qquad \int_{\rho(t-s,\, y-x) \geq 2\rho(\sigma-s,\, \xi-x)} \int |k(s, x; t, y) - k(\sigma, \xi; t, y)| \, dt \, dy \leq B$$

for all (σ, ξ) *and* (s, x) *in* $R \times R^d$, *then*

$$(1.17) \qquad \|Kf\|_{L^p(R \times R^d)} \leq \frac{4p^{p+1}}{(p-1)^p} (M_\infty)^{1-(2/p)} (M_2)^{2/p} \|f\|_{L^p(R \times R^d)}$$

for $2 < p < \infty$, *where*

$$(1.18) \qquad M_\infty = 2^{d+2} (|\{(t, y): \rho(t, y) \leq 2(d+1)\}|^{\frac{1}{2}} A + 2B)$$

and

$$(1.19) \qquad\qquad M_2 = 4A.$$

Proof. In view of Theorem A.1.5, all that we have to show is that

$$(1.20) \qquad \|Kf\|_{\mathscr{P}(R \times R^d)} \leq (|\{(t, y): \rho(t, y) \leq 2(d+1)\}|^{\frac{1}{2}} A + 2B) \|f\|_{L^\infty(R \times R^d)}.$$

To this end, let $f \in L^\infty(R \times R^d) \cap L^2(R \times R^d)$ be given and suppose that $Q = \{(t, y): s - \delta^2 \le t < s + \delta^2 \text{ and } x_j - \delta \le y_j < x_j + \delta, 1 \le j \le d\}$. We want to estimate $(|Kf - (Kf)_Q|)_Q$. Define

$$S = \{(t, y): \rho(t - s, y - x) \le 2(d + 1)\delta\}, \qquad \varphi = f\chi_S, \quad \text{and} \quad \psi = f - \varphi.$$

Then

$$(|Kf - (Kf)_Q|)_Q \le (|K\varphi - (K\varphi)_Q|)_Q + (|K\psi - (K\psi)_Q|)_Q.$$

Note that

$$(|K\varphi - (K\varphi)_Q|)_Q \le (|K\varphi - (K\varphi)_Q|^2)_Q^{\frac{1}{2}} \le (|K\varphi|^2)_Q^{\frac{1}{2}}$$

$$\le \frac{A}{|Q|^{\frac{1}{2}}} \|\varphi\|_{L^2(R \times R^d)} \le A\left(\frac{|S|}{|Q|}\right)^{\frac{1}{2}} \|f\|_{L^\infty(R \times R^d)}$$

and

$$\frac{|S|}{|Q|} = |\{(t, y): \rho(t, y) \le 2(d + 1)\}|.$$

Next, observe that if $(\sigma, \xi) \in Q$ and $(t, y) \notin S$, then $\rho(t - s, y - x) \ge 2\rho(\sigma - s, \xi - x)$. Hence, if we use the fact that

$$(|K\psi - (K\psi)_Q|)_Q \le \frac{2}{|Q|} \int_Q |K\psi(\sigma, \xi) - K\psi(s, x)| d\sigma d\xi,$$

then (1.16) yields:

$$(|K\psi - (K\psi)_Q|)_Q \le \frac{2}{|Q|} \iint_Q |K\psi(\sigma, \xi) - K\psi(s, x)| d\sigma\, d\xi$$

$$\le \frac{2}{|Q|} \iint_Q d\sigma\, d\xi \iint_{R \times R^d \setminus S} |k(\sigma, \xi; t, y) - k(s, x; t, y)|\, |f(t, y)|\, dt\, dy$$

$$\le \frac{2\|f\|_{L^\infty(R \times R^d)}}{|Q|} \iint_Q d\sigma\, d\xi$$

$$\times \int_{\rho(t - s, y - x) \ge 2\rho(\sigma - s, \xi - x)} |k(\sigma, \xi; t, y) - k(s, x, t, y)|\, dt\, dy$$

$$\le 2B\|f\|_{L^\infty(R \times R^d)}.$$

Combining these, we arrive at:

$$(\,|\,Kf - (Kf)_Q|\,)_Q \le (\,|\{(t,\,y): \rho(t,\,y) \le 2(d+1)\}|^{\frac{1}{4}}A + 2B)\|f\|_{L^\infty(R \times R^d)},$$

and so (1.20) is now proved. $\square$

A.2. Proof of the Main Estimate

We now have made all the preparations necessary for the proof of (0.4). What remains is to see how to apply these preparations. In order to eliminate unnecessary sub- and superscripts during the actual proof, we begin by establishing the notation which will be used throughout this section.

Let $c: [0, \infty) \to S_d$ be a measurable function which satisfies (0.1) for some $0 < \lambda < \Lambda < \infty$. Extend c to R by taking $c(-s) = c(s)$, $s \ge 0$, and set $C(s, t) = \int_s^t c(u)\,du$ for $s \le t$. Define

$$g(s, x; t, y) = \chi_{(0,\,\infty)}(t - s)[(2\pi)^d \det C(s, t)]^{-\frac{1}{2}}$$
$$\times \exp\left[- \frac{\langle y - x, C(s, t)^{-1}(y - x)\rangle}{2}\right]$$

on $(R \times R^d)^2$. For $f \in C_0^\infty(R \times R^d)$, define

$$Gf(s, x) = \int_s^\infty dt \int_{R^d} g(s, x; t, y)f(t, y)\,dy$$

and

$$G^*f(t, y) = \int_{-\infty}^t ds \int_{R^d} g(s, x; t, y)f(s, x)\,dx.$$

In order to prove (0.4) for all $1 < p < \infty$, it is sufficient to show that for all $1 \le i,\ j \le d$:

(2.1)
$$\left\|\frac{\partial^2 Gf}{\partial x_i\, \partial x_j}\right\|_{L^p(R \times R^d)} \le C_d(p, \lambda, \Lambda)\|f\|_{L^p(R \times R^d)}$$

and

(2.2)
$$\left\|\frac{\partial^2 G^*f}{\partial y_i\, \partial y_j}\right\|_{L^p(R \times R^d)} \le C_d(p, \lambda, \Lambda)\|f\|_{L^p(R \times R^d)}$$

for all $2 \leq p < \infty$, where $C_d(p, \lambda, \Lambda)$ depends only on d, p, λ, and Λ. Indeed, a simple duality argument shows that (2.2) for a given p implies (2.1) for the conjugate p' of p (i.e., $(1/p) + (1/p') = 1$). Since there is no difference between the derivations of (2.1) and (2.2), we will restrict our attention to the proof of (2.1) for $2 \leq p < \infty$.

For $0 < \varepsilon < 1$ and $1 \leq i, j \leq d$, define

$$
\begin{aligned}
k_{ij}^{(\varepsilon)}(s, x; t, y) &= \chi_{[\varepsilon, 1/\varepsilon]}(t - s)\frac{\partial^2 g}{\partial x_i \, \partial x_j}(s, x; t, y) \\
&= \chi_{[\varepsilon, 1/\varepsilon]}(t - s)[((C(s, t))^{-1}(y - x))_i \\
&\quad \times ((C(s, t))^{-1}(y - x))_j \\
&\quad - \delta_{ij}(C(s, t)^{-1})_{ii}]g(s, x; t, y).
\end{aligned}
$$

Clearly

$$
\iint |k_{ij}^{(\varepsilon)}(s, x; t, y)| \, dt \, dy < \infty
$$

for all $\varepsilon > 0$, $1 \leq i, j \leq d$, and $(s, x) \in R \times R^d$. Define

$$
K_{ij}^{(\varepsilon)}f(s, x) = \iint k_{ij}^{(\varepsilon)}(s, x; t, y)f(t, y) \, dt \, dy
$$

for $f \in L^\infty(R \times R^d)$. Our first lemma shows that the proof of (2.1) reduces to proving

$$(2.4) \qquad \sup_{0 < \varepsilon < 1} \|K_{ij}^{(\varepsilon)}f\|_{L^p(R \times R^d)} \leq C_d(p, \lambda, \Lambda)\|f\|_{L^p(R \times R^d)}$$

for all $1 \leq i, j \leq d$ and $2 \leq p < \infty$.

A.2.1 Lemma. *If* $f \in C_0^\infty(R \times R^d)$, *then* $K_{ij}^{(\varepsilon)}f$, Gf, *and all spacial derivatives of* Gf *are continuous and rapidly decreasing on* $R \times R^d$. *Moreover,*

$$(2.5) \qquad \widehat{\frac{\partial^2 Gf}{\partial x_i \, \partial x_j}}^x(s, \xi) = -\xi_i\xi_j \int_s^\infty e^{-\langle \xi, C(s, t)\xi \rangle/2}\hat{f}^x(t, \xi) \, dt$$

and

$$(2.6) \qquad \widehat{K_{ij}^{(\varepsilon)}f}^x(s, \xi) = -\xi_i\xi_j \int_{s+\varepsilon}^{s+(1/\varepsilon)} e^{-\langle \xi, C(s, t)\xi \rangle/2}\hat{f}^x(t, \xi) \, dt.$$

(Here $\hat{\varphi}^x(s, \xi) \equiv \int_{R^d} e^{i\langle \xi, x \rangle}\varphi(s, x) \, dx$, $\xi \in R^d$, *for* $\varphi \in L^1(R \times R^d)$.*) In particular,* $K_{ij}^{(\varepsilon)}f \to (\partial^2 Gf/\partial x_i \, \partial x_j)$ *uniformly on* $R \times R^d$ *as* $\varepsilon \searrow 0$.

Proof. The first assertion as well as (2.5) and (2.6) are easy consequences of the spacial translation invariance of the operators G and $K_{ij}^{(\epsilon)}$. Finally, from (2.5) and (2.6), it is clear that

$$\widehat{K_{ij}^{(\epsilon)} f}\,^x(s, \cdot) \to \frac{\widehat{\partial^2 G}}{\partial x_i \, \partial x_j}\,^x (s, \cdot)$$

in $L^1(R^d)$ at a rate which is independent of $s \in R$. Hence the last assertion follows from the Fourier inversion formula. $\square$

We now turn to the proof of (2.4). The idea is to show that $k_{ij}^{(\epsilon)}$ satisfies the hypotheses of Theorem A.1.6 with A and B independent of $0 < \epsilon < 1$. The first step is the following.

A.2.2 Lemma. (*Proof communicated by E. Fabes*). *For all* $0 < \epsilon < 1$ *and* $f \in L^\infty(R \times R^d) \cap L^2(R \times R^d)$:

$$(2.7) \qquad \|K_{ij}^{(\epsilon)} f\|_{L^2(R \times R^d)} \le 2(\Lambda/\lambda^3)^{\frac{1}{2}} \|f\|_{L^2(R \times R^d)}.$$

Proof. Let $f \in C_0^\infty(R \times R^d)$. By Parseval's equality:

$$(2.8) \qquad \|K_{ij}^{(\epsilon)} f\|_{L^2(R \times R^d)} = \frac{1}{(2\pi)^d} \left(\int \|\widehat{K_{ij}^{(\epsilon)} f}\,^x(\cdot, \xi)\|_{L^2(R)}^2 \, d\xi \right)^{\frac{1}{2}}.$$

By (2.6),

$$|\widehat{K_{ij}^{(\epsilon)} f}\,^x(s, \xi)| \le |\xi|^2 \int_s^\infty e^{-(C_\xi(t) - C_\xi(s))/2} |\hat{f}^x(t, \xi)| \, dt$$

where $C_\xi(t) = \int_0^t \langle \xi, c(u)\xi \rangle \, du$, $t \in R$ and $\xi \in R^d \backslash \{0\}$. Changing variables, we get

$$|\widehat{K_{ij}^{(\epsilon)} f}\,^x(s, \xi)| \le |\xi|^2 \int_{C_\xi(s)}^\infty e^{(C_\xi(s) - t)/2} \frac{|\hat{f}^x(C_\xi^{-1}(t), \xi)|}{C_\xi'(C_\xi^{-1}(t))} \, dt$$

$$= \varphi * \psi_\xi(C_\xi(s)),$$

where $\varphi(t) = e^{t/2} \chi_{(-\infty, 0)}(t)$ and

$$\psi_\xi(t) = \frac{|\hat{f}^x(C_\xi^{-1}(t), \xi)| \, |\xi|^2}{C_\xi'(C_\xi^{-1}(t))}.$$

Note that

$$\psi_\xi(t) \le \frac{|\hat{f}^x(C_\xi^{-1}(t), \xi)|}{\lambda}$$

and so

$$\|\widehat{K_{ij}^{(\varepsilon)} f}^{\,x}(\cdot, \xi)\|_{L^2(R)} \leq \|\varphi * \psi_\xi(C_\xi(\cdot))\|_{L^2(R)}$$

$$\leq \frac{1}{\lambda^{\frac{1}{2}} |\xi|} \|\varphi * \psi_\xi\|_{L^2(R)}$$

$$\leq \frac{2}{\lambda^{\frac{1}{2}} |\xi|} \|\hat{f}^x(C_\xi^{-1}(\cdot), \xi)\|_{L^2(R)}$$

$$\leq \frac{2\Lambda^{\frac{1}{2}}}{\lambda^{\frac{1}{2}}} \|\hat{f}^x(\cdot, \xi)\|_{L^2(R)}.$$

It is evident from this and (2.8) that (2.7) holds. $\square$

The final step is to check that there is a B, independent of $0 < \varepsilon < 1$, for which (1.16) holds when k is replaced by $k_{ij}^{(\varepsilon)}$.

A.2.3 Lemma. *There is a constant* $B_d(\lambda, \Lambda)$ *such that*

$$(2.9) \qquad \int_{\rho(t-s, \, y-x) \geq 2\rho(\sigma-s, \, \xi-x)} |k_{ij}^{(\varepsilon)}(s, x; t, y) - k_{ij}^{(\varepsilon)}(\sigma, \xi; t, y)| \, dt \, dy$$

$$\leq B_d(\lambda, \Lambda)$$

for all $0 < \varepsilon < 1$, $1 \leq i, j \leq d$, *and* $(s, x), (\sigma, \xi) \in R \times R^d$.

Proof. Let

$$k_{ij}(s, x; t, y) = \chi_{(0,\infty)}(t - s)[(C(s,t)^{-1}(y-x))_i (C(s,t)^{-1}(y-x))_j - \delta_{ij} C(s,t)_{ii}^{-1}]$$

$$\times g(s, x; t, y).$$

Note that for $s < t$:

$$|g(s, x; t, y)| \leq \frac{1}{(2\pi\lambda)^{d/2}} \frac{1}{(t-s)^{d/2}} e^{-|y-x|^2/2\Lambda(t-s)}$$

and so

$$k_{ij}(s, x; t, y) \leq \frac{1}{(2\pi\lambda)^{d/2}} \frac{1}{(t-s)^{d/2}}$$

$$\times \left| \left(\frac{|y-x|}{\lambda(t-s)} \right)^2 + \frac{1}{\lambda(t-s)} \right| e^{-|y-x|^2/2\Lambda(t-s)}.$$

Clearly:

$$|k_{ij}^{(\varepsilon)}(s, x; t, y) - k_{ij}^{(\varepsilon)}(\sigma, \xi; t, y)|$$
$$\leq |k_{ij}(s, x; t, y) - k_{ij}(\sigma, \xi; t, y)|$$
$$+ |\chi_{[\varepsilon, 1/\varepsilon]}(t - s) - \chi_{[\varepsilon, 1/\varepsilon]}(t - \sigma)| \, |k_{ij}(s, x; t, y)|.$$

We will first show how to estimate

$$\int_{\rho(t-x, y-x) \geq 2\rho(\sigma-s, \xi-x)} |\chi_{[\varepsilon, 1/\varepsilon]}(t - s) - \chi_{[\varepsilon, 1/\varepsilon]}(t - \sigma)|$$
$$\times |k_{ij}(s, x; t, y)| \, dt \, dy.$$

Obviously, this comes down to looking at

$$\frac{1}{(2\pi\lambda)^{d/2}} \int_{\rho(t-s, y-x) \geq 2\alpha}$$

$$\times \int \chi_{(0, \infty)}(t - s)|\chi_{[\varepsilon, 1/\varepsilon]}(t - s) - \chi_{[\varepsilon, 1/\varepsilon]}(t - \sigma)|$$

$$\times \left[\left(\frac{|y - x|}{\lambda(t - s)}\right)^2 + \frac{1}{\lambda(t - s)}\right] \frac{\exp[-|y - x|^2/2\Lambda(t - s)]}{(t - s)^{d/2}} \, dt \, dy$$

$$= \frac{1}{(2\pi\lambda)^{d/2}} \int_{\substack{\rho(t, y) \geq 2\alpha \\ t > 0}} \int |\chi_{[\varepsilon, 1/\varepsilon]}(t) - \chi_{[\varepsilon, 1/\varepsilon]}(t - (\sigma - s))|$$

$$\times \left[\left(\frac{|y|}{\lambda t}\right)^2 + \frac{1}{\lambda t}\right] \frac{\exp[-|y|^2/2\Lambda t]}{t^{d/2}} \, dt \, dy$$

$$= \frac{1}{(2\pi\lambda)^{d/2}} \int_{\substack{\rho(1, y) \geq 2\alpha/t^{\frac{1}{2}} \\ t > 0}}$$

$$\times \int \frac{1}{t} |\chi_{[\varepsilon, 1/\varepsilon]}(t) - \chi_{[\varepsilon, 1/\varepsilon]}(t - (\sigma - s))|$$

$$\times \left(\left(\frac{|y|}{\lambda}\right)^2 + \frac{1}{\lambda}\right) \exp[-|y|^2/2\Lambda] \, dt \, dy$$

where $\alpha = \rho(\sigma - s, \xi - x)$. Since $|\sigma - s| \leq \alpha^2$, the last expression can be written as the sum of four terms, each of which is less than or equal to:

$$\sup_{T \geq 0} \frac{1}{(2\pi\lambda)^{d/2}} \int_T^{T+\alpha^2} \frac{1}{t} \, dt \int_{\rho(1, y) \geq 2\alpha/t^{\frac{1}{2}}} \left(\left(\frac{|y|}{\lambda}\right)^2 + \frac{1}{\lambda}\right) e^{-|y|^2/2\Lambda} \, dy$$

$$= \sup_{T \geq 0} \frac{1}{(2\pi\lambda)^{d/2}} \int_{T/\alpha^2}^{T/\alpha^2 + 1} \frac{1}{t} \, dt \int_{\rho(1, y) \geq 2/t^{\frac{1}{2}}} \left(\left(\frac{|y|}{\lambda}\right)^2 + \frac{1}{\lambda}\right) e^{-|y|^2/2\Lambda} \, dy$$

$$= \sup_{T \geq 0} \frac{1}{(2\pi\lambda)^{d/2}} \int_T^{T+1} \frac{1}{t} \, dt \int_{\rho(1, y) \geq 2/t^{\frac{1}{2}}} \left(\left(\frac{|y|}{\lambda}\right)^2 + \frac{1}{\lambda}\right) e^{-|y|^2/2\Lambda} \, dy.$$

Using the fact that $\rho(1, y) \leq 1 + |y|$, one can easily check that

$$\frac{1}{t} \int_{\rho(1, y) \geq 2/t^{\frac{1}{2}}} \left(\left(\frac{|y|}{\lambda} \right)^2 + \frac{1}{\lambda} \right) e^{-|y|^2/2\Lambda} \, dy, \qquad t > 0,$$

is bounded by a constant depending only on d, λ, and Λ.

Thus, it remains only to estimate:

$$\int_{\rho(t-s, y-x) \geq 2\rho(\sigma-s, \xi-y)} |k_{ij}(s, x; t, y) - k_{ij}(\sigma, \xi; t, y)| \, dt \, dy.$$

Note that:

$$|k_{ij}(s, x; t, y) - k_{ij}(\sigma, \xi; t, y)|$$

$$(2.10) \qquad \leq \chi_{(-\infty, t)}(s \wedge \sigma) |\sigma - s| \sup_{s \wedge \sigma \leq \tau < (s \vee \sigma) \wedge t} \left| \frac{\partial k_{ij}}{\partial \tau} (\tau, x; t, y) \right|$$

$$+ \chi_{(-\infty, t)}(\sigma) |\xi - x| \sup_{0 \leq \gamma \leq 1} |\nabla k_{ij}(\sigma, (1 - \gamma)x + \gamma\xi; t, y)|.$$

Given $\tau < t$, we have:

$$-\frac{\partial k_{ij}}{\partial \tau} (\tau, x; t, y) = [C(\tau, t)^{-1} c(\tau) C(\tau, t)^{-1}(y - x)]_i$$

$$\times [C(\tau, t)^{-1}(y - x)]_j g(\tau, x; t, y)$$

$$+ [C(\tau, t)^{-1}(y - x)]_i$$

$$\times [C(\tau, t)^{-1} c(\tau) C(\tau, t)^{-1}(y - x)]_j g(\tau, x; t, y)$$

$$- \delta_{ij}(C(\tau, t)^{-1} c(\tau) C(\tau, t)^{-1})_{ii} g(\tau, x; t, y)$$

$$+ \tfrac{1}{2} < y - x, C(\tau, t)^{-1} c(\tau) C(\tau, t)^{-1}(y - x) > k_{ij}(\tau, x; t, y)$$

$$+ \frac{1}{2} \frac{\dfrac{\partial}{\partial \tau} [\det C(\tau, t)]}{\det C(\tau, t)} k_{ij}(\tau, x; t, y)$$

Thus for $\tau < t$:

$$\left| \frac{\partial k_{ij}}{\partial \tau} (\tau, x; t, y) \right| \leq \frac{1}{(2\pi)^{d/2}} \left(\frac{\Lambda}{\lambda^{d/2+2}} + \frac{d! \Lambda^d}{2\lambda^{3d/2+2}} \right)$$

$$\times \frac{1}{(t - \tau)^{d/2+2}} e^{-|y-x|^2/2\Lambda(t-\tau)}$$

$$+ \frac{1}{(2\pi)^{d/2}} \left(\frac{5\Lambda}{2\lambda^{d/2+3}} + \frac{d! \Lambda^d}{2\lambda^{3d/2+2}} \right)$$

$$\times \frac{|y - x|^2}{(t - \tau)^{d/2 + 3}} e^{-|y-x|^2/2\Lambda(t-\tau)}$$

$$+ \frac{1}{(2\pi)^{d/2}} \frac{\Lambda}{2\lambda^{d/2+4}} \frac{|y - x|^4}{(t - \tau)^{d/2 + 4}} e^{-|y-x|^2/2\Lambda(t-\tau)}$$

$$\leq C_d(\lambda, \Lambda) e^{-|y-x|^2/2\Lambda(t-\tau)} \sum_{l=0}^{2} \frac{|y - x|^{2l}}{(t - \tau)^{d/2 + 2 + l}}$$

where $C_d(\lambda, \Lambda)$ depends only on d, λ, and Λ. We now switch to the parabolic polar coordinates described in the paragraph preceding Theorem A.1.6. Let $\sigma = s + \alpha^2\theta_0$, $\xi = x + \alpha\bar{\theta}$, $t = s + \rho^2\bar{\omega}_0$, and $y = x + \rho\omega$ where α, $\rho > 0$ and $(\theta_0, \bar{\theta})$, $(\omega_0, \bar{\omega}) \in S^d$. Assuming that $\rho > 2\alpha$ and that $\sigma \wedge s < t$, we have for $\sigma \wedge s \leq \tau < (\sigma \vee s) \wedge t$:

$$\left| \frac{\partial k_{ij}}{\partial \tau} (\tau, x; t, y) \right| \leq \frac{C_d(\lambda, \Lambda)}{\rho^{d+4}} \exp\left\{ -\frac{|\bar{\omega}|^2}{2\Lambda\left(\omega_0 - \dfrac{\tau - s}{\rho^2}\right)} \right\}$$

$$\times \sum_{l=0}^{2} \frac{|\bar{\omega}|^{2l}}{\left(\omega_0 - \dfrac{\tau - s}{2}\right)^{d/2 + 2 + l}}$$

Clearly

$$\exp\left\{ -\frac{|\bar{\omega}|^2}{2\Lambda\left(\omega_0 - \dfrac{\tau - s}{\rho^2}\right)} \right\} \sum_{l=0}^{2} \frac{|\bar{\omega}|^{2l}}{\left(\omega_0 - \dfrac{\tau - s}{\rho^2}\right)^{d/2 + 2 + l}}$$

is bounded so long as $(\omega_0 - (\tau - s)/\rho^2)^2 \geq \frac{9}{32}$. On the other hand

$$|\bar{\omega}|^2 + \left(\omega_0 - \frac{\tau - s}{\rho^2}\right)^2 = 1 - 2\frac{\tau - s}{\rho^2}\omega_0 + \left(\frac{\tau - s}{\rho^2}\right)^2$$

$$\geq \left(1 - \left|\frac{\tau - s}{\rho^2}\right|\right)^2 \geq \frac{9}{16}, \qquad \sigma \wedge s \leq \tau < \sigma \vee s,$$

since $\alpha < \rho/2$. Thus $|\bar{\omega}|^2 > \frac{9}{32}$ if $(\omega_0 - (\tau - s)/\rho^2)^2 \leq \frac{9}{32}$. Combining these remarks, we have

$$\sup_{\sigma \wedge s \leq \tau < (\sigma \vee s) \wedge t} \left| \frac{\partial k_{ij}}{\partial \tau} (\tau, x; t, y) \right| \leq \frac{\gamma_d C_d(\lambda, \Lambda)}{\rho^{d+4}}$$

if $\rho > 2\alpha$. Hence

$$|s - \sigma| \int_{\substack{\rho(t-s,\, y-x) \geq 2\rho(\sigma-s,\, \xi-x) \\ t > \sigma \wedge s}} \sup_{\sigma \wedge s \leq \tau < (\sigma \vee s) \wedge t} \left| \frac{\partial k_{ij}}{\partial \tau}(t, x; t, y) \right| dt\, dy$$

$$\leq \gamma_d C_d(\lambda, \Lambda) \alpha^2 \int_{2\alpha}^{\infty} \rho^{-3}\, d\rho \int_{S^d} J_d(\omega)\, d\omega$$

$$= \tfrac{1}{2}\gamma_d' C_d(\lambda, \Lambda)$$

where γ_d' equals γ_d times the surface area of $\{(t, y): \rho(t, y) = 1\}$. We have used here the fact that $dt\, dy = \rho^{d+1} J_d(\omega)\, d\rho\, d\omega$, where $J_d(\omega)$ is discussed before Theorem A.1.6 and $d\omega$ denotes surface measure on S^d.

By essentially the same argument, one can show that there exists $\bar{C}_d(\lambda, \Lambda)$ such that

$$\sup_{0 \leq \gamma \leq 1} |\nabla k_{ij}(s, (1 - \gamma)x + \gamma\xi; t, y)| \leq \frac{\bar{C}_d(\lambda, \Lambda)}{\rho^{d+3}}$$

when $\rho \equiv \rho(t - s, y - x) \geq 2\rho(\sigma - x, \xi - x)$ and $s < t$. Thus:

$$|x - y| \int_{\substack{\rho(t-s,\, y-x) \geq 2\rho(\sigma-s,\, \xi-x) \\ t > s}} \sup_{0 \leq \gamma \leq 1} |\nabla k_{ij}(s, (1 - \gamma)x + \gamma\xi; t, y)|\, dt\, dy$$

$$\leq \bar{C}_d(\lambda, \Lambda) \alpha \int_{2\alpha} \rho^{-2}\, d\rho \int_{S^d} J_d(\omega)\, d\omega$$

$$= \tfrac{1}{2}\bar{C}_d(\lambda, \Lambda) \int_{S^d} J_d(\omega)\, d\omega,$$

where $\alpha = \rho(\sigma - s, \xi - x)$. Combining this with the preceding, we see that the lemma now follows from (2.10). □

We have at last made all the calculations necessary for the proof of (0.4). We give a precise statement in the following.

A.2.4 Theorem. *Let* $A_d(\lambda, \Lambda) = 2(\Lambda/\lambda^3)^{\frac{1}{2}}$ *and let* $B_d(\lambda, \Lambda)$ *be as in Lemma A.2.3. Set*

$$M_\infty = 2^{d+2}(|\{(t, y): \rho(t, y) \leq 2(d + 1)\}|^{\frac{1}{2}} A_d(\lambda, \Lambda) + 3B_d(\lambda, \Lambda))$$

and

$$M_2 = 4A_d(\lambda, \Lambda).$$

Then, for $2 \leq p < \infty$, (0.4) holds with

$$C_d(p, \lambda, \Lambda) = \frac{4p^{p+1}}{(p-1)^p} (M_\infty)^{1-2/p} (M_2)^{2/p};$$

and for $1 < p \leq 2$, (0.4) holds with $C_d(p, \lambda, \Lambda) = C_d(p', \lambda, \Lambda)$ where p' is the conjugate of p.

Proof. As we pointed out before Lemma A.2.1, proving (2.1), and therefore (0.4) for $2 \leq p < \infty$, is equivalent to proving (2.4) for this range of p's. But (2.4) with the specified choice of $C_d(p, \lambda, \Lambda)$, $2 \leq p < \infty$, is immediate from Lemmas A.2.2 and A.2.3 together with Theorem A.1.6. To get the desired result when $1 < p \leq 2$, we have to prove (2.2) for $2 \leq p < \infty$; and as we said before, this is accomplished in the same way as we proved (2.1). It turns out that (2.2) holds with the same $C_d(p, \lambda, \Lambda)$ as (2.1), and this is the reason that we can take $C_d(p, \lambda, \Lambda) = C_d(p', \lambda, \Lambda)$ for $1 < p \leq 2$. $\square$

A.3. Exercises

A.3.1. Starting with Lemma A.1.1, check that if $(E, \mathscr{F}, \mu)$ is a probability space, $\{\mathscr{F}_n\}_0^\infty$ is a non-decreasing sequence of sub σ-algebras of $\mathscr{F}$, and $\{X_n\}_0^\infty$ is a sequence of integrable random variables such that X_n is $\mathscr{F}_n$-measurable and such that

$$\sup_{m \geq 0} \sup_{n \geq m} E[\,|X_n - X_{m-1}|\,|\,\mathscr{F}_m]$$

is bounded μ-almost surely by some constant $M < \infty$, then for all $\gamma < 1/2Me$:

$$E\left[\exp\left(\gamma \sup_{n \geq 1} |X_n - X_0|\right)\right] \leq C_\gamma(M)$$

where $C_\gamma(M) < \infty$ is a constant depending only on γ and M. In particular, if $\mathscr{F}_0 = \{\phi, E\}$ and $\mathscr{F} = \sigma(\bigcup_0^\infty \mathscr{F}_n)$, apply this observation to show that for integrable X with $\|X^\dagger\|_{L^\infty(\mu)} < \infty$:

$$E[\exp(\gamma X^*)] \leq C(\|X^\dagger\|_{L^\infty(\mu)})$$

when $\gamma < (2\|X^\dagger\|_{L^\infty(\mu)} e)^{-1}$. (The notation here is borrowed from Lemma A.1.2.) Of course, this proves that for such an X and γ:

$$E[\exp(\gamma |X - X_0|)] \leq C(\|X^\dagger\|_{L^\infty(\mu)}).$$

The preceding inequality is, for the present context, the John–Nirenberg [1961] inequality.

A.3.2. The analogues of Lemma A.1.1 and the preceding for continuous parameter families of random variables can be derived when some regularity condition is imposed on the family as a function of the parameter. This can be done by a limit procedure starting from the discrete case. It is interesting that if enough regularity is assumed then a direct derivation can be easier. For instance, let $(E, \mathscr{F}, \mu)$ be a probability space, $\{\mathscr{F}_t: T \geq 0\}$ a non-decreasing family of sub σ-algebras of $\mathscr{F}$, and $\theta: [0, \infty) \times E \to C$ a continuous, progressively measurable function. Suppose that there is an integrable function Y such that

$$\sup_{t \geq s} E[|\theta(t) - \theta(s)| \,|\, \mathscr{F}_s] \leq E[Y|\mathscr{F}_s] \qquad (\text{a.s., } \mu)$$

for all $s \geq 0$. Show that for all $\alpha, \beta > 0$:

$$\left(\sup_{t \geq 0} |\theta(t) - \theta(0)| \geq \alpha + \beta\right) \leq \frac{2}{\alpha} E\left[Y, \sup_{t \geq 0} |\theta(t) - \theta(0)| \geq \beta\right].$$

The proof can be accomplished by the same procedure as we used in the proof on Lemma A.1.1, only this time one should exploit continuity. Once one has this inequality, it is clear that the analogues of Lemma A.1.2 and the preceding exercise follow easily. In particular, if

$$\sup_{t \geq s} E[|\theta(t) - \theta(s)| \,|\, \mathscr{F}_s] \leq M \qquad (\text{a.s., } \mu)$$

for all $s \geq 0$, then

$$E\left[\exp\left(\gamma \sup_{t \geq 0} |\theta(t) - \theta(0)|\right)\right] \leq C_\gamma(M)$$

for all $\gamma \leq 1/2Me$. An interesting application of this result is the following observation due to Gaveau [private communication]. Let $a: [0, \infty) \times R^d \to S_d$ and $b: [0, \infty) \times R^d \to R^d$ be bounded measurable coefficients which determine a measurable Markov family $\{P_{s,x}: (s, x) \in [0, \infty) \times R^d\}$ of solutions to the martingale problem for a and b. Suppose that $f: [0, \infty) \times R^d \to \mathbb{C}$ is a measurable function such that

$$E^{P_{s,x}}\left[\left|\int_s^t f(u, x(u))\, du\right|\right] \leq M$$

for all $0 \leq s < t$ and $x \in R^d$. Show that for all $\gamma < 1/2Me$, $s \geq 0$, and $x \in R^d$:

$$E^{P_{s,x}}\left[\exp\left(\gamma \sup_{t \geq s} \left|\int_s^t f(u, x(u))\, du\right|\right)\right] \leq C_\gamma(M).$$

The idea is to set $\theta(t) = \int_s^t f(u, x(u)) \, du$ and check that:

$$E^{P_{s,x}}[|\theta(t_2) - \theta(t_1)| \, | \, \mathcal{M}_{t_1}] \leq M \qquad \text{(a.s., } P_{s,x})$$

for all $s \leq t_1 < t_2$. Note that if $f \geq 0$, this result proves that

$$\sup_{(s, x)} E^{P_{s,x}}\left[\int_s^\infty f(u, x(u)) \, du\right] < \infty$$

if and only if there is a $\gamma > 0$ such that

$$\sup_{(s, x)} E^{P_{s,x}}\left[\exp\left(\gamma \int_s^\infty f(u, x(u)) \, du\right)\right] < \infty$$

A.3.3. Let $(E, \mathcal{F}, \mu)$ and $\{\mathcal{F}_n\}_0^\infty$ be as in Lemma A.1.2. Note that the same argument plus an obvious change in notation enables one to prove Theorem A.1.3 when T is defined on functions from $(E, \mathcal{F})$ to a complex separable Hilbert space $\mathcal{H}_1$ into functions from $(E, \mathcal{F})$ into a second complex separable Hilbert space $\mathcal{H}_2$. That is, if $T: L^2(\mu; \mathcal{H}_1) \cap L^\infty(\mu; \mathcal{H}_1) \to L^1(\mu, \mathcal{H}_2)$ and T satisfies:

$$\|(Tf)_{\mathcal{H}_2}^\dagger\|_{L^\infty(\mu)} \leq M_\infty \|\|f\|_{\mathcal{H}_1}\|_{L^\infty(\mu)}$$

and

$$\|\|Tf\|_{\mathcal{H}_2}\|_{L^2(\mu)} \leq M_2 \|\|f\|_{\mathcal{H}_1}\|_{L^2(\mu)},$$

where $(\varphi)_{\mathcal{H}_2}^\dagger \equiv \sup_{n \geq 1} E[\|\varphi - E[\varphi | \mathcal{F}_{n-1}]\|_{\mathcal{H}_2} | \mathcal{F}_n]$ for $\varphi \in L^1(\mu; \mathcal{H}_2)$, then for $2 < p < \infty$

$$\|\|Tf\|_{\mathcal{H}_2}\|_{L^p(\mu)} \leq \frac{4p^{p+1}}{(p-1)^p} M_\infty^{1-2/p}(4M_2)^{2/p} \|\|f\|_{\mathcal{H}_1}\|_{L^p(\mu)}.$$

As an application of this remark, we outline the following derivation of Burkholder's inequality (cf. 4.6.12). Take $\mathcal{H}_1 = l^2(C)$ and $\mathcal{H}_2 = C$, and consider the operator $T: L^2(\mu; \mathcal{H}_1) \to L^2(\mu; \mathcal{H}_2)$ defined by

$$T(\{f_k\}_1^\infty) = \sum_1^\infty (E[f_k | \mathcal{F}_k] - E[f_k | \mathcal{F}_{k-1}]).$$

Check that

$$\|T(\{f_k\}_0^\infty)\|_{L^2(\mu)} \leq \|\|\{f_k\}_1^\infty\|_{\mathcal{H}_1}\|_{L^2(\mu)}$$

and that

$$E[|T(\{f_k\}_1^\infty) - E[T(\{f_k\}_1^\infty)|\mathcal{F}_{n-1}]|^2 | \mathcal{F}_n] \leq 4\|\|\{f_k\}_1^\infty\|_{\mathcal{H}_1}\|_{L^\infty(\mu)}^2.$$

Thus, by the preceding remark:

$$\|T(\{f_k\}_1^\infty)\|_{L^p(\mu)} \le \frac{4p^{p+1}}{(p-1)^p} 2^{1-2/p}4^{2/p}\| \{f_k\}_1^\infty \|_{\mathscr{H}_1} \|_{L^p(\mu)}$$

for $2 < p < \infty$. Next, show that $T^*: L^2(\mu) \to L^2(\mu; \mathscr{H}_1)$ given by

$$T^*f = \{E[f \mid \mathscr{F}_k] - E[f \mid \mathscr{F}_{k-1}]\}_1^\infty$$

is the adjoint of the map T. Also, check that

$$\|(T^*f)_{\mathscr{H}_1}^\dagger\|_{L^\infty(\mu)} \le 2\|f\|_{L^\infty(\mu)}.$$

Hence, for $2 < p < \infty$:

$$\| \|T^*f\|_{\mathscr{H}_1} \|_{L^p(\mu)} \le \frac{4p^{p+1}}{(p-1)^p} 2^{1-2/p}4^{2/p}\|f\|_{L^p(\mu)}.$$

Working by duality, conclude that for $1 < p < \infty$:

$$\|T(\{f_k\})\|_{L^p(\mu)} \le C_p\| \|\{f\}\|_{\mathscr{H}_1} \|_{L^p(\mu)}$$

and

$$\| \|T^*f\|_{\mathscr{H}_1} \|_{L^p(\mu)} \le C_p\|f\|_{L^p(\mu)},$$

where $C_2 = 1$, $C_p = 4p^{p+1}/(p-1)^p 2^{1+2/p}$ if $2 < p < \infty$, and $C_p = C_{p'}$, if $1 < p < 2$. Finally, observe that

$$T^*f = T^*(f - E[f])$$

and

$$TT^*f = f - E[f].$$

Thus

$$\frac{1}{C_p} \| \|T^*f\|_{\mathscr{H}_1} \|_{L^p(\mu)} \le \|f - E[f]\|_{L^p(\mu)} \le C_p\| \|T^*f\|_{\mathscr{H}_1} \|_{L^p(\mu)}.$$

This is essentially Burkholder's inequality. To get from it to the form in which the inequality is usually stated, let $(X_n, \mathscr{F}_n, \mu)$ be a martingale. Given $N \ge 1$, set $f = X_N$ and note that $X_n = E[f \mid \mathscr{F}_n]$, $0 \le n \le N$. Hence

$$\frac{1}{C_p} E\left[\left(\sum_1^N |X_n - X_{n-1}|^2\right)^{p/2}\right]^{1/p} \le E[|X_N - X_0|^p]^{1/p}$$

$$\le C_p E\left[\left(\sum_1^N |X_n - X_{n-1}|^2\right)^{p/2}\right]^{1/p}.$$

Since the middle term is a non-decreasing function of N, this now yields:

$$\frac{1}{C_p} E\left[\left(\sum_1^\infty |X_n - X_{n-1}|^2\right)^{p/2}\right]^{1/p} \le \sup_{n \ge 1} E[|X_n - X_0|^p]^{1/p}$$

$$\le C_p E\left[\left(\sum_1^\infty |X_n - X_{n-1}|^2\right)^{p/2}\right]^{1/p}$$

for $1 < p < \infty$. Note that the middle term is commensurate with $E[\sup_{n \ge 1} |X_n - X_0|^p]^{1/p}$ for p's in this range, but that it no longer is when $p = 1$. A beautiful fact, discovered originally by B. Davis [1970], is that Burkholder's inequality remains true when $p = 1$ after one brings the "sup" inside the expectation. For an elegant derivation of Davis's inequality, the reader is referred to A. Garsia [1973].

A.3.4. The reason for our working with parabolic scaling can be most easily appreciated if one goes back to the kernel k in (0.6). One can easily check that $k(\lambda^2 s, \lambda x) = (1/\lambda^3) k(s, x)$ for any $\lambda > 0$. This type of homogeneity meshes perfectly with the homogeneity of the metric ρ introduced before Theorem A.1.6 and is the reason why ρ is the natural metric with which to study such kernels.

In connection with the preceding paragraph, suppose that $k: R \times R^d \backslash \{0, \mathcal{O}\} \to \mathbb{C}$ is a smooth function satisfying $k(\lambda^2 s, \lambda x) = \lambda^{-d-2} k(s, x)$ for all $\lambda > 0$. Let $k^{(\varepsilon)}$, $\varepsilon > 0$, be given by

$$k^{(\varepsilon)}(s, x) = \chi_{[\varepsilon, 1/\varepsilon]}(\rho(s, x)) k(s, x).$$

Show that there is a B for which

$$\int_{\rho(t-s, y-x) \ge 2\rho(\sigma-s, \xi-x)} |k^{(\varepsilon)}(s - t, x - y) - k^{(\varepsilon)}(\sigma - t, \xi - y)| \, dt \, dy \le B$$

for all $(\sigma, \xi), (s, x) \in R \times R^d$ and $\varepsilon > 0$. Assume in addition that there is an A such that

$$|\widehat{k^{(\varepsilon)}}(\sigma, \xi)| \le A$$

for all $\varepsilon > 0$ and $(\sigma, \xi) \in R \times R^d$. Show that

$$\|k^{(\varepsilon)} * f\|_{L^p(R \times R^d)} \le C_p \|f\|_{L^p(R \times R^d)}, \qquad 1 < p < \infty,$$

where C_p is independent of $\varepsilon > 0$ and depends only on A and B as well as p.

Bibliographical Remarks

Chapter 1. The study of weak convergence of probability measures on complete separable metric spaces is carried out in detail in the works of Prohorov [1956] and Varadarajan [1958] and [1966]. One can find a convenient treatment of the topic in the books of Parthasarathy [1967] and Billingsley [1968].

For a detailed study of the theory of martingales the reader is referred to the books of Doob [1953], Meyer [1966] and Neveu [1975].

The compactness criterion for Markov processes that we have presented in terms of martingales is a modification of the results contained in Prohorov [1956]. Standard references for the extension theorems of Kolmogorov and Tulcea are Kolmogorov [1956] and Doob [1953].

Chapter 2. The regularity of paths in the Brownian motion case goes back to Wiener [1923]. Other basic references for results on the regularity of sample paths are Slutzky [1949]; Dynkin [1952] and Kinney [1953] for Markov processes, and the more recent work of Garsia, Rodemich and Rumsey [1970], for general stochastic processes.

Chapter 3. For partial differential equations there are several sources. For example Protter and Weinberger [1967] on the maximum principle, Friedman [1964] for the existence and properties of solutions to partial differential equations with Hölder continuous coefficients and Ladyženskaja, Solonnikov and Ural'ceva [1968] for the existence and uniqueness of solutions in Sobolev spaces. For degenerate parabolic equations, the result we have proved is taken from Oleinik [1966]. These results, at least when the diffusion matrix $\{a^{ij}\}$ has a smooth square root, have been proved probabilistically in Gikhman and Skorohod [1973] and McKean [1969].

Chapters 4 and 5. The classical reference for the theory of stochastic integration is Itô [1951]. By using the theory of martingales the proofs were streamlined by Doob [1953]. Other references for stochastic integrals are Kunita and Watanabe [1967], Strassbourg seminar notes by Meyer [1974–75] and McKean [1969]. The existence of Lipschitz continuous square roots for smooth coefficients $\{a^{ij}\}$ was proved in a paper by Phillips and Sarason [1967].

Chapter 6. The idea that one should look for distribution uniqueness for solutions of stochastic differential equations goes back to some papers by Girsanov [1960] and [1961]. The martingale formulation and the systematic use of the techniques of this chapter were initiated in Stroock and Varadhan [1969]. The existence of solutions to the martingale problem is a reformulation of the generalized existence theorem for Itô's equations proved by Skorohod [1961]. The technique of random time change goes back to Volkonskii [1958] and was used by him and by Itô and McKean [1965] to study one dimensional diffusions. The Cameron–Martin–Ginsanov formula was derived in a particular case in Cameron and Martin [1953]. Girsanov [1960] and Skorohod [1957] have made computations of Radon–Nikodym derivatives for stochastic processes.

Chapter 7. Tanaka [1964] and Krylov [1966] constructed a Markov semigroup (without proving uniqueness) for an arbitrary set of time homogeneous elliptic coefficients assuming for smoothness only the continuity of the second order terms. Some of the results proved in this chapter have been extended to diffusions on manifolds in Yang [1972].

Chapter 8. The results of this chapter have their origin in Watanabe and Yamada [1971].

Chapter 9. Most of the estimates on the transition probabilities that we derive in this chapter can also be derived by using standard techniques in the theory of partial differential equations. See for instance the works of Fabes and Riviere [1966] and Ladyženskaja, Solonnikov and Ural'ceva [1968].

Chapter 10. When the dimension of the space is 1, Feller obtained complete results concerning conditions for explosion. These can be found in Itô–McKean [1965]. In the multidimensional case the main results are due to Khasminskii [1960]. They may be found in the book by McKean [1969].

Chapter 11. The idea of proving limit theorems for processes goes back to Erdös and Kac [1947] and Donsker [1951]. The papers by Prohorov [1956] and Skorohod [1956] were very important in the development of the subject. The technique we use to prove limit theorems is very close in spirit to the proof of Trotter [1959] of the central limit theorem. These same martingale techniques have been used effectively to study the central limit problem for random variables on a Lie group in Stroock and Varadhan [1973].

Chapter 12. The problem of Markov selections was first solved by Krylov [1973].

Chapter 13. The results on singular integrals and their application to partial differential equations go back to Calderon and Zygmund [1952] in the elliptic case and Jones [1964] in the parabolic case. The results on singular integrals go back even further to Mihlin [1938] and Giraud [1932].

There is, of course, a considerable amount of material on Markov processes in general and diffusion processes in particular that has not been covered in this book. The expository articles by Dynkin [1962] and Williams [1974] contain excellent references. The books by Dynkin [1965], and Gikhman and Skorohod [1974], [1975], [1978] have a much wider scope than is attempted here. The two volumes by Friedman [1976] deal with several applications of stochastic integral equations besides developing the theory. Stratanovich [1968] and Lipster and Shiryayev [1977] and [1978] deal with the problems of control and filtering of Markov processes.

For a more detailed study of one dimensional diffusions Itô and McKean [1965] and Dynkin [1959] are good sources. For diffusions with boundaries Watanabe [1971], Sato and Ueno [1965], Wentzel [1959] and Stroock and Varadhan [1971] are some of the relevant references.

Bibliography

Alexandrov, A. D.
1963 The method of "supporting" mapping in the study of the solutions of boundary problems. Outlines Joint Symp. Partial Differential Equations (Novosibirsk, 1963) 297–302.

Billingsley, P.
1968 Convergence of Probability Measures. New York, John Wiley.

Burkholder, D. and Gundy, R.
1970 Extrapolation and interpolation of quasilinear operators on martingales. Acta Mathematica **124**, 249–304.

Calderon, A. P. and Zygmund, A.
1952 On the existence of certain singular integrals. Acta Math. **88**, 85–139.

Cameron, R. H. and Martin, W. T.
1953 The transformation of Wiener integrals by nonlinear transformations. Trans. Amer. Math. Soc. **75**, 552–575.

Davis, B.
1970 On the integrability of the martingale square function. Israel Jour. Math. **8**, 187–190.

Donsker, M. D.
1951 Four papers in probability. Mem. Amer. Math. Soc. No. 6.

Doob, J. L.
1952 Stochastic Processes. New York, John Wiley.

Dynkin, E. B.
1952 Criteria for continuity and lack of discontinuities of the second kind for trajectories of a Markov stochastic processes. Izv. Akad. Nauk SSSR. Ser. Mat. **16**, 563–572.
1959 One dimensional continuous strong Markov processes. Theor. Prob. and Appl. **4**, 1–52.
1960 Foundations of the Theory of Markov Processes (English translation). Oxford–London–New York–Paris, Pergamon Press.
1962 Markov processes and problems in analysis. Proc. Int. Cong. of Math. Stockholm, 36–58.
1965 Markov Processes, Vols. 1, 2. Berlin, Springer-Verlag.

Erdös, P. and Kac, M.
1946 On certain limit theorems of the theory of probability. Bull. Amer. Math. Soc. **52**, 292–302.

Fabes, E.
1966 Singular integrals and partial differential equations of parabolic type. Studia Math. **XXVIII**, 6–131.

Fabes, E. and Rivière, N. M.
1966 System of parabolic equations with uniformly continuous coefficients. Jour. Analyse Math. **XVII**, 305–335.
1966a Singular integrals with mixed homogeneity. Studia Math. **27**, 19–38.

Fefferman, C. and Stein, E. M.
1972 H^p spaces of several variables. Acta Mathematica **129**, 137–193.

Feller, W.
1936 Zur Theorie der Stochastischen Prozesse [Existenz und Eindeutigkeitssätze]. Math. Ann. **113**, 113–160.

Freidlin, M.
1967 On small (in the weak sense) perturbations in the coefficients of a diffusion process. Theory of Prob. and Appl. **12**, 487–490 (English Translation).

Friedman, Avner
1976 Stochastic Differential Equations and Applications, Vols. 1 and 2. New York, Academic Press.

Friedman, A.
1964 Partial Differential Equations of Parabolic Type. Englewood Cliffs, NJ, Prentice Hall.
1969 Partial Differential Equations. New York, Holt, Rinehart and Winston.

Garsia, A.
1973 Martingale Inequalities. Reading, MA, W. A. Benjamin.

Garsia, A., Rodemick, E., and Rumsey, H., Jr.
1970–71 Indiana Univ. Math. Journal **20**, 565–578.

Gihman, I. I. and Skorohod, A. V.
1973 Stochastic Differential Equations. New York, Springer-Verlag.
1974, 1975, 1979 The Theory of Stochastic Processes, Vols. I, II, and III. Berlin, Springer-Verlag.

Giraud, G.
1932 Sur une extension de la théorie des équations intégrales de Fredholm. C. R. Acad. Sci. Paris, **195**, 454–456.

Girsanov, I. V.
1960 On transforming a certain class of stochastic processes by absolutely continuous substitution of measures. Theory of Prob. and Appl. **5**, 285–301.
1961 On Itô's Stochastic Integral Equation. Soviet Mathematics, Vol. 2, pp. 506–509.

Halmos, P. R.
1950 Measure Theory. New York, Van Nostrand.

Hille, E. and Phillips, R. S.
1957 Functional Analysis and Semigroups. Providence, American Math. Soc.

Hormander, L.
1960 Estimates for translation invariant operators in L_p spaces. Acta Math. **104**, 93–140.

Itô, K.
1951 On stochastic differential equations. Mem. Amer. Math. Soc. **4**, xxx–xxx.

Itô, K. and McKean, H. P., Jr.
1965 Diffusion Processes and Their Sample Paths. Berlin, Springer-Verlag.

Jacod, J. and Yor, M.
1976 Étude des solutions extrêmales et representation intégrale des solution pour certains problèmes de Martingales. C. R. Acad. Sciences, Paris T. 283 Series A, pp. 523–525.

John, F. and Nirenberg, L.
1961 On functions of bounded mean oscillation. Comm. Pure Appl. Math. **14**, 785–799.

Jones, B. F.
1964 A class of singular integrals. Amer. Jour. of Math. **86**, 441–462.

Khasminskii, R. Z.
1960 Ergodic properties of recurrent diffusion processes and stabilization of the solution of the Cauchy problem for parabolic equations. Theor. Prob. and Appl. **5**, 179-196.

Kinney, J. H.
1953 Continuity properties of sample functions of Markov processes. Trans. Amer. Math. Soc., **74**, 280-302.

Kolmogorov, A. N.
1931 Uber die Analytischen Methoden in der Wahrscheinlichkeitsrechnung. Math. Ann. **104**, 415-458.
1956 Foundations of the Theory of Probability. New York, Chelsea.

Krylov, N. V.
1966 On quasi diffusion processes. Theor. Prob. and Appl. **11**, 373-389.
1973 The selection of a Markov process from a Markov system of processes (Russian). Izv. Akad. Nauk SSSR, Ser. Mat. **37**, 691-708.
1974 Certain bounds on the density of the distribution of stochastic integrals. Izv. Akad. Nauk, **38**, 228-248.

Kunita, H. and Watanabe, S.
1967 On square integrable martingales. Nagoya Math. Journal **30**, 209-245.

Kuratowski, C.
1948 Topologie, Vol. 1. Warzawa, Monografie Matematyczne.

Ladyženskaya, O. A., Solonnikov, V. A., and Ural'ceva, N. N.
1968 Linear and quasilinear equations of parabolic type. Providence, Amer. Math. Soc. Translations of Math. Monographs, No. 23.

Lévy, P.
19-8 Processus Stochastiques et Mouvement Borwnien. Paris, Gauthier-Villars.

Lipster, R. S. and Shiryayev, A. N.
1977, 1978 Statistics of Random Processes, Vol. I, Theory, and Vol. II, Applications. Berlin, Springer-Verlag.

McKean, H. P.
1973 Geometry of Differential Space. Ann. Prob. **1**, 197-206.

McKean, H. P., Jr.
1969 Stochastic Integrals. New York-London, Academic Press.
1975 Brownian Local Times. Advances in Mathematics, Vol. 16, pp. 91-111.

Meyer, P. A.
1966 Probability and Potentials. Waltham, MA, Blaisdell.
1974-75 Séminaire de Probabilités, Université de Strasbourg. Springer-Verlag, Lecture Notes No. 511.

Mihlin, S. G.
1938 The problem of equivalence in the theory of singular integral equations. Mat. Sbornik N.S., **45**, 121-140.

Neveu, J.
1975 Discrete Parameter Martingales (translated by T. P. Speed). Amsterdam, North-Holland.

Oleinik, O. A.
1966 Alcuni risultati sulle equazioni lineari e quasi lineari ellitico—paraboliche a derivate parziali del second ordine. Rend. Classe Sci. Fis. Mat., Nat. Acad. Naz. Lincei, Ser. 8, **40**, 775-784.

Parthasarathy, K. R.
1967 Probability Measures on Metric Spaces. New York, Academic Press.

Phillips, R. S. Sarason, L.
19xx Elliptic Parabolic Equations of Second Order. J. Math. and Mech. **17**, 891-917.

Prohorov, Yu. V.
1956 Convergence of stochastic processes and limit theorems in probability theory. Theor. Probability Appl. **1**, 157-214 (English Translation).

Protter, M. H. and Weinberger, H. F.
1967 Maximum Principles in Differential Equations. Englewood Cliffs, NJ, Prentice-Hall.

Pucci, C.
1966 Limitazioni per soluzioni di equazioni ellitiche. Annali di Matematica pur ed applicata, Series
IV, **LXXIV**, 15-30.

Rivière, N. M.
1971 Singular integrals and multiplier operators. Arkiv für Matematic **9**, 243-278.

Sato, K. and Ueno, T.
1965 Multidimensional diffusions and the Markov process on the boundary. Jour. of Math., Kyoto
Univ. **4**, 529-605.

Skorohod, A. V.
1956 Limit theorems for stochastic processes. Theor. Prob. and Appl. **1**, 261-290.
1957 On the differentiability of measures which correspond to stochastic processes, I, Processes with
independent increments. Theor. Prob. and Appl., **2**, 407-432.
1961 On the existence and uniqueness of solutions for stochastic differential equations. Sibirsk. Mat.
Zur. **2**, 129-137.
1961 Existence and uniqueness of solutions to stochastic diffusion equations. Sibirsk Math. Zh. **2**,
129-137.

Slutzky, E. E.
1949 Some statements concerning the theory of random processes. Publications of Middle Asian
University, Mat. Ser. (5) **31**, 3-15.

Stratanovich, R. L.
1968 Conditional Markov Processes and Their Applications to the Theory of Optimal Control. New
York, American Elsevier Publ. Co.

Stroock, D. W.
1973 Applications of Fefferman-Stein type interpolation to probability theory and analysis. Comm.
Pure Appl. Math. **XXVI**, 477-496.

Stroock, D. W. and Varadhan, S. R. S.
1969 Diffusion processes with continuous coefficients, I and II. Comm. Pure Appl. Math. **XXII**,
345-400, 479-530.
1970 On the support of diffusion processes with applications to the strong maximum principle.
Proceedings of Sixth Berkeley Symp. on Mathematical Statistics and Probability, Vol. III,
pp. 333-360.
1971 Diffusion processes with boundary conditions. Comm. Pure Appl. Math., **XXIV**, 147-225.
1972 On degenerate elliptic-parabolic operators of second order and their associated diffusions.
Comm. Pure Appl. Math. **XXV**, 651-714.
1973 Limit theorems for random walks on Lie groups. Sankhyā **35**, Ser. A, 277-294.

Sussman, H.
1977 An interpretation of stochastic differential equations as ordinary differential equations which
depend on the sample point. Bull. Amer. Math. Soc. **83**, 296-298.

Tanaka, H.
1964 Existence of diffusions with continuous coefficients. Mem. Fac. Sci. Kyushu Univ., Ser. A., **18**,
pp. 89-103.

Trotter, H. F.
1959   An elementary proof of the central limit theorem. Archiv der Mathematik **10**, 226–234.

Varadarajan, V. S.
1958   Weak convergence of measures on separable metric spaces. Sankhyā **19**, 15–22.
1965   Measures on topological spaces. Transl. Amer. Math. Soc., Ser. II, **48**, 161–228.

Volkonskii, V. A.
1958   Random time changes in strong Markov processes. Theor. Prob. and Appl. **3**, 310–326.

Yang, W. Zu.
1972   On the uniqueness of diffusions. Z. Wahr. Verw. Geb. **24**, 247–261.

Watanabe, S.
1971   On stochastic differential equations for multidimensional diffusion processes with boundary conditions. Jour. of Math., Kyoto Univ. **11**, 169–180.

Watanabe, S. and Yamada, T.
1971   On the uniqueness of solutions of stochastic differential equations. Journal of Math. of Kyoto Univ. **11**, 155–167.

Wentzel, A. D.
1959   On boundary conditions for multidimensional diffusion processes. Theor. Prob. and Appl. **4**, 164–177.

Wiener, N.
1923   Differential space. J. Math. Phys., **2**, 131–174.
1930   The homogeneous chaos. Am. J. Math. **60**, 897–936.

Williams, D.
1974   Brownian motions and diffusions as Markov processes. Bull. Lond. Math. Soc. **6**, pp. 257–303.

Yamada, T. and Watanabe, S.
1971   On the uniqueness of solutions of stochastic differential equations, II. J. Math. Kyoto Univ. **11**, 553–563.

Subject Index